Springer Biographies

The books published in the Springer Biographies tell of the life and work of scholars, innovators, and pioneers in all fields of learning and throughout the ages. Prominent scientists and philosophers will feature, but so too will lesser known personalities whose significant contributions deserve greater recognition and whose remarkable life stories will stir and motivate readers. Authored by historians and other academic writers, the volumes describe and analyse the main achievements of their subjects in manner accessible to nonspecialists, interweaving these with salient aspects of the protagonists' personal lives. Autobiographies and memoirs also fall into the scope of the series.

Giulia Pancheri

Bruno Touschek's Extraordinary Journey

From Death Rays to Antimatter

 Springer

Giulia Pancheri
INFN Frascati National Laboratory
Frascati, Roma, Italy

ISSN 2365-0613   ISSN 2365-0621 (electronic)
Springer Biographies
ISBN 978-3-031-03828-0   ISBN 978-3-031-03826-6 (eBook)
https://doi.org/10.1007/978-3-031-03826-6

Front cover credits: Family Documents, E. Amaldi The Bruno Touschek legacy Vienna 1921 - Innsbruck 1978, CERN-81-1, content available under CC-BY-3.0, and INFN-LNF Multimedia.
Back cover credits: Sapienza University of Rome–Physics Department Archives and INFN-LNF Multimedia.

This Springer imprint is published by the registered company Springer Nature Switzerland AG
The registered company address is: Gewerbestrasse 11, 6330 Cham, Switzerland

This book is dedicated to my family,
Yogi, Neelam and Amrit Srivastava
and
to the memory of Elspeth Yonge Touschek

Preface

"Signorrina, dobbiamo guadagnarrrci il pane e burro", and then, "Dobbiamo amministrrarre le corrrezioni rradiative". This is what Bruno Touschek, with his strong rounding of the letter "r" told me one day in Frascati, 54 years ago.[1]

I do not have a clear memory of when I first saw Bruno Touschek, although I know I attended his course on statistical mechanics at the University of Rome, and loved it. But I remember very well the day, when, two months after graduating in physics from the University of Rome, I arrived at the Frascati National Laboratories, some thirty kilometres southeast of the city, and joined the theoretical physics group Touschek was assembling. I was aware of something extraordinary, which I may have dreamed of, but never expected would really happen to me, first to become a physicist and then to do so under the guidance of a great scientist.

It was May 1966. Touschek's office was a corner room on the first floor of the Edificio delle Alte Energie, the "High Energy building", looking down to a wide plaza where the main building housed the Frascati electron synchrotron, a particle accelerator which, in 1959, had given Italy the entrance to the very exclusive club of countries with the know-how to build particle accelerators.

The synchrotron building was at the end of a short road, flanked by Roman pines, coming straight from the entrance gates of the Laboratories. Outside the gates, across the street, a modern-day futuristic accelerator, ADONE, was under construction. The road, 'Strada del Sincrotrone' as it was then called, now Via Enrico Fermi, separated the old from the new. Envisioned in December 1960, after Touschek saw that matter–antimatter colliders could be built and would work as he had dreamed, ADONE in 1966 was still to be fully mounted, but it was a reality and it would be completed and ready to start in less than three years.

On that day in May, Touschek took me to visit what is—still called—the ADONE building. Inside, the large circular hall was almost empty. Or perhaps it was not empty, but it appeared such to me. On the floor, there was a scattered number of magnetic components, dipoles and quadrupoles, which were lifted and put in place

---

[1] *Young lady, we must earn our bread and butter* and then *"We must administer the radiative corrections"*.

by a large crane, which Touschek pointed to me with great pride. In his mind, he could see the machine which would be built, a circular accelerator where matter and antimatter, electrons and positrons created in the laboratory, would circulate in opposite directions almost at the velocity of light, and collide at yet unseen center of mass energies. I had no such vision, and knew very little of the long road which had led to the construction of this machine, but I did partake of Touschek's enthusiasm and pride.

Indeed, I had no clear idea of the importance of what he had done and was trying to accomplish, nor of the adventurous, at times dangerous and tragic, life he had led before arriving in Rome in December 1952. In the years which followed, I would come to know parts of it. Only much later, after he had passed away, I decided I needed to know more about it. As Touschek's fame dimmed, machines such as the Large Hadron Collider, following the same principle as the one he made to work in the 1960's, became more and more popular, even in the public's mind. The decision to recount the story of Bruno in a book was taken. The story would start from his early days in Vienna before the war until he settled in Rome, and tell the experiences and studies which led him to devise AdA, a proof-of-concept of a new type of machine, the electron–positron colliders, which became the norm of particle physics probes in the second half of twentieth century physics, and more.

Brookline, US   Giulia Pancheri
September 2021

Acknowledgments

This book would not have existed without the help and collaboration of the late Elspeth Touschek, Bruno Touschek's widow, to whose memory this book is dedicated. Through her, an invaluable cache of personal and family documents, from which this book was born in its present form, was opened. I am also very grateful to Francis Touschek, Bruno's son, for his friendship and collaboration.

Much of the present writing and research work comes from my friend Luisa Bonolis, from the Max Planck Institut für Wissenschaftsgeschichte (MPI WG), in Berlin. Luisa Bonolis' archival and bibliographical research constitutes the historical backbone of this book (Bonolis 2021). We started working together in 2009 to prepare an updated biography, which became our joint 2011 article in the *European Physical Journal for History*. Then, after Elspeth Yonge Touschek passed away, the project for a book about Touschek's life arose. Unfortunately, by the time we were ready to start the actual writing, Luisa had to renounce accompanying me to the very end, because of other professional engagements, but continued helping with bibliographic searches and reading of the various chapters, suggesting venues of investigation, and corrections. She generously gave me permission to use parts of papers we had written together (Bonolis and Pancheri 2018; Pancheri and Bonolis 2018; Bonolis and Pancheri 2019; Pancheri and Bonolis 2020), and from which four chapters of this book have taken shape. Further collaborative work is present in Chap. 3, which was part of an article that should have been written with Pedro Waloschek, but was never finished because he passed away during our work.

For hospitality during the writing of this book, I thank the Center for Theoretical Physics (CTP) of the Massachusetts Institute of Technology (MIT), the Laboratoire de l'Accélérateur Linéaire (LAL) in Orsay and the Frascati National Laboratories (LNF) of the Istituto Nazionale di Fisica Nucleare (INFN). I acknowledge encouragement and help from LNF staff, the computing center, and the Documentation Services. I am grateful to LNF friends now retired, Orlando Ciaffoni, Claudio Federici, and Vincenzo Valente, whose interest for Touschek's work allowed conservation of many important documents, which I used and consulted.

I am most grateful to Irma Wulz, from the IKG, Israelitische Kultusgemeinde, in Vienna, for information and links which have provided information about Bruno

Touschek's family and his studies at the University of Vienna. Thanks are due to the Archives of the Deutsches Museum, Munich, Arnold Sommerfeld papers, for copies of the correspondence between Arnold Sommerfeld and Bruno Touschek, and to the Edoardo Amaldi and Bruno Touschek papers collection Archives, from the Physics Department of Rome Sapienza University, where Touschek and Amaldi's official correspondence is maintained. I am grateful to Giovanni Battimelli, from the History of Physics group at Sapienza University for sharing with me information and insight about Touschek's papers in the Archives. I thank the University of Glasgow Archives & Special Collections, University collection, GB 248 DC 157/18/56 (UGA) and Special Collections, for consultation and permission to use material about Touschek's years as a graduate student and Nuffield Lecturer. Special thanks to Antonella Cotugno from the Sapienza University of Rome–Physics Department Archives for providing copies of the Max Born-Bruno Touschek correspondence Churchill Archives Centre, The Papers of Professor Max Born of Churchill College, Cambridge University (CHA). I am grateful to Jens Vigen from CERN Documentation Service for help and advice about photographs and material from CERN archives and publications. A special thank is due to the Library staff at INFN Frascati National Laboratories, Lia Sabatini, Davide Cirillo, and Antonino Cupellini, who have assisted me with hard to find articles.

For proofreading the entire book, data checking, detailed and overall suggestions, I am most grateful to Galileo Violini, Director Emeritus of the Centro International de Fisica, Bogotá, a friend with whom I shared attending classes and preparing exams when we were students at the University of Rome. His criticism and observations led to many corrections and accompanied me in the final preparation of the manuscript.

As I started the book project, much support arrived from friends and colleagues. Lluis and Nuría Oliver were among the friends who first encouraged me. Thanks to their invitation to give a talk at Orsay, I had my first contact with LAL for what concerned Touschek's story in Orsay, and in 2013, Frascati and Orsay organized a celebration for 50 years having occurred since first collisions were observed in AdA at Orsay. In 2018, Lluis, formerly at the Institut National de Physique Théorique, painstakingly read and inserted corrections in the first version of the Introduction. This happened before COVID-19, when the book had not been written, only imagined and planned. Thus, the Introduction does not look quite like the one he worked on and whose hand corrected copy he mailed to me from Paris. The parts of the Introduction I retained, after the book took shape during the last three years, have followed the spirit and much of Lluis' corrections.

I thank my good friend Per Osland, from the University of Bergen, for many conversations about this project and first alerting me to the Norwegian film-maker and writer Aashild Sørheim's interest in Rolf Widerøe's life (Sørheim 2015, 2020). He was instrumental in my contacting Aashild, whom I thank for providing information on her project about Widerøe, in the early phase of this writing. I remember warmly my conversations with Klaas Bergmann from the University of Kaiserlautern, who patiently listened to my descriptions of Touschek's life, asking questions, which focused my research. For friendship and interest during my frequent visits to MIT CTP, I thank Earle Lomon, who helped me to clarify many points of Touschek's story, and Scott Morley for his support during my visits and afterwards. I thank Ruggero Ferrari, from University of Milan, whose criticism and interest since our first discussions as both MIT CTP visitors in 2008, sharpened and encouraged me in the project about Touschek's work.

Many thanks are due to Wolf Beigelböck, whose passion for the history of physics led him to create the *European Physical Journal for History*, where the germ of this book appeared as an article with Bonolis, the first one in its first issue (Bonolis and Pancheri 2011). Beigelböck had welcomed our proposal for the article, and followed its preparation, as an informal referee. I remember spending with him a fall 2010 afternoon in Cambridge, Massachusetts, going over his corrections of the manuscript, line by line, trying to answer some questions, whose answers neither Luisa, nor myself, knew at that time.

I am grateful for the encouragement I received from the late Professor Peter Schuster, and the members of the History of Physics Conference (HoP) group he was instrumental in creating. In particular, I also thank Prof. Edward Davies, from the University of Cambridge, chair of this group, which provided an avenue to present my project to other physicists with similar interests in the history of physics. The 2016 EPS History of Physics Conference in Pollau, Austria, September 2016, organized by Peter Schuster shaped and focused my work towards presentation to an audience which did not only include particle physicists. From all of them, I received encouragement and stimulating questions. Through them, an invitation to the 2017 Bristol Workshop on the History of Colliders enforced my belief in Touschek's essential contribution to Europe's development of particle colliders.

Crucial contributions for a complete and correct description of AdA's adventure have come from two of its protagonists, the late Carlo Bernardini (1930–2018), from Universty of Rome, and Jacques Haïssinski, from Laboratoire de l'Accélérateur Linéaire in Orsay. I was a young physicist in Frascati when I first met Carlo Bernardini. Our friendship continued through the years and we continued exchanging notes and observations until a few weeks before his death, in Rome, in 2018. I am grateful to Carlo for his affectionate and inspiring company through this project, and to his wife Silvia Tamburrini, who would occasionally provide some anecdotes or forgotten details about AdA's times. Both Carlo Bernardini and Jacques Haïssinski gave important contributions to three docu-films about AdA, through the interviews often quoted in this book. After Carlo Bernardini passed away, Jacques Haïssinski became the reference point for all things about AdA. I thank him for the criticism and suggestions he generously provided, and still provides me, to this day. His voice has given me the courage to continue writing about things I did not know enough.

My first physics teacher, Giorgio Salvini (1920–2015), has been a major protagonist of the Italian accelerator adventure, first with the Frascati electron synchrotron, then AdA and ADONE. I owe to him not just my first introduction to the hard work required to become a physicist, but also constant friendship and advice. During his last years, he shared with us, Luisa Bonolis, my husband Yogi Srivastava and myself, his life memories, partly found in this book. For all this, I am most grateful to him and Costanza, his wife for many years.

The input for the book did not only come from physicists. When I started my long search for the origin of Touschek's great accomplishments and causes of his early death, I was quite ignorant of many events surrounding the Holocaust and their impact on Bruno's life. It is thanks to conversations with some close American friends, in particular, Dorit Straus and Norman Krasnow, that I slowly reached a better understanding of Touschek's early years and war experience. It is through our bi-annual visits between Rome and Manhattan, regularly held before COVID-19, that I learnt much from Dorit and Norman. Their point of view, opinions and suggestions were often present while writing this book: for this and their continuing friendship, I am most grateful.

Encouragement and useful suggestions came from a young friend, Elena Spaventa, formerly from University of Durham, now at Bocconi University, who shared with me her experience with copyright laws and publishing rights. I thank Elena for her very useful advice, and interest in my book project, often proffered over her excellent roast beef and Yorkshire puddings.

Important contributions to the book's writing come from the interview for three docu-films about Touschek and the making of AdA, all three directed by Enrico Agapito, and co-authored by Luisa Bonolis, or Giulia Pancheri, or both. I thank Enrico Agapito for providing the full registration of all the interviews, which allowed transcriptions in this book and contributions to AdA and Touschek's story.

My family has provided me with moral and intellectual support throughout the preparation of this book which took more than 15 years. My son Amrit Paolo Srivastava has contributed to the work related to Touschek's life project since its earliest days. In 2005, he contributed to the English translation of the first docu-film by Agapito and Bonolis *Bruno Touschek and the Art of Physics*, later he was companion and photographer during one visit to Elspeth. In 2016, he was with me in Paris during the preparation for a docu-film about the impact of AdA on the development of France' large accelerators. Our visit to the Musée des Arts et Métiers enlightened my understanding of the French scenario which Touschek stepped into when AdA was brought to Orsay. He has painstakingly transcribed and translated into English the many movie interviews from Italian and French, and other relevant documents cited in the text. My daughter Neelam has provided me with continuos encouragement, in addition to frequent reading and corrections of the initial drafts. Finally, my husband Yogi has been the source of inspiration and advice on many theoretical physics questions, in addition to precious criticism.

Part of the material presented in this book is based on Touschek's letters to his family, or to Arnold Sommerfeld. These letters, the majority of which was typewritten, are in German. As such, the present text is based on translations by Luisa Bonolis, Amrit Srivastava and, occasionally, myself. I thank Luisa Bonolis for preparing a chronological summary of all the letters, which were transcribed by Amrit Srivastava one by one from the photographs of the original German text, and then translated into English. I am indebted to Cinzio and Antonio Ianiro for their help in graphical elaborations of Bruno Touschek's drawings included in the letters.

I thank many librarians across the world, who helped me trace copyright permission and photographs, from the Niels Bohr Library at the American Institute of

Physics (AIP), the University of Illinois, University of Göttingen, University of Vienna, the Max Planck Institute in Berlin and others.

I also wish to thank people and friends whose contribution did not ultimately come into this book, but who helped me in the search for images or texts. One of them is my friend Rohini Godbole from the Indian Academy of Sciences, another is Olga Shekhovtsova, from Kharkiv, KIPT, who helped with information about the origin of Russian names. For help with maps of Rome in the '20s and '30s, I thank Giorgio Capon, my colleague from the Frascati National Laboratories, and Teresa Calvani, both among my high-school friends from some sixty years ago.

In addition, I wish to thank, chapter by chapter, friends and colleagues, who contributed to this book with readings, criticism, documents and memories relating to different aspects of Touschek's story. I apologize if this list may not be as complete as it should: my age and the long time spent on the book carry the responsibility of any missed acknowledgment.

Thanks for Chap. 2

In addition to Irma Wulz, I wish to thank Claudia Wulz, from the University of Vienna and a colleague physicist, for facilitating and encouraging me to contact the IKG. I am grateful to my friend Galileo Violini for providing helpful advice and information about the Nuremberg laws.

Gianni Battimelli, from the University of Rome, has provided much information and advice. Together with Michelangelo De Maria and Giovanni Paoloni, he collected Touschek's office papers after Bruno passed away, and visited Touschek's family. He has been a great help, recently contributing the little portrait of Oskar Weltmann he had found among Touschek's paper in the Rome University Physics Institute. This portrait has shed light on the personal tragedy that his uncle's suicide represented for Bruno, a young 13 years old at the time.

I am grateful to Michael Eckert from the Deutsches Museum for sending me copies of Sommerfeld's correspondence between Arnold Sommerfeld and Bruno Touschek, and would also like to thank Luisa Bonolis for putting us in contact.

I thank Heinrich Mitter from University of Graz for providing information about Touschek's gymnasium years, and Herbert Poesch, from the University of Vienna, for help and archival information about the standing of half-Jewish students in the University of Vienna, in particular about Touschek, and for providing documents about Touschek's university expulsion in 1941.

I am grateful to our close family friend Steven Saletan, from New York, who has followed this long effort and was kindly appreciative after reading the first chapter, encouraging the telling of a story I felt insecure to recount, more than once.

Thanks for Chap. 3

In this chapter, where the first year of Touschek's life in Germany is detailed, the major source comes from the letters Bruno Touschek wrote to his father after he left Vienna. The letters used for this chapter cover the period February 26, 1942 (arrival in Munich) to February 13th, 1943, in total of over 40 letters, mostly typewritten. Thanks are due to Luisa Bonolis and Amrit Srivastava for clarifications about chronology and translations.

Thanks for Chap. 4

This chapter is closely based on Pedro Waloschek's posthumous book *Death Rays as Life-Savers during the Third Reich*, in the English version posted at DESY, and his earlier work on Widerøe's life, entitled *The infancy of particle accelerators*, both translated from German by his daughter Karen. Pedro Waloschek, who passed away in 2012, gave an important contribution to this and the following chapter, as he sent to me and Luisa Bonolis a copy of Widerøe's unpublished 1943 article, which is at the core of Widerøe's 15 MeV project. The discovery that the article was never published offered the first clue that Touschek's war years had only been barely understood, and sent Bonolis and me on a detective search of Touschek's importance in Widerøe's betatron project. I had known Pedro in Frascati during the 60s and through his frequent visits in the years to come. In 2011, we also started writing together a paper, which would connect Bruno's war years with ongoing and war-related *death-ray* projects of interest to the military, and shortly before his death, Pedro sent to me and Luisa his last English version of the *death-ray* book. I am most grateful to Pedro's memory for having shared with us his knowledge of those years and providing a guide to understand Touschek's work on the betatron project. My thanks also go to Karen Waloschek, who graciously gave permission to use parts of her father Pedro's writings.

I thank Saverio Braccini from the University of Bern, who sent me a copy of the biographical memoir about Fritz Houtermans, which he had edited together with Antonio Ereditato and Paola Scampoli, from Edoardo Amaldi's unpublished notes.

Thanks for Chap. 5

Almost completely based on Touschek's letters to his father, from mid-February 1943 until June 1945, this chapter owes much to the letters' translator, to Luisa Bonolis' chronological summaries and to our joint research efforts to clarify the relationship between Bruno Touschek and Rolf Widerøe during the betatron early planning and construction.

Thanks for Chap. 6

This chapter is based on an earlier article posted in the ArXiv with Luisa Bonolis (Bonolis and Pancheri 2019). Thanks are due to her scholarship and her corrections, which were incorporated in the present version.

Thanks for Chap. 7

It is thanks to Robert MacLaughlan and my daughter Neelam Srivastava, from Newcastle University, and Maria Daniella Dick, from the University of Glasgow, that this chapter could take the present form. Traveling from Newcastle-upon-Tyne to Glasgow through Edinburgh, gave me a better understanding of Touschek's years in Scotland. Together, we drove through the streets near the University to see the places where Touschek lived during five years in Glasgow, including 11 University Square, the house for University Professors, whose first occupant had been Lord Kelvin. I am also very grateful to Neelam for accompanying me to visit the University Archives and assisting in searching and photographing the existing documentation about Bruno Touschek.

I thank Michael C. Gunn from Birmingham University for contributions about his parents, including memories of their Tyrol vacation.

I am grateful to Luisa Bonolis for kindly agreeing to my use of the content of our joint posting about this period of Touschek's life (Pancheri and Bonolis 2020).

Thanks for Chap. 8

For the historical layout of this chapter, I owe much to Luisa Bonolis, who was also very kind in reading its final version and providing comments and corrections. For the transcriptions of Salvini's movie interviews, I had Amrit Srivastava's precious help. I also thank Gianni Di Giovanni, from the Frascati National Laboratories, for contributions from the Laboratories photographic archives.

Thanks for Chap. 9

I am thankful to Peter Minkowski from the University of Bern for the invitation to talk about Bruno Touschek, as it gave me the occasion to meet a professor from the Mathematics department working on the well known Cini-Touschek transformation, reinforcing what Nicola Cabibbo always pointed out, namely that Touschek's contribution went beyond accelerator physics.

I thank my dear friend Guido Cosenza, formerly from the University of Naples, for sharing his memories and knowledge of the period Touschek used to visit Naples, where Touschek also met his future wife Elspeth.

From the University of Naples, I am happy to acknowledge Salvatore Esposito's help with information about Caianiello, and the Theoretical Physics Institute in Naples.

Thanks for Chap. 10

In this chapter, I have tried to definitely establish AdA's priority in the development of electron-positron storage rings, and I am grateful to Jacques Haïssinski for his very careful reading of the chapter and for providing a number of important observations, in addition to asking very insightful questions.

I am also grateful to Vincenzo Valente, one of my former colleagues from the Frascati National Laboratories, who gave me a copy of the list of seminars held at Frascati from 1959–1960. Such a list allows to pinpoint the date of Wolfgang Panofsky's visit to Frascati in October 1959, which can be taken as the starting point for AdA's creation. Valente, who has written extensively about the history of the Laboratories and its protagonists, has also provided a copy of the minutes of the February 17th, 1960 meeting in Frascati, where a proposal for starting an electron – positron experiment was launched by Touschek. In this meeting, the Laboratories' scientific council solicited the preparation of a realistic plan to be soon submitted for approval.

I also thank Gianni Battimelli, for reading this chapter and providing some useful criticism, and Francesco Calogero, from the University of Rome, for an e-mail exchange about this period.

I am obliged to Luisa Bonolis, who gathered important parts of the archival material used in this chapter when she prepared the 2004 docu-film *Bruno Touschek and the Art of Physics*, also based on contributions given at the 1987 Touschek Memorial Lectures, given in Frascati. Bonolis's earlier correspondence with Nicola Cabibbo and Raoul Gatto has allowed to clarify the nebulous early genesis of AdA. Also, many parts of this story would have remained foggy, had it not been for Elspeth Yonge Touschek's generosity in letting us photograph Bruno Touschek's letter to his father.

Often mentioned as an unforgettable learning experience, Touschek's lectures on Statistical Mechanics were first given in the academic year 1959–60. I thank Paolo Camiz and Carlo Di Castro, both formerly from the University of Rome, for their memories of attending Touschek's lectures.

Thanks for Chap. 11

This chapter is based on joint research with Luisa Bonolis for more than ten years (Bonolis and Pancheri 2018) and the posting prepared to remember Touschek's passing in 1978. In this chapter, unlike the others, Touschek's voice is absent. The letters to his father, which have gifted us with unexpected insight into Touschek's life and how he arrived to conceive AdA, stop in November 1960, to briefly reappear in 1969, as shall be seen. Thus, about the glorious years AdA was in Orsay, only official documents from the University of Rome or Laboratoire de l'Accélérateur Linéaire archives speak for Bruno. From LAL, I gratefully acknowledge Jacques Haïssinski's many useful suggestions and help in providing archival documents on the exchanges between Rome and Paris, which led to AdA's transfer to France.

In addition to documents from Orsay and Rome Sapienza University, and personal memoirs such as (Marin 2009) or Bernardini's (Bernardini 2006), this chapter is enriched by interviews with both French and Italian scientists and technicians, obtained during a visit to Orsay in May 2013, or in Frascati in June 2013. I thank Enrico Agapito for providing the original movie shots. Thanks are due to François Lacoste and Giuseppe Di Giugno for providing photographs and memories, both in interviews and, particularly, through a personal visit to Di Giugno's lovely house, in spring 2019, where Di Giugno showed us some mementos from his pioneering contribution to digital music.

Thanks for Chap. 12

This chapter is based on a posting with Luisa Bonolis (Pancheri and Bonolis 2018) For this chapter, which follows the previous one without the solution of continuity, Jacques Haïssinski provided memories, archive material and a copy of his 1965 doctoral thesis, which constitutes the most complete document about AdA's operation in Orsay. I am grateful to Mario Fascetti, a Frascati technician, who had also worked for the electron synchrotron. Fascetti shared hilarious stories and personal memories of AdA's period in Orsay, in addition to letters and photographs, which he had carefully kept for over 60 years.

I am grateful to Franco Buccella, formerly from the University of Naples, for a description of how his thesis on single bremsstrahlung calculation in electron–positron collisions, done jointly with Guido Altarelli came about, and for sending me his photograph for inclusion on the book. I also thank Monica Pepe-Altarelli, from CERN, for photographs of her late husband Guido.

Thanks for Chap. 13

Many friends and colleagues contributed to the preparation of this chapter. Rinaldo Baldini, from LNF, Giorgio Bellettini from the University of Pisa, and Luciano Maiani, from the Sapienza University of Rome, contributed their memories about ADONE's period and the hectic days of the J / Ψ discovery. Mario Greco, from Roma Tre University, was most helpful in reading the chapter and providing further details to what he himself had already written about this. I am also grateful to Greco for clarifying points about the formation of the Frascati theory group. I owe to Franco Buccella a contribution about the student disruptions at the University of Rome, as described in this chapter.

Contents

Abbreviations

ACO	Anneau de Collision d'Orsay
AdA	Anello di Accumulazione
ADONE	A larger and better AdA
BTA	Bruno Touschek Papers in Sapienza University of Rome–Physics Department Archives
CERN	Conseil Éuropéen pour la Recherche Nucléaire, European Organization for Nuclear Research
CZ	Chemischen Zentralblatt
D.S.I.R.	Department of Science and Industrial Research, also DSIR
DAFNE	Double Accelerator for Nice Experiments
EAA	Edoardo Amaldi Papers in Sapienza University of Rome–Physics Department Archives
ETH	Eidgenössische Technische Hochschule (Zürich)
eV	electron Volt, unit of energy in elementary particle physics equivalent to the energy an electron gains traveling through a potential difference of 1 Volt.
Family Documents	All photographs, letters, drawings and any other material which was photographed by Luisa Bonolis, Amrit Srivastava or the Author, courtesy of Mrs. Elspeth Yonge Touschek, and reproduced with permission by Francis Touschek
INFN	Istituto Nazionale di Fisica Nucleare
LAL	Laboratoire de l'Accélérateur Linéaire, Linear Accelerator Laboratory
LNF	Laboratori Nazionali di Frascati, Frascati National Laboratories
MVRA	MegaVoltResearchAssociation
NEBB	Norsk Elektrisk og Brown Boveri
RLM	ReichsLuftfahrtMinisterium
RSUPDA	Sapienza University of Rome–Physics Department Archives
SIF	Società Italiana di FIsica
SPEAR	Stanford Positron Electron Asymmetric Ring

| VEP-1 | Russian electron-electron Collider |
| VEPP-2 | Russian electron-positron Collider |

List of Figures

Chapter 1
Introduction

This is the story of Bruno Touschek, the scientist whose vision and leadership led to the development of a new type of elementary particle accelerator, the matter-antimatter colliders. The aim of this book is to firmly establish Touschek as one of the major protagonists of 20th century physics, by describing his life and his scientific accomplishments, against the background of the technological and scientific roads which led to the reconstruction of European science after the destructions brought by World War II.

In the 1960's, and until his premature death in 1978, Bruno Touschek was very famous within the community of accelerator physicists. Unfortunately the dramatic events of his youth, which had led him to leave his native Austria in 1942, took a toll on his health and he died before seeing the advances in particle physics brought in by accelerators inspired by the first successful particle-antiparticle collider, AdA, *Anello di Accumulazione*, Storage Ring in English, the machine he invented and built. Thus, while in the years following his death, great discoveries were made with proton-antiproton or proton-proton colliders, he did not live to see them.

The interest in Touschek's life and the relevance of AdA, are connected with a wider interest in the long road which led to the formulation of the Standard Model of elementary particles. The road had started more than 100 years ago, but its main successes have appeared when colliders became a working tool for search and discovery. This landmark can be placed at the so called November Revolution, which took place in 1974, when a new type of quark, the *charm*, almost unsuspected until that time, was discovered in the USA, and then confirmed in Frascati with ADONE, a machine that Touschek had proposed in November 1960, after AdA's initial successes. The November Revolution was followed by a period of great discoveries, based on the construction of large infrastructures for particle acceleration, such as the colliders with the necessary particle detectors. Such were the discoveries of new quarks (Augustin et al. 1974; Aubert et al. 1974; Bacci et al. 1974; Abe et al. 1995; Abachi et al. 1995) and of the $W-$ and Z-bosons (Arnison et al. 1983a), the carriers of the electro-weak force. All along, from and through these

G. Pancheri, *Bruno Touschek's Extraordinary Journey*, Springer Biographies,
https://doi.org/10.1007/978-3-031-03826-6_1

experimental and engineering feats, an extraordinary theoretical construction was built, aiming at understanding all the forces which keep matter together. At the turn of the millennium, the missing link of the Standard Model was the Higgs boson, a particle theoretically envisioned in 1964 (Higgs 2010). Discussions around its search had begun at CERN in the mid 1980s. Carried through by three Director Generals, (Llewellyn Smith 2015), the search culminated in 2012 with the discovery of the Higgs boson at the CERN Large Hadron Collider (LHC) (Aad and ATLAS Collaboration 2012; Chatrchyan et al. 2012). In the LHC, which started operations in 2008, protons were made to circulate in opposite directions in a new machine built in the LEP tunnel. Colliding practically at the velocity of light, protons can break up and their hidden constituents, quarks, anti-quarks and gluons, can annihilate, creating the Higgs boson, the particle at the heart of the Standard Model.

Ever since CERN announced the discovery of the Higgs boson on July 4th 2012, the question often comes up of how did the road to particle colliders start. This book will describe the roads and pathways which led Bruno Touschek to propose the construction of AdA, ancestor of the LHC.

By now, the number and type of sources about the early times of the particle colliders and Bruno Touschek's life constitute a non negligible collection, but, written in different languages, and in different media format, are not fully available. The idea behind this book has been to collect and merge in a single narration the existing published and unpublished literature about AdA's construction, and provide a novel contribution to the history of particle colliders, drawing on a number of unpublished primary sources, in a unified description of apparently disconnected parts of Touschek's life.

Major Sources

Two are the main public sources about Bruno Touschek's life and scientific accomplishments: the Physics Department Archives of University of Rome Sapienza, https://archivisapienzasmfn.archiui.com, and the INFN Frascati National Laboratories http://w3.lnf.infn.it/multimedia/. The Rome University archives include all the papers found in Touschek's office after his death, and more material which Edoardo Amaldi, his biographer and friend, gathered and put together to write Bruno's biography (Amaldi 1981). Other documents were given by his wife Elspeth and son Francis Touschek. The saving of Touschek's office papers is a fortunate occurrence, with literally saved them from trash. While Touschek's office was being cleared, Amilcare Bietti, the professor who occupied it at the time, alerted two young physics historians of the department, Gianni Battimelli and Michelangelo De Maria, to come and see what could be usefully kept. Pulled out of the big trash bin used to clear professors' offices after their leaving, these document have allowed much understanding about Bruno's scientific life and his genius. They have also marked the beginning of the extensive collection of scientists' papers kept in the Rome Physics Department archives.

As indicated in text and footnotes of this book, Frascati's contribution mostly comes from internal reports, starting from early days of the Laboratories, the first of them available on line, LNF-53-057, dates from 1953. From these reports, the path followed towards conception and realization of the accelerators which brought Italy to play a leading role in post war particle physics can be traced. Until the 1990's, the laboratories used to have their internal printing services. In addition, through the years, books and pamphlets were published about the history of the laboratories, often in form of conference proceedings, now part of the Frascati Physics Series. Some crucial documents about the decisions leading to the construction of AdA come from this source, through Vincenzo Valente, the physicist in charge of the Frascati library services for many years and who extensively wrote about the laboratories (Valente 2007).

Other public sources of relevance to Touschek's scientific life have come from the Laboratoire de l'Accélérateur Linéaire in Orsay, where AdA proved the feasibility of electron-positron storage rings as important tools to probe elementary particles and their world. Published documents and archival documents such as letters exchanged between Italy and France comes from Laboratoire de l'Accélérateur Linéaire's archives thanks to Jacques Haïssinski.

The first major tribute to Touschek's life and accomplishments is due to Edoardo Amaldi, mentor and friend, who visited him at the Hôpital de La Tour near Geneva in 1978, and collected his memories of early times, to include them in the most comprehensive biography of Bruno's scientific life written up to now (Amaldi 1981). As Amaldi reports in the notes he took at the time:[1]

> At 1 pm I went to visit Bruno Touschek at La Tour Hospital, in Meyrin [...] Bruno has started telling me a part of his life and I have taken some notes, which I reproduce here in an immediate attempt to re-arrange what he told me in his Italian, so extraordinary in its precision and clarity. [...]

Amaldi's work was soon followed by its Italian translation (Amaldi 1982). Not long after Touschek's death, Giovanni Battimelli, Michelangelo De Maria and Giovanni Paoloni gathered Touschek's papers which had originally been in his office, before leaving for CERN, and catalogued in a later publication, *Le carte di Bruno Touschek*, Bruno Touschek's Papers in English, published in 1989 by Università La Sapienza, Centro Stampa Ateneo (Battimelli et al. 1989).

In 1985, Fernando Amman, co-author of the proposal to build AdA, and ADONE's director, reconstructed the early times of electron-positron colliders (Amman 1985). Touschek's memory was still much alive in Italy and, in 1987, the Frascati National Laboratories decided to honor Touschek's memory through the institution of yearly *Bruno Touschek Memorial Lectures*. The event included three lectures by John Bell on *Quantum Mechanics: an inexact science*, and a number of contributions from eminent scientists, among them Burton Richter, Carlo Rubbia, Simon van der Meer

---

[1] Translated from Italian, typescript dated Geneva, 28 February 1978, Amaldi papers Sapienza University of Rome–Physics Department Archives, Box 524, Folder 6, reproduced with permission, documents provided for purposes of study and research, all rights reserved.

(Greco and Pancheri 2004).[2] In the years following his death, Touschek was often remembered by Carlo Bernardini, his close friend and collaborator in the AdA adventure. Bernardini kept alive his friend's memory in many occasions, but, with some exceptions (Bernardini 1989), most of this writing was in Italian, and Touschek's scientific figure started to be forgotten outside Italy. One example can be found in an article about *The Rise of colliding beams*, in which Burton Richter, 1976 Nobel Prize winner for the discovery of the J/Ψ together with Samuel Ting, mentions AdA as a 'scientific curiosity, that contributed little of significance to the development of colliding beams' (Richter 1992). The genesis of AdA's success as outlined in this book will hopefully contradict Richter's statement.

In the 1990's, the publication of Rolf Widerøe's autobiographical account (Waloschek 1994) brought more information about the war period and the construction of the Hamburg 15 MeV betatron, including Touschek's involvement. In 1993, there came the dismantling of ADONE, the better and bigger electron-positron collider proposed by Touschek in November 1960. A special celebration day was organized to record the history and meaning of ADONE, bringing new memories of AdA's early days (Bernardini 1997; Cabibbo 1997). ADONE was closed in order to leave the floor to DAFNE, a new generation, high luminosity electron-positron collider. The change, which was going to take place in both the physics and the involved technology, opened the road to a historical reflection about AdA. In 1998, on the twenty-year anniversary of Touschek's death, an international conference was organized in Frascati, where hitherto unpublished memories about AdA were contributed by Carlo Bernardini (Bernardini 1998), Jacques Haïssinski (Haïssinski 1998), and others.

Then, in 2003, 25 years since Touschek's death, Carlo Bernardini suggested a docu-film about Touschek. He had been much affected by Bruno's loss and knew of Bruno's exceptional life and talents. He was firmly convinced that Touschek's life story should be brought to book or movie form. Luisa Bonolis, historian of physics and author of movie documentaries and oral history works, took up the challenge. The movie, *Bruno Touschek e l'arte della fisica* was made by Bonolis and Enrico Agapito, director of many outreach movies, and with my collaboration. It includes lengthy interviews with scientists from University of Rome and Frascati Laboratories. Such interviews hold a unique historical interest, as pertinent to the birth of particle colliders, but are not fully available, since only portions have been used in the movie. This difficulty is partly overcome by their quotations in this book. The making of the movie set Bonolis into her long archival research about Touschek's life (Bonolis 2021) and led to new and extensive contributions in English journals (Bernardini 2004; Bonolis 2005). Bonolis' research also led to her first contacts with the late Elspeth Yonge Touschek, who gave her access to a number of personal documents

---

[2] Because of unforeseen circumstances, the regular occurrence of these events was interrupted in 1988, and resumed only in 2009. A consequence of this interruption was that the transcription of contributions presented at the Lectures did not appear in print until the year 2004, thanks to the painstaking work by Luigina Invidia, at the time in charge of preparing the Frascati Laboratories publications.

and photographs. During one of Bonolis' visits, Elspeth also mentioned the existence of letters written by Touschek to his father, but the letters remained in some unopened folder.

These letters, which Elspeth made available to Bonolis and me in the years 2008–2009, have been a major inspiration for writing this book. They were used to prepare an updated version of Touschek's life, which was proposed for publication in a new section of EPJ, the *European Journal of Physics*. I had just retired from my position as Director of Research at the Frascati National Laboratories, and had the time to work on a new project. What I really wanted was to understand how Bruno Touschek had arrived to conceive and build AdA and why he died so young, before seeing the fruits of his genius. The new interest in colliders brought in by the start of LHC and the Higgs search, was also giving a special momentum to such work. With Bonolis, we prepared and published a lengthy article, including the more recent literature findings, and a choice of selected material from the letters. The EPJH article (Bonolis and Pancheri 2011) clarified and updated many aspects of Touschek's life and his impact on the history of colliders, but the letters were telling much more than what we could understand at the time. They are extensively used in this book, with the permission of Francis Touschek.

When Elspeth brought to us the carton boxes with the letters from Bruno to his father, a new perspective about Touschek's life was opened. Franz Touschek had carefully gathered in chronological ordering more than 200 letters from his son. The first letter was from 1934, the last from 1971, the year Bruno's father passed away. Yellowed by the many years, the letters were in German, often typewritten on very thin paper, due to the war shortages, or even scraps of used envelopes. They occasionally included some drawings by Bruno, such as one about bombed out Berlin buildings in 1942, or his summer adventures, when he was a student in Glasgow. Bruno's *persona* was suddenly coming to life. One could see how much could still be said about his road towards particle physics, and one could now understand the tragedies which led to his self-destruction and early death. Many questions about Touschek's life could now be answered and pathways understood. Thirty years had passed since his death, and, with Bonolis, we decided it was time to bring his accomplishments to the attention of the new generation of particle physicists working at the LHC. Many details, unknown at the time of Amaldi's biography, were checked and corrected, by consulting the letters or more recent bibliography, one such being Carlo Bernardini's autobiographical memories in *Fisica Vissuta* (Bernardini 2006).

The 2011 EPJH article included research about the 1962 transportation of AdA from the Frascati Laboratories to the Laboratoire de l'Accélérateur Linéaire (LAL) in Orsay. This transfer represents a major step in AdA's contribution to understand how a storage ring functions. New informations came thanks to Jacques Haïssinski, whose doctoral thesis documents how AdA was made to successfully operate in Orsay. Copies of letters from the archives of LAL much clarified the exchanges between the two laboratories and a new docu-film was prepared in 2013. Including in this book parts of the interviews with French scientists shows how the French-Italian collaboration arose and succeeded in the Herculean effort to prove AdA's feasibility ahead of the American rising competition. So it is the inclusion of material from Pierre

Marin's book about accelerators (Marin 2009), which describes how the collaboration started and developed. Without LAL's letters and Marin's book, its beginnings would have remained foggy. This is shown by the fact that, during the preparation of the EPJH article, we had asked Carlo Bernardini when the French scientists had first come to Frascati to see AdA. He could not remember the circumstance except in very general terms. The interest in knowing the precise date is that it highlights the role played by CERN as forger of new ideas from the spreading of scientific information. Indeed, the Orsay documents mention Marin's visit to occur in July 1961, immediately following a conference in Geneva where Touschek had announced that Frascati was building two storage rings, AdA, already in operation, and ADONE, bigger and more beautiful, proposed in November 1960. The documents also complement and partly correct Marin's narration.

Of particular relevance to AdA's story, is that the book brings to light new circumstances surrounding Bruno Touschek and Rolf Widerøe's joint work on the construction of a 15 MeV German betatron. This experience underlies Bruno's later understanding of the ways in which accelerated electrons behave in a storage ring. This capacity was pivotal to propose the building of AdA and believe in its feasibility. Combined with Bruno's formation as a theoretical physicist in Göttingen, where he was Heisenberg's assistant for six months, and in Glasgow, where a so far unpublished correspondence with Max Born highlights his direct connections with the great scientists of the first half of the century, Bruno would thus be able to make the perfect synthesis between theory and experiment which created the first electron-positron collider.

Appearing in this book for the first time is how AdA's proposal was born. In much of the current narration, Touschek's suggestion and AdA's approval appear as sudden events, springing from Touschek's genius in a beautiful March day in 1960. Careful studies of published and unpublished documents, including Bruno's letters to his father during this period, tell otherwise. The genesis of the first electron-positron collider owes much to Touschek's genius, but also to the presence in the University of Rome of a formidable group of theoretical physicists, among them Nicola Cabibbo, one of Touschek's first students, and Bruno's colleague Raoul Gatto. In addition, the availability of advanced accelerator technology in Italy, through the Frascati electron synchrotron, and in France, through the Orsay linear accelerator, allowed for AdA to be envisaged and work. Such synthesis will be highlighted through published and unpublished material, such as interviews and e-mail exchanges with the protagonists of the AdA adventure.[3]

The books I have mostly consulted, and drawn from, are many, but a few of them are relevant to more than one chapter. These are the CERN Yellow Report by Amaldi (Amaldi 1981), a daily companion through the entire writing, Widerøe's autobiography compiled by Pedro Waloschek (Waloschek 1994), and Waloschek's book about *death-rays* published first in German and, after his death, in English (Waloschek 2012), which was an eye-opener about why Touschek could work for

---

[3] See also interviews and personal recollections in Bonolis and Melchionni 2003; Bonolis 2008.

the military and why he was arrested. I would also like to add Pierre Marin's story about particle accelators (Marin 2009), from which the image of AdA in 1961 as a *vrai bijou* always comes to my mind, when I see the little accelerator shown to visitors on the grounds of the Frascati Laboratories.

In addition to the existing literature, the letters written by Bruno Touschek to his family are foremost among the sources on which this book is based. They are quoted, with date and place, whenever possible, as footnotes, and refer to the episodes being narrated. Of special interest, but not only, are those related to the years 1943–1945, during which he worked with Widerøe on the betatron and those during the Glasgow period, two periods both relatively unknown so far. On the archival side the main sources are the personal papers of Bruno Touschek, Edoardo Amaldi, Giorgio Salvini, Nicola Cabibbo kept in Sapienza University of Rome–Physics Department Archives. Further sources are Werner Heisenberg's papers at the Max-Planck-Gesellschaft Archive in Berlin, Arnold Sommerfeld' and his son Ernst's papers at the Archives of the Deutsches Museum, Munich, and the Max Born-Touschek correspondence, copy of which was only recently brought to the Rome Sapienza University Archives from the Churchill Archives, in Cambridge University.

Other relevant sources for the reconstruction of the AdA story are Carlo Bernardini's personal papers and recollections, and Jacques Haïssinski's memories, generously shared before and during the writing of this book. Jacques Haïssinski's 1965 *Thèse d'Etat* was another constant companion during the writing about AdA and is an invaluable document.

Different aspects of the trajectory leading to AdA's construction have appeared in three documentary films on the life and science of Bruno Touschek, with contributions from the Istituto Nazionale di Fisica Nucleare (INFN) and the Institut National de Physique Nucléaire et de Physique des Particules (IN2P3). Starting in 2003 with *Bruno Touschek e l'arte della fisica* and a corresponding 2005 English version, by E. Agapito and L. Bonolis, the story was continued 10 years later on the occasion of the 50 year anniversary since collisions were first observed in AdA. This docufilm enlarged the view to the development of colliders in France, and appeared in 2013 in both an English http://www.lnf.infn.it/edu/materiale/video/AdA_in_Orsay.mp4 and an Italian version, http://w3.lnf.infn.it/ada-anello-di-accumulazione/, by Enrico Agapito, Luisa Bonolis and this author. A further documentary, in French, on AdA and its contribution to the development of accelerators in France was completed in 2017, entitled *60-ans-d'exploration-de-la-matière avec des accélérateurs des particules*, https://webcast.in2p3.fr/video/60-ans-dexploration-de-la-matiere, by Enrico Agapito and the Author, with Jacques Haïssinski's collaboration. The interviews, on which these docu-films are based constitute a second set of primary sources, still unpublished in written form.

Touschek's Publication Body

Among the sources, Bruno Touschek's published output plays an important role. At the time of this writing, not much of Bruno Touschek's work is freely available through the web. A full publication list, can be found in Amaldi's work at https://cds.cern.ch/record/135949?ln=en, (Amaldi 1981, 55). This list, with some corrections and additions, is also to be found in the already mentioned book *Le carte di Bruno Touschek* (Battimelli et al. 1989, 54). Although in Italian, this collection constitutes an invaluable sources of information about Touschek's public writings, in addition to reproduction of various drafts and handwritten notes and providing a guide through Bruno Touschek's archived material in Sapienza University of Rome–Physics Department Archives.

Outline of Chapters

The book was written mostly following a close chronological order. Thus chap. 2 includes Touschek's early years in Vienna, where he was born, the only son of the marriage between Camilla Weltmann, from a Jewish family with artistic ties to the Viennese Secession movement, and Franz Xaver Touschek, an Austrian Army officer. In this chapter, many dramatic occurrences influenced forever Touschek's life, among them his mother's death, his maternal oncle Oskar's suicide, a disruption of his high-school studies for what Amaldi calls racial-political reasons, attempts to emigrate to England, studies at the University of Vienna, his expulsion in 1940, and refusal of his readmission request. The correspondence with Arnold Sommerfeld leading to Touschek's move to Germany at the end of February 1942 closes the chapter. In Chap. 3 a close description of Touschek's first year in Germany draws from his letters home. Early difficulties with lodging in Hamburg, first experiences with heavy Allied bombing will appear here for the first time. An hour by hour description of the fateful encounter with Mira Hatschek, which will lead to his working at Löwe Opta and learning of Widerøe's betatron proposal, makes clear the move to Berlin. The chapter closes with a letter where Touschek tells his parents of having recently read a *sehr dummen Artikel*, a silly article, and how Touschek's boss at Löwe Opta was excited about it, to the point of mentioning Heisenberg and possible military interest.

In order to understand why such an article could be of war-related interest, Chap. 4 moves to Rolf Widerøe and the events surrounding the construction of a 15 MeV betatron. A very brief outline of the developments of particle accelerators, inclusive of Widerøe's formulation of the betatron's working principle in his PhD thesis in 1928, places Kerst's 1941 announcement of the first successful operation of a betatron as the prime reason for Widerøe's article submission, and its adoption by the German military. The role played by *death-ray* projects as life-saviours against the background of forced labor and deportation is highlighted. A suggestion about

Touschek's involvement in the approval of Widerøe's proposal is advanced. More details about Touschek's participation in Widerøe's project follow in Chap. 5. Also based on his letters home are descriptions of heavy bombing in Berlin, and the calls for forced labor he received through 1944 and 1945, which explain his arrest by the Gestapo in March 1945. The well known episode concerning the failed attempt to kill him in the last days of the war is recalled.

Two chapters, Chaps. 6 and 7, are dedicated to Touschek's formation as a theoretical physicist. Bruno is shown in Göttingen, where he obtained his Diploma in Physics in 1946, and in Glasgow, where he received the PhD in physics in 1949 and enjoyed a regular university life, after the losses and wounds inflicted by the discrimination and constant treat to his life during the last years of the war. In both chapters, Touschek's voice from the letters describes friendly episodes and, so far unpublished, details of correspondence with Werner Heisenberg and Max Born.

The move to Rome in December 1952 and what was happening in Europe with the creation of CERN, and in Italy, with the construction of the Frascati National Laboratories, are followed through Chaps. 8 and 9. The narration moves then to focus on the genesis of AdA. A succession of international conferences, held between 1956 and 1959 and attended by Touschek, shows a clear path which explains the proposal he will put forward in 1960. The move to AdA's initial successful operation and the move to Orsay are covered in Chaps. 11 and 12. The stage is now set for Chap. 13 where the last part of Touschek's life, both his golden years and his rapid decline, occur. His creation of a theoretical physics group in Frascati, and the construction of ADONE are an important legacy, established during these years. ADONE's operation was unfortunately delayed by a labor strike, and other difficulties arose in the university, starting in 1968. Touschek's drawings, some of which are reproduced in this chapter, are a testimonial of life in the University of Rome during those years. The chapter covers the events surrounding the 1974 J/Ψ discovery and this book basically stops there, with only a brief description of Touschek at CERN in 1977–1978.

Why to Stop Here?

Before leaving the way to the next twelve chapters of the book, I would like to explain the choice not to describe Touschek's last years. One reason is that I was not living in Italy during this period, and I could add no new information about what was happening at the Laboratories or at the University. Conversely, these years are extensively and thoughtfully described by Amaldi (Amaldi 1981), both at the beginning and at the end of his work. His report can be downloaded from CERN and nothing needs to be added to what he chose to report from his closeness to Bruno or from letters he received about Touschek's death, and which are maintained in the Sapienza University of Rome–Physics Department Archives. The little I want to include about Touschek's last years, can be found in the Epilogue at the end of this book.

Part I
The Years of Learning: From Austria and Germany to the UK

Chapter 2
The Vienna Years: 1921–1942

*The time here goes by very slowly. The visa from England could
already come.*
Die Zeit hier vergeht sehr langsam. Das Visum aus England
könnte schon kommen.
Bruno Touschek's letter to his parent, from Rome, 1939.

Abstract Except for summer vacations in Tyrol or in Italy, the first 21 years of
Bruno Touschek's life were spent in Vienna, where he was born on February 3rd,
1921, the only son of a mixed marriage. The annexation of Austria to Germany in
1938, the *Anschluß*, interrupted his Gymnasium studies, but he still managed to pass
his *Matura* in February 1939. After a brief vacation in Rome and some attempts to
emigrate to England and study chemistry in Manchester, he returned to Austria and
was enrolled to study physics at the University of Vienna, but was expelled in May
1940 because of his Jewish origin from the maternal side. Following an unsuccessful
renewed application in 1941, he decided to continue his studies in physics by moving
to Germany, under the protection of Arnold Sommerfeld. In February 1942, he left
Austria for Munich, not knowing when or whether he would return to the country of
his birth.

Bruno Touschek came from Vienna. His life crossed Europe both in space and time.
He saw discrimination and suffered profound personal losses in his first years, and,
through World War II, survived death and followed the dream of becoming a physicist.
In this he succeeded, and gave to Europe and the world the first matter-antimatter
collider, a particle accelerator which opened the way to fundamental discoveries in
particle physics. This he did in 1960, by joining theory and experiment, up-to-date
accelerator technology and imagination. He thought the unthinkable. But the effort
to survive family losses, expulsion from the university and the tragedies of the war
he was witness to, took their toll and he died young. His story starts in this chapter.

© The Author(s), under exclusive license to Springer Nature Switzerland AG 2022 13
G. Pancheri, *Bruno Touschek's Extraordinary Journey*, Springer Biographies,
https://doi.org/10.1007/978-3-031-03826-6_2

2.1 Family History

Bruno Touschek was born in Vienna, on February 3rd 1921, the only son of a mixed religion marriage. His mother, Camilla, came from a well-to-do Jewish family. She was the youngest of five children of Josefine Brammer and Leopold Weltmann.[1] The eldest was named Adele, the second one was Frida, then, in birth order, there came Oskar, Ella and Camilla, Fig. 2.1. Oskar, who was born on May 21st, 1885, was a medical doctor and a professor of University of Vienna. He was also a painter of some renown, but died rather young, only 46 years old, on May 19th, 1934, as recorded in the Lutherische Stadtkirche of Vienna. In fact, Oskar, as his two sisters Adele and Ella, had converted to the Protestant faith in 1911,[2] conversion to the Christian faith for assimilated Jews being rather frequent in those years of growing anti-semitism (De Waal 2012). Ella married (and later divorced) the architect Emmanuel Joseph Margold, assistant to Joseph Hoffmann, the founder of the *Wiener Werkstätte*, the Vienna Workshop movement. Of Adele, the eldest one, nicknamed Ada, we shall hear more later, through Bruno's story. In later years it was said that Bruno gave the name AdA to the ground-breaking particle accelerator he invented in 1960, in memory of the aunt who had just passed away at 84 years of age.[3]

Touschek's maternal grandfather, Leopold, was born in Pressburg, now Bratislava, on May 16th, 1843, and passed away on January 25th, 1908.[4] At the time that Leopold passed away, the family lived in Schottenring 15, but later moved to Ruckergasse.[5]

Camilla's mother, Josefine, née Brammer, died in the Holocaust. She had a tragic life, as she lost her second daughter Frieda still only 12 years old, and then lost her husband when her youngest, Camilla, was only 9 years old. By 1934, two more of her five children had died, Camilla in 1930 and Oskar, suicide, in 1934. Josephine's younger brother M.H. Brammer had also passed away in 1929.[6]

Camilla was born in Vienna on November 30th, 1889 and changed her religion in order to be able to marry Bruno Touschek's father, Franz Xaver Touschek, who was a Staff Officer in the Austrian Army, and fought on the Italian front during the First

---

[1] About being Jewish in Vienna see, for instance, https://germanics.washington.edu/hans-neurath.

[2] Courtesy of Vienna Jewish Archives, see https://data.matricula-online.eu/de/oesterreich/wien-evang-dioezese-AB/wien-innere-stadt-lutherische-stadtkirche/STB54/?pg=17.

[3] In fact, the name was proposed by Giorgio Ghigo, one of scientists who built AdA, see Chap. 10.

[4] He is buried in the Jewish Cemetery in Vienna, where one of his daughters, Frieda, is also buried. Death announcements can be found at http://anno.onb.ac.at/cgi-content/anno?apm=0&aid=nfp&datum=19080126&seite=25&zoom=2 and http://anno.onb.ac.at/cgi-content/anno?apm=0&aid=nfp&datum=18950310&seite=18&zoom=2, courtesy of Vienna Jewish Archives.

[5] The address in 1904 is found at the Lehman register https://www.digital.wienbibliothek.at/wbrobv/periodical/pageview/170300, courtesy of Vienna Jewish Archives, the later address is given in Camilla's religion conversion record, and it also appears in a postcard she sent to her mother in 1916.

[6] See   http://anno.onb.ac.at/cgi-content/anno?apm=0&aid=nfp&datum=19291203&seite=19&zoom=2.

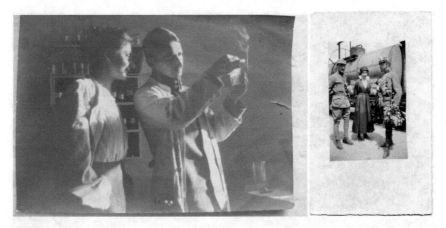

Fig. 2.1 At left, Bruno's aunt Ella and uncle Oskar. At right a photo postcard sent by Camilla Weltmann to her mother, dated 1916. Family Documents, © Francis Touschek, all rights reserved

World War (WWI), (Amaldi 1981).[7]. Franz Touschek was born in January 1884.[8] Camilla's conversion is recorded as having taken place on March 2nd, 1915, and the marriage may have followed sometimes in 1916.[9] There exists a postcard sent by Camilla to her mother in 1916, shown in Fig. 2.1, with a photograph which could have been taken on her wedding day. This card comes from a collection of Touschek family documents, such as photographs, letters, drawings and other material reproduced with permission from Francis Touschek.[10] Camilla and Franz lived in Josefstädterstrasse 81, in Vienna VIII, nearer to the city center, some 4 kilometer from her mother's home in Ruckergasse. Bruno was born in 1921.

Bruno's ties with Italy and Rome, where he would later spend a major part of his life, were established very early, since Camilla's eldest sister Adele married an Italian businessman and went to live in Rome. For a well-to-do family, as were

---

[7] In the first draft of notes taken following conversations and letters exchanged with Touschek in 1978, Amaldi writes 'Maggiore dello Stato Maggiore' of the Austro-Hungarian empire. Carlo Bernardini, professor at the University of Rome, and one of Touschek's closest friends told the Author that Franz Touschek was a 'General' in the Austrian Army. Bruno Touschek in an early draft of his CV writes 'Maggiore (Major) in KaiserSchützen', which partly checks with what Amaldi writes since the KaiserSchützen were three regiments of the Austro-Hungarian mountain infantry, which distinguished themselves on the Italian front.

[8] He died at 87 years old, according to the information from Israelitische Kultusgemeinde, and the burial took place on June 30th, 1971. Buried in the same grave, there are his first wife Camilla, his second wife Rosa, aged 84, and his sister Mathilde Schreiber, aged 90, aunt Thilde, frequently mentioned in Touschek's letters to his parents,

[9] The record of the conversion is kept at https://data.matricula-online.eu/de/oesterreich/wien/01-unsere-liebe-frau-zu-den-schotten/01-61/?pg=196, where her birthdate and family address are also recorded, courtesy of Vienna Jewish Archives.

[10] These family documents were photographed courtesy of Mrs. Elspeth Yonge Touschek, by Luisa Bonolis, Amrit Srivastava or the Author, during visits to Elspeth Yonge Touschek in 2008–2009.

Fig. 2.2 Bruno and his mother Camilla Touschek from their passport. At left, Bruno in a later passport photograph. Family Documents, © Francis Touschek, all rights reserved

the Weltmanns, it would have been natural for the sisters to exchange visits, and for Camilla to bring her little boy along when visiting Rome. Thus we see him in Fig. 2.2, as a bright and happy child in his mother's arms, in a page of the passport, which may have been issued in view of a visit to Camilla's eldest sister Adele.[11]

Camilla shared with her siblings a strong artistic interest, and participated in the exceptional atmosphere of the Vienna Secession movement (Amaldi 1981). She was a good painter, according to her son,[12] and passed on to Bruno her artistic bent and interests, as one sees in an early drawing by Bruno, shown in Fig. 2.3, entitled *Supreme War Council, in Russia* and dated March 31st, 1927. This drawing, made when he was only 6 years old, is remarkable both in the subject, probably influenced by political discussions at home, and in its graphic realization. While it is not surprising that he would be good at drawing at such age, the subject, the structuring of the group of military men and how differently each subject is depicted, all this goes beyond the average child's artistic attitudes. This drawing foreshadows many sketches and vignettes which, in later life, would delight family and friends, either enclosed in his

---

[11] To travel there at an early age, a young boy needed to be registered with the passport of the travelling parent.

[12] This is mentioned in one of Touschek's CVs in Bruno Touschek Papers, Sapienza University of Rome–Physics Department Archives, https://archivisapienzasmfn.archiui.com, all rights reserved.

Fig. 2.3 Front and back of a drawing by Bruno, when he was six years old, *Oberster-Kriegsrat*, Supreme War Council. Family Documents, © Francis Touschek, all rights reserved

letters home during the war and in the post-war years, or drafted on any piece of paper he could find and occasionally make available to his Rome University colleagues.[13] In his home in Vienna, there hung paintings by Egon Schiele, some of which followed Bruno to Rome in later years.

Life darkened as the 1920's gave way to the 1930's, both for young Bruno and his surroundings. Bruno's mother, who had become seriously ill when Bruno was still young, passed away, on April 27th, 1930, leaving her husband Franz to look after their young boy.[14] Four years later, his maternal uncle also died.[15]

In those years, the political scene started changing dramatically. In particular, for Viennese assimilated Jews, 'the years of comfort and eminence came to an abrupt end'.[16] Bruno's father resigned from the Army and went into the reserve in 1932, finding a job in an employment agency. This may have been prompted by the fact that it would have been quite difficult for an Army officer to take care of a young boy, but it may also have been a consequence of the political situation in Austria and the growing anti-semitism, that put him in the difficult position of having a son from his

---

[13] Once, when Luisa Bonolis and myself visited Touschek's widow Elspeth, we were told that she used to search for these pieces of paper in Bruno's pockets, upon his return home from the University. She would then iron them out carefully and collect them in folders. This is how many of Touschek's drawings commenting Rome University life and the 1968 unrest, have survived.

[14] Information about Camilla Touschek's death date was obtained from Archive of the Jewish Community of Vienna, the Israelitische Kultusgemeinde (IKG), with the kind assistance of Irma Wulz, see https://data.matricula-online.eu/de/oesterreich/wien/10-st-anton-von-padua/03-30/?pg=61.

[15] See an article about the death of Univ.Prof. Dr.med. Oskar Weltmann: http://anno.onb.ac.at/cgi-content/anno?aid=kvz\&datum=19340518\&seite=5\&zoom=33\&query=\%22dr.\%2Beltmann\%2210\&ref=anno-search,courtesyofIKG.

[16] From a *New Yorker* article about a recent biography of Stefan Zweig, https://www.newyorker.com/magazine/2012/08/27/the-escape-artist-3.

Fig. 2.4   At left Bruno, in Rome, ca. 1932, Family Documents, © Francis Touschek, all rights reserved, in aunt Ada's garden in Via di Villa Sacchetti, 2021 photograph by A. Srivastava, all rights reserved

marriage to a Jewish girl.[17] Not long after, in or before 1934, Franz also remarried, with Rosa Reichel. A letter Bruno wrote to his father and stepmother, dated July 23rd, 1934, suggests that they took a vacation following the wedding, leaving Bruno in Vienna, perhaps with his grandmother. Franz and Rosa had no children of their own, and Rosa was always very fond of Bruno.[18] Since 1934 until his father's death, the letters of young Bruno to Rosa and Franz, occasionally addressed to 'Lieber Vater, lieber Mutsch', or more often 'Lieben Eltern', are an invaluable source of information on Touschek's complex and adventurous life.

Throughout the 1930's, Touschek regularly visited his aunt Ada in Rome during his summer vacations. Figure 2.4 shows Bruno in the garden surrounding her apartment in via di Villa Sacchetti, in posh Parioli neighbourhood. He may have been 11 or 12 years old, and is shown lovingly holding a dog in his arms. Bruno was always very fond of pets. In Glasgow, where he spent 5 years after the war, he took care of a cat and then a dog, neither of whom unfortunately survived the harshness of the Glasgow winters.[19] Likewise, later in Rome, where he had settled in 1953, he had a dog as well as two cats, whom he named after famous physicists, Planck and Pauli (Amaldi 1981, 15).

---

[17] According to Amaldi, Bruno felt himself responsible for having been a burden to his father's career (Amaldi 1981, 4).

[18] In later years, when Bruno was in Glasgow in 1948, Rosa worried of not receiving his letters for two months, even going so far as to inquire about him with Philip Dee, the Chair of the Physics Department (Pancheri and Bonolis 2020, 26).

[19] Letters to parents in 1948 and 1949.

2.2 The School Years

Bruno started high-school in 1931. He entered the Piaristengymnasium in the eighth district (G8, now bG8), in the academic year 1931–1932, and continued until 1938, when the annexation of Austria to Germany, the *Anschluß*, changed everything for Vienna Jews (Shirer 1941; Zweig 1943; De Waal 2012; Offenberger 2017).

The Piaristengymnasium was, still is, a public high-school, with a classic academic curriculum, eight years of Greek and four of Latin studies, with the reputation of being the best academic public Gymnasium in Vienna, and even beyond. Bruno's memories of his professors from these years are described in a typewritten note he drafted many years later, as follows:[20]

There was in Vienna at the time – in 1938 – my teacher in German and Latin language. His interest was comparative philology, comparing "equus" and "caballus", the strange habit English people have of giving Saxon names to the animal (ox, steer), and Latin names to their meat (beef), and also Grimm's law and the futility of vowels. All things which interested me at the time and still do.

There was the religious studies teacher [...], whose unctuous manner led to an entire generation of atheists - including a bishop's nephew, who became a surgeon, and whose almost psychoanalytic deafness was probably due to having been exposed to such false sounding voice.

Then there was the math and physics professor, a university graduate, but also a carpentry teacher and the owner of a cinema house, who would make us build mechanical and electrical models, and was also teaching us the theory of shooting, being obsessed with war machines, I don't know why. None of us became an artillery officer, but I am still interested in how to throw a tennis ball, or build something mechanical to take the place of a boy catching a flying ball.

Finally, there was the botanics and zoology professor, one meter and a half tall, who, possessing like me a Slavic name, was therefore convinced of the superiority of the German race.

There was another math professor, with a long beard [...] who never succeeded in finishing a calculation on the blackboard. Instead of the usual 'quod erat demonstrandum', he would then say 'errare humanum est' until one us would say 'Sir, you are very human'. However, there is something which stayed with me from these messy lectures: that the right to make mistakes is what distinguishes the scientist from the faithful.

---

[20] The note was written in Italian and translated by the Author. Its date is unknown, but its placement among other documents, in Mrs. Elspeth Yonge Touschek's possession, suggests it may have been written in 1971.

When the German troops entered Austria in March 1938, Bruno was in the 7th year of his studies and would soon start preparing for the *Matura* examination, which normally would have taken place in June 1939, at the end of his eighth year, and which would have allowed him to attend the university. He was suddenly aware of how the school itself was changing. In the above mentioned memoirs, he says:

```
One who was short and fat was the Professor of Greek. [...]
When the Nazis arrived in 1939 he decided to take
measurements of the students' skulls. It turned out that I
was the only 'Nordic' student - with a Jewish mother and a
Czech-German father - which proves that one can learn even
from indifferent teachers.21
```

During this time the first interruption of Bruno's studies took place, and his life changed course. With the *Anschluß*, German laws were applied, in particular concerning the proportionately large population of Jewish students in the most prestigious of the Vienna high-schools (Beller 1989). From the Piaristengymnasium, these students were moved to another newly created Jewish-only school. As for Bruno, his half-Jewish status allowed him to continue at the Piaristen for some time, but he had to leave it in 1938, before taking the final exam. Still he managed to find a way to finish and take the 1939 *Matura*.[22] How this happened is not completely clear. One account is given by Edoardo Amaldi, who collected Touschek's memories of this period and then transcribed and translated them into English with Bruno's assistance. A second comment comes from Heinrich Mitter, who had met Touschek in person in the '60s and did some research on Bruno's studies in Vienna.[23] The third comment is from Klaus Gottstein, who met Touschek in Göttingen, after the war.[24] Both Mitter and Gottstein's comment came to us through e-mail exchanges after the publication of (Bonolis and Pancheri 2011), where the story of Touschek's high-school interruption follows Amaldi's description. Such interruption was a very traumatic event in Bruno's memory and it is worth analyzing it in detail, since Gottstein and Mitter's comments enlarge, but also contradict Amaldi's description, leaving some questions unanswered.

---

[21] The year 1939 does not sound right, as the Nazis arrived in 1938. And in 1939 Touschek had already left the Piaristengymnasium.

[22] This end-of-school examination, prerequisite to university admission in Germany is called *Abitur*, in Italy *maturità*.

[23] Heinrich Mitter, professor of Theoretical Physics at the University of Graz, had met Touschek in Tübingen, where Mitter was professor at that time and had invited Touschek to give a seminar. He had known of Touschek much earlier. Firstly because his colleague Paul Urban told many stories about Touschek. Next he had encountered Touschek's name through the (chiral) transformation named after him (Cini and Touschek 1958; Touschek 1959), having collaborated for quite some years with Heisenberg in Göttingen and Munich on the nonlinear spinor theory, for which the transformation was very important. Personal e-mail communication to the Author.

[24] Professor Klaus Gottstein (1924–2020), Emeritus Professor of the Max Planck Institute, started as a particle physicist, later becoming involved in issues at the interface of science and politics, as in his work about The Amaldi Conferences. His views about the controversial question of Heisenberg and Bohr can be seen in (Gottstein and Chao 2002).

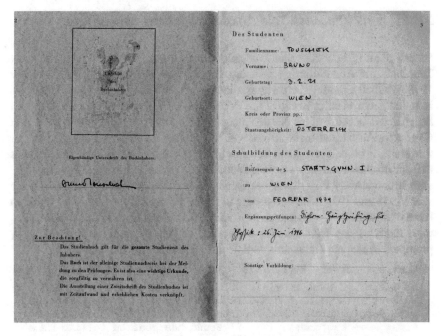

Fig. 2.5 Copy of Touschek's student book from Göttingen University, noting his graduation from high-school in February 1939, Family Documents, © Francis Touschek, all rights reserved

According to Amaldi (Amaldi 1981, 3), Touschek 'was told that he could no longer attend school because [...] his mother was Jewish' (Amaldi 1981, 3). About the final exam, Bruno's CV indicates that he graduated from the Schottengymnasium in Vienna. Mitter writes to have found records proving that Touschek remained at the Piaristen until December 1938, and then took and passed his exam at the Staatsgymnasium on February 28th, 1939, a date and a school confirmed by Touschek's student book from Göttingen University, Fig. 2.5 (Bonolis and Pancheri 2019). Gottstein's comment complements Mitter's and refers to a question, which can highlight Bruno's characteristic reaction to authority, namely that Bruno could not have been expelled by the Piaristengymnasium for purely racial reasons. So why was Touschek not allowed to finish his studies at the Piaristen and what happened in the months between December 1938 and February 1939? This is a time when Bruno's life was disrupted and his recollections, as reported by Amaldi, can give a glimpse of Vienna in the aftermath of the *Anschluß* until the start of WWII, in September 1939, and the impact on Bruno's life from such dramatic changes.

To look for an indication as to what happened in the months between December 20th, 1938, when he was dismissed from the Piaristen, and February 28th, 1939, when he passed his final exam, one can first turn to Amaldi, then to Mitter and Gottstein.

2.2.1 What Touschek Told Amaldi About His School Years

On the point of how Touschek was able to obtain his *Matura*, let us check what Touschek told Amaldi during their last conversation in Geneva, in late February 1978. We shall reproduce here a translation of the typewritten notes Amaldi prepared in Italian, after visiting Bruno at the Hôpital de la Tour, where Touschek had been hospitalized, because of his fast declining liver condition.[25] Shortly after, Bruno left Geneva for Austria, where he arrived with a CERN car in March. He was staying near Innsbruck, in his beloved Tyrol, where he had hoped to recover. Amaldi sent his notes to Bruno to check, and they were then returned, with some corrections. This exchange, with Bruno already very ill, may explain some confusion of dates pertaining to the years 1938 and 1939, which remained in Bruno's biography by Amaldi, published three years after Bruno's death on May 25th, 1978, (Amaldi 1981). Such confusion, most likely arising from Bruno's corrections or lack of them, is what originally prompted Mitter and Gottstein's comments to (Bonolis and Pancheri 2011).

> Geneva, February 28th, 1978
>
> At 1 p.m., I went to visit Bruno Touschek at the Hôpital de La Tour in Meyrin. I had already gone to visit him yesterday with Sergio Fubini, but I wanted to talk to him once again. After some exchanges, [. . .] Bruno has started telling me a part of his life, and I took some notes, which I reproduce here in an immediate attempt to re-arrange what he told me in his Italian, so extraordinary in its precision and clarity. [. . .]
>
> As Summer 1938 started Bruno was finishing his 8th [actually 7th, Author's note] class of the Piaristen Gymnasium, one year away from the *Matura* exam, when he was told he could no more attend the school as he was of mixed heritage, since his father Franz Xaver Touschek, a Major in the Austrian Army [*nello Stato Maggiore dell'Esercito Austriaco* in original notes], had married Camilla Weltmann, who was Jewish.[26] He thus stopped attending school but he had many friends, whom he would meet

---

[25] From Edoardo Amaldi Papers in Sapienza University of Rome–Physics Department Archives, https://archivisapienzasmfn.archiui.com, all rights reserved.

[26] In (Amaldi 1981), one reads "Bruno had attended school in Vienna and, at the beginning of Summer 1937, he had completed the 8th class of the Piaristen Gymnasium, that is a year before the Abitur (state examination), when he was told he could no longer attend school because he was of mixed blood as his mother was Jewish." We note that at beginning of the summer 1937, Bruno would have finished the 6th year of school and would have been entering the 7th in the following fall semester, as he had entered the Gymnasium in the academic year 1931–1932, with Bruno being, at the time 10 years old. This was checked by Mitter and recently by the Author, *via* correspondence with Piaristen Gymnasium. Amaldi then continues, and the next strange date comes as he writes that "[. . .] the Abitur examination of 1938 was brought forward to February so that a large number of young men could would be immediately available as junior officers", and then "Thus he passed his state examination and in February 1938 he went to Rome for the 'school-leaving holiday' according to the tradition of the bourgeoisie of that period." and "Bruno had thought of of studying engineering in Rome and so he began to attend the first two-year course in engineering in the spring 1938."

in the *cafés*, and who kept him informed of what was happening. With the beginning of the war in sight, the Austrian High Command was already making preparations. In particular the *Abitur* examination of 1938 [actually 1939, Author's note] was brought forward to February, so that a large number of young men would be immediately available as junior officers. A friend who attended another school, suggested that he sat at the exam at a different school as an external student without making any mention of his real position. Bruno took his advice and sent his application to the Director of Education of the SchottenGymnasium. He was allowed to take the exam and passed it very well in all subjects except Greek, where he was declared at first as 'nicht genünged' (insufficient), but the decision was later changed into 'genügend', as an order had arrived to pass large numbers of young men who would soon be needed as reserve officers.

Thus he passed his state examination [...] in February 1938 [actually 1939, Author's note] and then he went to Rome for the 'school-leaving holiday'.[27]

What Amaldi writes, namely that Bruno found a way to take the exam by enrolling in the Schottengymnasium (officially, Schottengymnasium der Benediktiner in Wien), is also confirmed in one of Touschek's CVs kept at University of Rome.[28] The Schottengymnasium was a private catholic school, which was closed after the Nazi took over, and it appears that the students were officially enrolled in the Staatsgymnasium, which granted the *Matura*, as the available official documents show in Touschek's case.

2.2.2 Comments About the Austrian School System and Touschek's *Matura*

Here follows what Prof. Heinrich Mitter, from University of Graz, who knew Touschek in the 1960's and has done some research about Touschek's school years, wrote to the Author on June 8th, 2011.

In order to clarify the facts concerning Touschek's school education it seems appropriate to start with some remarks on the school system.[29]

This date should be read as '1939', as seen from Bruno's 1939 letters from Rome, which we shall discuss later.

[27] Translated from Italian by the Author from original photographed by Luisa Bonolis, courtesy of Sapienza University of Rome–Physics Department Archives, https://archivisapienzasmfn.archiui.com, all rights reserved.

[28] Bruno Touschek Papers, courtesy of Sapienza University of Rome–Physics Department Archives, https://archivisapienzasmfn.archiui.com, all rights reserved.

[29] The school system in Austria in the 20th century (before, during and after the time of the Nazi government, even up to now) started for children at the age of 6 years and involved four years of elementary school (Volksschule) and then (according to the choice of the parents) either four years

Most of the Austrian schools were and are state schools managed by the federal republic ("Bundesrepublik") without tuition. Before 1938 also private high-schools existed, which were managed by the Catholic Church (mostly by religious orders). They charged tuition. The Matura had to be passed in presence of a state commission. The Nazis closed these schools in the first weeks after the "Anschluß". The catholic schools were re-established in 1945.

One of the most famous catholic schools was and is the "Schottengymnasium" managed since 1807 by Benedictine monks inside their monastery (founded 1155) in central Vienna (address Freyung 6). After the school was closed by the Nazis the classrooms were used by the Staatsgymnasium Wien 1 (SG 1).

As far as Touschek's high-school studies are concerned, I have obtained copies of Touschek's year certificate for the 8th class and his MZ (Maturazeugnis), as well as the corresponding "Hauptprotokoll" (Main Report) on the Matura 1939 from his last school Staatsgymnasium Wien 1 (SG 1). From these documents, the relevant dates of his school years can be reconstructed. I have also obtained some additional information on the political situation, in particular with respect to Jewish students, from school histories. The following facts have emerged.

Touschek entered the Gymnasium in the 8th district (G8) in 1931 (school year 1931/32), aged 10 years. When the Nazis came to power on March 13th, 1938, he was in the 7th class (1 year before Matura). The director of G8, who was a firm Christian democrat, was fired immediately and replaced by a fanatic Nazi. According to the German laws, racially Jewish students were not allowed in high-schools. If there were few Jewish students in a school, they could finish the running school year ending in summer 1938. In Vienna, however, a considerable fraction of the high-school students were racial Jews. In April 1938 some rapid transfers took place, by means of which a few exclusively Jewish high-schools were created, were the students could go on for some time. Thus, at G8 all 42 Jewish students were transferred to the now Jewish G2 in April 1938. Touschek was, however, not transferred: according to the laws he was a "1st class half-breed" (one of the parents was Jewish) and these were allowed in high-schools until 1942. So Touschek finished the 7th class at G8 and stayed there even for the first trimester of the 8th class, for which he obtained a certificate dated Dec. 20th, which is quoted in his subsequent records. After the Christmas vacations he changed to SG1, where he finished the 8th class in February 1939 and passed the Matura in the same month. The MZ is dated Feb. 28th, 1939. It contains the remark, that Touschek "had begun his studies at G8 in 1931/32 and continued at SG1 without interruption in Jan.1939". The topics (marks) for the written part of the Matura exam were German (good), Mathematics (good), Latin (good) and Greek (sufficient). For the oral exam he had chosen Mathematics (very good) and Latin (good).

Thus, according to Mitter, Bruno was allowed to continue the school year after March 1938 and start the next trimester in September. While the mixed up dates in (Amaldi

---

of main school (Hauptschule, ages 10–14) or eight years (ages 10–18) of high-school. The latter type required some entrance exam and the parents could choose between different curricula; the most frequently offered ones were Gymnasium (Latin for 8 years, Greek for 4 years,) Realgymnasium (a modern language for 8 years, Latin for 4 years,) and Oberrealschule (no ancient languages). Usually the school year started early in September and ended in the first half of July. It was partitioned into two semesters until 1938, afterwards into three trimesters. The students obtained semester (resp. trimester) certificates and year certificates. After passing all eight classes successfully, the students had to pass the Matura exam, which involved written and oral parts in a number of topics. The corresponding certificate (Maturazeugnis, MZ) contained also the marks of the year certificates from the 5th to the 8th class and was/is a necessary requirement for any university studies. From H. Mitter's note.

1981) are easily clarified as probably due to Bruno's traumatized memories of that period, the question remains as to which motivation could be used for his leaving the Piaristen in December, since Bruno was not a so-called *first class Jew*?

A suggestion, as well as a confirmation of Mitter's, words can be found in a 2012 correspondence from Prof. Klaus Gottstein, who had met Bruno Touschek in Göttingen after the war. In this correspondence, Gottstein comments upon information given in (Bonolis and Pancheri 2011, 20–21).

Pages 20/21: The information given about Touschek's high-school experience is strange. It is true that the discriminating laws of Nazi Germany became valid also in Austria after the annexation of Austria by Germany in March 1938. However, in Nazi Germany, "Half-Jews" were expelled from high-schools (gymnasiums) and excluded from the Abitur (matura) only in and after 1942. I myself, also considered to be "half-jewish" by Nazi terminology, attended a Berlin gymnasium and obtained my Abitur in 1941 without any problem. I just noticed by chance in the Internet that the Austrian "half-jewish" professor of medical history Philipp Engel also attended a regular school in Vienna until 1942. Thus, it is likely that the reason for Touschek's abandoning the Piaristen Gymnasium was not related to the general racial laws introduced by the Nazi Government but to the arbitrary action of an anti-semitic school director. Otherwise the Schottengymnasium would not have been allowed to accept him for graduation either. Moreover, according to your report, Touschek was able to enroll at the University of Vienna in October of 1939, attended courses and passed exams there. This would have been impossible if the later racial school laws of the Nazis had already been in force in Vienna in 1938 and 1939. Touschek was expelled from university life in May 1940, at a time when this happened to "Half-Jews" all over Germany.

Gottstein confirms that Bruno could not have been expelled purely in relation to the racial laws, and suggests an arbitrary action by an overzealous school director, an hypothesis resonating from Touschek's reminiscences of his Professor of Greek, who decided to take measurements of the students' skulls when the Nazis arrived. Thus Bruno, who clearly made fun of that professor, was probably forced to leave the school for a combination of reasons, at the least a disciplinary action motivated by Bruno's intolerance of authority and racial motives.

2.2.3 1939: Dreaming of England from Rome

A war was approaching, timing was crucial, choices had to be made. In 1938, Bruno's grandmother had moved to Rome to live with her daughter Ada. Having lost two of her children, Oskar and Camilla, and with daughter Ella living abroad as well, it was a natural move for her to leave Vienna, escaping both the danger impending on Viennese Jews, and joining her eldest daughter. But the Italian city was not a safe haven for Jews either. In Italy, 1938 marks the year when the Mussolini government forged its alliance with Hitler's Germany, marked by the Führer's visit to Rome in May. At the end of the year, following the example of the Nuremberg Laws, the Italian Parliament promulgated the infamous *Leggi razziali* [Racial Laws], the anti-semitic laws which canceled the civil liberties of Italian citizens of Jewish origin, such as excluding them from higher education, and stripping them of their assets.[30] And thus, in 1939, probably as a consequence of these changed circumstances, her daughter Ada convinced Josefine to return to her homeland (Amaldi 1981, 13).

It is quite possible that Bruno was the one who accompanied his grandmother to Rome in Spring 1938. We have no record of it, except in a letter he wrote home on March 16, 1939, where he makes the comment

```
Rome without the Führer is quite nice.
```

This comment seems to imply that he was in Rome during Hitler's visit, which took place on May 3rd, 1938 (Cavendish 2008).

Through 1938, the thought of leaving Vienna matured in Bruno's mind. During the violence and disruptions, which followed the *Anschluß*, the feeling that he could be a hindrance to his father's life and professional advancement brought the decision to leave and study abroad. His father would soon be asked to return to active service in the Army, as his expertise and professional contacts would be valuable in the war preparations undergoing in the German world. Franz refused, probably for political reasons, which he shared with his son, and Bruno appreciated it. But, at the same time, as he told Amaldi, he felt his father was sacrificing his career in solidarity with his son, exposed to discrimination by this same government which wanted Franz's loyalty: 'At the height of the racial persecution, Bruno suffered at the thought that he was a burden on his father, and that he was unintentionally damaging his career as well as his private life, as a living testimony of his first marriage to a Jewish girl' (Amaldi 1981, 4).

Indeed further dangers and tragedy could be foreseen. Bruno started considering where to go and continue his studies. As reflected by the results of his final exams, *very good* in Mathematics, only *good* in Latin and barely *sufficient*—if not worse—in Greek, his interests were clearly oriented towards science.[31] An option was offered by the University of Rome, endowed with excellent science and engineering schools. This option was relatively inexpensive, as he could live with aunt Ada and her husband G. Vannini in their nice apartment. While waiting for his final exam, he inquired around, and came also to consider emigration to England, and study Chemistry at the University of Manchester.[32] He learnt about the Society of Friends, an organization established in Vienna and run by the Quakers, which was engaged in helping people wanting to emigrate from Austria, by assisting them to obtain entry visa to Great Britain.[33]

---

[30] A moving memory of what happened to an Italian Jewish family during the war can be found in (Camiz 2018). Paolo Camiz, who became a successful physicist and a musician, in 1959 attended Bruno Touschek' course on *Statistical Mechanics*, at University of Rome. This was the first year Touschek was teaching this course, which was to have a long life and influenced many students.

[31] We rely here on Prof. Mitter's note.

[32] A recent article on the UK newspaper *The Guardian*, formerly *The Manchester Guardian*, recalls the 200 ads which saved Jewish children by asking for foster care or proffering household help. Similar ads were also placed in other UK newspapers, but appeared most frequently in *The Manchester Guardian*.

[33] Another important organization which helped relocation of German scientists persecuted by the Nazi regime, was the Society for the Preservation of Science and Learning (SPSL). The SPSL

Shortly after passing his *Matura* exam at the end of February, Bruno went to Rome. There was no time to ponder a decision. It was 1939, and emigration to England for Austrian citizens of Jewish origin was already difficult.[34] A letter from the Society of Friends in London had sent Bruno a questionnaire to be filled, but it seemed that this road was closed, as Bruno, who had already turned 18 years old five weeks before, would not qualify.[35] By March 11th, Bruno had already gone to the British consulate in Rome and spoken to the consul, who was very kind, but not immediately very helpful. From the British consul, he learnt that there was an English College in Rome, which might perhaps be of assistance in his emigration efforts.[36] Bruno went there, but probably because of his family history or the experience with the infamous high-school teacher of religious study, he turned around without further pursuing this road. He had previously written to the British Ministry of Education, and was expecting an answer. At the consulate, they assured him that the Ministry always replied. Perhaps, he thought, their letter had arrived and had been put into aunt Ada's mail box, which hung in her garden, with its back door always left open. He whimsically mused that his letters could have escaped the mail box and, taking flight with the March wind, scattered away to find a resting place in a wastepaper basket in the nearby park of Villa Borghese, just outside the north side of Rome's ancient walls.[37]

In March, Rome can be at its best. The spring light fills the sky and wakes up the city, which winter usually puts in a semi-dormant state. Nowhere else in the city as in Villa Borghese, one can see the progression of nature. Trees and parks get a fuzzy dressing from the new leaves, taking over the old ones, still on the branches, which never get to be really bare, as winter is mild in Rome. Aunt Ada's apartment was a short walk from the Villa Borghese and the *zoo*, which Bruno visited often.

---

succeeded in collecting large sums of money to help scientist to move and relocate, and to inspire awareness to the plight of scientists in Germany and Austria, mostly Jewish, but not only (Zimmerman 2006).

[34] Sigmund Freud had emigrated to London in 1938, almost at the last moment, as also described in https://www.thejc.com/lifestyle/features/how-a-nazi-saved-sigmund-freud-1.13679 but, even for him, it had not been possible to take all his family away.

[35] The Quaker organizations were very active in helping Jews to emigrate from Germany and Austria, and other nazi occupied countries. One of the most well known action was the so called Kinder Transport for children younger than 17 years of age. From wikipedia: '10, 000 children, the majority of whom were Jewish, were brought to Britain from Germany, Austria, Czechoslovakia and Poland to escape persecution by the Nazis between 1 December 1938 and 1 September 1939. What came to be known as the Kindertransport was the result of the combined efforts of Jewish and Quaker organisations in successfully persuading the British government, in the days after Kristallnacht in November 1938, to ease its immigration restrictions for refugee children. The children were permitted to enter Britain on temporary visas without their parents if a guarantee of 50 pounds per child were provided to cover the costs of care, education and re-emigration from Britain once the war was over. If the children were over 14, they were to be found work in agriculture or domestic service.'

[36] The Venerable English College of Rome is a Catholic institution for the training of priests destined to the missions of England and Wales. It was originally founded in 1362 as the English Hospice, a place for pilgrims from England and Wales, on the site of the present College in the center of Rome.

[37] Letter to parents, March 11th, 1939.

Fig. 2.6 Photograph from the Pincio terrace in Villa Borghese, where Bruno used to go for walks, 2021 photograph by A. Srivastava, all rights reserved

He used to walk from his aunt's place to the Pincio Terrace, the highest point in the park, from where one can see the city beyond the river Tiber, towards the Vatican, and, just below, Piazza del Popolo, with the obelisk at its center, the twin churches on the viewer's left and, on the right the Church of Santa Maria del Popolo, which houses two Caravaggio paintings, Fig. 2.6.

He also went to the University of Rome and attended a course on 'Mathematical Analysis' for students of science and engineering, taught by Francesco Severi, a renowned mathematician (Roth 1963), later remembering this experience as having been one of 'more enthusiasm than success'.[38] Although anxiously waiting for the visa to go to England, he was also young, and could enjoy the city and its many attractions. He would go fencing in the American club, went to parties where eligible girls were shown out by their mothers, visited Venice and, impatiently, listened to his uncle Nino's lessons about how to make money. Aunt Ada ran a strict household, and Bruno was soon worried about having to ask for his everyday expenses, and it did not take long for him to start feeling constrained by her checking even how much electricity was consumed by his reading at night.[39]

---

[38] See Bruno Touschek's CV in Amaldi Archives.
[39] Letters to parents March 16th and 20th, 1938.

He had registered at the University of Manchester, and was waiting to hear if he had been enrolled. For this, he needed proof that he had passed his *Matura*, alternatively he would have to pass the English *Matriculation*, and asked his father to please send his school leaving certificate with the next post.[40] Fearful that the papers could not arrive in time from Vienna, he started preparing for the physics 'matric', as it was colloquially called, thus rapidly improving his English as well.

The papers from Austria arrived, and one obstacle was eliminated. Around March 20th, he was still quite hopeful that the plan for England would go through and listed the progression of this plan, as follows:[41]

```
1. I do not need to pass a graduation exam in England.
2. I have enrolled at the University of Manchester, and
I am probably registered there as a student.
3. The Society of Friends (Manchester not London)
handles the entry and all matters for study assistance.
4. My visa to England is expected to come this week or
next week.
5. Travel question not yet clarified.
```

Clearly, in late March, all still depended on getting the visa, and, were this granted, handling the expenses and cost of travel. The question of handling the money was no more a minor one, however. Bruno's family in Vienna was now experiencing financial difficulties. The worsening situation for all Vienna Jews affected Josefine, and her money. Bruno was aware of this and was very worried about his much loved grandmother, who had returned to Vienna, and must have felt lonely and unsafe. As for Bruno's father, Franz could only rely on his Army pension, as he probably saw business opportunities drying up, both because of the impending war preparation and because of his decision not to re-enter the Army.

As the days passed, Touschek was feeling more and more restless, finding it difficult to go along with the political ideas of his uncle and the money worries of his aunt.[42] As it would happen many times in his life, the uncertainty about his future would be hard for him to bear, and the world around him would be laid bare with its cracks and pains. When this happened, he would want to leave, and search for a way out. In mid-March he had liked Rome and enjoyed walking around as a carefree tourist, without a guide, talking to girls and visiting the zoo, a few steps away from aunt Ada's home.[43] But as time slowly passed, his mood became darker and he felt it had been a mistake to go to Italy. Years later, in Göttingen, he remembered those anxious days in Rome. In 1946, waiting to know about his future, and responding to

---

[40] Letter to parents, March 11th, 1939.

[41] Letter to parents, on March 20th, 1939.

[42] Undated 1939 letter to parents. In this letter Bruno writes that his uncle keeps telling him how much he had earned through the Abyssinian War, and drily comments 'as if this were so noble and good'.

[43] Letter to parents on March 16th, 1939.

his parents' pressure to return to Vienna, he would regret not to have gone to England before the war, calling it 'the first mistake' (Bonolis and Pancheri 2019, 30).[44]

At the end of March, Bruno went back to Vienna.[45] Had the visa been granted? We do not know. He had done all he possibly could to get away. He was in Rome on March 11th, 1939, and remained in the city until the end of the month. During this time, he had enrolled to study chemistry at University of Manchester, applied to enter the UK, and attended courses at the University of Rome, waiting day by day for the visa to arrive from the British consulate. But then he returned to Vienna, just as his grandmother had done. From there, in 1942, Josefine would be tragically deported to the extermination camp of Theresienstadt, where she died in March 1943.[46] Bruno's destiny, would be different, although his life did come very close to a dramatic ending. But this is part of Touschek's legend and will be recounted later.

2.3   Trying to Study Physics: Expulsion from the University of Vienna in May 1940

On September 1st, 1939, German troops invaded Poland, and the war, which had been under preparation on all sides for quite sometime, started. Bruno, like so many others, had seen it coming, especially during his month in Rome. Unlike the annexed Vienna, Rome was the center of national political decisions, in particular about whether Italy would join Germany or not in an impending war, and the city was full of foreign journalists and rumors. This had in fact been one of his worries during the three weeks or so he had been waiting day by day for permission to go to England. He could see the war would be a major thwart for his plans, because England would be closed in the event of war. But at the same time he did not want to stay in Italy, perhaps because of money, or disagreements with his aunt and uncle.

In September, he was enrolled to study physics at the University of Vienna. According to an official University document Bruno was in fact registered as a regular student at the Faculty of Philosophy from the winter term 1939/40 until the first trimester 1940.[47] During this first year, effectively only two trimesters, Bruno soon proved to be an exceptional student. He attended and passed the examination for the required ten courses, Physics Laboratory, Differential Equations, Mathemat-

---

[44] His regrets about not having gone to England were expressed in a letter to his parents from Göttingen, on 16th December 1946.

[45] In addition to the three letters from Rome on March 11th, 16th and 20th 1939, there is one more letter, dated only 1939, and then there are no letters from Bruno to his father and stepmother until February 1942, from Munich. Given Bruno's habit to write very frequently, he must have returned to Vienna not much longer than one week, or so, past March 20th.

[46] https://www.holocaust.cz/databaze-dokumentu/dokument/96485-weltmann-josefine-oznameni-o-umrti-ghetto-terezin/, courtesy of IKG Archives. Josefine's probable deportation date is given as 1941 in Amaldi's (Amaldi 1981), but Shoah database gives the date as of 27.8.42, courtesy of G. Violini.

[47] Document courtesy of Prof. H. Posch, University of Vienna.

ics Pro-Seminar, Rational Mechanics, Theory of Functions, Seminar in Theoretical Physics, Statistical Theory of Heat, Exercises in the Theory of Functions, Physics Laboratory II, Thermodynamics. In the Mathematics Pro-seminar, where students would give presentations of their choice, he discussed the Markov double sum series, in verses (Amaldi 1981, 4).

Soon however, the normal course of Bruno's life was disrupted once more and he was forced to take a different path. On May 8th 1940, the "Reichsminister für Wissenschaft, Erziehung und Volksbildung", declined the application of Touschek (who was a so-called 1st class half-breed) for a continuation of his studies.[48] Touschek's expulsion from the University at the end of the 1939–1940 academic year is also confirmed by the Archives of the Israelitische Kultusgemeinde. After a renewed application, a conditional registration was granted for the winter term 1941/1942, but in the dangerous political situation rapidly engulfing Vienna in 1940–1941—the first years of the war—professors had to show their loyalty to the government and its anti-semitic policy, and it was uncertain whether Bruno would be able to continue his studies. Luckily, he found support and friendship in Paul Urban, a young faculty member, tolerated in his career, but not encouraged because of suspected lack of political loyalty.[49] As Urban wrote to Amaldi, after Touschek's death,[50]

Having been dismissed from my state employment (State Railroads) on March 13th 1938 [the day of annexation of Austria] because of my 'hostile' behaviour [...] under the existing regime, my position at the University also became very uncertain. My political positions were so well known that I could not become a 'Dozent' but I had to content myself with the title of 'Dr. habil.' in order to prevent me from coming into contact with students through the teaching. [...]

Urban was assistant to the German Professor Erwin Fues, who had been a graduate student of Arnold Sommerfeld, one of the early founders of quantum mechanics (Eckert 2013a).[51] At the Institute for Theoretical Physics, Fues had replaced Hans

---

[48] See also the article https://www.tandfonline.com/doi/pdf/10.1080/14623528.2019.1634908? needAccess=true which gives information on mixed marriage Jews (Beller 1989).

[49] Paul Urban (1905–1995) was Professor of Theoretical Physics at University of Graz from 1948 until retirement. His obituary by Fritz Rohrlich in the July 1995 issue of *Physics Today*, https:// physicstoday.scitation.org/doi/pdf/10.1063/1.2808109, mentions in particular his 'dangerous' train rides to Poland during the war to help 'destitute Jewish physicists' and his efforts after the war to prevent 'former Nazis to regain important positions.' See also (Plessas 1995).

[50] From Edoardo Amaldi Papers, courtesy of Sapienza University of Rome–Physics Department Archives, https://archivisapienzasmfn.archiui.com, all rights reserved.

[51] Arnold Sommerfeld (1868–1951) formulated the first semi-classical description of the atom, the famous Bohr-Sommerfeld model, where for the first time the fine structure constant $\alpha = 1/137$ was introduced. Through his many extraordinary students, he was a major influence in the development of the great theoretical physics school of the first half of the 20th century. Among his students and post-doctoral researchers there were many Nobel prize winners, such as Peter Debye, Wolfgang Pauli, Werner Heisenberg, Linus Pauling, Max von Laue, Isidor Rabi and Hans Bethe, although himself never won the Nobel Prize, even if he was nominated many times. He obtained his Habilitation in Göttingen under Felix Klein, and in 1906 became Professor of Theoretical Physics in Munich. He was an early strong supporter of Einstein's theory of relativity, and opposed the anti-semitic laws through which a large number of scientists were dismissed from German Universities. Under new

Thirring, who had been sent into forced early retirement for political reasons in 1938.[52] Not much caring for political issues, Fues had accepted Urban as assistant.[53] Urban was generous and courageous and could not avoid helping his students. In the letter to Amaldi, he also says:

> I took Bruno with me, to give him the possibility, together with a few similar cases, such as Koch, Fränkl, etc., to work with me and use the library undisturbed. He immediately appeared to me to be a talented person, but rather difficult because he was self-opinionated.

Recognizing the intellectual promise of young Bruno, Urban suggested that he read, with his assistance, the major treatise on atomic and quantum physics, Arnold Sommerfeld's Atombau und Spektrallinien (Eckert 2013b).[54] Having carefully gone through it, Bruno spotted some small errors in the first volume and discussed them with his Professor of Mathematics, Edmund Hlawka, who suggested that Bruno write directly to Sommerfeld.[55] In Fig. 2.7 we show a photograph of Paul Urban on the left and Edmund Hlawka on the right.

Sommerfeld replied and invited Touschek to read the second volume, as well, which Bruno did, as one can see in the preface to the second edition of this second volume, where Sommerfeld thanks Touschek for a critical revision of the text, as shown in Fig. 2.8.

And thus, from the correspondence between young Touschek and Sommerfeld, there unfolded the events which ultimately led to Bruno's proposing and constructing the first electron-positron collider in the world, AdA.[56]

---

laws made by the regime, he should have resigned in 1936, and suggested his former pupil Werner Heisenberg to become his successor. Opposition to Heisenberg kept him in his post until 1939, when a candidate more loyal to the Nazi regime was found. Sommerfeld then retired, but returned to his Chair in 1945, after the war.

[52] Hans Thirring (1888–1976) served as assistant, professor, and head of the institute for theoretical physics of the University of Vienna until his forced retirement in 1938 after the Anschluss. He was a pacifist and a socialist. He is known for the prediction of the Lense-Thirring frame dragging effect of general relativity in 1918. After the end of World War II, he was reinstated and became dean of the philosophical faculty in the years 1946–1947.

[53] Information about Paul Urban was kindly contributed by Professor Heinrich Mitter, from University of Graz, where Urban was Professor at the Institute for Theoretical Physics since 1947 until retirement. Mitter had obtained his doctoral degree in Graz, with a thesis supervised by Urban, and was for some time Urban's assistant, much later in 1976 becoming his successor.

[54] Sommerfeld's treatise had many editions: Atombau und Spektrallinien (Friedrich Vieweg und Sohn, Braunschweig, 1919 (1. Auflage), 1921 (2. Auflage), 1922 (3. Auflage), 1924 (4. Auflage), 1929 (Atombau und Spektrallinien. Wellenmechanischer ErgŁnzungsband), 1931 (Atombau und Spektrallinien. Band 1. 5. Auflage), 1939 (Atombau und Spektrallinien. Band 2. 2. Auflage), 1944 (Atombau und Spektrallinien. Band 1. 6. Auflage),1951 (Atombau und Spektrallinien. Band 1. 7. Auflage + Atombau und Spektrallinien. Band 2. 3. Auflage des wellenmechanischen ErgŁnzungs-bandes, information courtesy of Luisa Bonolis. It is not known which edition, Auflage, Touschek studied, but he probably read the first volume, Band, of the fifth edition, which came out as Volume 1 in 1931, and was followed by a second volume in 1939.

[55] Edmund Hlawka (1916–2009) made notable contributions to the theory of numbers (Amaldi 1981).

[56] The Sommerfeld-Touschek correspondence includes letters maintained at the Deutsches Museum, Archive.

Fig. 2.7 Paul Urban (1905–1995) on the left, reprinted by permission from Springer Nature: (Plessas 1995), copyright © 1995, Springer Verlag. Edmund Hlawka (1916–2009) on the right, courtesy Archive of the University of Vienna, picture archive, Originator: Foto Karl Winkler, Wien Signatur: 106.I.1073 - 1955

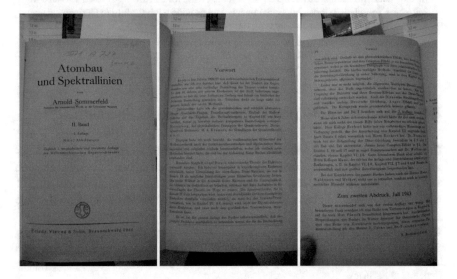

Fig. 2.8 Front page of *Atombau und Spektrallinien* and acknowledgment of Touschek's contribution in the Preface to the second edition of the second volume. Photograph courtesy of Luisa Bonolis

After his expulsion, Bruno's presence in the University became very difficult, and pressure grew on Urban and his Principal Theodor Sexl, who was always called on the phone from the Second Institute for Experimental Physics and requested to take measures against Urban, should the latter still be gathering with students who were not of pure race (Amaldi 1981, 8).[57] In the end, the temporary solution was that Bruno and other Urban's *protégés* (as he calls them in the letter to Amaldi) met at Urban's home, where their minds were nurtured by Paul's tutoring and the books he borrowed from the library, while his mother 'provided them with food.'

On December 1st 1941, Bruno applied again to be admitted at the university as an auditor, but his application was refused. This is the answer that he received on January 17th, 1942:[58]

> Based on the opinion of the Dean of the Philosophical Faculty, the conditional enrollment required by the hybrid I Grades Bruno Touschek as guest listener for the winter semester 1941/42 can immediately be revoked. Touschek is to be informed of this by submitting the application enclosures.

Nothing more could be done. But Bruno was not alone. What happened in fall and winter 1941–1942 and which would decide Bruno's future fate was described by Paul Urban in his long letter to Amaldi (Amaldi 1981, 7):

> At the time, I was working on the [...] *tunnel effect*. I wanted to give a lecture on this subject at the seminar led by Sommerfeld. As is well known, Sommerfeld had been sent away from his institute and had been replaced by Prof. Müller (hydrodynamics). Professor Clusus allowed the old Sommerfeld to hold a small seminar at the Institute of Chemical Physics at the University of Munich at Amelienstrasse, where his friends and admirers could meet and discuss with him. Taking advantage of my seminar, fixed for 24 November 1942, I wanted to introduce Touschek to those important people and provide a job for him in Germany, since his presence in Vienna had already been difficult.

> Thus we went to Munich together and I gave my lecture in the presence of Sommerfeld and other famous physicists (Touschek took care of the projection of the slides).

2.4   Bruno Touschek, Paul Urban and Arnold Sommerfeld

Urban had started planning for his seminar soon after the summer, writing to Sommerfeld in mid October, proposing to give a seminar in November.[59] A visit to Munich and a correspondence between Sommerfeld, Touschek and Urban followed, part of which has survived.[60] Urban and Touschek's visit to Munich was successful. Possible

---

[57] This is mentioned in Urban's long letter to Edoardo Amaldi, which follows Amaldi's 1978 message to all the people who had been Touschek's friends or mentors, asking them to share with him their memories of Bruno, Edoardo Amaldi Papers in Sapienza University of Rome–Physics Department Archives.

[58] Courtesy of Prof. Posch, University of Vienna.

[59] Letter from Urban to Sommerfeld on October 17th, 1941, Edoardo Amaldi Papers in Sapienza University of Rome–Physics Department Archives, box 524, folder 4.

[60] Courtesy of the Deutsches Museum, Archive, there are letters from Touschek to Sommerfeld on December 20th and 31st, 1941 and January 12th, 1942, letters from Urban to Sommerfeld on

ways for Bruno to go to Germany and continue to study were discussed with Sommerfeld, who was already well disposed towards Bruno, following the exchanges about his treatise. The plan had a number of problems, such as ensuring some form of university attendance for Bruno and adequate financial support to live on his own in Germany. The name of Paul Harteck as a possible first contact point in Hamburg was proposed.[61] Bruno said that his father could perhaps help him financially during the first period.

Back in Vienna, Urban continued his efforts to help Bruno in leaving Austria and furthering his physics studies in Germany, where it could be hoped that Bruno's last name would not betray his Jewish origin.[62] Urban wrote to Sommerfeld thanking him for the hospitality and sending greetings for the coming birthday on December 5th.[63] Sommerfeld, shown with Pauli in Fig. (2.9), answered immediately, replying to the greetings[64] and proposing that Touschek could attend Harteck's and Lenz's excellent lectures at the University of Hamburg.[65] To express his gratitude, both for the invitation to give his seminar, and for the expected support for his protégé, Urban carefully selected some wine to compliment Sommerfeld for the coming Christmas holidays. In the meanwhile, Sommerfeld had also found a possible position for Touschek, in Hamburg, and wrote directly to Bruno to inquire where it could be acceptable to him, and, probably, what kind of financial support Bruno's family could provide.[66]

At this point, something went temporarily wrong. When Touschek responded to Sommerfeld about the envisaged employment in Hamburg, he also started a discus-

---

January 4th, 1942 and from Sommerfeld to Urban on January 9th, 1942, while Sommerfeld's to Touschek have been lost. The Deutsches Museum, Archive maintain two more later letters from Touschek to Sommerfeld, September 28th, 1945 and October 5th, 1950. The Amaldi Archives, University of Rome, maintain the Amaldi - Urban correspondence, which includes two long letters by Urban to Amaldi in 1980 and copies, provided by Urban, of letters between Urban and Sommerfeld concerning Touschek: Urban to Sommerfeld on October 17th and Sommerfeld to Urban on December 2nd, 1941. The letter from Urban to Sommerfeld on December 2nd, 1941, mentioned in (Bonolis and Pancheri 2011), is from Deutsches Museum, Archive, and is part of material directly sent to the Author, courtesy of the Deutsches Museum, Archive.

[61] At the time Paul Harteck (1902–1985) was Professor of Chemical Physics at University of Hamburg. He had been Sommerfeld's student, and was a member of the *Uranverein*, the German atomic project, established by the German government in 1939 to explore the possible use of atomic energy. As such, he was one of the internees in Farm Hall after the war, together with Heisenberg (Bernstein 2001; Cassidy 2017).

[62] The name Touschek is from the Sudetenland, in former Czechoslovakia.

[63] Archivio Amaldi, box 524, f.4 (036).

[64] Letter from Sommerfeld to Urban on December 2nd, 1941, Edoardo Amaldi Papers, courtesy Sapienza University of Rome–Physics Department Archives, https://archivisapienzasmfn.archiui.com, all rights reserved. For these letters, source is indicated in (Bonolis and Pancheri 2011).

[65] Wilhelm Lenz (1988–1957) had been a student of Arnold Sommerfeld, and was Professor of Theoretical Physics at University of Hamburg, since 1921. His major scientific accomplishment is the formulation of the Ising model (Brush 1967).

[66] This letter has been lost, but it must have been sent prior to the one by Touschek on December 20th, 1941. This is so, because Sommerfeld mentions its content in his January 9th letter, 1942 to Urban, adding to have been the first to write to Touschek, who had replied on December 20th, 1941.

Fig. 2.9 Arnold Sommerfeld (at left) and Wolfgang Pauli in Geneva, October 15th, 1934. Reproduced courtesy of the Pauli Archive, CERN (PAULI-ARCHIVE-PHO-041). At right, the January 9th, 1942 letter sent by Sommerfeld to Urban, in which he recommends love between the "children", courtesy of the Deutsches Museum, Archive

sion about a physics problem, which was debated between him and Sommerfeld in two letters. But, much to his dismay, Touschek's argument had some errors. Acknowledging them, he wrote to Sommerfeld on December 20th:[67]

> Vienna, December 20th, 1941
>
> Dear very honoured counsellor!
>
> First of all I would like to thank you very much for the kind treatment of the Hamburg affair. Then I have to apologize, 1) because of the stupid signs in my letter 2) because of its shortness.[68]
>
> ...
>
> I ask you not to be angry because of the sloppy last letter. I really intended to bother you with it as little as possible. ...
>
> I would also like to thank you once again for the trust shown to me.

---

[67] Translation from the original German into English of the quoted correspondence between Urban, Sommerfeld and Touschek, was done by the Author. In this letter, Touschek starts with *Sehr geehrter Herr Geheimrat!*, a very honorific term used for highly distinguished university professors, courtesy of L. Bonolis.

[68] Bruno appears to refer to some sign mistakes he had made in his calculation.

A few days later, he used the last day of the year to apologize once more for his mistake.[69] Given the times, it was a rather unusual procedure by the part of a student, such as Bruno, to start discussing a physics problem with one of the great theoretical physicists of the time, especially since it turned out to be partly wrong. When Bruno, totally unaware of having possibly committed a sin of *lèse-majesté*, shared with Urban this correspondence with Sommerfeld, Urban was both embarrassed, in relation to Sommerfeld, and annoyed with Bruno. On January 4th, perhaps feeling that he was being excluded by the exchange, Urban wrote to Sommerfeld:[70]

> Thank you for your letter of December 27th, which I only got to read today because I was away for a few days. I am pleased that you approved of the wine and I confess that its selection was based on a rather large number of tastings.
>
> [...]
>
> I am somewhat astonished that Mr. Toušek wrote to your Excellency without my knowing about it. [...]

But the times were not such as to allow some ruffled feathers to get between a well-meaning teacher as Urban was, and his young and promising student, in danger of losing his education or even risking deportation. As for Sommerfeld, he had not minded or noticed Bruno's unusual and brash behavior. He replied immediately to Urban, Fig. (2.9):[71]

> I would be sorry if you and Mr. Touschek were to be alienated because of a fault of mine. For my part, I had first written to Touschek about an inquiry from Hamburg (industrialists were looking for assistants). On my request, he sent me an answer on 20.XII. That he also wrote about the scattering problem on this occasion I did not see as wilfullness or even insincerity against you, as can also be seen from the fact that I sent him greetings to you and Prof. Sexl. I very much hope that under these circumstances you can revise your suspicions against T. It is clear that you have the better knowledge in these questions. I would consider it a privilege if you let him continue to work on the tricky question.
>
> [...]
>
> With the admonition 'Children love each other' I remain your old
>
> [Sommerfeld]

Urban had been really upset by Touschek's approaching Sommerfeld on his own and discussing a physics problem he himself was engaged in, with Touschek making some mistakes in his calculations, and all this without Urban having been neither involved nor informed. But Sommerfeld's letter had its effect and Urban forgave Bruno, who had felt heavy remorse for his 'immature behaviour'. In the meanwhile, one of the people Sommerfeld had approached in Hamburg, his former student Günther Jobst, wrote to Bruno a friendly letter, which did not promise anything definite,

---

[69] Courtesy of the Deutsches Museum, Archive, letter from Bruno Touschek to Arnold Sommerfeld, December 31st, 1941.

[70] January 4th letter from Urban to Sommerfeld, Deutsches Museum, Archive.

[71] Letter from Sommerfeld to Urban, January 9th, 1942, Deutsches Museum, Archive.

but had an encouraging effect. At the same time, Bruno could write to Sommerfeld that his father was ready to partially support him during the initial stay in Germany.[72]

Thanks to Sommerfeld's generosity, peace returned between Bruno and Urban, and preparations started for Bruno to leave Vienna. As he was leaving, Urban recommended that Bruno look into a problem, which was receiving great attention among theoretical physicists, the so called *double-β-decay* (Goeppert-Mayer 1935; Wick 1935; Maiorana 1937). As Urban wrote to Amaldi, 'later he obtained his diploma on the basis of a paper on this theme'.[73]

Thus it happened that on February 24th, 1942, Bruno boarded the train from Vienna to Munich and left Austria for Germany. The train was of a very modern type, electrically powered, with cars which were upholstered even in the third class wagons, where Bruno had his ticket. It moved silently across the snow covered land, stopping in Salzburg and then crossing into Bavaria.[74]

---

[72] Letter from Bruno Touschek to Arnold Sommerfeld on January 12th, 1942, courtesy Deutsches Museum, Archive.

[73] Urban's statement to Amaldi is not quite correct, since Touschek's Diploma from University of Göttingen was on the theory of the betatron. Urban may refer to work Touschek did while in Göttingen but published in 1948 (Touschek 1948b). Touschek worked on the double $\beta - decay$ process in the months following the Diploma he obtained in June 1946. In the summer, he was shortly in the UK and, in a letter to his parents, dated November 11th, 1946, he mentions having had a 'paper in my pockets' on this subject upon his return from there.

[74] Letter to parents from Munich, on February 26th, 1942, the day he had gone to see Sommerfeld. The date of 24th for leaving Vienna can be determined from the detailed account of his arrival in Munich, the search for the hotel, which did not end until 6 a.m., and then going to the theatre, presumably on the day after his arrival in Munich.

Chapter 3
Bruno Touschek's Extraordinary Journey to Munich, Hamburg and Berlin: 1942–1943

After all, I think the account of the wonderful journey is pretty unique in the history of my life so far.
Immerhin glaube ich, dass der Bericht von der wunderbaren Reise ziemlich einzigartig in der Geschichte meines bisherigen Lebens dasteht.
Bruno Touschek's letter to a friend, Berlin, November 29th, 1942.

Abstract Bruno left Austria for Germany in February 1942, to follow his call to study physics, naively hoping his Jewish origin on the maternal side could go unnoticed to the Gestapo. After first visiting Sommerfeld in Munich, he went to Hamburg, attending classes at the University, while doing odd jobs to earn his living. At the end of the year, unsatisfied with his position at the Studiengesellschaft für Elektronengeräte, he moved to Berlin, where he started to work at Löwe Opta. This move, which ultimately led to his encounter with Widerøe and learning about electrons' way in a betatron, came to be through a rather chancy encounter, in Berlin, on the S-bahn, on November 12th, 1942. Asking a girl for directions, they discovered to be both from Vienna, both physics students with Hans Thirring, and having heard of each other through the link of being both half-Jews and working in Germany. This casual event, which opened Bruno's extraordinary journey, was always remembered by Bruno, even appearing in Amaldi's biography, as a turning point. Mira Hatschek worked for Löwe Opta and opened the way for Bruno's being hired there as well. The next turning point in his life took place in February 1943, when he came across an article submitted to the *Archiv für Elektrotechnik* by a Norwegian scientist, Rolf Widerøe, and, from that time on, his life followed a road which would ultimately lead him to construct the first matter-antimatter collider, an elementary particle accelerator which opened the way to modern day high energy physics.

When Bruno boarded the train in Vienna on February 24th, 1942, his life changed for ever. As the train left Austria and entered Germany, he could not know that he would never return to live in his native city. He longed for it during the tragic war years, and went back many times to visit his family, both during and after the war, but when

© The Author(s), under exclusive license to Springer Nature Switzerland AG 2022   39
G. Pancheri, *Bruno Touschek's Extraordinary Journey*, Springer Biographies,
https://doi.org/10.1007/978-3-031-03826-6_3

he arrived in Munich, he started a life journey that would take him away, to follow his call for physics and personal freedom. This journey sees him traveling through Europe, both in space and time. He was in Germany during the war and witnessed the early post-war efforts to restart Germany from the ashes, then moved through most of Europe, first to Great Britain, then to Italy, France, Italy again, and Switzerland. He lived through the reconstruction of Europe and was one of the protagonists of particle physics history in the '60s and early '70s. He returned to Austria to die, in 1978.

Bruno's five years in Germany started with a visit to Arnold Sommerfeld in Munich.

3.1 Sommerfeld's Day in Munich

Munich was not immediately friendly to Bruno. His original hotel reservation was not honoured at his arrival, other hotels were full. This was probably the first time that Bruno was travelling alone to a totally unknown place, without relatives or colleagues whose experience to count on, but finally at 6 a.m. in the morning, a place in the hotel was found, and the odyssey was over. Later in the day, he called at Sommerfeld's office and was told to come on the following morning. Bruno was looking forward to meet again the great scientist and start a new life, of work and study, completely on his own, in a great country, which was still confident to win the war. The losses and the heavy bombing of the Northern cities would come later, in that same year, but for the time being the future may not have looked so grim to the 21 year old.

He was young. Easily bouncing back from the lost night, he toured the city, enjoyed some local food and went to the theatre to see *The clever Viennese* [Die kluge Wienerin], at the time a very popular play in Vienna.[1]

As agreed, at 10 a.m. on February 26th, the 'Sommerfeld day' started.[2] After discussing some physics points, the plan for Bruno's move to Hamburg was examined. There were two separate problems to solve: foremost for Bruno's continuing physics studies, was how to attend classes at the University in Hamburg, and, on a more basic level, how to earn a living. The first one was solved by Sommerfeld writing letters to his former students Paul Harteck, Wilhem Lenz and Hans Georg Möller, recommending that they allow Bruno to attend undisturbed their lectures. As for the second question, Sommerfeld had already written to another of his former students, Günther Jobst, through whom a job was now to be offered to Bruno at the Studiengesellschaft für Elektronengeräte, a laboratory which built electronic material, some of

---

[1] The Viennese author Friedrich Schreyvogl (1899–1976) wrote 'Die kluge Wienerin' in 1941, a highly successful comedy, which in Vienna alone had 64 performances at the Deutsches Volkstheater in 1941–1942, from https://de.wikipedia.org/wiki/Friedrich_Schreyvogl.

[2] This is how Bruno refers to his day at Sommerfeldf's institute in a letter from Munich to his family on February 26th, 1942, where the first days in Munich are described in detail.

which was of interest to the war industry.[3] Sommerfeld had also asked Jobst to help Bruno and support him so that he could have some free time to study.

Before turning to sleep, he dutifully reported his first days in Germany to his parents in Vienna in a February 26th letter. The day had been fruitful, the road ahead was reasonably clear. These may well have been some of the last care-free days in Bruno's life before the end of the war and the Göttingen days in 1946. His last thoughts in the letter were for his grandmother Josefine and aunt Thilde.[4] He was aware of impending changes, as the train trip to Hamburg was known to be not as pleasant as the one he had just taken leaving Vienna. Intermediate stops and change of train were scheduled, and the train had no dining car. Unlike Munich, that was not heavily involved in industrial and military production, and whose distance from Great Britain lessened the probability of aerial attacks, Hamburg had been a regular target of air raids since the start of the war. Going to Hamburg was akin to start feeling what the war meant for German civilians and civic life.[5]

3.2 Hamburg: 1942

The first months in Hamburg are of unique interest in Bruno's life. These early months gave to Bruno the foundation of his future as a physicist: in the industrial firm Studiengesellschaft für Elektronengeräte, where he had a job as recommended by Sommerfeld, he learnt to work with his hands, to use a drill, to weld, to see things with an engineer's eye, to make circuits work, while at the University he learnt theoretical physics not just from books, as he had been forced to do in the last year in Vienna, but through the lectures of some great physicists such as Harteck and Lenz. This unique experience forged Bruno into the exceptional scientist who would unite theory and experiment by conceiving and constructing AdA, an experience reinforced between 1943 and 1945, when Bruno worked with Rolf Widerøe on the betatron, at the same time following lectures both at University of Hamburg and Berlin.

---

[3] The laboratory was controlled by the Dutch firm Philips, which also controlled C.H.F. Müller, as mentioned in the publication by the U.S. Department of Commerce Bibliography of Scientific and industrial records, pag. 152, issued shortly after the end of the war, on January 11th, 1946. This report included the survey of German firms in US occupied European zone. It is of note that Widerøe's betatron would be constructed at C.H.F. Müller in 1944 by a group which included Bruno Touschek. The laboratory was building 'drift tubes' of interest for high frequency communications. According to (Amaldi 1981) Touschek worked there for a long time, but Bruno's letters to his father indicate that he resigned from employment at the Studiengesellschaft in fall 1942 and moved to Berlin.

[4] Frequent mention of greetings to the grandmother in Bruno's letters home in the first half of 1942 indicate that, when Bruno left Vienna in February, she was still living free in Vienna.

[5] See https://en.wikipedia.org/wiki/Bombing_of_Hamburg_in_World_War_II with dates of raids on Hamburg, starting with September 10/11 1939. For a timeline of the war from the UK point of view, see https://www.historic-uk.com/HistoryUK/HistoryofBritain/World-War-2-Chronology/.

Bruno reached Hamburg on the evening of March 1st, a Sunday. The train ride was actually not as bad as he had anticipated. Thanks to the compartment door which was hard to open and discouraged unwelcome fellow travellers, he could reach Hamburg undisturbed. Made wise by his initial misadventure in Munich, when he had not been able to find sleeping accommodation until 6 a.m., he had booked a hotel recommended by the people at Sommerfeld's institute, the Reichshof Hotel. Noise from draining pipes gave him a bad night, but morning brought a good breakfast, mini-rolls with butter and marmalade, practically a main meal, and he was then ready, albeit a bit drowsy, to spend the day on a tour of the city before engaging in his visit to the Studiengesellschaft für Elektronengeräte, (S.f.E.), the firm where he hoped to get some employment, through Sommerfeld's recommendation. His first impressions of Hamburg were not entirely favourable, as his Viennese ear was not accustomed to the harder Hamburg accent. He mused that the people were polite but reserved. As for the food, he marveled at the strange idea of serving unpeeled potatoes.

After a first day during which he acquainted himself with the city and its inhabitants, he moved on to serious business, visiting Dr. Jobst, and the grounds of the S.f.E.[6] In this first encounter, Bruno met Günther Jobst, who had been Sommerfeld's student in Munich, and was very kind to him. They discussed Bruno's salary and the kind of work he could do or would be expected to do. It was agreed that Bruno would work initially in the high frequency laboratory. As it turned out, this employment would only last until the fall, not much, but still as much as Bruno would possibly want to take on, as he was keen to start attending lectures at the University in the next semester and would not be able to combine full time employment at the Laboratory with University attendance and the required study time. He had also envisaged working for Sommerfeld, with whom he was keeping close contact. Sommerfeld was preparing an Italian edition of his treatise and wished Bruno to assist him. Bruno had spent enough time in Italy to know Italian quite well, and could thus help checking the language, in addition to being throughly familiar with the physics content.[7]

After a tour of the firm's grounds chaperoned by a Dr. Busse, Bruno was invited to join Jobst for lunch, a distinction which created some resentment among the other employees, as he was later made aware through another fellow worker, a Viennese painter, working there as a technical draftsman. Hearing this, Bruno got himself

---

[6] G. Jobst was a pioneer of electronic tubes, see note in https://onlinelibrary.wiley.com/doi/pdf/10. 1002/phbl.19570130105, in *Physikalischer Blätter* on the occasion of Jobst' 60th birthday (Hahn et al. 1954). About the Studiengesellschaft für Elektronengeräte, see http://www.jogis-roehrenbude. de/Roehren-Geschichtliches/Hiller/Hiller.htm, and the Final Report on the Investigation of the X-Ray Industry in Germany, reported by Caperton B. Horsley on behalf of the U.S. Technical Industrial Intelligence Committee, August 11, 1945, published by Combined Intelligence Objectives Sub-Commitee (CIOS), G-2 Division, SHAEF (Rear) APO 413. (file CIOS-XXVIII-31). At pag. 27 it is said that the Studiengesellschaft für Elektronengeräte functioned as the engineering and research department of the Philips Valvo-Werke in Hamburg, which manufactured radio-tubes. Philips also controlled the C.H.F. Muller Aktiengesellschaft, where Rolf Widerøe's betatron was to be built, as described elsewhere in the report.

[7] Actually the Italian edition of Sommerfeld's treatise never appeared in print, most likely because of the war engulfing all activities, and the changed post war scenario.

ready for a fight, something which held a certain appeal for him all through his life, as we shall see in many instances, as he was unable to suffer offense or challenges which he felt to be undeserved. In this case, this may have been just a typical office gossip, and nothing happened.

Once the visit to his future place of employment was over, Bruno turned to look for accommodation, as he obviously could not afford to spend too many days in the hotel. It turned out that accommodations were harder to find than he expected, and he decided to get a broker to take care of it. This freed him to go to the University on Friday.

At Hamburg University, among his referrals there were Paul Harteck and Hans Georg Möller. Paul Harteck, a chemical physicist and a Viennese by birth, was the Director of the Department of Physical Chemistry, and was well known for work on heavy water, the chemistry of deuterium compounds, artificial radioactivity and neutron physics. He was a member of the German nuclear project. Earlier in 1939, he had become aware of the war time possibilities of nuclear energy and made contact with the German government. In September, when the German nuclear project, the so called *Uranverein*, was created with Werner Heisenberg and other major German scientists, Harteck joined it. At the end of the war, Harteck would be captured by the British and kept *incommunicado* for six months at an isolated country house in Great Britain, Farm Hall, together with Heisenberg and other famous German physicists such as Otto Hahn, Max von Laue, Carl von Weizsäcker (McPartland 2013; Cassidy 2017).[8] But all this was still away in the future, after Germany would be defeated, after millions dead in the concentration camps, on the battle front, and in the old cities destroyed by bombing. In early 1942, when Bruno reached Hamburg, almost care-free, the heavy bombing of Germany had not yet started: it was going to happen very soon, in a couple of weeks in fact.[9]

Paul Harteck was extremely courteous and kind to the young Austrian. He advised Bruno on how to behave in his next visit to Prof. P. P. Koch in order to gain access to his Institute, where Bruno would attend some classes.[10] Harteck recommended Bruno to start with a snappy *Heil Hitler!* and then ask, and answer to, intelligent questions about the Professor's main claim to fame, a color photometer.[11] Jokingly,

---

[8] Werner Heisenberg had been awarded the Nobel Prize in 1932 for 'the creation of quantum mechanics, the application of which has, inter alia, led to the discovery of the allotropic forms of hydrogen', Max von Laue in 1914 for 'his discovery of the diffraction of X-rays on crystals'. Otto Hahn received the news of having being awarded the 1944 Nobel Prize in Chemistry for 'his discovery of the fission of heavy nuclei.', while under isolation in Farm Hall in 1945. He was refused travelling to Stockholm to receive the prize, but he was allowed to write a letter of acceptance. The scientists' conversations were secretly recorded by the British military. According to Hans Bethe (Bethe 2000), these transcripts prove that Heisenberg's group was not working on an atomic bomb.

[9] The first bombing of the civilian population of Hamburg took place on January 14th, 1942, but it was a relatively light engagement, as from https://ww2db.com/event/timeline/1942/. Ultimately 75% of Hamburg would be destroyed.

[10] Peter-Paul Koch has been awarded his doctorate as a student of Wilhelm Röntgen. He was an enthusiastic Nazi follower, and, after the war, committed suicide, in October 1945, see https://peoplepill.com/people/peter-paul-koch.

[11] *Koch gehen: zackig Heil Hitler machen!*, in letter to parents on March 7th, 1942, from Hamburg.

Fig. 3.1 Hamburg University: © UHH/Schell at left and, at right, Paul Harteck in 1948, photograph
from Bundesarchiv, Bild 183-2005-0331-501/Unknown author/CC-BY-SA 3.0, from wikipedia

Harteck asked Bruno: 'You don't know what this is?', a question he himself answered
by immediately adding: 'Neither do I'. To be sure, Bruno studied with great detail the
instrument developed by Koch, based on a high sensitivity method for the photometric
analysis of X-ray plates. Thus, when examined, Touschek was able to answer very
competently and was successfully admitted to the Institute (Amaldi 1981, 4). After
seeing Harteck, Bruno's next successful visit was to Hans Georg Möller, an older
Professor, Director of the Institute for Applied Physics, who made Bruno at ease.[12]
Thus, the enormous respect which all German physicists held for old Sommerfeld,
coupled with Bruno's clearly extraordinary intellectual qualities, opened to the young
Austrian a new life in Germany, in which he would find work and could continue
his physics studies, skirting death – but surviving – and learning through war times
(Fig. 3.1).

     To complete the day, and his first week in Hamburg, Bruno went to the theatre to
see a recent play by Gerhard Hauptmann, *Iphigenie auf Delphi*, but did not like the
adaptation, which was totally non classical, and was disturbed by the actors' pronun-
ciation, in the spoken theatrical German, which sounded 'barbaric' [barbarischen]
compared with the Viennese way.[13] The play finished early in the evening, and he
went to the theatre restaurant, where he came across some German chess club mem-

---

[12] From https://de.wikipedia.org/wiki/Hans_Georg_M%C3%B6ller: The Institute for Nanostruc-
ture and Solid State Physics (INF) was founded in 1925 under the old name Institute for Applied
Physics (IAP) at the Jungiusstrasse location, and Hans Georg Möller was appointed as its first
director. The focus of the work at that time was on high-frequency technology, the applications of
the cathodic tube [electron tube] and later on radar technology.

[13] Gerhart Hauptmann (1862–1946) was awarded the 1912 Nobel Prize in Literature 'primarily in
recognition of his fruitful, varied and outstanding production in the realm of dramatic art' from
https://www.nobelprize.org/prizes/literature/1912/summary/. *Iphigenie auf Delphi* was written in
1940-1941 and was premiered in Berlin on November 15, 1941. It seems that the adaptation was
placed during pagan times wth a Christian undertone. See also (Ziolkowski 1959). William Shirer
in (Shirer 1941, 179) comments about Hauptmann having once been 'an ardent Socialist and a great
playright', and having now become 'a nazi and a very senile man'.

bers. Never one to shy away from a challenge, and clearly already a good chess player himself, he joined them playing six games, of which only one was lost and one was a draw. Fortified by this success, the week ended with a decision to join the chess club.[14]

The problem of finding affordable accommodation was not immediately solved. For many years, such searches would be a constant problem in Bruno's life. After leaving the security of the Vienna home and the warm embrace of his maternal and paternal family (grandmother Josefine, his parents and aunt Thilde), no place could give him the security he had lost. His need for constancy and comfort led him often to conflicts. In the next ten years, through the war and after, in Germany and in Scotland, Bruno often changed home, quarrelling with his landlady or landlord. Occasionally a place was found to be satisfactory, not often anyway, and in such case, he would include detailed drawings in his letters home. Bruno's quest for a home was never fully settled. In Glasgow where he spent five years after the war, he changed his residence four or five times. Even in Rome where he lived for twenty five years, married, raised his own family, and had a secure position, he still moved through four different places.

In Hamburg, Bruno's house hunting took an extra week, and only on March 14th, he could leave the hotel and move to a pension in Alsterdammstrasse,[15] where house services were provided for a number of guests, as it was the custom at the time, and food could be obtained through food stamps.[16] In an undated letter, with header 'Reeperbahn', a street in Hamburg red-light district,[17] he described his room in the Pension. His room could not be locked, as the door had a purely aesthetic handle, and its furnishing consisted of a previously two foot legged table, balancing at the end of the bed in 'statically undetermined equilibrium', a bed enjoying ownership of four fragile legs, and numerous nails on the walls, intended both for hanging items of clothing and for tearing them, as Bruno drily commented to his parents. The bathroom was shared with the proprietor, a generosity which costed Bruno the fast disappearance of his shaving cream in the black market. In the living room, in front of the mirror, was the dirty brown bronzed plaster statue of a bare-breasted lady, exhibiting five half fingers on each hand. One cannot but admire the courage of Bruno, or marvel at his resilience: only two months earlier he had abandoned the Vienna home where paintings by Egon Schiele would hang in his bedroom, together with those by other Vienna artists, including his uncle Oskar, but he still could describe his tragically changed surroundings with some sarcastic detachment, even humor.

Dinner or meals during week ends were at the Pension, paid by the food stamps of his ration card. Daily regular lunch presented some difficulties, as he dismissed

---

[14] Letters to parents on March 4th and March 7th, 1942.

[15] The Alster is a right tributary of the Elbe and is Hamburg's second most important river.

[16] Food stamps had been established in Germany just before the break of the war, in late August 1939 (Shirer 1941) and were part of everyday life in Germany during-the-war.

[17] A 1943 German movie, Grosse Freiheit Nr. 7, vividly depicts life in a locale in Hamburg red-light districts. It was one of the first Agfa color movies by the production house TERRA. The movie ignores the contemporary existence of the war, including shots in the Hamburg harbor which required masking the warships in the background.

the simple food at the modest restaurant where the S.f.E. had struck a deal for the employees. Losing his food stamps, at it happened in one of his first two weeks in Munich, he could have starved. Resourceful as he was, he found an excellent place, the Alstereck restaurant, in fact a very expensive one on the Alster Promenade where he was offered free meals for a whole week by the very kind Viennese proprietor.[18] On Thursday evenings he played chess, and went fencing on Monday and Wednesday with a Viennese master, who knew the teachers Bruno had in Vienna and Rome.[19]

3.2.1   The Bombing of Lübeck

Soon, however, life in Hamburg started darkening. On the night of March 28th, two weeks after he had moved to the Lembke pension, Bruno got his first fire-alarm, when Hamburg was flown over several times by British bombers between 10:30 and 3:00 in the night. It was not the first bombing attack the harbour city had suffered, but it was the first to heavily hit the civil population, not only military targets.[20]

The night of March 28th was one of almost full moon, just before Palm Sunday. Bruno stood at the window of his room in the Pension and watched the shooting of the Flak (Flugabwehrkanone), the German anti-aircraft defense. In a letter to his parents, he observes that an air raid may look like fireworks, except that, of course, bombing, which might seem more human than medieval killing and torturing, is much more deadly in terms of the sheer number of human lives. On the night in question, he reports that the city of Lübeck was almost without anti-aircraft gun, probably as a result of their detachment to other more strategically valued targets. The English knew that, of course, and, that night, left Lübeck almost completely in ashes. The train station was badly hit, famous old churches erased with entire

---

[18] Letter to parents on March 14th, 1942, address is Hamburg, c/o Lembke, Alsterdamm 36, and undated letter, with Reeperbahn as sender's location. When compared with previous and later letters, this letter can be dated between March 14th and April 3rd.

[19] Bruno took fencing lessons both in Vienna and Rome, during his pre-war visits. Bruno mentions his Viennese teacher's name as being Targler, possibly Leopold Targler, the author of a 1913 book *Das Fechten mit der Stoß - und Hiebwaffe in sportlicher und moderner Auffassung* [Fencing with a rifle and bat in a sporty and modern way]. By 1908, Vienna had 24 fencing clubs, most of them fencing according to the Italian school. Berlin and Prague had only three each, as from https://hiebfechtkunst.wordpress.com/2011/08/25/la-schiabola-e-la-vostra-penna-the-expansion-of-the-italian-saber/. In Rome, the fencing Academy was located in Via del Seminario, next to the ancient Roman Pantheon, in the center of the city. Fencing instructions, originally destined to the upper classes or the military, were revived in the 1950's by the booming Italian movie industry. Famous fencing pupils included major Hollywood stars of the time, such as Errol Flynn.

[20] See article about *Bombing of Hamburg, Dresden, and Other Cities 28 Mar 1942 - 3 Apr 1945*, as in https://ww2db.com/battle_spec.php?battle_id=55. Hamburg would ultimately be bombed on seventeen occasions, resulting in destroying 75% of the city.

Fig. 3.2 Lübeck cathedral burning after the 1942 bombing, https://commons.wikimedia.org/wiki/
File:Bundesarchiv_Bild_146-1977-047-16,_L, Bundesarchiv, Bild 146-1977-047-16 / CC-BY-SA
3.0

streets. Firefighters came from all neighbouring cities - including Hamburg. The
count, according to Bruno, was 600 dead, 2000 missing.[21] It was the first attack on
Lübeck and the people were still on the streets, when it started. The narrow streets
of the old city did the rest. A large part of Lübeck's people were left with only a
night shirt and a pair of slippers, as Bruno heard from a man who had in fact come to
Hamburg to buy socks, and whom he got to know in the Dammtor restaurant. People
in Hamburg could see Lübeck burning some 70 kilometers away, Fig. 3.2.[22]

3.2.2   A Racial Slur and Search for a New Home

Bruno had planned to remain in the Lembke's pension for about two weeks and then,
on April 1st, move to a small apartment leased to him by a relative of his landlord. But
his departure was hastened by an incident which can shed light on the reason why,
three years later, around March 15, 1945, just as the Western forces were approaching

---

[21] In (Hopkins 2008), a number of casualties as high as 2000 civilians is mentioned as reported by
(Grayling 2006). See also the graphic novel by Wayne Vansant, (Vansant 2013), (Eberle 2007), and
Bombing of Lübeck on March 28th, 1942, from wikipedia.

[22] Letter sent on Good Friday (April 3rd 1942).

Hamburg, he was arrested by the Gestapo, and in April sent to probable death to the Kiel concentration camp (Amaldi 1981; Bonolis and Pancheri 2011).

On Sunday, the day after the air attack on Hamburg, he had gone to his landlord and asked for a sandwich, which he thought was his due, since he had given him all his food stamps soon after arriving. Bruno had then learned, by chance, that the man was maintaining a very flourishing black market activity with his pensioners' food stamps, certainly Bruno's. Other pensioners seemed less concerned about the quality and quantity of the food, not so Bruno, who was young, hungry, and still unused to see basic rights, such as his deserved worker's food, to be taken away from him. Thus a discussion ensued, during which the landlord made an offensive comment about Bruno's parentage, 'Abstammung' is the word used by Bruno in recounting the incident to his parents, no other explanation is given. The landlord had apparently learned of Bruno's 'Abstammung', at the Police Station where he had gone for the usual report of new Pension guests. Bruno, in telling the incident to his parents, does not specify the type of offense which was given, but very likely it referred to his being half-Jewish. A scuffle followed, or, rather, Bruno, young and fast, as fitting to the good fencer he was, reacted by hitting the landlord, not just once, but twice or thrice. The landlord threatened to call the police, and Bruno challenged him to do so, well knowing that the landlord had a bad conscience because of his black market activities and wouldn't dare to do it. Bruno left the place immediately, and, as a parting shot, told the landlord that he could get his rental dues from the police.[23]

This episode shows Bruno's courage and refusal to accept insults or injustice of any kind. Perhaps trivial, as we do not really know the nature of the offense, it hints to something very relevant for Bruno's life in the next three years, namely that the Hamburg police in March 1942 knew that Bruno was half-Jewish. So, why was he not arrested between March 1942 and 1945? Because, as we shall see, his clear capacities for scientific work, could be usefully employed in Germany's scientific and technological war efforts, as was the case for other half-Jewish scientists and engineers (Waloschek 2012), as we shall see in Chap. 4.

The header in the letter, in which Bruno related this episode and its follow up with the search for new accommodation, indicates the Studiengesellschaft für Elektronengeräte, as this was the place where he spent his next few nights as a fire-guard. At first, after leaving the Pension, and being Sunday afternoon, the only place he could think of was an hotel-by-the-hour, where by handing over two English books to a Norwegian girl, he was then left undisturbed for the night. On the following day, he committed himself to the fire watch service at the Studiengesellschaft. He slept for a few nights in the guard room, with the only duty to make calls if a roof fire developed. He was given a steel helmet for his own protection and was even paid, but poorly, for his watch duty.

Once the immediate sheltering needs were taken care of, he then went door-to-door to inquire about possible rooms to rent, until, finally, one lady was sufficiently

---

[23] Letter to parents on April 3rd, 1942.

Fig. 3.3 Drawing included in one of Bruno's April 1942 letters, describing his temporarily homeless status, the steel helmet for his new fire-watching job, desire for bed and food, all of his 'desiderabilia'. Family Documents, © Francis Touschek, all rights reserved

impressed by his good manners, and a made-up referral. Thus, on April 3rd, he secured new accommodation.[24]

April went by, and Bruno's luggage, which had been sent from Vienna, with many little things he needed, such as toilet necessaries, arrived at the Pension. To retrieve it from there was not a trivial affair, but after some shouting from the resentful landlord and help from a kind passer-by, the big heavy trunk was lifted into a tram, and reached Bruno's new home.

This month's adventures were humorously depicted in a drawing he sent home, shown in Fig. 3.3: the fire-fighter helmet, the big trunck he had to carry by himself from the Pension, the suddenly missed bed, a sorely missed nail file, etc., calling all these things his *desiderabilia*.

His new address was in Fröbelstrasse 14, not far from the University of Hamburg, where he had started following some lectures. At the University, Bruno had been warmly welcomed by Wilhelm Lenz, who also invited him to his Colloquium.[25] He thus came to know various professors from other German universities, who would come to Hamburg to listen to some of these seminars, such as Hans Jensen, well

---

[24] He would ring the bell and say that he had heard of a room for rental from Dr. So and So, figuring that someone could be impressed by Bruno knowing a doctor, who had recommended their place to him. April 3rd, 1942 letter.

[25] In 1919, the Physikalisches Staatsinstitut (founded in 1885) was integrated into the newly established Universität Hamburg. Then, in 1921, Wilhelm Lenz was appointed to the first professorship for theoretical physics. One of his first assistants was Wolfgang Pauli. Indeed, it was here that Pauli first formulated his exclusion principle for which he was later conferred the Nobel Prize in Physics. See also https://www1.physik.uni-hamburg.de/en/th2/ueber-das-institut.html.

 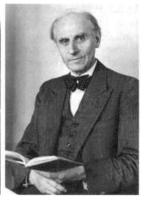

Fig. 3.4 At left, Markus Fierz, Wolfgang Pauli and Hans Jensen, Copenhagen Conference, 1934. Photograph by Paul Ehrenfest, Jr., courtesy AIP Emilio Segré Visual Archives, Weisskopf Collection, photos.aip.org. At right, Wilhelm Lenz from University of Hamburg Physics web site

known theoretical physicist from the University of Hannover.[26] On April 30th, Bruno had received his first invitation to Lenz' Colloquium on the occasion of a lecture on superconductivity. This was for him a good introduction to the Colloquium, as he had studied the subject and knew it well. All in all, this was the best day he had had so far, as he wrote to his parents. After the Colloquium, Lenz took him to a cosmological seminar at the Institute of Mathematics, by Prof. Beckmann. One of the speaker's points was addressing the question of how far do you get without the theory of relativity. He would not use the word 'relativity', of course, but, to be understood and at the same time save himself from a possibly anti-Semitic physics listener, he called it 'general theory of gravitation'. In the discussion, when the point about relativity came up, Beckmann said: 'You can talk about it—maybe even think about it', which impressed Bruno to the point of writing it to his parents.[27] The dinner was extended with a stroll the three of them took together along the Alster Basin. It was beautiful: the stars were shining over the water and there was no sound of 'radio tubes'.[28] In Fig. 3.4 we show photographs of Jensen and Lenz.

---

[26] J. Hans D. Jensen 1907–1973) shared half of the 1963 Nobel Prize for Physics with Maria Goeppert Mayer 'for their discoveries concerning nuclear shell structure', the other half of the prize was awarded to Eugene P. Wigner for unrelated work. Jensen obtained his Ph.D in 1932 in Hamburg, and then became scientific assistant at the Institute for Theoretical Physics of the University of Hamburg. He became Professor of Theoretical Physics at the Technische Hochschule in Hannover in 1941, and in 1949 was appointed Professor at the University of Heidelberg.

[27] The directive by Himmler was that it was allowed to teach modern physics, as long as one did not mention Jewish authors (Bernstein 2004). Kurt Gottstein, who worked under Heisenberg for twenty years (1950–1970) notices however that Heisenberg disregarded this directive in the 1944 edition of his book *Die phylikalischen Prinzipien der Quantentheorie*, where Einstein's name was mentioned 5 times, in addition to other Jewish scientists.

[28] Letter to parents on May 3rd, 1942.

Bruno was also working hard at the Studiengesellschaft, and had the prospect of taking two weeks vacation for the Pentecost festivity, on May 24th. The opportunity was there for him to go home and see his family. But money was becoming a serious problem. He had to continuously borrow from his employer, and occasionally had to ask his father for help. However, another visit could not be expected to take place for another year, and he was worried about his grandmother Josefine, as shown in the correspondence with his father during the first half of 1942, and his constant requests for her letters or news. It is known that she died in the concentration camp of Theresienstadt on March 10th, 1943, having been deported in August 1942, and she was still in Vienna in February 1942, when Bruno had left. Thus he managed to find the money, and join his family.[29]

While Bruno was learning to live by himself in precarious conditions, away from his family, and with very little money, the war continued with unstoppable pace, towards more and more deadly outcomes.[30]

3.2.3   Depression

The record of Bruno's letters to his father between May and June 1942, points to Bruno taking the proposed vacation and visiting Vienna and Munich.[31] In June, there were negotiations with the Studiengesellschaft for ending the apprenticeship period and entering in a new contract. At this time, he experienced a bout of depression. This was certainly because of overwork and the strain of the first three months of hardship in a completely new type of life. The return to Hamburg, after visiting Sommerfeld in Munich, made him compare his present status with the expectations he had held: where had his future as a physicist gone? Why did he spend his days in the S.f.E. without having the time to study what he loved, relativity most of all? Something similar had happened in September 1939, probably following the start of the war and the definite acknowledgment of the impossibility to emigrate and study in England. Very likely and, if so, foremost in causing his depression, he may have been worrying about his grandmother's deportation. After his mother's death in 1930, followed by his uncle Oskar's suicide in 1934, his grandmother's fate would be one more tragic loss affecting his emotional life.

Bruno was now more on his own that he had ever been. The stark difference with the living conditions he had been used to, through most of his life, the lack of prospects for his studies and life in general as the war moved on, were sufficient reason to despair and be discouraged. But his friends were watching and looking after him. He received comforting letters from Hans Thirring, 'like a father', and knew that Sommerfeld was concerned about him. In Vienna, Urban was also trying to help, and recommended

---

[29] Lack of letters between May 3rd and June 7th, points to Bruno having been in Vienna during the two weeks around Pentecost.

[30] See http://www.bbc.co.uk/history/worldwars/wwtwo/ww2_summary_01.shtml.

[31] A handwritten June 7th letter mentions travelling and Munich.

Bruno as a reviewer for articles submitted to the prestigious *Chemisches Zentralblatt*, himself renouncing to do it – and to the small *honorarium* – in favour of the unknown Touschek.[32] Bruno was normally critical and sometimes even bothered by Urban's attention, but he was now grateful for the help and appreciative of Urban's concern. Bruno found no difficulty in reviewing. He was amazed on how fast he could do it, such as seven (!) reviews in half a day.[33] The work for the *Chemisches Zentralblatt* prepared him for work for other journals, most fatefully, as we shall see, as assistant to Editor of the *Archiv für Elektrotechnik*, a few months later.

During the summer months, also to assuage his anger at the working conditions and poor salary with the firm, he kept himself active, not only with fencing, but also swimming and playing soccer. But he also had a number of small accidents, as he was first hit while fencing, and then a soccer injury kept him home with a swollen knee. He was also unsatisfied with his living place, as his room did not have a suitable working desk, and in August decided he had to make a change, and move to a better place. This time, he felt happy enough about the prospective new place to ask his father to send him some of the pictures hanging in his room in Vienna, such as the two paintings by Schiele, a torso and a Kosak head, one painting by his uncle Oskar, and the like. Luckily Bruno's father did not follow his son's requests, since these precious things would not have survived the hard times ahead in Bruno's life, the move to Berlin and the bombings of the city, the imprisonment by the Gestapo in 1945, the post-war times in occupied Hamburg.[34] Letter after letter, he also sends greeting to his grandmother, inquires about her, asks his father to let him know how is she doing.

3.2.4 Jobless

His depression slowly eased because of the prospects of improved living conditions in the next month. It was now late August, and, in the evening, from his window, he could see children walking and singing, holding little paper lanterns on long sticks.[35] The view inspired a drawing, shown in Fig. 3.5, next to which he commented that he could just be like one of these children, walking through a wood and holding the lantern of his life, while humming, whistling and thinking. In September, he moved to the new apartment, but the unhappiness and strain of the previous months, finally

---

[32] The Chemisches Zentralblatt is the first and oldest journal published in the field of chemistry. It covers the chemical literature from 1830 to 1969 and describes therefore the birth of chemistry as a science, in contrast to alchemy. From wikipedia.

[33] Letter on July 18th, 1942.

[34] Some of these paintings remained in possession of the Touschek's family long after the end of the war.

[35] It is a tradition of German speaking countries for the children to carry such paper lanterns on a long stick, singing *Laterne, Laterne, Sonne, Mond und Sterne*, which Bruno paraphrased in his drawing as *Vaterne, Laterne*, with the two children looking like twin brothers. Such a tradition is typically held on St. Martin's day in November, but it can also occur in the summer.

took their toll. While waiting for his tram home, in mid September, he caught cold and was so ill with high fever that his kind new landlady had him brought to the hospital, where they diagnosed him with pneumonia. This was clearly a mis-diagnosis, since the fever was down the next morning and an X-ray did nor confirm pneumonia. He was back at home in a few days, and recovered after resting and eating well enough to rapidly gain weight under the kind attentions of his landlady. She was a young woman from Vienna, with a fiancé from the Baltic countries, an Officer in the Wehrmacht. The landlady and her fiancé were both very kind to him, and the three of them often played chess together in the evening.

He proudly described the new lodgings to his father including a detailed floor map in his letter, Fig. 3.6.

The new place was very nice, but expensive. To afford paying for the higher rent, it was necessary to earn quite a different salary than at the Studiengesellschaft, and Bruno, who had gained grounds at the firm and was now giving mathematics lessons to his colleagues, started looking around for a more rewarding position. This did not appear too unrealistic: he had heard that a girl he knew from Vienna, Mira Hatscheck, was earning almost twice his salary by working at Löwe Opta in Berlin, for a type of work not too different from the one at the Gesellschaft.[36] He looked for openings at Philipps, at the C.H.F. Müller factory, all of them involved in electronic type work for the military, as Germany was intensifying its scientific efforts. At this time, he was not only worried about money, but also concerned about passing his life doing things which did not interest him, rather than studying what he really wished to. As he writes, 'under all circumstances I want to avoid coming to a laboratory, the work of which I am not interested in after a few months.' Finally, he decided to leave the Studiengesellshaft, before the notice advance period. This turned out to be a temporary disaster.[37]

---

[36] Letter to parents from Hamburg, September 6th, 1942.

[37] Letter to parents from Hamburg on October 10th, 1942.

Fig. 3.6 The plan of Bruno's room in Schlüterstrasse where Bruno moved into on September 1st, 1942, from one of his drawings. Family Documents, © Francis Touschek, all rights reserved

Suddenly, in October, he found himself without a job, with a nice room and a kind landlady but no money to pay for the rent, and in need of his father's help to go through the month. He did have some back up reserves, but they were soon finished as the many job options he was envisaging did not become actual offers. He finally asked his father for money, which was sent together with kind words of encouragement. Bruno was not alone, however. A bank check was sent by Aunt Ada from Rome, to help him out in those difficult days. His friends worried as well, and tried to find solutions to his temporary joblessness. One colleague from the firm, Dr. Busse, tried to get him some loose employment position with the Reichsluftfahrtministerium (RLM), the Minister of Aviation of the Reich, in the Air Force Laboratory in Gräfelfink near Munich, while Lenz from Hamburg University, as well as Sommerfeld from Munich and Thirring from Vienna, were doing their best to help him. At the end, while his friends may have been helping all along, he found the solution by himself, as it shall be seen.

In October he frequently visited the Hamburg employment office, but no offers came from this. A trip for a position at the Carl Zeiss firm in Jena or in Berlin, also with the RLM, was envisaged but he actually had no money to spare and finance an exploratory trip. To keep some balance in such a difficult situation, he found solace in reiterating the reasons for the apparently rash decision to resign: that he could not spend his life in a job which could not possibly interest him, such as the one at the Studiengesellschaft. He also confessed to his landlady the difficult position he was in, and the 'confession' partly relieved him of his anxiety.[38]

The situation was so worrisome financially, that he even considered going back to Vienna. It would have been a retreat, however, although it had a certain appeal: to be back home, relieved of financial worries, his old friends and protectors to discuss physics with. It was no more than a fantasy, he had had to leave Vienna, returning now was out of question. He could only move forward. And this is what he did.

Bruno was resilient and had great internal resources, and a passion for physics which would be his strongest drive. With no new position to show up, he became impatient, and decided to push for employment possibilities in Berlin, while, at the same time, collect the money due to him from his referee work with the *Chemischen Zentralblatt*. And thus, on November 12th, he left for Berlin, without knowing whether he would have the money to return to Hamburg. It was to be a very short trip, barely two days, but, as it turned out, it was very eventful. Reporting this trip to his friend Peter on November 29th, he wrote that '...*dass der Bericht von der wunderbaren Reise ziemlich einzigartig in der Geschichte meines bisherigen Lebens dasteht*', '... the account of the wonderful journey is pretty unique in the history of my life so far.' And, in fact, this trip changed the course of Bruno's life.

3.3 To Berlin: New Friends and Work at Löwe Opta Radio

He left on November 12th with the Hamburg to Berlin express train, the fastest in the world at the time, and directly went to collect his dues from the *Chemischen Zentralblatt*, in Sigismond Strasse.

This first trip to Berlin turned out to be of critical importance in Bruno's future, as it ultimately put him in contact with Rolf Widerøe's betatron project. It lasted less than two days and it is worth reporting it in detail, as it it highlights the elements of chance in a great life, such as Bruno's. There are two records of the events, one in (Amaldi 1981, 4), and a very detailed one, almost by the hour, in a letter Bruno wrote to his friend Peter.[39] According to Amaldi's narration 'Once, in a train in Berlin he [Bruno] met a girl, M. Hatschek, – she too was half Jewish – who worked in a factory which had changed its name from *Löwenradio*, typically Jewish, to that of *Opta*. Miss

---

[38] Letter to parents, from Hamburg, October 28th, 1942.

[39] Letter to Peter, November 29th, 1942. The letter, which Bruno told his friend could be shared with his parents, was obviously given to Bruno's father, and was among the cache of letters which Elspeth showed to L. Bonolis and the Author.

Fig. 3.7 A present day
image of the Kranzler Coffee
shop, where Bruno, on
November 12th, 1942, met
with Mira, from https://en.
wikipedia.org/wiki/Caf%C3
%A9_Kranzler#/media/File:
Berlin_-_Caf%C3%A9_
Kranzler1.jpg, author
WladyslawTaxiarchos

Hatschek introduced Touschek to the management, …..'. This description attributes
the encounter to pure chance, being so in Bruno's memory to the end of his days.
In fact, Bruno knew of a 'Miss Hatschek', as a student of Hans Thirring, working
at Löwe Opta, and a *mischling*, half-Jewish like himself. When he was envisioning
a possible move to Berlin, Bruno had considered getting in touch with her.[40] Now,
thanks to the letter to his friend Peter, the picture of him as engaging with some
girl he didn't know and, through her, finding a job, can be enlarged and its details
seen through the eyes of the young Bruno, almost as things happened. Infact, Bruno
did meet Mira by chance, and this happened in the Berlin S-Bahn on his way to
the *Chemischen Zentralblatt*, and on the same day of his arrival. Asking a girl for
directions, he soon discovered her to be the same Mira Hatschek he knew about.
Before each went to their separate ways, they agreed to meet later, at the Kranzler
coffee shop on Kurfurstendamm, (Fig. 3.7).[41]

When Bruno and Mira met again in the afternoon, the possibility for Bruno to
work at Löwe Opta was discussed and he took the decision to inquire whether there
was a free position for him. Bruno had intended to speak to Prof. Schottky in the
evening and inquire about the status of his application for a job at Siemens, but
this was postponed in view of the decision to go to Löwe Opta on the following
morning.[42] Mira promised to find him accommodations for the night, and he went
to have dinner in a Russian restaurant, where another chance encounter took place.
While waiting to call Mira, and find out where he would spend the night, he started
talking with two girls, Monika and Maria, who would remain his friends for a long
time.

---

[40] Letter to parents on October 28th, 1942.

[41] The Kranzler was a famous meeting place in Berlin. It had been founded in 1825. Destroyed in
1945, it was rebuilt after the war, and further transformed after the fall of the Berlin Wall in 1989.
It remains one of the main landmarks for social life in the city.

[42] Walter Schottky (1886–1976) is known for important contributions to the physics of semi-
conductors. After graduating from University of Berlin, and appointments at Jena, Wurzburg and
Rostock University, at the time of Bruno's' planned visit he was at Siemens laboratories in Berlin.

Fig. 3.8 Header of a letter sent by Bruno to his family in 1943, where the previous name of the firm is still prominently shown. Family Documents, © Francis Touschek, all rights reserved

Things were moving very fast. On November 13th he was hired at Löwe, starting on November 23rd as laboratory assistant.[43] In the afternoon, he went back to Hamburg to arrange for his move to Berlin. And thus, from this trip, Bruno's extraordinary adventure in the world of accelerator physics started.

Returning on the 20th to Berlin, he found reasonable accommodations at Sponholzstrasse 43, and then went to meet the new friends, whom he had rather casually come to know at the restaurant, where he had gone to have dinner, while waiting for Mira during his previous trip to Berlin. They were all half-Jewish like him. Before Bruno's work became too absorbing, as it started to be later in January, they would spend week-ends discussing psychoanalysis and philosophy, dreaming of starting a school for *mischlingen*, the mixed-race children as they were called (Raggam-Blesch 2019).[44]

His work at Opta Radio started on November 23rd, when he went to the office and met Dr. Karl E. Egerer and Dr. Tomfohr, whom he was told had to be thanked for his employment. The firm, which had previously been called Löwe Radio from his Jewish owners, had now changed its name into *Opta Radio*, as one can see from the header of a letter Bruno sent to his family later in 1943, reproduced in Fig. 3.8.[45] At the time, Bruno did not know why he had obtained this job, but mentions that Dr. Tomfohr, who hired him, was married to a Jew - and this could have been the reason.[46] Tomfohr was later dismissed, on the ground of being accused of espionage, according to what Bruno wrote in January to his father. *A posteriori* one can see

---

[43] In the November 29th letter to his friend Peter, Bruno makes a pun on the name of the firm in Hamburg, the Studiengesellschaft für Elektronengeräte, by saying that his position at Löwe Opta would be similar to the one he had at the Stupidiengesellschaft für Elektronenverräter, *Stupid Society on the electronic traitors*.

[44] In this letter to his friend, on November 29th, 1942, we see Bruno using the term for mixed marriage children, which hardly ever appears in the letters to his father.

[45] Löwe Radio, later Opta Radio, was founded in 1923 by the two brothers Sigmund and David Löwe as a radio manufacturing company, with advanced research in integrated circuits, and started research in television in 1929.

[46] Bruno in a December 1942 letter seems not to believe this to be reason, but offers no other explanation as to why it would have been Tomfohr to hire him. Infact, if the Hamburg police knew

two good reasons for Bruno to be hired at Opta: Egerer, the director of the section where Bruno was going to be employed, was a Viennese, a connection which helped Bruno in many occasion. Egerer had been Hans Thirring's student, as Mira Hatschek had also been. The other reason was unknown to Bruno in those days: the firm was providing material of interest to the RLM, and Egerer was one of the scientists trusted by General Milch, at the time one of the most powerful persons at the Luftwaffe (Waloschek 2012).[47] In October, when Bruno had left the Studiengesellschaft für Elektronengeräte and had been looking for a similar position elsewhere, he had started contacts with some officers from the RLM, who were known to be interested and promoting war related high-frequency research. These attempts had not been successful, but Bruno's capacities and personal details may have been known to be of interest to Egerer, via the RLM. The Viennese connection, namely referrals among other Viennese in Hamburg or Berlin, played often a role in Bruno's getting work, and even, as we have seen, free meals in a fancy restaurant.

At the office, Bruno kept very much to himself, trying to avoid attracting attention or getting into conflict. His attitude towards his new boss was to support his ideas, if it did not require additional studying. Thus, initially, he was able to go along well with Egerer, and was also rather happy, having found books he could take home to study, such as Wigner's book on group theory (Wigner 1931).[48] However, after some time, something started worrying Bruno: Dr. Egerer keeps inquiring with me about physics things that don't interest me and that have nothing to do with the business [of the firm]. These things appeared to be related to 'elektrische Maschinen' [electrical machines], which he had little desire to read about at home. Coupled with a warning, which Egerer gave to Bruno in the first days of his employment, namely to 'consider this position as a war effort', something more than just normal work appeared to be expected of Bruno at the Opta.[49]

Bruno's instinct was right. There was more afoot with Egerer and Opta Radio than appeared at first sight. In 1943, he would find this out, and would slowly become more and more involved on secret projects financed by the RLM, Reich Ministry for Aviation. These projects were part of a war effort directed to manufacture the so-called *death-ray* machines (Waloschek 2012).

Bruno was half-Jewish, it would probably have been known to Opta, which had research contracts with various Ministries.

[47] General Field Marshal Erhard Milch (1892–1972), at that time Director of Air Armament (Generalluftzeugmeister) and Air Inspector General (Generalinspekteur) of the Luftwaffe, was one of the most powerful man under Hermann Goering (1893–1946) in the Third Reich.

[48] Eugen P. Wigner (1902–1995) was awarded one half of the Nobel Prize in Physics 1963 for 'his contributions to the theory of the atomic nucleus and the elementary particles, particularly through the discovery and application of fundamental symmetry principles'. The original German version of Wigner's book on group theory was translated into English in 1959 (Wigner 1959).

[49] November 29th letter to Peter, and undated December 1942 letter to parents from Berlin.

3.4 February 1943: Egerer's Crazy Plans

Bruno stayed in Berlin for Christmas and New Year. It was the first time in his life
that he had spent the Christmas holidays away from home, and would not be the last.
But he was not sad. He was with his new friends, and had a good time. On January
2nd, 1943 he sent greetings for his father's birthday, and a card with funny drawings,
which reflected his good spirits. The card had some *coriandoli* glued to it, remnants
of a holiday party at the Studiengesellschaft, Fig. 3.9.

The work at Opta Radio however was somewhat different from what Bruno had
expected. It involved administrative work for the laboratory, such as dealing with
orders and providers, even managing jealousy between various departments. It also
required some work for Bruno's boss Dr. Egerer in a 'secret laboratory', whose details
remain unspecified.

Fig. 3.9 Bruno's self
portrait in the bottle,
with a friend, Family
Documents, © Francis
Touschek, all rights reserved

Once more, busy as he was at Opta and with work for Egerer's secret laboratory, Bruno felt that his physics interests were neglected because of the many hours spent at the firm, and mused that his intellectual work was becoming limited to the referee work he was doing for the *Chemischen Zentralblatt*. As a matter of fact, he was learning more than he thought. His work was appreciated so much that he soon received commissions from other scientific journals as well, occasionally to referee articles whose subject went beyond his actual knowledge, such as one, bordering on philosophy, on *Epistemology and Modern Physics*. This did not not discourage him, quite the opposite: what he did not know, he would learn. Thus, when asked to referee the work on epistemology and modern physics, he immersed himself in Russell, Poincaré and the other modern natural philosophers.

Bruno's capacity and willingness to learn on his own all kinds of subjects had come early to him. Since his first being prevented from taking the Austrian *Matura* with his fellow students of the Piaristengymnasium in 1939, he had adapted to become his own teacher, a habit which he had to perfect when his studies were once more interrupted by the expulsion from the University of Vienna in May 1940, and afterwards, as he found himself alone in Hamburg, in 1942. When he was asked to referee for the *CZ*, most of the articles he was assigned to read were about physics, but he would do his theoretical work so well that, at a certain point, he was listed among the reviewers as a Graduate in Chemistry. The thing embarrassed him quite a lot, and, following his father's advice, he requested a correction to be published in the journal.[50] He was also doing work for some journals in which Egerer was involved, the *Elektrotechnische Zeitschrift* and the *Archiv für Technisches Messen*. Most importantly, as we shall see, Karl Egerer was Editor of the scientific magazine *Archiv für Elektrotechnik*, where an article proposing the principle of a new type of accelerator, a betatron as it is now called, had been published in 1928 by the Norwegian scientist Rolf Widerøe (Widerøe 1928).

Germany was not winning the war. In January, the war arrived directly in Berlin, with day-time fire-bombing for the first time.[51] The aerial superiority of the Anglo-American forces terrified the civil population, while military resources were diverted to the Russian and African fronts. In the ongoing search for offensive-defensive weapons against the devastating bombing, some military officers at the RLM had been focusing on devices which could allow the extraction of electromagnetic rays powerful enough to hit enemy planes, block their engines or kill the pilots, the so-called *death-ray* weapons. Such devices were of two types. One was the so-called *X-rays gun*, whose construction was pursued by Ernst Schiebold, Professor of Physics

---

[50] Undated letter, placed between two letters dated February 2nd and February 3rd, 1943.

[51] A prelude to the devastating November and December 1943 raids came from the De Havilland Mosquito, which hit the capital on January 30 1943, the tenth anniversary of the Nazi party coming into power, the *Machtergreifung*. That same day, both Göring and Goebbels were known to be giving big speeches that were to be broadcast live by radio. At precisely 11.00 am, Mosquitoes of No. 105 Squadron arrived over Berlin exactly on time to disrupt Göring's speech. Later that day, No. 139 Squadron repeated the trick for Goebbels. From https://en.wikipedia.org/wiki/Bombing_of_Berlin_in_World_War_II. See also https://www.airforcemag.com/article/0708bigb/ about the strategy adopted by the British government about bombing of Berlin.

at the University of Leipzig, who would propose it to the Luftwaffe in April 1943. Work for such device would take place in Leipzig, but part of the device would be built at the Luftwaffe's Gross Ostheim Research Station.[52] A description of the proposed work at the Research Station can be found in (Waloschek 2012, 63–93), where one also learns that in 1943 Bruno was doing theoretical calculations for Ernst Schiebold. This might throw some light onto Bruno's work with an otherwise unspecified firm 'F.G.' and Bruno's efforts to obtain permission to work there from Opta Radio, namely from Egerer, as mentioned in his February 1943 letters.[53] Details about F.G. are missing, most likely because of secrecy. In such case, 'F.G.' could in fact stay for *Forschungsstelle Gross Ostheim*, and refer to perspective work for the secret weapon proposed by Schiebold to the Luftwaffe. However, Amaldi's biography never mentions activity for this research station.

Another type of *death-ray* producing devices financed by the RLM was supposedly based on the betatron, Widerøe's great invention in accelerator physics, (Waloschek 1994; Sørheim 2020).[54]

Touschek worked for two years, from 1943 until 1945, on one such project, which was directed by Widerøe himself, a fact of great import for Touschek's future contribution to accelerator physics. How did he get involved in this project, which at the time would be totally useless as a war weapon, but which gave a pivotal contribution to the birth of electron-positron colliders, eighteen years later? The clue to how the Norwegian connection, or, if one prefers, the Norwegian road to high energy physics, entered into Bruno's story, can perhaps be found in one of Bruno Touschek's letters to his father, which is reproduced in its original German text in Fig. 3.10 (Bonolis and Pancheri 2011) and whose most relevant parts read as follows:

```
Bruno Touschek, Berlin-Friedenau, Sponholzstrasse 43.
February 15th 1943

[...]

Following a lecture by [Hans] Thirring [in Vienna, added by
the Author] about the cyclotron, I casually made the remark
to Sexl and Urban that with such high energies relativistic
```

---

[52] This research station had been installed adjacent to the Gross Ostheim military airport near to the city of Aschaffenburg.

[53] Letters to family in which F.G. or FG is mentioned are dated as follows: (i) February 2nd, 1943, (ii) undated but placed after the February 2nd in the sequence kept by Mrs. Touschek, likely following Bruno's father ordering, (iii) February 10th, 1943, (iv) February 15th,1943.

[54] A betatron is a circularly shaped device to accelerate electrons to typical 1–100 Million electron Volt energies, first proposed in 1928 by the Norwegian Rolf Widerøe. It was in use for particle physics research until shortly after the war when a new type of accelerator, the *synchrotron* was proposed as able to accelerate particles to higher energies. Betatrons continued to be used for medical purpose, where the range of particle energies is optimal for biological applications. Rolf Widerøe, the inventor of the betatron principle, was instrumental in promoting and spreading its medical use through Europe. The betatron is typically a glass tube, in which electrons are kept on a circular orbit by a fixed magnetic field, while being accelerated by an electric field, which is generated by a changing magnetic field flux through the circular area enclosed by the glass tube. The device is based on Faraday's Law and the action of the Lorenz force.

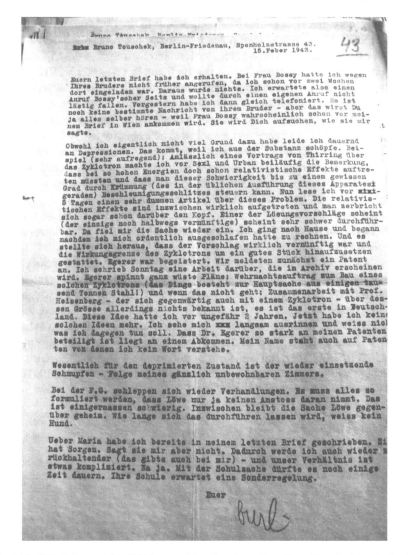

Fig. 3.10 BT's February 15th 1943 letter to his father, from (Bonolis and Pancheri 2011). Family Documents, © Francis Touschek, all rights reserved

effects would have to occur and that this difficulty can be controlled to a certain extent [...]

Now, I read a very stupid article about this problem 5 days ago. The presence of relativistic effects has been confirmed in the meantime and we are even racking our brains over it. One of the proposed solutions [...] seems difficult to implement. Then I thought again about it. I went home and started to calculate, after I had slept well. Egerer was

thrilled. We first got a patent. I wrote a paper about it on
Sunday that will appear in the Archive. Egerer spins very
wild plans: a Wehrmacht order to build such a cyclotron (the
thing mainly consists of a few thousand tons of steel!) and
if that doesn't work: cooperation with Prof. Heisenberg -
who is currently working on it with a cyclotron - however
its size is not known, it is the first in Germany. I had
this idea about two years ago. Now I have no such ideas
anymore. I see myself slowing down and do not know what to
do about it. [...] Dr. Egerer is so heavily involved in my
patents because of an agreement. My name is also on patents
that I don't understand.

[...]

At the F.G. negotiations drag themselves again. Everything
has to be formulated in such a way that Löwe just takes no
offense at it. That is somewhat difficult. In the meantime,
the Löwe matter remains secret from anyone. How long this
can be done nobody knows.

[...]

This letter is very difficult to understand, seemingly more like a series of notes
than a coherent narrative. Since Bruno talks of being depressed, and specifically
attributes the cause to excessive mental straining, it could be dismissed, were it not
for its mentioning a '*dummen*' article which Bruno had read a a few days earlier
and for an explicit address to '*wüste Pläne*' by his boss Egerer in the context of
particle accelerators, the cyclotron in particular.[55] Next, Bruno talks about a possible
interest from the Wehrmacht and Heisenberg. To understand or at least interpret the
meaning of such letter, it is necessary to go back to earlier times, the development
of elementary particle accelerators in the 1930's and early 40's and to the role held
by the Norwegian scientist Rolf Widerøe in such developments. We shall thus try to
answer the question as to who was the author of the article which excited Egerer and
led him to involve the Wehrmacht and even interest Heisenberg.

---

[55] Cyclotrons were particle accelerators invented by Ernest Lawrence (1901–1958) to accelerate
protons (Lawrence and Livingston 1931) or ions (Lawrence and Livingston 1932) along a circular path, https://web.archive.org/web/20060203005231/http://www.aip.org/history/lawrence/first.
htm. Lawrence was inspired in its invention by Rolf Widerøe's proposal for the betatron and his
successfully operated linear accelerator (Widerøe 1928), a device which accelerated particles along
a linear path. He was awarded the 1939 Nobel prize 'for the invention and development of the
cyclotron and for results obtained with it, especially with regard to artificial radioactive elements',
from https://www.nobelprize.org/prizes/physics/1939/lawrence/biographical/. As particle accelerators to high energies, cyclotrons were superseded first by betatrons and then by synchrotrons.

Chapter 4
The Road from Norway: Rolf Widerøe and the Strahlentrasformator

...our dreams had no limits. [1943]
Rolf Widerøe, Europhysics News, Vol. **15** *(1984) 9*

Abstract In 1943, Bruno Touschek and Rolf Widerøe's lives crossed paths in Germany, working on a secret project financed by the Aviation Ministry of the Third Reich. The project was set in motion in 1941 by Donald Kerst successfully building a working betatron, a novel type of particle accelerators, able to accelerate electrons to much higher energies than so far attained. Kerst's article in *The Physical Review* prompted the Norwegian Rolf Widerøe, who had formulated the betatron's working principle in 1928, to put forward his own proposal in an article submitted in September 1942 to the German scientific journal *Archiv für Elektrotechnik*. German accelerator science surrounding Rolf Widerøe's proposal were of interest to the military, and were also an opportunity for German scientists to employ their Jewish colleagues in projects of national interest, thus shielding them from forced labor and deportation. Such projects included betatrons and giant X-rays installations able to produce electromagnetic rays intense enough to damage enemy's planes, the so called *death-rays*. An overview of the secret *death-ray* projects during WWII in Germany shows why Widerøe's article led to the proposal reaching the Aviation Ministry, in the person of General Erhard Milch, Director of Air Armament (Generalluftzeugmeister) and Air Inspector General (Generalinspekteur) of the Luftwaffe. Circumstance surrounding Bruno Touschek's review of Widerøe's article, can explain how a classified project to build this betatron was set in motion.

There are things which happen by chance, others by wilful choices of the people involved. For chance to lead to further improvements in knowledge or benefits to society, both must occur. Chance is at the core of the conception of AdA through the encounter of Bruno and Rolf Widerøe. One was a 22 year old half-Jewish Austrian trying to continue his physics studies in Germany in the dark shadows of World War II (Amaldi 1981; Bonolis and Pancheri 2011). The other was a Norwegian, a family

G. Pancheri, *Bruno Touschek's Extraordinary Journey*, Springer Biographies,
https://doi.org/10.1007/978-3-031-03826-6_4

man, a successful engineer in his early forties, who went to Germany following the dream of his youth, the construction of a new type of particle accelerator, a *Strahlentransformator* (Widerøe 1928), and it was a machine to accelerate electrons, the β-rays first discovered in the late 19th century. But he had not succeeded to make it work.[1]

Chance lies in how Touschek came to read Widerøe's 1943 article, on the one hand, and on how Rolf came to write this article, after many years during which he had seemingly put aside his dream (Waloschek 1994; Wiik 1998; Sørheim 2020). We have seen how Bruno happened to read Widerøe's article, according to his biographer, namely through his job at Löwe Opta, secured by the chance encounter with Mira Hatschek on a train in Berlin (Amaldi 1981, 4). As for Widerøe, in this chapter we shall see how the last issues of an American journal to reach the University of Trondheim in Norway before the end of the war, led Widerøe to go back to his dream.

Of course, chance is a concept often borne by ignorance of the events behind the occurrence of some facts. War events were behind the last issues of *The Physical Review* reaching Trondheim when it did, and Bruno's work for Löwe Opta could have occurred anyway, sooner or later, as he had already thought of looking for Mira, when leaving Hamburg for Berlin, in November 1942. But, in Bruno's 1978 conversations with Amaldi in the Hôpital de La Tour in Geneva, he remembered meeting Mira as a chance encounter. In any case, one can see that the two roads which led Bruno and Rolf to work together owe their crossing to a background of events all naturally occurring. In other words: the essence of chance is a time coincidence.

The transformation of these dreams into great accomplishments, comes from courage and genius. When Bruno, a would be theoretical physicist, and Rolf, with a successful particle accelerator described in his doctoral thesis—a linear accelerator first in the world of that type—came together with their dreams, it can be said that the future of high energy physics started its long road to the successes of the 20th century. Rolf saw these successes, Bruno, albeit the younger by almost twenty years, did not. Chance staged Bruno's casual encounter with Mira Hatschek on a train in Berlin - and thus Bruno's work at Löwe-Opta. Chance made Kerst's article to appear in the last issue of *The Physical Review* which reached Trondheim, in Nazi occupied Norway, in Fall 1941. The dreams of the two scientists did the rest.

4.1 Particle Accelerators

Particle accelerators were developed in the last century, not long after Ernest Rutherford bombarded a gold foil (the *target*) by means of a beam of α-particles (the nuclei of the Helium atom, the *projectiles*) in 1911.[2]

---

[1] Such type of machine was later called *betatron*, a name said to have been chosen in a department contest, after Kerst' first success in building such machine (Pender and Warren 1943), according to wikipedia.

[2] Ernest Rutherford (1871–1937) was a New Zealand born British scientist. His famous experiment was carried out at the University of Manchester, by Rutherford's undergraduate student Ernest

Rutherford's experiment opened the way to the discovery of the atomic structure, which is now described as a positively charged nucleus, surrounded by negatively charged electrons, orbiting at a distance hundred thousand times the nucleus' dimensions. Previous to this experiment, the atoms, the unbreakable constituents of matter according to the ancient Greeks, had been modelled as a 'plum pudding', with the electrons as the 'plums' embedded in a 'dough' of positive matter. Rutherford's experiment, with some α-particles scattered away from their initial trajectory at very large angles, contradicted this picture and, in 1913, Niels Bohr, Rutherford's student in Cambridge, presented a model of the atom as an infinitesimal replica of a solar system, with an electrically charged nucleus around which negatively charged electrons circulated through Coulomb force attraction. The experiment implied that a sufficiently energetic particle, such as the α-particles emitted by a radio-active nucleus, could break the atom, travel through the vacuum between the electrons and the nucleus, and then be scattered away, by the strong repulsive force acting between the positively charged nucleus and the just as positively charged α-particles.

At the time of Rutherford's experiment in 1911, the source intensity was selected according to the radioactive material producing the α-particles. The path was controlled by slits and shields. The next step in the study of the basic constituents of matter was to control the particle sources and their energy. Varying the sources and increasing their momentum would allow deeper penetration into matter, and production of isotopes. To proceed further took time and it is only after the interruption of WWI that new developments occurred. A step in reaching higher acceleration was obtained through electrostatic accelerators, such as the Van de Graaf generator operational in 1931, or the Cockcroft-Walton voltage multiplier in 1932, both called by the name of their inventors.[3] But first came Rolf Widerøe's linear accelerator, subject of his PhD thesis in Aachen, and, before this, the idea of the betatron, which Widerøe had developed in 1922 at the University of Karlsruhe (Widerøe 1928; Widerøe 1947).

Marsden, and with the collaboration of the German physicist Hans Geiger, of the Geiger-Müller tool for measuring radioactivity. A nice simulation can be found at https://micro.magnet.fsu.edu/electromag/java/rutherford/.

[3] For a concise history of particle accelerators see https://www.britannica.com/technology/particle-accelerator#ref364994. See also https://en.wikipedia.org/wiki/Cockcroft%E2%80%93Walton_generator: 'The Cockcroft-Walton voltage multiplier is an electric circuit that generates a high DC voltage from a low-voltage AC or pulsing DC input. It was named after the British and Irish physicists John Douglas Cockcroft and Ernest Thomas Sinton Walton, who in 1932 used this circuit design to power their particle accelerator, performing the first artificial nuclear disintegration in history. They used this voltage multiplier cascade for most of their research, which in 1951 won them the Nobel Prize in Physics for 'Transmutation of atomic nuclei by artificially accelerated atomic particles'. The circuit was discovered in 1919, by Heinrich Greinacher, a Swiss physicist. For this reason, this doubler cascade is sometimes also referred to as the Greinacher multiplier. Cockcroft-Walton circuits are still used in particle accelerators....It was also one of the early particle accelerators responsible for development of the atomic bomb.'

4.1.1   Roads from the North of Europe and Elsewhere: Linear Accelerators, Beam Transformers, Cyclotrons and Induction Accelerators

One of the roads to modern day particle accelerators comes from the Nordic countries. From Sweden one traces the first proposal for a linear accelerator (Ising 1924).[4] Around the same time, a young Norwegian studying engineering in Germany, Rolf Widerøe, thought of a circular accelerator. He did not succeed in making his idea into a working instrument, which he called a *ray trasformer*, but the principle behind it would later inspire Donald Kerst in the USA to build the first operational betatron (Kerst 1940, 1941; Kerst and Serber 1941).

Widerøe is one of the founders of accelerator science. He was successful in the construction of the first linear accelerator. His fascination with particle accelerators started very early, as he was still in his high-school days. Passion for science arises as a beacon to a young man or a young women, when they are still in their teens, a call which remains for life, often through difficulties and disappointments. Such a passion is what drove Bruno Touschek away from Vienna in 1942, to continue his physics studies in Hamburg and Berlin, where he would meet Widerøe and start the long road towards building the first ever electron-positron collider.

Rolf Widerøe holds a special place also in the history of particle colliders, which are accelerators in which particles are made to collide head-on. He appears in Touschek's notes, when he was proposing the construction of AdA to the scientists of Frascati National laboratories, in Italy. He acknowledged Widerøe's insight when he wrote: The first time I heard of colliding beams was from Rolf Widerøe.[5] This conversation had occurred in Hamburg, in 1943 during the dark times of the Second World War. In his autobiography Widerøe mentions the conversation as taking place after an autumn vacation with his family in Norway, at Lake Tyndall. At the time, Norway had been occupied by the Germans since 1941, and Widerøe was just starting a classified project to build a 15 Million electron Volt (MeV) betatron,[6] commissioned by the Aviation Ministry of the Reich, the Reichsluftfahrtministerium, R.L.M. or RLM. How this happened is a story only partially known.

---

[4] See https://indico.cern.ch/event/532397/contributions/2170633/attachments/1343755/2025070/Alesini_LINEAR_ACCELERATOR_handouts.pdf.

[5] Undated typescript, probably prepared at the end of February 1960, and before the official AdA proposal on March 7th, Bruno Touschek Papers in Sapienza University of Rome–Physics Department Archives.

[6] Electron Volt (eV) is a unit of energy used in elementary particle physics, and corresponds to the energy gained by an electron passing through a potential difference of 1 Volt.

4.1.2 Rolf Widerøe and His Dreams

Widerøe's dream was to build particle accelerators, (Sørheim 2015, 2020). This is a theme which runs through Rolf Widerøe's autobiography, compiled by Pedro Waloschek (Waloschek 1994), published a few years before Widerøe passed away at 94 years of age. In Rolf Widerøe's honor, a symposium was held at the Oncological Hospital of Oslo, and a talk was given by Bjørn H. Wiik, the late Director of the Deutsches Elektronen Synchrotron laboratory from 1993 until 1999 (Wiik 1998), inclusive of a concise summary of the history and working principle of particle accelerators. Wiik, who was himself a Norwegian and knew Widerøe very well (Kienle 1999), tells about the young Rolf having to decide about his University studies in 1919, and becoming fascinated by articles about the nascent science of elementary particles accelerators, which appeared in magazines and newspapers.

Rolf Widerøe was born in Oslo in 1902, one of three children of a well-to-do middle class family, and, at eighteen, was a keen reader of popular scientific magazines, where the nascent science of atomic physics and of particle accelerators, was presented to a public eager to discover a better world to be built over the ruins of the old regimes. In the years before and after WWI, the reading public was keen to follow the successes of atomic physics: early discoveries, such as the X-rays and their property to penetrate human bodies as shown by the famous photograph of Mrs. Röntgen's hand, or Eddington's 1919 experiment proving Einstein's general relativity hypothesis, had a large impact on the popular perception of what science could do.

Less popular, but still frequently appearing in science magazines, were descriptions of what scientists were doing with newly found particles beams, the α particles, and the β and γ-rays, emitted by the newly discovered phenomena of natural and artificial radioactivity.[7] In 1911, Ernest Rutherford's experiment in Cavendish Laboratory in Cambridge had shown that the atoms had an inner structure and could be penetrated through proper use and control of these particles. But the new science was not only pure research. Beneficial use of radioactivity had been also evident through the war, thanks to the dedicated activity of Marie Curie, with the assistance of her daughter Irène, driving a hospital ambulance to apply radiotherapy to wounded soldiers. In those years nuclear physics had not yet shown its deadly face and gave hopes for a better future.

Bohr's description of the structure of atoms was easily grasped by scientific magazines readers. After Rutherford's experiment, physicists looked to other atomic constituents, electrons or protons, to be used a as more penetrating probes into the structure of matter. The science of particle acceleration was starting. Rolf Widerøe

---

[7] Newly discovered elementary 'particles' at the beginning and the turn of last century were named from the first letters of the Greek alphabet. Discovered in that order, α-particles were soon found not to be elementary at all, but to be ionized Helium nuclei, four times heavier than Hydrogen, β-rays were then identified as consisting of electrons, particles with a definite mass value, roughly 2000 times lighter than than the Hydrogen atom. The definition of *rays* is more appropriate for γ-rays, which correspond to electromagnetic radiation of a given frequency. The habit of using Greek names for elementary particles, in addition to Latin names, is still followed.

told his biographer Pedro Waloschek how he had imagined the vast possibilities offered by electromagnetism to control the motion of charged elementary particles, when he was still in his late teens. He had then decided to study engineering, a field which was less esoteric than physics and more grounded in building equipment or new devices. Through these, he could envision finding a path to the mysteries of matter. When it came to choose university studies, Rolf's father thought that technical training at Norwegian institutions at the time were not as excellent as what was available in Germany, where he himself had studied. Following his father's advice, Rolf left Norway and went to study at the University of Karlsruhe, in Southern Germany.

4.1.3   A Transformer to Accelerate Charged Particles

It is in Karlsruhe that Rolf developed the idea of accelerating charged elementary particles by way of a changing magnetic field.

The first suggestions to increase the energy of charged particles had been to make them move through an electric field, gaining an amount of energy proportional to the potential difference and the particle's charge. The idea was put forward by the Swedish physicist Gustav Ising (1883–1960), Fig. 4.1, (Ising 1924).

Another way for charged particles to gain energy is through a time-changing magnetic field—based on a phenomenon known as Faraday's law of electromagnetism. The magic of Faraday's law allows to transform an alternating current (AC voltage) into a direct one (DC), by means of the changing magnetic field flux, generated by the AC current through a loop wire. This is the principle on which the well known

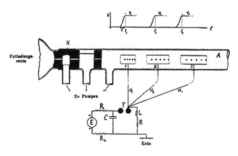

Fig. 2.13

Ising's proposal for a linear particle accelerator. The high-frequency field is supplied by a discharge across the spark gap F; K is the cathode; a_1, a_2, a_3, connections to the drift tubes. Ising, *Kosmos, 11* (1933), 171.

GUSTAF ISING

Fig. 4.1   Gustav Ising, courtesy of Swedish Biographical Dictionary (SBL), https://sok.riksarkivet.se/sbl/Presentation.aspx?id=11994, and his proposed accelerator (Ising 1933)

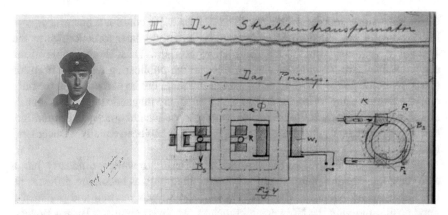

Fig. 4.2 The eighteen year old Rolf Widerøe, from the collection of the ETH-Bibliothek, published on Wikimedia Commons as part of a cooperation with Wikimedia CH, licensed under https://creativecommons.org/licenses/by-sa/4.0/deed.en. At right a reproduction of Fig. (2.2) of his sketches for the proposed Strahlentrasformator from (Widerøe 1928, 22)

transformers, which give DC currents to our households, operate, and which led Widerøe to propose a new type of particle accelerator.

In Fig. 4.2 we show a photograph of Widerøe in his high-school days, and the original Widerøe's figure illustrating his point. His idea was that instead of a loop wire, as in the traditional transformer shown at left in the sketch, one could put electrons in a glass ring shaped tube, in which they would be constrained to circulate by the force exerted by a fixed magnetic field. The electrons circulating in the closed area defined by the glass tube, shown at right, would be subject to exactly the same laws as those of a current in closed coil, and, again thanks to Faraday's law, they could be accelerated if a changing magnetic field would act through the area enclosed by the glass tube.

The principle of such a device, appropriately called beam or ray transformer, *Strahlentrasformator*, was envisioned by Widerøe during his years as an engineering student at at Karlsruhe.

Beautiful as the idea was in principle, Rolf was however unable to build a working device.[8] But Rolf's dream to build a particle accelerator would not easily dissolve. As he could not obtain a doctorate on the basis of a device whose feasibility could not be proven, he went to the Aachen Polytechnic and proposed to build a different type of accelerator, the linear electron accelerator, such as envisaged by the Swedish physicist Gustav Ising. This time, Rolf succeeded and the first linear electron accelerator was built. Rolf Widerøe was awarded his PhD and published the results in the

---

[8] The difficulties came with keeping stability of the electron's path, since electrons could be scattered out of their fixed orbit by their collision with residual gas particles in the glass tube. Contemporary vacuum techniques were unable to eliminate the residual gas to the extent needed for the device to work, and Widerøe's professor, an expert in such techniques, could not be convinced to award a doctorate just for the idea.

German scientific journal *Archiv für Elektrotechnik*. He could not however ignore his dream of a ray transformer, as its working principle, later to be known as the *Widerøe's condition*, was too beautiful not to be right. Thus, the principle of his unsuccessful beam transformer was included in the publication resulting from his thesis (Widerøe 1928). He did not know it at the time, but he had not been the only one to envisage using a circular device with a changing magnetic field flux to accelerate elementary charges: Ernest Walton, at the Cavendish Laboratory in Cambridge, had also independently thought of such a scheme, but had been equally unsuccessful (Walton 1929).[9]

With his doctorate, in 1928 Wideroe went to Berlin, to work for industry, but, a few years later, as the political and economic situation was becoming difficult, he left Germany and went back to Norway. In 1934 he was married and obtained a good position joining the Norwegian branch of the Swiss company Brown Bovery, 'Norsk Elektrisk og Brown Boveri', NEBB. He was not working on building particle accelerators, but would occasionally follow the field, which was in the meanwhile taking giant steps. Driven by the unparalleled successes in atomic, nuclear and particle physics of the 1930's, mostly obtained through cosmic ray studies, accelerator science was galloping along, with the United States taking first stage. There, at the University of California at Berkeley, the first circular accelerator was built in 1930.

4.1.4 Ernest Lawrence's Merry Go-Round

In 1930, far away from Norway, in the USA, in Berkeley, a young American physicist, Ernest O. Lawrence, the son of a Norwegian immigrant, took a giant step in particle accelerator development when he morphed the linear accelerator—Ising proposed and Widerøe built—into a circular machine able to accelerate charged particles, such as protons or ions, to higher energies, through a combination of electric and magnetic fields (Lawrence and Livingston 1931, 1932; Heilbron and Seidel 1990). The new machine was called a 'cyclotron', and Lawrence used to refer to his invention, for which he won the Nobel Prize in 1939, as a 'proton merry go-round'.[10] Charged

---

[9] The paper says: 'The present paper contains an account of some work done in an attempt to produce high speed electrons, using an indirect method suggested by Sir Ernest Rutherford. Although it was not found possible to make the method work, yet a description of it may be of interest on account of the importance attached to any method of obtaining fast electrons.'

[10] See https://www2.lbl.gov/Science-Articles/Archive/early-years.html. See more details in https://history.aip.org/history/exhibits/lawrence/first.htm. The step from linear accelerators to circular ones marks a fundamental change in the history of particle accelerators: while linear accelerators needed longer and longer installations for the particles to gain higher energies, circular paths could now be used with great savings in the space needed for such installation in the pursuing of higher energies. In fact, as particles would be made circling many times while at each turn getting a new kick in their acceleration, gaining more energy, they would remain constrained to a circular area, of smaller dimensions than those needed to reach the same energy with a linear accelerator. As the particles increased their energy, the path radius would increase, until the maximum sustainable by such mechanism (Bethe and Rose 1937).

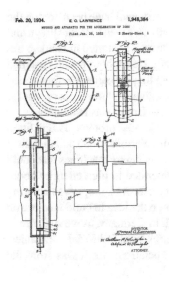

Fig. 4.3 Figures from the cyclotron patent Ernest Lawrence submitted to the US patent office in 1934, from https://patents.google.com/patent/US1948384, and one portrait held by Niels Bohr Library & Archives, AIP Emilio Segré Visual Archives, Physics Today collection, id: Lawrence Ernest Orlando A5

particles, kept on a circular path by a constant magnetic field, would be accelerated to higher energies by an electric field giving a kick at each passage from one half of circle to the next. The particles thus accelerated would then be extracted from the cyclotron and directed to hit a target with the material to be investigated. In Fig. 4.3, we show a photograph of Lawrence and his sketch of the cyclotron from the 1934 patent.

The higher energies reached by the bombarding particles in a cyclotron allowed production of nuclear isotopes, and gave way to visions of tapping an energy source more powerful than anything else in existence, the nuclear force which keeps together the components of the atomic nucleus.[11] After the first cyclotron was built in 1931 at the Radiation Laboratory of University of California at Berkeley, now called 'Lawrence Radiation Lab', it was Frédéric Joliot-Curie, at Collège de France in Paris, who first picked up the lead to have a cyclotron in Europe in 1937. Built in Switzerland, it was installed at Collège de France in Paris and became operational during the war.[12] In those years, the construction of a cyclotron was also underway in

---

[11] For the role played by Lawrence and the cyclotron in the Manhattan project, see https://history.aip.org/exhibits/lawrence/bomb.htm.

[12] See https://publishing.cdlib.org/ucpressebooks/view?docId=ft5s200764&chunk.id=d0e16658&toc.depth=1&toc.id=d0e16658&brand=ucpress for a detailed history of cyclotrons. The first working cyclotron outside the United States came to life not in a great center of nuclear physics in

Copenhagen, through Niels Bohr's efforts, and could very well have been among the subjects discussed during the now famous 1941 Heisenberg's visit to Bohr (Frayn and Butler 2017). In Germany, the construction of a cyclotron was started in Heidelberg. As head of the German nuclear project, the *Uranverein*, Heisenberg would have had an obvious interest in cyclotrons and it is not surprising that his name was mentioned by Karl Egerer, when discussing with Bruno Touschek about Widerøe's proposal for a new type of particle accelerator in the *dummen* article Bruno was reviewing.

Both the American and the French cyclotrons accelerated protons, as this fitted well in the development of nuclear physics, in Italy with Fermi's group and in France with Frédéric Joliot and his group. But the interest in developing machines that could accelerate electrons was also present. Electrons accelerated to high energies could produce powerful X-rays of both pacific and military use. As Europe was entering into a new war, in the United States as well as in Europe the idea to build an electron accelerator following Wideroe's idea was considered. In Germany, as we shall see in detail in the next sections, a number of different groups from industrial companies, such as Siemens, tried to obtain funding to build a betatron. But it was from the United States that the first success came.

4.2   Scientific Articles and *Death-Ray* Proposals

With a series of three articles which appeared in *The Physical Review* through 1940 and 1941, Donald Kerst first, and then Kerst with his student Robert Serber, showed how they had succeeded to build a circular machine in which electrons could be accelerated to an energy of 2.3 Mega eV (MeV) and kept circulating for hours (Kerst 1940, 1941; Kerst and Serber 1941).[13] These articles, reaching Europe in 1941, set in motion events which would open the path to electron-positron colliders. In Norway they prompted Rolf Widerøe to return to his dream of building a betatron, and led to the project where Touschek learnt the art of making accelerators. In Italy, the articles caught the attention of two professors of physics at University of Milan, who thought it important enough to suggest to their student Giorgio Salvini to study it. Salvini, stationed with the Italian Army in the Northeast, prepared his thesis and obtained his Laurea in Fisica in 1943 on the theory of betatrons. The effect of this would surface 10 years later, in postwar Italy, when Salvini became the first director of the future National Laboratories of Frascati, where an electron synchrotron successfully operated in 1959, and, in March 1960, AdA's proposal was approved.

While this was still in the future, let us see how Kerst's articles affected Rolf Widerøe's life and future scientific work.

Europe, but at the Institute for Physical and Chemical Research (Riken) in Tokyo. The Swiss firm who made the magnet for the Paris cyclotron was Oerlikon of Zurich.

[13] One Mega corresponds to one million.

Fig. 4.4 The situation in Europe at the end of 1941. In blue, countries occupied by the Third Reich or allied, as was the case of Italy, from https://commons.wikimedia.org/wiki/File: Second_world_war_europe_1941-1942_map_en.png

4.2.1 An Issue of the Physical Review Arrives in Trondheim

The two articles in which Kerst announced to have built a working betatron were published in *The Physical Review* issue of July 1st, 1941. What happened next in Oslo is better left to the description by Roald Tangen, from Oslo University, in a 1993 letter to Pedro Waloschek (Waloschek 1994, 60).

At the time Tangen describes, Norway was partially occupied by the Germans, who would take it over completely in December.[14] We show in Fig. 4.4 the situation of Europe at the end of 1941.

Tangen remembered that in 1941 access to American journals, such the *The Physical Review*, had stopped, after the 1940 Norway occupation by the Germans. Thus, when Tangen, who was working on a small van de Graaf generator, was invited to give a lecture about particle accelerators in Oslo, he was initially unaware of the more recent developments about betatrons. The invitation had come from the Oslo Physics Association and had to take place in December 1941. However, some scientific information would still be unofficially exchanged among scientists, and, coincidentally, a few days before Tangen's trip to Oslo, the one issue of *The Physical*

---

[14] The Germans started their occupation of Norway in April 1940. See https://time.com/5885434/ nazi-norway-history/ for a review of a recent book by Despina Stratigakos on *Hitler's Northern Utopia: building the new order in occupied Norway*. Part of such plans included the creation of a New Trondheim, next to the old one, where Tangen's reading of Kerst' article took place in 1941 (Stratigakos 2020).

Review with Kerst's article reached him, by what he calls a misterious way.[15] Tangen immediately saw the interest of Kerst's announcement and included it in his lecture. When he delivered it, he mentioned Kerst's reference to a paper by an author he did not know at the time, but whose name, Widerøe, pointed to a Norwegian. Widerøe, still and always interested in following physics news and, in particular, those about particle accelerators, was in the audience. In fact, it was not really a coincidence, as Rolf, after a few years in Berlin, was then living in Oslo, with a position with NEBB. Rolf had kept his interest in physics and its developments and was a member of the Association, which he had contributed to financially support (Waloschek 1994, 57). It was thus quite natural for Rolf to be in the audience. He could not know it at the time, but this conference changed the course of his life, as he embarked in an adventure which would take him to Hamburg and Berlin, and to build a betatron for the Germans, during the last two years of the war. It also led to an official inquest over charges of collaborationism when he returned to Norway in 1945.

The inquest would change Widerøe's life for ever, and make him leave Oslo for Switzerland, after the war. The events surrounding the inquest are hardly mentioned in Widerøe's autobiography, and it took an article from the oncological ward of the Norwegian Radium Hospital in Oslo to resurrect them (Brustad 1998). Brustad's article was interestingly entitled *Why is the Originator of the Science of Particle Accelerators so Neglected, Particularly in his Home Country?* This article started author Aashild Sørheim's search for a full understanding of the motives and occurrences behind Rolf Widerøe's choice to work for the German military - occupying Norway against the people's will - while his brother was in prison in Germany (Sørheim 2015, 2020). Rolf Widerøe had passed away in 1996, having been bestowed with many honors, in Germany and elsewhere, for his contributions to science. The wounds of war were definitely over, and Brustad brought up the story of Widerøe's war period in Germany. The answer to Brustad's provocative question focuses on the charge for collaborationism and the related post-war inquiry.

Brustad's article opened up a much more complex story that what had been generally known at the time, in particular the connection with the infamous German *death-rays*, which, as it turns out, never came up to their fame. The subject had fascinated Pedro Waloschek, Widerøe's biographer, who published an extensive description of death-rays projects in Germany entitled *Todesstrahlen als Lebensretter* (Waloschek 2004). The interest in the subject led him to start on its English version entitled *Death-Rays as Life-Savers in the Third Reich*, with his daughter Karen's assistance.[16] The English version only appeared after his death (Waloschek

---

[15] Kerst had submitted his famous papers on April 18th, 1941, reporting the operation of a 2.3 MeV betatron (Kerst 1941), and citing Widerøe's, Walton's and Jassinski's papers, apparently on a request by the editor, with the corresponding theory appearing in a second paper (Kerst and Serber 1941), both papers in the July issue. According to Wolfgang Paul (Waloschek 1994, 156), this was the last issue of *The Physical Review* which arrived legally in Germany, and illegally in occupied Norway (Trondheim), mailed as an ordinary letter.

[16] Pedro Waloschek sent a copy of his last version of the book to this Author and Luisa Bonolis, one month before his death, following a correspondence about a possible collaboration about Touschek and Widerøe's war-related work. After Waloschek's death, the book was posted at DESY, upon

2012). Thanks to Waloschek's work and the more recent work in (Sørheim 2020), an almost complete description of the war events surrounding Widerøe's betatron is now available, and Bruno Touschek's years in Germany can also be made clearer.

As for Touschek, Amaldi's narration of this period does not clearly mention that the betatron project was connected to the development of *death-rays* to help the German military effort.[17] Bruno also gave rather fantastic explanations, both to his friends in Rome, in later years, as well as to his biographer, as to why he was arrested by the Gestapo in March 1945. All this clouded the importance of this period in Bruno Touschek's life and formation as a physicist, which are surrounded by partial neglect. However, a different narration emerges from the letters to his father during the period of construction of the betatron, and, by combining Waloschek's work with the letters, a coherent picture emerges.

4.2.2   How Rolf Widerøe Went Back to His Dream

It had been December 1941, when Rolf Widerøe attended Tangen's conference and learnt that an American physicist, in Illinois, had built the machine he had dreamed of almost twenty years before, without succeeding to make it work.

In those late December days of 1941, while Bruno Touschek, in Vienna, was planning his departure for Germany, Rolf, in Oslo, reflected on his old work on the beam transformer. 1941 had been a difficult year for both. Bruno had been denied his renewed application to attend the University and forced to give up all his dreams to study physics in Vienna. Rolf's brother Viggo, a pilot and active participant in Norway's resistance against the German occupying forces, had been imprisoned in May 1941, and sentenced to death for helping Norwegian fighters to flee the country. The sentence had then been converted to ten years imprisonment and he was then in Akershus Fortress, near Oslo, waiting to be transferred to the infamous Hamburg-Fühlsbuttel prison in Germany. The two brothers had been very close, Viggo, younger to Rolf, was a charismatic figure, and his fate obscured family life, already tragically altered when, in 1937, the youngest of the Widerøe's brothers, Arild, had died in an aerial accident.

4.2.3   The Article Which was Never Published

In the months following Tangen's conference in Oslo, Rolf went back to his ray transformer. Kerst had now demonstrated that Widerøe's principle could work. As

---

request by Karen Waloschek. Quotes from (Waloschek 2012) are included in the present book, by kind permission from Karen Waloschek.

[17] Amaldi mentions the secrecy surrounding the project, but there is no mention of how direct the connection was to the war effort, as he writes: 'The proposal [Widerøe's] was kept secret because of its possible applications. Such secrecy appears today and would have seemed to me (and to many others) rather curious even at that time. [...] Furthermore, it was already clear that the betatron could be employed only as a source of X-rays used manly for medical purposes.' (Amaldi 1981, 5).

the inventor of the idea, Rolf felt confident that a machine with energy quite higher than in Kerst's betatron could be feasible and spent the next 8 months preparing an article where he proposed a 15 MeV betatron.[18] His idea was to submit it to the journal where his original paper had appeared. In September, his article was ready and he submitted it to the *Archiv für Elektrotechnik*, where it was received on September 15, 1942.

At that time, Bruno Touschek was still in Hamburg, working for the Studiengesellschaft, but would soon resign from this position, moving to Berlin with a new job at Löwe Opta. During the previous months he had supplemented his income working as a referee for a number of scientific journals. The articles he would review ranged from chemistry to physics, to technical matters, to epistemology. If he did not know the subject, he would make himself read about it as much as possible. He was very good at this, and fast. His capacity as a reviewer may have helped him in being hired at Löwe Opta, where his future boss, Karl Egerer, was also *Editor-in-chief* of the *Archiv für Elektrotechnik* (Waloschek 2012, 96), and in the confidence of the Luftwaffe General Milch.

And this is how, in February 1943, while assisting Egerer at Löwe Opta, he came across Widerøe's article. There is no proof that the article Bruno mentioned to his parents, calling it a *dummen Artikel*, was Widerøe's submitted paper. Still, the most reasonable explanation for Widerøe's proposal coming to the attention of the military, is that a link between Widerøe and the RLM was established because Bruno Touschek called Egerer's attention to the submitted article. Many trails link Egerer and the German military interest in betatrons (Waloschek 2012). At the time, Egerer's frequentations with Milch had brought him to know of other betatron projects being aired in the months between the end of 1942 and early 1943. Egerer himself later claimed to have suggested—to General Milch—to include Widerøe's idea in an existing project (Waloschek 2012, 96). It is thus very likely that the fist step towards Widerøe's 15 MeV betatron, was taken at Löwe Opta, when Bruno drew Egerer's attention to the *dummen* article he was reviewing. About Egerer and Touschek, Widerøe says: 'It was probably Egerer who brought Touschek to us in Hamburg' (Waloschek 1994, 90).

Contrary to what appears in his memoirs, Widerøe's article was never published. The proofs exist and the betatron he proposed was built, but the pages of the *Archiv für Elektrotechnik*, where the article should be found, do not include an article by Widerøe, but something different, of no import to this story. Were a scholar, who had read Widerøe's autobiography, to go to the Deutsches Museum in Munich to consult volume 37 of the year 1943, and search for the article between pages 342 and 355, as indicated by Widerøe in his 1994 autobiography, she would not find it. She could then ask to consult the bounded volumes from other years, 1942, 1944, or even 1945,

---

[18] Electrons in Kerst' betatron had a maximum energy of 2.3 MeV. New projects in the USA aimed at reaching 100 MeV.

perhaps thinking that Widerøe made a typo, but this would not bring the article back to the journal where it had been destined.[19]

Why was the article never published, but also why did Rolf Widerøe hide this fact to his biographer? The answer can now easily be found, because Widerøe's proposal was accepted and the project was classified as war secret, and the secrecy was a proof that the charges of collaborationism, laid by the Norwegian judiciary against Widerøe, were, at least partially, founded (Waloschek 2012; Sørheim 2020).

There is a *hyatus* in Widerøe's story after he submitted his article for publication in September 1942. In his autobiography, he does not mention hearing anything about the fate of his article for quite sometime. Then something happened, in Oslo, in the spring (Waloschek 1994, 61), probably March or April, 1943. At this time, two or three German officers came to NEBB and approached him, asking if he would come to the Grand Hotel with them as they wanted to discuss something with him. Norway had been under occupation by the Germans since 1940, and one could not refuse the invitation. In the Grand Hotel, they asked him to return to Berlin with them, and, in his autobiography, he remembers to have been flown there two days later. He accepted because they mentioned this to be of relevance to help Rolf's brother Viggo, in prison in Rendsburg jail, in Germany, and in poor conditions. In Berlin, he was made aware of how his proposal for the 15 MeV betatron was of interest to the Luftwaffe, and things for its construction were set in motion.

A *posteriori* the chain of events following the arrival of Rolf Widerøe's article to Touschek's attention can be traced. Egerer was right in becoming excited when Touschek had spoken to him of the article he was reviewing. The article was in fact of interest to the German military, and fitted well with the current state of betatron research development in Germany. Egerer was aware of current *death-ray* projects under examination by the Luftwaffe, such as one by Ernst Schiebold from Leipzig, about powerful X-rays from giant anti-cathodes, and of two other betatron proposals, under discussion at around the same time. He contacted General Milch, while Bruno, either on his own or, more likely, prompted by Egerer, wrote to Widerøe and transmitted his objections to the treatment of the electron's orbit in the article (Amaldi 1981, 5).

4.3 Betatrons and *Death-Rays* Projects in Germany

The German interest for the betatron was pre-existent to the war, and it became of urgent concern in early 1943, when the German military was searching for a *miracle* weapon to turn back the tide of its losses on both the Russian and the African fronts. Indeed, after Kerst's article appeared in 1941, as many as three projects were put in motion, one by Siemens, one by a small company owned by Heinz Paul

---

[19] This is what happened to L. Bonolis in 2011, when together with the Author, they decided to check the relevant issue of the *Archiv für Elektrotechnik*, while their (Bonolis and Pancheri 2011) article was put to print. Personal communication by L. Bonolis.

Schmellenmeier in Berlin and, later, the one by Widerøe. Both industry and government were interested in supporting these projects: on the one side there were two industrial firms, Siemens (at Erlangen and Berlin), and Philips - which controlled the C.H.F. Müller Company in Hamburg -, on the other there was the government, through funding from the Reich's Research Council and the Luftwaffe, from the Aviation Ministry (RLM).[20] What started these different projects shows a combination of societal and human motives: in one case, Schmellenmeier aimed at protecting his friend Richard Gans, a well known scientist of Jewish origin, who could otherwise have been sent to hard labor camp, whereas Siemens projects were directed to industrial and medical exploitation. Widerøe's project was on the other hand more closely connected to *death-ray* production, favoured, among others, by Ernst Schiebold, from the University of Leipzig. In the following we shall briefly discuss all these projects, following Rolf Widerøe's last work (Waloschek 2012) and an article prepared by Waloschek, with Bonolis and the Author, about the connection between Bruno Touschek, Rolf Widerøe and the *death-ray* projects.[21]

4.3.1 The Siemens Project

The Siemens project was the natural development of the pioneering work by Max Steenbeck of Siemens, who had patented in 1933 a proposal for the construction of an apparatus named Elektronenschleuder ("electron catapult"), together with Rüdenbeck.[22] This was a prototype of the machine later to be called betatron by Kerst.[23] However, the electron beam produced by the experimental apparatus that Steenbeck developed in strict secrecy at Siemens-Schuckert in Berlin was so weak, that it was decided to stop the project. But after Kerst reported on his first successful betatrons, Steenbeck went back to his original idea and convinced Siemens to accept

---

[20] The Siemens project, after the successful commissioning of a 6 MeV betatron in 1944, was stopped in 1945 because the US-troops confiscated it. Work was resumed in 1947 in Göttingen, only limited to research on radio-biology and radiation-therapy, because nuclear physics in Germany was strictly forbidden at the time.

[21] The article preparation was interrupted by Rolf Widerøe's death in March 2012, and was never completed. Parts of the draft are reproduced here, courtesy of Luisa Bonolis and with Karen Waloschek's permission.

[22] A patent from Rüdenberg, R. and Steenbeck, M., Elektronenschleuder, German-Patent (Siemens) 656–378, was submitted on March 1, 1933, published on Febr. 4, 1938, referred to as [STEEN1933] in (Waloschek 2012).

[23] Kerst did not mention Steenbeck's 1933 patent in his articles, being probably unaware of it. A general review on the development of betatrons in the Siemens centers of Berlin and Erlangen during the period 1934–1976 can be found in (Osietzki 1988; Busch and W. Bautz 2005) and (Kaiser 1947). See also Widerøe's reply to Kaiser about his own invention of the betatron (Widerøe 1947). Actually, already in 1927/28 Steenbeck had developed the basic ideas for an accelerator which corresponded exactly to the cyclotron proposed in 1931 by Lawrence, but he did not publish anything on this work (Waloschek 2012, 175).

the proposal for a 6-MeV betatron put forward by an X-ray specialist, Ing. Konrad Gund.

Gund, an electrical engineer who was working in the laboratories for electro medical applications of the Siemens-Reiniger-Werke in Erlangen, was put in charge of developing the "electron-catapult". In May 1942 he presented a preliminary design for such a device, and after a series of tests performed in autumn, which confirmed its scientific and medical perspectives, the Luftwaffe requested the construction of a betatron in October 1942. In a letter dated January 12, 1943, Siemens was granted a contract for such research which was declared of war interest. The works to be performed were classified as secret (see footnote 29 in Busch and W. Bautz 2005) and it was established that Hans Kopfermann from Göttingen would act as supervisor. Support for the plan came from Werner Heisenberg. In February 1943 Steenbeck expressed his great regret to Kopfermann that previous research results had not yet been published, so that now it would appear as if Kerst had built the first operational betatron. For this reason Steenbeck was allowed to report about this activity, and especially about his previous results, in the journal *Die Naturwissenschaften* (Steenbeck 1943).[24]

Gund's betatron at Siemens had been planned aiming at medical applications, and was also used by physicists Hans Kopfermann and Wolfgang Paul for scientific measurements in nuclear structure. Kopfermann and Paul, who were planning to build a small betatron and use it for research aims, offered their help to test the already built Siemens machine. Wolfgang Paul himself told Pedro Waloschek (Waloschek 2012, 179): "I started learning the required impulse-electronic and some nuclear physics and we mounted the required apparatus in Göttingen." During the summer of 1944 they were able to start the first measurements, on "the time structure of the X-rays produced in an internal target, the angular distribution and the energy of the electrons, both from their range and using photo-effect on beryllium and deuterium." It is thus no wonder that support for the project came also from Heisenberg, as *per* footnote 30 in (Busch and W. Bautz 2005). Gund's betatron was the first to be operational in Germany and the first to be used to perform scientific research work.[25]

---

[24] On Thursday, 27 April 1944 the commissioning of the first betatron developed for medical purposes was done in Erlangen, and the Luftwaffe asked that a 20 MeV betatron should be built, as per footnote 33 in (Busch and W. Bautz 2005). When Gerlach became the successor of Abraham Esau as Commissioner for Nuclear Physics Research within the Research Council of the Reich (Reichsforschungsrat), in summer 1944 three orders with very high priority were issued to Dr. Max Anderlohr, director of the Siemens-Reiniger-Werke in Erlangen, for the construction of a 5–MeV betatron, for the research and development on a 20–25 MeV betatron, and for the planning and preliminary design work for a 100 MeV betatron (Osietzki 1988). By that time it was well known that a 100 MeV betatron, largely designed by Kerst, was being built in the United States by W. F. Westendorp and E. E. Charlton at General Electric Company, in Schenectady (Westendorp and Charlton 1945). The machine was completed and fully operational in 1945.

[25] Information about betatrons in Germany during the war used to be found in the FiAT, CIOS and BIOS reports based on joined British—American investigations, covering German Science and Industrial Institutions, covering a large number of topics. According to the CIOS report August 1945 (footnote in p. 7), two betatrons had recently been constructed and were being tested at the Siemens-Reiniger plant in Erlangen: the first was the just described 6 MeV device, the second

4.3.2   The Schmellenmeier's Project and Richard Gans

The case of Schmellenmeier's project was completely different, and fits with a policy adopted by many German scientists, who, although critical of the Nazi regime, had not left the country, often for genuinely patriotic reasons, and were trying to be of help to their fellow scientists of Jewish origin. To avoid their being sent to forced labor or concentration camp, scientists such as Richard Gans or promising young men such as Bruno Touschek, would be offered employment in some type of technical or scientific job, with their contribution to the firm's activity labelled as of 'interest to the country'.

Heinz Paul Schmellenmeier owned a small company, the 'Entwicklungslaboratorium Dr. Schmellenmeier', which had contracts with the Army Ordnance Office (Heereswaffenamt). In late March 1943 the idea of proposing the construction of a small betatron was discussed by Heinz Paul Schmellenmeier, with Hans D. Jensen and Fritz Houtermans.[26]

This idea had been strongly advocated by Houtermans, Schmellenmeier's close friend, who was worried about the fate of Richard Gans, a well known theoretician and a specialist in the field of magnetism, who continuously risked of being deported to a concentration camp, because of his Jewish origin.[27] Gans had the proper experience for tackling the task of calculating the electron orbits and the complicated magnetic

---

one, more recently constructed, was a 7 MeV machine. At that time plans were still being made to construct a 15 MeV betatron. Recently, as of June 2021, these reports are no longer accessible by the public. Most of the material about betatrons can be found in (Kaiser 1947). Presently available to this author as file *BIOS-Miscell-77.pdf*, as well as file *BIOS-201*, CIOS Item n.21, where a visit to C.H. Müller factory was reported, dated October 1945. In the interview of one De. Fehr, it s said that the 15 MeV betatron built by a Norwegian scientist 'had been experimented with for the Luftwaffe with the hope (?) of obtaining a death-ray for anti-aircraft work'. Since none of the scientists on charge were present at the factory in October 1945, this may just have been the official explanation given to justify the project as war-related and valuable. On the other hand, the war-related value of Widerøe's betatron is dismissed both in Amaldi's biography and in (Waloschek 1994).

[26] Houtermans (1903–1966), the son of a Dutch banker, was brought up in Vienna by his mother Elsa Wanek. After obtaining his Ph.D. at Göttingen in 1927, he had been Gustav Hertz's assistant at the Technische Hochschule in Berlin. A convinced communist, he emigrated with his wife after Hitler came to power in 1933, first in Great Britain and then to the Soviet Union. Arrested and tortured during the Great Purge in 1937 he was handed over to Gestapo after the Hitler-Stalin Pact of 1939. Through Max von Laue he was released in 1940 and got a job in Manfred Baron von Ardenne's industrial laboratory. In 1944 he took a position at the Physikalisch-Technische Reichsanstalt as nuclear physicist (Amaldi 2012). He became a close friend of Touschek through all the post-war years.

[27] Richard Gans (1880–1954) (von Reichenbach 2009) had gone to Argentina in 2012, where he had been the third director of the Institute of Physics of the National University of La Plata, returning to Germany in 1925. In 1933, as a Jew, Gans had lost his position of Professor of Theoretical Physics in Königsberg and director of the Second Institute of Physics. After finding work, as a "privileged non-Aryan", at AEG Research Institute in Berlin and later at Telefunken and GEMA, a company developing radar devices, in 1943 he was employed in the private laboratory run by Manfred von Ardenne, but he was obliged to leave and was sent by the 'Labor Deployment Department of Jews' to work clearing bomb sites. At that time he was 63 years old and was too weak to do such heavy physical work. For this reason he was in great danger of being deported.

fields for the development of a betatron, so that Houterman's idea was to transfer Gans to Schmellenmeier's company under the pretext of his being absolutely necessary to a project of great importance for the war. Thus the idea of a betatron.[28] Several well-known academics were asked to provide advice and support for the proposal; these included Max von Laue, Walter Friedrich, Richard Becker, Werner Heisenberg and, later, Walther Gerlach, who was a leading scientist of the 'Uranverein', the group of scientists involved in the development of the German nuclear program. It is also to be remarked that Walther Gerlach, at the time appointed by Göring to be Commissioner for Nuclear Physics in the Reich's Research Council, had been Gans' student in Tübingen, and had become a great friend of his former teacher after becoming assistant to Paschen at the Institute of Physics.[29]

For such plan to work, it needed to be absolutely appealing to the Air Ministry, and a strong motivation had to be found for proposing such a machine. Betatrons could reach electron energies which were considered high for the time, so that this suggested the revival of the old idea of using them to make strong bundles of X-rays. This was the *death-ray* idea, originally formulated by the British inventor Harry Grindell-Matthews in the early 1920s, and also suggested by Nicola Tesla (Hecht 2019; Ford 2013).

In order to get financial support for his project, Schmellenmeier claimed that the expected radiation could be of great significance to the conduct of the war, and in fact suggested that it could be used to pre-ionise the engines of aircraft, causing ignition and ceasing of functioning.[30] At the time every effort was being made to prevent the intrusion of American and British aircraft into German airspace, so that it was not difficult to catch the interest of the Luftwaffe. In order to meet military requirements, the range needed by X-rays produced by such devices should be of the order of Kilometers and the radiation to be fairly hard in order to penetrate through several centimeters of metal and other materials to reach the engine's interior. Physicists of the caliber of Jensen and Houtermans must have been fully aware that such performance was impossible to achieve with available technologies, not perhaps so the Luftwaffe.

In any case, even if the proposers—Schmellenmeier included—were aware of its scientific and technical weaknesses, the project was still worth pursuing, being interesting from a purely scientific point of view, as it required calculations which only a few physicists in the world sufficiently familiar with the theoretical tools could execute. In addition, suitable equipment would have needed to be built in order to confirm the calculations, and this made the project of constructing a betatron to possess high scientific relevance. This last point was particularly important given the

---

[28] Waloschek remarks that Schmellenmeier, Houterman and Jensen were of communist ideas, and shared ideals of solidarity due to their beliefs (Waloschek 2012).

[29] An extensive correspondence between Gans and Gerlach is preserved in the Archives of the Deutsches Museum in München.

[30] According to a report drafted by Schmellenmeier (Waloschek 2012), "the bundled, highly penetrant radiation could be used to pre-ionise the engines of aircraft, which would cause the ignition to cease functioning, the machines would be unable to continue their flight and would therefore come into the flak area."

scientific authority held by the supporters of the project, Heisenberg and Houtermans, among them, but what mostly mattered was to help Gans to avoid being drafted to forced labor, and obtain the military support.

At the time, the plan's capacity to produce a beam of enough hard X-rays to be used as *death-rays* for war aims was relegated to the background, but the project was sufficiently realistic to justify a serious consideration, and on 1st May 1943 Schmellenmeier submitted to the Reich's Air Ministry a proposal to construct a 1.5 MeV betatron. Although no concrete answer was given until later in August, a positive outcome was that by end of May 1943 Gans was sent to compulsory work at Schmellenmeier's laboratory, where the latter enrolled him as scientific labourer.[31]

4.3.3   *Schiebold's* Death-Ray *Project*

In Spring 1943, while the above mentioned machines were being developed by Schmellenmeier in Berlin and by Siemens-Schuckert in Erlangen, a project with an apparently much higher war potential, a real *X-ray gun*, was put in motion in Berlin by Ernst Schiebold, a well known mineralogist and expert in X-ray technology, who ran a semiprivate Institute for X-ray materials research which had a special position at the University of Leipzig, where he had been professor until 1941.[32]

While there is no known connection between the two German led betatron projects and Widerøe's proposal, this is not so in the case of Schiebold's and Wideroe's projects. At a certain moment, the two projects run in parallel. Both projects were classified as highly secret, for some time even sharing the same personnel. This was the case for Touschek, who is known to have worked for both, Schiebold's and Widerøe's.

From the beginning, Schiebold's project was presented as "a new means to fight the enemy" (Waloschek 2012) and was surrounded by a higher degree of secrecy than Schmellenmeier's or Gund's.

Shiebold's proposal, aiming at "the defence of the country in the Total War," was put forward on 5th April 1943 to the Berlin office of General Field Marshal Erhard Milch, Director of Air Armament (Generalluftzeugmeister) and Air Inspector General (Generalinspekteur) of the Luftwaffe. At the time, Milch was one of the most powerful men under Hermann Göring in the Third Reich, and could be directly interested to a project aiming to produce "an additional defensive weapon to combat and annihilate the crews of enemy aircraft and ground troops by means of X-rays and

---

[31] Young Klaus Gottstein, who was classified as a "half-Jew", was at the time already in Schmellenmeier's team. He has testified how several new attempts were made to send Gans to a concentration camp, but each time Schmellenmeier successfully used his contacts to prevent this from happening. In a correspondence with this author and L. Bonolis in May 2012, Gottstein made clear that Gans' life was saved by Schmellenmeier's betatron project.

[32] In 1943, Schiebold was professor for X-ray physics and non-destructive materials testing at the Technical University of Dresden and at the time collaborated with industry to run his institute in Leipzig (Waloschek 2012).

electron beams."[33] Schiebold was well known internationally as a specialist in the mentioned fields, as well as in their application.[34] Schiebold believed that obtaining the required harmful effects over distances of several kilometres would have been technically feasible if "radiation of sufficient penetration power and intensity is used" (Waloschek 2012, 65). In order to attain such hard X-rays it would be necessary to accelerate electrons applying a very high tension to the X-ray tube—Schiebold mentioned "several million Volts" as desirable, but at the time no generator could achieve such high voltages with sufficient beam intensity, so that he remarked that an electric tension of at least 500,000 Volt would be necessary.[35] Within this first proposal Schiebold recommended the participation of five specialists (a high voltage engineer, a design engineer, an X-ray engineer, a thermodynamics engineer and a gun construction specialist) and in particular he made a list of possible candidates: the first one was Richard Seifert, the owner of an important X-ray devices factory in Hamburg, 'R. Seifert & Co'.[36] Schiebold and Seifert were very good friends since the time both had been students in Leipzig, and shared interest in X-ray equipments and their application. Moreover, Seifert's company in Hamburg was one of the main providers of X-ray equipments and measuring instruments to Schiebold's institute for X-ray materials research. All these circumstances well explain their good relationship and Seifert's presence in Schiebold's project.

Following Schiebold's proposal of 5th April, a series of official meetings took place with Seifert and, on the 20th, Schiebold was presenting his ideas in person to the office of General Milch. On that occasion, a power of attorney for the realization was issued personally by General Milch for Schiebold, the engineer, and Luftwaffe captain, Ludwig Fennel and Theodor Hollnack, both men of confidence of the RLM.[37] Schiebold's friend, Richard Seifert, who also lead an official

---

[33] Ernst Schiebold, "For an additional defensive weapon to combat and annihilate the crews of enemy aircraft and ground troops by means of X-rays and electron beams" [Vorschlag eines zusätzlichen Kampfmittels [...], copy of the original document, dated April 5th 1943 (10 pages) preserved in the Sächsisches Staatsarchiv, Leipzig (Waloschek 2012, 215).

[34] Schiebold considered two possibilities: (a) Radiation concentrated on a relatively small target; (b) Radiation spread out over a wide cone in space. Whilst 'concentrated radiation' (a) could be directed at low-flying aeroplanes or ground troops in the open-air, 'spatial radiation' (b) would be preferable as a defensive weapon for the protection of entire regions or towns from aerial attack.

[35] It is to be remarked that after completing his first manuscript Schiebold sent it to Arnold Sommerfeld in Munich, who was rather negative, mainly because of his political and ethical views as regards that particular type of research. Heisenberg, too, was consulted. It is not clear what was his opinion.

[36] According to a report prepared on behalf of the U.S. Technical Industrial Intelligence Committee (CIOS) immediately after the war (The X-ray industry inGermany, Item No. 1.9 & 21, File no: XXVIII 31, August 11, 1945, document available at http://www.cdvandt.org/cios-xxviii-31.htm "Seifert was the largest manufacturer of industrial X-ray equipment in Germany [...] Seifert purchased tubes from Siemens, Müller, A.E.G. and manufactured only a few pump connected tubes [...] the total number of employees was about 350 [...]." Siemens and Müller of Hamburg were also described in the same section, "Ultra High Voltage Equipment".

[37] Hollnack worked under General Milch, and he might also be directly interested in X-ray tests to be performed on metal products built by his firm in Essen for the use of the aircraft industry. Schiebold had become acquainted with Hollnack, who was a businessman and administrator, on

working group on electro-medical applications, was nominated technical adviser of the project. The top secret project started by General Milch thus included Schiebold as scientific leader, Seifert as expert in industrial X-ray devices, Theodor Hollnack as administrator reporting to the RLM, the financing agent of the project, and Fennel, probably representing the Luftwaffe. Both Seibert and Hollnack would later be part of the team which built Widerøe's 15 Mev betatron.[38]

For his project, Schiebold was aware that it needed much more powerful X-ray sources than those provided by accelerators already at disposal. This is what made Widerøe's article particularly interesting, as it also contained the proposal for building a 100 MeV machine—meaning one able to produce a really high energy X-ray beam. Widerøe's proposal may have been circulating since March or even February. It was thus put forward at the right moment and in the right context. It is conceivable that, already in April if not before, the proposal had been made known to General Milch through Karl Egerer, and the idea surged in Berlin of discussing both Schiebold and Widerøe's projects. This is how Widerøe's own recollection of the events of spring 1943 in Oslo comes into light. Namely, the planned April 20th meeting might have triggered the visit by the German Air Force officers approaching Rolf Widerøe in Oslo (in March or April 1943) asking him to come to Berlin at a very short notice. The importance of the arguments discussed and the place, would warrant the airlift from Norway. Once in Berlin, Widerøe was told about the plans to build betatrons (Waloschek 1994, 49). During this meeting, a connection with Schiebold's project could have been aired. As the war ran its course, Schiebold's project run into difficulties, General Milch lost his influence, but Widerøe's project continued, with Touschek's full commitment. Eventually Widerøe's project took over, while Shiebold's was later dismissed.

4.4 Widerøe's Article and the *Death-Ray* Projects

One can now ask again the question of who made the Luftwaffe aware of the potentialities of Widerøe's betatron project in the first place. Barring, as said before, a direct address by Widerøe to the Luftwaffe, of which there is no evidence, the first initiative was clearly born inside Löwe Opta after the submission of Widerøe's (later

---

the occasion of the financing by the Air Force Ministry of a research project on materials testing. During these discussions, which took place since 1942, Schiebold mentioned the idea of developing powerful radiation sources as weapons and eventually prepared the project submitted in April 1943 (Waloschek 2012, 89)

[38] A letter written by Schiebold in December 1943, clarifies the strong connection between the X-ray gun and Widerøe's project. In describing to Walther Georgii, director of the German Research Institute for Glider Flight (Deutschen Forschungsanstalt für Segelflug), all the actions taken by that time, Schiebold included an order placed with the C.H.F. Müller company for an X-ray generator tube, as well as an order worth 150,000 RM placed with the same C.H.F. Müller company "for the development and construction of a 15 megavolt ray transformer of the 'Widerøe type'. Delivery expected in mid-1944." (Waloschek 2012).

unpublished) article and following Touschek's interest about relativistic effects in calculating the electron's orbits in Widerøe's machine. Egerer himself wrote this to Ernst Schiebold on January 2, 1945 (Waloschek 2012, 96), namely to have been the one to suggest to General Milch the inclusion (or merging) of Widerøe's ideas in Schiebold's project. And, from Touschek's letter to his parent on February 13th, 1943, one can argue that it was Touschek who brought Egerer's attention to Widerøe's proposal.[39]

As soon as Widerøe's proposal became known within Milch's circle at the RLM, other trusted consultants, already aware of Schiebold's proposal and of existing betatron projects, caught its potentialities in the context of producing very high energy X-rays and went into action. As Pedro Waloschek learnt [40] (Waloschek 2012, 96), such was the case of Richard Seifert also involved in calling Widerøe's to Germany. His firm in Hamburg had longstanding good contacts with a big X-ray-tubes and radio-valves factory called 'C. H. F. Müller', also known locally as 'Röntgenmüller', which supplied X-ray-tubes for Seifert's materials testing devices. This factory had been founded in 1865 by the glass-blower C. H. F. Müller and had a tradition of excellence. In 1943, the company was part of the Philips group and, as such, was in competition with Siemens, where Gund was constructing the 6 MeV betatron. It is thus understandable why Seifert's firm was definitely interested in any new development concerning betatrons.[41] As a matter of fact, the Luftwaffe's betatron, namely Widerøe's betatron, was finally built in Hamburg, at the C. H. F. Müller Co. (Waloschek 1994, 74): "After a little while we realised that the best place to build the betatron was at the big X-ray-tubes and radio-valves factory C. H. F. Müller". [42]

To place Touschek's work in its perspective, we anticipate here that, in mid 1943, Widerøe's project received its go-ahead, and that, by fall, active planning was put in place. In Schiebold's documents, archived at the University of Leipzig there is a contract—written following Schiebold's instructions—dated October 19, 1943, exactly specifying Widerøe's duties and rights during his work in Germany

---

[39] This is also confirmed in (Amaldi 1981, 5).

[40] This was told to Waloschek by Mr. Willi Markert, who in 1943 was the commercial director of Seifert's factory.

[41] According to a report prepared on behalf of the U.S. Technical Industrial Intelligence Committee (CIOS) immediately after the war (The X-ray industry inGermany, Item No. 1.9 & 21, File no: XXVIII 31, August 11, 1945, document available at http://www.cdvandt.org/cios-xxviii-31.htm, Philips actually controlled and owned a great number of organizations in the whole world, among which C. H. F. Müller in Hamburg. The Studiengesellschaft für Elektrongeräte, where Touschek worked during his first year in Germany, functioned as the engineering and research department for the just mentioned radio tube plant.

[42] According to the first biographical notes taken by Amaldi on 28th February 1978, in Geneva, at that time Touschek "continued to work for Müller and for Seifert & Co. The latter built the cannon for the injection of electrons in the betatron", from E. Amaldi, typescript with handwritten notes, Amaldi Archive, Physics Department, Sapienza University of Rome, Box 524, Folder 6. However, Amaldi did not include this information neither in the English nor in the Italian version (Amaldi 1981, 1982).

(Waloschek 2012, 109).[43] Any proof of the existence of this contract except in Schiebold's papers could not be found; Widerøe never mentioned such contract nor its negotiations to his biographer Pedro Waloschek. However, the existence of formal agreements about Widerøe's project is reflected in a letter Touschek wrote to his parents, around that time:[44]

> Yesterday I signed my death sentence. A very formal declaration
> of secrecy provided with a hundred oaths and threats. I've as
> well fabricated a theory—which is quite nice. As you can see I'm
> back in high favor with E[gerer]. I would like I could get more
> salary.

A few days later, on November 6, Widerøe wrote a report entitled "Vorschläge über eine möglichst rasche Durchfurung von Konstruktion, Bau und Aufstellung des Strahlentransformators", namely "Proposal for the fastest possible execution of the design, construction and installation of the ray-transformer". In this report Widerøe did not mention Schiebold, and later he was clear in recalling that he disapproved the latter's proposal, as most experts did at the time (Waloschek 2012, 109).[45] He divided the work to be done in three groups: (a) Preliminary tests and afterwards construction of the 15-MeV beam-transformer; (b) Preliminary tests and afterwards construction of the 200-MeV beam-transformer; (c) Planning and installation of the test-station Gross-Ostheim. The Gross-Ostheim Station (also known as Forschungsstelle Grossostheim) was adjacent to the military airport near the city of Aschaffenburg and under the command of the Luftwaffe Research Department. The main purpose of the center was the development and test of the X-ray-tubes with big anodes proposed by Schiebold.[46]

The 200-MeV apparatus, which was in Widerøe's November 6, 1943 working program, was the outgrowth of the preliminary design work he had carried out for the 100-MeV machine, described in his first unpublished article. Its final destination was in principle the center of Grossostheim, where it should have been installed and used as a source of high energy electrons. It was exactly the same site where Schiebold's 'X-ray gun' should have been built. This aim vanished with the dismissal of such an unrealistic project; on the other hand the realization of a 100-MeV machine was also an illusion at the moment. Nevertheless it absorbed a lot of preparatory work, which undoubtedly represented a most important theoretical challenge for the young Bruno Touschek, a fundamental step in his self-taught and on-the-field formation. The design of the 200-MeV unit was in fact judged by Kaiser "the most interesting and remarkable development due to the Widerøe group" (Kaiser 1947).

---

[43] The contract also states that this service was "ordered by the Reichskommissar for the occupied Norwegian regions." It was also stated that he was free to travel whenever and wherever he wanted.

[44] *Gestern habe ich mein Todesurteil unterschrieben. Sehr formale mit hundert Eiden und Drohnungen ausgestattete Geheimhaltungserklärung.*, in letter to parents on October 29th, 1943.

[45] A copy of such document is preserved at the archives of the ETH in Zürich, Hs 903:48.

[46] About Grossostheim Research Station, see also http://www.grossostheim-im-krieg.de/html/reportage_02.html.

4.4.1 Touschek and His Contribution to the *Death-Ray* Projects

A direct connection between Schiebold's and Widerøe's projects is present through Bruno Touschek, who later prepared some notes on X-rays production for Schiebold. Since the beginning of his work at Opta, Touschek had been involved in research on cathode ray tubes, and from some of his letters home one learns that he dealt with issues regarding some undefined 'electric' machines and physics of electrons in general. He also frequently mentions works and contracts to be discussed with an undisclosed firm *F.G.*. In light of the plans for Schiebold's devices to be built at the Forschungsstelle Grossostheim, F.G. is very likely to be a short hand, secret, pointer to such installation. In the next chapter, we shall show an application for a patent by Touschek and Egerer, dated March 22nd, 1943, in which a *Cathode ray tube for voltage limitation* is described. He continued to work on similar problems as confirmed by a letter to his parents on May 31st, 1943:

> Egerer torments me constantly with a theoretical study on the
> inertia of cathodes.

One does not know whether—and to which extent—this problem was related to Schiebold's plans. In any case all this background work at Opta—and the previous work at the Studiengesellschaft in Hamburg—gave him the necessary competence to handle the question related to Schiebold's idea that it would be possible to build an X-ray tube having a concave cathode, with electrons focused on the anode emitting X-rays in a narrow bundle. With sufficiently high voltage and giant anticathodes, Schiebold argued, it would then be possible to achieve high radiation intensities at long distances.[47]

Indeed, one of the main technical problems in the production of Schiebold's "X-ray gun" that should handle such large amounts of power, regarded the risk of damage—even of melting—of the small focal spot on the anticathode on which accelerated electrons producing X-rays would concentrate. The solution proposed by Schiebold to this problem, to enlarge the area of the anticathode, was criticized,[48] and, since the beginning of 1944, doubts about Schiebold's project feasibility became

---

[47] Schiebold proposed to enlarge the focal spot, and thus he began to consider areas with a radius of 10 cm, and his later proposals foresaw huge anticathodes of up to 2 meters in diameter which he thought would allow for much improved cooling. Tubes with such extended focal region and run at voltages of several million Volts would be the heart of what Schiebold refered to as "X-ray guns."

[48] Gerlach, who had doubted the feasibility of Schiebold's proposal, showed to his friend Helmut Kulenkampff the documents related to the extremely secret project. Kulenkampff was an expert in the production of X-rays and saw that Schiebold had applied two well known relations in a range in which they were no longer valid (Waloschek 2012, 121). He wrote an explanatory letter to General Milch criticizing both the efficiency in transforming energy of fast electrons into X-rays and the range of X-rays produced by anticathods of big dimensions. On February 22, 1944, Kulenkampff sent a letter to all members of the Board of Trustees for the Luftwaffe-Research-Centre Grossostheim where he showed concrete doubts about the efficiency of the emission of X-rays and about the range of radiation produced by anticathods of such dimensions.

apparent.[49] In the meantime it appears that Touschek had been entrusted with the delicate task of the calculations related to such giant anthicatodes, which were a main component of Schiebold's scientific program. In a document of 2 pages entitled "Zur Frage der Emission flächenformiger Antikathoden", Touschek calculated the intensity of the radiation produced by an anticathode of 1 meter diameter at a distance of 30 meters—as well as of 3 km—with electrons of 200 MeV energy.[50]

None of this is openly mentioned in Amaldi's biography of Bruno Touschek, probably because the connection to the Nazi war effort and planning for deadly weapons was not an activity easily justified in 1978, requiring long explanations about a partly forgotten past. Or perhaps Bruno had not believed in Schiebold's project as a realistic war weapon possibility, and ignored its implications. Or he was kept unaware of them.

All this makes clear how and why it became quite natural to consider Widerøe's concrete betatron project alongside (if not to embed into) Schiebold's proposal of an utopian weapon for Hitler's 'total war'. Since the very beginning, the Luftwaffe became interested in building a 'high-energy' betatron in connection with the military *death-ray* project proposed in April 1943 by Schiebold, and we recall that Widerøe was in fact contacted in Oslo by officials of the Luftwaffe in that same period. Schiebold's project required electrons accelerated to high energies, so that eventually the use of betatrons as sources became one of the main aims of the plan. The two projects were strictly connected through people such as Egerer, Seifert and Hollnack, all involved in the production of X-ray devices and thus having the highest interest in deriving benefit from the ongoing war for getting funds and improve their trading activity, and preparing for future exploitation of the market. This explains why the 15-MeV betatron project was born under secrecy. Once Widerøe's proposal became the object of active RLM consideration, Widerøe's article, and all patents related to betatrons submitted in that period, became secrets of war. The same fate fell on

---

[49] The members of the Board created to judge and control Schiebold's activity were: Egerer, Abraham Esau, Fennel, Friedrich Geist (colonel of the Luftwaffe), Gerlach, Georgii, Heisenberg, Heuser, Hollnack, v. Lossbert, Seifert and F. Tamms (designer of the big hall required to place the high voltage generators required by Schiebold's project).On May 4, 1944, when his project was beginning to be judged as not feasible, Schiebold wrote a report in which he mentioned the construction of the 15-MeV beam transformer at C.H.F. Müller in Hamburg. On July 12 Schiebold gave an example for the use of a 200 MeV beam transformer—clearly referring to Widerøe's for the moment "utopical" proposal—in connection with an X-ray gun equipped with a two-meter anticathode (Waloschek 2012, 137).

[50] Touschek's calculations were included as attachment to a final letter written on September 20, 1944 by Schiebold to Gerlach where the former tried to defend his scientific position also mentioning theoretical computations done for him by a "Dr. Touschek." A copy of this undated document was provided by the late Pedro Waloschek to Luisa Bonolis and through her to the Author. This undated document is contained in a microfilm preserved by Schiebold's family, while a copy (B. Touschek, Berechnung für grossflächige Anode, AZ 037077/448, NES 70-71) is among Schiebold's papers stored at the University Archive in Leipzig.

a second article submitted to the *Archiv für Elektrotechnik* by Wideröe on July 12, 1943 in which he presented accurate computations and constructions details for a 200 MeV betatron. This article was declared a secret of war since the very beginning.[51] Unfortunately, this also happened to all the work Bruno did for the project.

---

[51] The design of the 200 MeV machine was prompted by news that the General Electric was constructing a 100 MeV betatron (Kaiser 1947, 2).

Chapter 5
Bruno Touschek and Widerøe's Betatron in 1943–1945: The Dark Years

I am currently working on a very interesting problem - that has roughly the order of magnitude of the world record.
Ich arbeite gegenwärtig an einer sehr interessanten Problem das so ungefähr die Grössenordnung Weltrekord besitzt.
Letters to parents, November 29th 1943.

Abstract During the last two years of WWII, Bruno Touschek was in Germany, working on the secret betatron project directed by the Norwegian Rolf Widerøe and financed by the Reichsluftfahrtministerium, the Aviation Ministry of the Reich. He was moving between Berlin, where he continued working for Löwe Opta, and Hamburg, where the betatron was being built at the C.H.F. Müller factory. He also attended University classes by famous physicists such as Max von Laue and Hans D. Jensen, and experienced the heavy fire-bombing of both Hamburg and Berlin. In March 1945, as the English and American forces were approaching Hamburg, Bruno and Widerøe transferred the working betatron to Wrist. On his return to Hamburg, Bruno was arrested by the Gestapo and brought to the infamous Fuhlsbüttel prison.

In 1943, the war came to Berlin (Shirer 1950). Military losses in the East, with the disastrous Russian campaign, and the African retreat in the South combined with heavy bombing of Berlin and other German cities, by the Royal Air Force during the night and the American during the day, signalled a turning point. By January Bruno was in Berlin, where his search for learning and studying what could be more central and closer to his interests, was leading him nearer and nearer to the dark heart of the war.

Since the previous fall, after resigning from the Studiengesellschaft in Hamburg, Bruno had tried to find a position with other firms. He could not formulate clearly what he wanted to become, busy as he was with finding money for his day-to-day needs, decent accommodations, and adjusting to a working environment. Life in Vienna, since he had been denied access to the University and his academic hopes

G. Pancheri, *Bruno Touschek's Extraordinary Journey*, Springer Biographies,
https://doi.org/10.1007/978-3-031-03826-6_5

had been crushed, had taken an ugly turn, but in Vienna he had still counted on a secure home, with good meals and care for his clothes, a room with lovely paintings, many friends and no serious financial worries yet. There, surrounded by the affections of his family, from both maternal and paternal sides, he had advanced in his physics studies enough to feel confident to discuss problems with Arnold Sommerfeld, one of the great German theoretical physicists. The change from Bruno's 1941 life in Vienna and the year in Hamburg which had followed, had been enormous.

When Bruno had moved to Germany, he held the hope that he could follow courses at the University, and find a position to finance his living costs. Instead, he had to continuously borrow money from his father, request advances from his boss, and fight for his intellectual space with colleagues he did not particularly like in a working environment he despised. He could have succumbed, given up his plans to study, and thus be lost. But it did not happen. He did put aside the promised work with Sommerfeld, which was impossible to maintain while adjusting to the new life, but he did not give up. When the situation in Hamburg became unbearable, he jumped out of it, plunging into the unknown, aided by new friends and his will to do better things. One stares admiringly at Bruno's courage in breaking out of Hamburg and into the Berlin adventure, as he did without even have the money for the return trip to Hamburg. He was following a survival path, and it is in Berlin, where his life as a physicist began. There, in 1943 Bruno started to become aware of his capacities as a theorist, and, during the next two years, Bruno learnt how a particle accelerator is constructed from scratch, calculating the electron orbits and preparing technical reports. He officially joined Widerøe's betatron project in fall 1943, and his letters home from this period show a young man moving along with stark determination, to follow his study and calculate whatever he is asked to do, for Schiebold's death X-ray project or Widerøe's betatron, all the while trying to save his books from the fire bombs engulfing his room.

All this came about after February 1943, when he had been discussing, with his boss Karl Egerer, what he first thought to be a *sehr dummen* article submitted to the *Archiv für Elektrotechnik*.

5.1   Secrets Works in Early 1943

In February 1943 Touschek held only an academic interest in particle accelerators, and his discussion with Thirring and Sexl had been concerned with relativistic effects on the electron's orbit and not in building an accelerator.[1] He had been intrigued by some aspects of relativity, which he had studied and was fascinated by, and the discussions were mostly of theoretical interest.

During this time, there are traces of Bruno's involvement in the X-ray project planned by Ernst Schiebold at the Grossostheim Research station. The hint comes from mention of contacts with the never defined firm called *F.G.*, appearing in

---

[1] Letter to father on February 15th, 1943.

his letters starting from October 1942. At this time, he had also tried to obtain a position with the well know research center directed by Manfred von Ardenne in Berlin-Lichterfelde, but nothing came out of it.[2] Contacts with F.G. had intensified after he moved to Berlin and was working with Löwe Opta. In one of his letters, F.G., often simply FG, is described as an important military research institute.[3] As we have seen in Chap. 4, it is possible that FG referred to the Luftwaffe research center Forshungstelle Grossostheim, in Bavaria near Fliegerhorst Grossostheim, the Luftwaffe air base. The abbreviation used by Bruno may have been just a short cut in his letters or related to the secrecy surrounding the projects under way at the base.

In January and February 1943, negotiations with F.G. intensified in connection with some work, which could be carried out without interfering with the one at Löwe Opta. The work involved was related to his theoretical capacity, based on his already well developed mathematical knowledge, instilled by the Vienna school - Hans Thirring among them – but also from his natural ability and inclination. In February he had a contract almost in his hands. It required only theoretical work, which Bruno estimated to be about 1 hour a day, requiring timely delivering of whatever would be requested, and weekly meetings to report on the work. This is what Bruno had been trying to obtain since his Hamburg days: a work where he could learn something interesting, freedom to study and then attend some University classes. It was very appealing, providing Bruno with some extra income.

At first Egerer seemed to agree. As things would later show, Bruno was wrong about this. His extra work would clash with Egerer, possibly because of professional jealousy, more likely because he feared Touschek would leave Löwe Opta, where Bruno's duties included not only scientific work (Amaldi 1981, 5), but administrative and editorial work for the various scientific journals which Egerer was involved in. He also had Bruno doing calculations of various kind for the many patents he submitted, a most common practice for researchers in private firms to establish their scientific priority. As a compensation for Bruno's collaboration in matters not directly related to Opta, Egerer would include Bruno's name in some of his patents. Often, he spent time fantasizing about grandiose plans to make money. Once, Egerer even started a plan to submit a patent having for subject the well known twin-paradox by Albert Einstein. Egerer bet that registration for a patent was possible. However, the hope to win a research contract from the RLM for space navigation had one major drawback, which they both thought quite funny, namely that the corresponding theory came from a Jew.[4]

The continuing interest from F.G., the referee work for different journals, the consideration in which Egerer clearly held him, returned to Bruno some of the self

---

[2] Manfred von Ardenne (1907–1997) was a German researcher, applied physicist and inventor. In 1923, at the age of 15, he received his first patent for an electronic tube for applications in wireless telegraphy, and prematurely left the Gymnasium to pursue the development of radio engineering with the entrepreneur Siegmund Löwe, who became his mentor. He attracted high level scientists to work in his facility, such as Fritz Houtermans in 1940, from https://en.wikipedia.org/wiki/Manfred_von_Ardenne.

[3] On February 2nd 1943, letter to his parents.

[4] Letter to parents, no date, photography n. P1030059, in Author's personal archive.

Fig. 5.1 Oskar Weltmann, Touschek's maternal uncle, a medical doctor and a painter, self portrait, Family Documents, © Francis Touschek, all rights reserved. At right, Oskar Weltmann in a later photograph found by Giovanni Battimelli among Bruno Touschek Papers, reproduced courtesy of Sapienza University of Rome–Physics Department Archives, https://archivisapienzasmfn.archiui. com, documents provided for purposes of study and research, all rights reserved

confidence in the theoretical capacities he thought to have lost. During the past year in Hamburg, he had not been able to go through with the work with Arnold Sommerfeld. Although this was a natural consequence from having been so busy, first settling in Hamburg, an unfamiliar city, then starting to work at the Studiengesellschaft and fighting with his boss or with other colleagues, this inability to follow up with his exchange with Sommerfeld had occasionally given him pangs of guilt. In Berlin, he felt better and, when, in January, Sommerfeld's son Ernst came to visit Bruno, his worries about the relationship with Sommerfeld senior were assuaged.[5] They soon became good friends and remained such for many years after.

On February 3rd, Bruno turned 22 years of age. As it often happens, the birthday made Bruno think about his hopes of the previous year and the reality in front of him. He reflected on his passion to become a scientist rather than an artist, family traits embodied in his uncle Oskar, whom he had taken for a model. A self portrait by Oskar Weltmann is shown in Fig. 5.1, together with a photographic portrait, found among Bruno's Rome Institute papers, after his death, witness to his long-held attachment to his uncle.[6]

---

[5] Letter to parents, February 8th 1943.

[6] The small photographic portrait, 10x5 cm, came to the Author's attention courtesy of Giovanni Battimelli, curator of Touschek's papers for Sapienza University of Rome–Physics Department Archives (Battimelli et al. 1989).

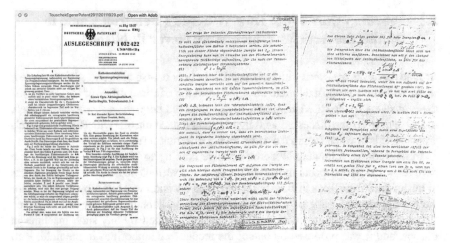

Fig. 5.2 At left, a patent for *Kathodenstrahlröhre zur Spannungsbegrenzung*, Cathode ray tube for voltage limitation, submitted by Bruno Touschek and Karl Egerer on March 22nd, 1943, registered after the war in 1953. At right, two pages of calculations by Touschek, probably done around the same time, courtesy of Pedro Waloschek

Bruno saw that some of his hopes had not worked out, at least not as yet. He had hoped to be able to secretly work on a dissertation, but this plan crashed against the regulations in force at the University, such as a diploma examination with some constraints about the number of semesters etc. which had to be taken before passing to the doctorate. None of this could be done of course, and Bruno realized that the University of Berlin was as much under the Nazi rules as Vienna had been.[7] Looking back, the past year appeared to Bruno as having been wasted, as he longed for having time to sit in a library and sift through books or magazines, and having a job where he could prepare himself for continuing his studies with a financially secure position after the war. This was the deeper background of his protracted negotiations with F.G., as he could not see such possibilities with his work at Löwe Opta.

Unfortunately, Opta did not appear to let him go so easily and he worried to miss the opportunity.

In the meanwhile, Bruno was working with Egerer on problems related to X-ray emission, and was co-author of a patent they submitted together on March 22nd, 1943, shown in Fig. 5.2. It is quite possible that during this same period he did some work for Schiebold as well. A two page typescript from Schiebold's papers is also reproduced at right.

---

[7] Letter to parents on February 8th and 10th 1943.

5.2   February-July 1943: How Widerøe's Betatron Came to Be and the Articles Which Were Never Published

Around this time, the event which changed the whole course of Bruno's life occurred, when Widerøe's proposed article travelled from the editorial office of the *Archiv für Elektrotechnik* to the RLM attention. If, as suggested here, the *sehr dummen* article Bruno reviewed (and discussed with Egerer) was indeed the one with Rolf Widerøe's proposal, this is when Bruno's destiny took a different turn.

In the months following Widerøe's article coming to Touschek's attention, we have seen that the proposal moved to the desks of various authorities and scientists in Berlin until it surfaces in July 1943, when the decision was taken by the RLM for the construction of Widerøe's 15 MeV betatron in Hamburg. As the article aimed at building an accelerator with a power not yet reached by any such machine in Europe, and it was proposed by one of the recognized authorities in the field, Rolf Widerøe, it was just what the RLM was looking for, a scientific project with real war like potential.[8] While Schiebold's x-ray proposal was moving along what actually became a tortuous path, Widerøe's proposal was immediately seriously considered and carefully examined. Not many scientists were able to evaluate its feasibility. As we know, in the case of Shiebold's proposal, among others, Sommerfeld had been consulted, and it cannot be excluded that, in the course of approval of Widerøe's proposal, Sommerfeld suggested Touschek as a reliable referee and that Bruno's criticism was at the heart of the decision to approve it or not.

From March to June, an exchange of letters must have taken place between Bruno and Widerøe. None of this correspondence is known to have survived, nor is there any mention of this exchange, or Widerøe's name, in Bruno's letters home, until June 17th, 1943. But Amaldi explicitly writes that Bruno wrote to Widerøe, after reading his proposal, and that the latter replied. Bruno tried to convince Widerøe to change some parts of the article. On June 17th, he finally succeeded. The article could thus be published. Rolf prepared a second article for an even more powerful machine. At the RLM the decision was set in motion.

5.2.1   More About Interest for Betatrons as Death Rays and How They Found Their Way in the RLM

When Widerøe's proposal had been brought to the attention of Egerer, it had soon moved forward to Milch. While Widerøe's article certainly did not mention war application or death ray emission, its potentiality for war purpose fell within current discussions at the RLM. As also seen in Chap. 4, apart from Shiebold's X-ray project, the RLM had also been approached by Schmellenmaier to support the construction

---

[8] Even though Widerøe had not worked on accelerator research since his PhD days, his work had been referred to in Kerst's article, a proof of recognized expertise in the field, at least in the view of the German military.

of a 1.5 MeV rheotron, as it was still called at the time.[9] The idea had come from two of his friends, Fritz Houtermans and Hans Jensen from Hannover.[10] Both were already quite well known physicists, and sharing socialist, if not communist, ideals.

From (Waloschek 2012, 40):

It was around 11 pm on a late March Friday in 1943 when Heinz Schmellenmeier received an agitated telephone call from his friend Fritz Houtermans who later that night even came to call on him. This, however was not unusual for Houtermans temperamental style of life. He had just heard that Richard Gans had been set to work clearing bomb sites and was insistent that something had to be done immediately.' Houtermans proposed that Schmellenmeier requests that Gans be transferred to his company, as his contribution to an important project being conducted there was indispensable. A sufficiently important idea, scientifically sound, but of national, if not war-related relevance, had to be conjured up. Houtermans and his friend Hans Jensen came up with the idea that the construction of a betatron could be a project of war interest and help Richard Gans, who had the necessary scientific credentials for the idea to be credible. To justify the proposal he submitted to RLM on April 2, 1943, Schmellenmaier mentioned the potential for such machine to emit a type and intensity of radiation which could be of 'great significance to the conduct of the war'.

Thus, hope for such projects to produce miracle weapons surged through the RLM in spring 1943. Either for desperation rising through the rubble of the burning German cities, or from the deadly military losses in Russia and North Africa, Widerøe's proposal was bound to attract the military interest and ultimately be approved.

While Bruno, perhaps unaware of the article's interest to RLM, was working on convincing Widerøe to change some of his calculations, at the RLM the project was given serious consideration, prompting the dispatch of the German officers to Oslo (in the spring, as Widerøe writes).

At the time, how and when Widerøe's proposal acquired official status was buried as a war secret. The two protagonists of this story, Rolf and Bruno, did not disclose too many details about the RLM involvement in the betatron project. There are contradictions and silence in the story Widerøe told to his biographer Pedro Waloschek. Most of what is now known comes from (Sørheim 2020) or Waloschek himself (Waloschek 2012). Touschek's story as he told it to Amaldi at the end of his life, is simply shorter, but his letters to the family during the last two years of the war, tell much more, as we shall soon see.

In Widerøe's case, the circumstances just described have a dramatic central place in his life, which is clearly divided between events before and after the article submission, or more precisely, before and after the fateful conversation with the German officers in spring 1943, in Oslo. Even considering the many honors and successes which surrounded him in his life after he left Norway in 1946 to work in Switzerland, it is quite normal that he would have preferred not to remember what led to the postwar inquest for collaborationism. The inquest subjected him to personal humiliation, including a heavy financial loss and was one obvious reason to leave his native country and work abroad. However, it is still worth asking why did Widerøe hide the fact

---

[9] This is the type of accelerator later called 'betatron'. The name was given by Kerst, who had called it 'induction accelerator' in his original articles.

[10] For Houtermans' adventurous life see footnote in Chap. 6 and (Amaldi 2012).

that his article for the construction of the 15 MeV betatron was never published. He must have known about this, careful as he always was in recording his patents. He did not have many scientific publications in his name, and of course this article would be proof that he had proposed and built the 15 MeV betatron.[11] Acknowledging that the article had not been published meant that he knew the proposal had become a secret project, financed by the German military. In such case, he had to admit to have been aware of the military interest in the project, which, at least officially, could only be related to the development of a secret weapon. He is likely not to have worried about the military exploitation of his project, knowing quite well that the potentiality of the betatron was clearly insufficient for making a serious weapon, and being mostly interested in building the machine for its scientific interest, and perhaps envisaging medical application. In addition, he had his brother's well being to consider and would go along with the RLM interest. Be as it may, hiding of this fact obfuscates the accuracy of his autobiography. This is especially true, concerning submission of his patent for storage rings, to which we shall turn later.

The concision of Bruno's description of the events in 1943, as reported by Amaldi, has a different but no less dramatic origin. Amaldi (Amaldi 1981, 5) writes:

> [...] Dr. Egerer [was] working on the development of Brown's small tubes (i.e. cathodic oscillographs) for television. Egerer was also Chief Editor of the scientific magazine *Archiv für Elektrotechnik*. Egerer had Bruno's help in this work too and it was thus that, at the beginning of 1943, Touschek heard of a proposal presented by Rolf Widerøe to construct a 15 MeV betatron. The proposal was kept secret because of its possible [war-like] applications. [...] Reading Widerøe's proposal Touschek had the impression that the relativistic treatment of the stability of the [electron] orbits contained some mistakes. He wrote to Widerøe, who replied and invited him to go and work with him when, towards the end of 1943, he was ordered to build a machine of this type.

This is a short but simplified version of the events which took place between February, when Touschek read Widerøe's proposal, and November 1943, when Bruno signed his contract with RLM, but one should remember that, when these recollections were written down by Edoardo Amaldi, between February and March 1978, Bruno Touschek was at the Hôpital de la Tour in Geneva, already very ill. On February 28th, he had started telling Amaldi about his life in Germany during the last two years of the war. He would soon be taken to Austria, by a CERN car, and pass away in less than two months, at only 57 years of age. There was no time for lengthy interviews, and no need for in-depth reconstruction.

As shown in (Waloschek 2012) and partly in (Waloschek 1994), much more went on before Widerøe's betatron proposal would become an actual secret project financed by the RLM. However, the manoeuvring of scientists and military persons surrounding it, in those days of 1943, became clear to Bruno only later in the year. In

---

[11] Similar considerations were expressed by Pedro Waloschek in a personal e-mail communication to Luisa Bonolis and this Author, in late 2011, expressing doubts about the soundness of some of Widerøe's claims to scientific priority. Pedro Waloschek was close to the end of his life, when he sent these comments, which close with a question: 'I could add more, but why?'. He would pass away four months later, and may have felt free to say things he had thought about, but not put down in writing.

the first months of 1943, he had been worrying more about his studies and, later, his survival under the bombing of Berlin. But it is of relevance to spend time on what happened during those 1943 months, as these events, constituting a pivotal moment in his life, made Bruno the physicist who could see in his mind the ways and modes electrons move inside an accelerator (Bernardini 2006).

5.2.2   Beating the *Norweger* Point by Point

Between February and June, Bruno was writing to Rolf discussing relativistic effects in the proposed betatron. However, in Bruno's letters home from March through May, there is no mention of the correspondence between them. It is possible that Bruno knew little of what was happening at RLM about Widerøe's idea, and did not think it worth writing about it to his father. But it could also be due to the secrecy, which immediately surrounded Widerøe's proposal, and of which Bruno must have been made aware. In any case, his criticism of the article reached Widerøe. By early May, Rolf was correcting his proofs (Waloschek 1994), which may have been passed on to Bruno. It is only on June 17th, that 'my Norwegian', *Mein Norwerger*, suddenly appears. On that day, Bruno triumphantly wrote to his parents that he had beaten his Norwegian point by point along the line and that the article in print would have to be modified to include Bruno's suggestions before printing. This is the first time that Widerøe, still referred to only as a 'Norwegian', is mentioned in Touschek's letters to his parents and fits well with the secrecy surrounding the project, which was still running in parallel or together with Schiebold's.

Following his first article, where plans for a 100 MeV induction accelerator had been indicated, Widerøe had, in the meanwhile, worked on an even more powerful machine, and in July was preparing submission of another article. In June, Bruno, like Rolf, was still convinced that Widerøe's original article, once corrected of the errors he had pointed out, would be published, but the article did not appear, although this was not discovered (or made public) until very recently by L. Bonolis and this author. Bruno also believed that the work he had done during these months, in connection with the problem of radiation emission among others, would be published in one of the RLM journals, and that it could be part of a dissertation. None of this could happen. The betatron project was considered of military interest and neither Widerøe's articles nor Touschek's work could be published. Instead of Widerøe's, an article by a different author with a different subject, was published.[12]

Following Bruno's letter to his parents, one can place the approval of the construction of Widerøe's betatron between June 17th, when Bruno expected to see the article published, and early August when specific plans for starting the project were put in place. It was decided that the betatron was going to be built in Hamburg and then brought to the Luftwaffe research station in Grossostheim for operation and mea-

---

[12] The article, published where there should have been Widerøe's, is entitled *Einphasig gespeister meßphasenschieber aus drehtransformator und drehspannungsscheider*, (Beindorf 1943).

depicted item: "Röntgenröhrenfabrik Müller" / X-ray tube factory Müller, Hamburg, (west elevation)
source: photo collection of PMS, Hamburg (unknown photographer)
date: ca. 1930

Fig. 5.3 C.H.F. Müller building

surements. The official reason, to develop in Hamburg the first phase of Widerøe's project, was strategically motivated as the city had undergone a devastating bombing in late July, and it could perhaps be spared from further destruction in the months to come (Waloschek 2012).[13] These raids were part of what the Anglo-American Command referred to it as *Operation Gomorrah*.[14] They started on July 24th and lasted 7–8 days, destroying a large part of the city, with as many as 40000 casualties in the single night of July 27–28.[15]

The choice of Hamburg for the first stage of the project had also another reason. Seifert, who had promoted Schiebold's project, had also been involved in evaluating Widerøe's proposal. One of the crucial components of the betatron was the glass tube, in which the accelerated electrons would be made to circulate, and Seifert was director of C.H.F. Müller in Hamburg, Fig. 5.3, with proven experience in the construction of glass tubes.

5.3 University at Last!

In the spring, while busy writing to Widerøe, Bruno had also actively pursued his physics studies. He was adapting to life in Berlin. In April, he had changed house and liked the family he was staying with. Later on, he would be disappointed by the

---

[13] As a matter of fact, Hamburg was again devastated in late 1944 and early 1945.

[14] See https://www.bbc.com/news/uk-england-43546839.

[15] See Bomber Command record for July 1943 at https://webarchive.nationalarchives.gov.uk/20070706055428/http://www.raf.mod.uk/bombercommand/jul43.html.

food, as it always happened in those days, but remained in this place, at Eislebener Strasse n. 4, for longer than in other places.

During the first months of employment with Opta, relations with Egerer were good. In April there were no contrasts yet. Work at Opta could bring moments of friendship and relaxation. During the mandatory overtime from 5 to 7 pm, Bruno could take care of light duties, such as letter or report writing, things he would refer as his *Schmonzes*, an Yiddish term to describe non important things. In this period, yiddish terms would occasionally appears in Bruno's letters home. Through April and May, many of the letters were written from the Opta office, while coffee was brewing on the Bunsen burner next door, and visitors or friends could drop by.

Bruno's outlook for continuing on his physics studies was also less bleak than it had appeared in the winter. He was told that there was a good probability that in the following semester some university arrangements could be made for him, basically around three mornings a week at the university. However, after the many disappointments of the past months, he was not too optimistic, and, in April, he would gauge the probability for this to happen to be no more than 10%. In fact it was much higher and, between April and May, Bruno university engagement was set in motion.

After returning from a few days vacation in Vienna around Easter time, which he took hoping his absence from the office would not be noticed, he wrote to his parents to be busy reviewing a manuscript, which is likely to have been Widerøe's.[16] Not long after, Bruno was at the university attending lectures. The first time was on the occasion of von Laue's seminar on May 13th. He had become noteworthy.

Bruno had fought for a place among the likes of Schiebold and Egerer, had survived, and was surviving, many obstacles. The time had arrived for him to be accepted as a student, albeit unofficially. It was not something so easy to do. There was danger in protecting non-Aryans, and Bruno's Jewish origin was known to the authorities. But Bruno had proved to have the intellectual and mental strength to go back to physics under those increasingly difficult times. This capacity had been shown by Bruno's correspondence with Widerøe about the submitted article, while the project was under scrutiny for its soundness. As in the case of Schmellenmaier or Schiebold's project, some of the best German scientists, such as Max von Laue, at the time Chairman of the Prussian Academy of Sciences, or Werner Heisenberg, had been consulted about the proposed X-ray projects. Bruno's name, either through Egerer or Schiebold, both involved in these plans, is likely to have come to these scientists' attention in favourable terms. Mostly, Sommerfeld, the time-honored teacher, consulted by his former students, was certainly advocating that Touschek be allowed to restart his university path.

In mid May, probably encouraged by Sommerfeld or Thirring, with whom Bruno was in regular contact through his frequent trips to Vienna or *via* telephone calls,

---

[16] In a letter to parents on May 3rd, 1943, he writes to be officially reading a manuscript, but that, since he already knew exactly what was in it, this would be a labor saving. This is no proof that it was Widerøe's, but it hints to be a manuscript which Bruno had already known about unofficially. About Widerøe correcting his proofs, see also (Waloschek 1994, 65).

Bruno went to the University, where von Laue was holding a course of seminars in theoretical physics on superconductivity. He approached him before the lecture, and asked for his permission to attend. As expected (Bruno would not have approached him, had he not known it already), von Laue had nothing against it and Bruno was admitted to a very distinguished circle. He lively participated to the discussion which followed the seminar, and then went home with a mathematician who even invited him for dinner. The day after was a Friday, and Bruno went to other lectures. He first attended a nuclear physics lecture by Joseph Mattauch (from Vienna), who also agreed to Bruno sitting through it. During the lecture Bruno contributed to a discussion about some error the professor had made. Mattauch's lecture was followed by a colloquium at von Laue and Heisenberg's Seminar. Bruno had hardly sat down that von Laue came up to him and asked very detailed information about his educational background. 'Mattauch does the same', as Bruno proudly wrote to his parents.[17]

Von Laue was impressed by Bruno and the week after, during the next seminar, asked him to come and see him, as he wanted to give him a task. Such offer was just what Bruno had hoped would happen. Perhaps this could lead to a dissertation, the ultimate result of his period in Germany. And one week later, there came in the mail a large envelope with special prints from M. von Laue. It was getting serious and he was excited by the opportunity.

On May 24th, he was feeling that he was 'overloaded to the point of maddening' as he was then still working at Löwe Opta, engaged with the F.G., probably meaning Shiebold and Widerøe's proposals, reviewing for the *Chemischen Zentralblatt*, some undisclosed work for von Ardenne, and the physics assignment by von Laue. To get by with the money, he was also giving some private lessons. All this was also happening during heavy bombing of Berlin by the Allied, often making it difficult for him to work.

Apart with the rather unexpected task for von Laue, Bruno also held hopes for a dissertation in connection with his careful study of Widerøe's work. When a confrontation with Widerøe had occurred on June 17th, Bruno had expected to be on the way for publication of his own work. Bruno had written:[18]

> [. . .] Today was a meeting at the R.L.M. In which he [his Norwegian] had to agree with me point by point. A work in print on the theory of a certain device will be changed quite substantially due to my remarks and I will also appear in the *Archiv for high frequency technology and electroacoustics* in March. Only part of my work, of course, for reasons of confidentiality.

> The complete work (actually a cycle of 3 works, the first has already been written, the second has been calculated, the third is only finished in my brain, but since it should mainly contain tables, it requires little mind but a lot of work) in the form of RLM reports. Perhaps a publication in the journal for aviation research is possible.

---

[17] Letter to parents May 15th, 1943.

[18] June 17th, 1943 letter.

```
The people around me are considering the plan to use my work
as a secret dissertation now. But I do not think that
anything will come of it.
```

Bruno's pessimism was well founded, nothing came out of that work during the war times. Still, the June 17th meeting sealed the fate of Widerøe's project and Touschek's studies. After this, a plan for Widerøe's betatron to be built began to be worked out and by the end of October 29th Bruno would be hired by the RLM to be part of the secret project directed by Widerøe.

5.4 Colliding Clouds in the Northern Sky and Darkening Times

In July Rolf Widerøe had been busy preparing the second article about a possible 200 MeV betatron, while discussing with RLM about feasibility and construction of the 15 MeV machine. Once the decision had been taken that the betatron would be initially constructed in Hamburg, Widerøe, always the practical man, organized his future time in the hanseatic town, where he now had agreed to spend the necessary time. He rented a room to live and was provided with adequate office space, while Bruno went to Austria in mid-July to spend his summer vacation with his family.

After preparation for work in the months to come, and agreement on whatever conditions should be in his contract with RLM, Widerøe was now free to go back to Norway for a long delayed summer vacation with his family. The vacation gave him time to think, his mind wandering through dreams of particle accelerators, and new ideas came up. Clouds colliding in the clear light of a Northern summer sky evoked particle collisions. He could imagine charged particles colliding head-on in a new type of accelerating machine. It was an exciting, totally new concept (Waloschek 1994, 81).

5.4.1 How Bruno Missed the Point

Back to Germany to start work on the betatron, Rolf prepared the details of his idea about colliding particles. He consulted Bruno about it, but failed to impress him. As Widerøe says (Waloschek 1994, 82): "After I returned to Hamburg, I spoke with Touschek about my ideas and he said they were obvious, the type of thing that most people would learn at school (he even said *primary school*) and that such an idea could not be published or patented." In his memories, Widerøe placed this conversation in 1943, in Hamburg (Waloschek 1994), after he had returned from the vacation in Norway. Bruno did not know enough about particle accelerators at that time to appreciate the novelty in Widerøe's idea, or did not care, and discarded it as trivial. Years later, when the time came for him to understand the revolutionary

Let me first explain why a storage ring is an important
instrument, particularly when fed with electrons and positrons.
The first suggestions to use crossed beams I have heard during
the war from Widerøe, the obvious reason for thinking about
+ throws away   them being, that one ~~wasted~~ a considerable amount of energy
by using 'sitting' targets - most of the energy being wasted
to pay for the motion of the centre of mass. If one wants
to study electrodynamics one should turn to ...

Erteilt auf Grund des Ersten Überleitungsgesetzes vom 8. Juli 1949
(WiGBl. S. 175)

BUNDESREPUBLIK DEUTSCHLAND

AUSGEGEBEN AM
11. MAI 1953

DEUTSCHES PATENTAMT
PATENTSCHRIFT
№ 876 279
KLASSE 21g GRUPPE 36
W 687 VIIIc / 21g

Dr.-Ing. Rolf Wideröe, Oslo
ist als Erfinder genannt worden

Aktiengesellschaft Brown, Boveri & Cie, Baden (Schweiz)

Anordnung zur Herbeiführung von Kernreaktionen

Patentiert im Gebiet der Bundesrepublik Deutschland vom 8. September 1943 an
Patentanmeldung bekanntgemacht am 18. September 1952
Patenterteilung bekanntgemacht am 26. März 1953

Kernreaktionen können dadurch herbeigeführt werden, daß geladene Teilchen von hoher Geschwindigkeit und Energie, in Elektronenvolt gemessen, auf die zu untersuchenden Kerne geschossen werden. Wenn die geladenen Teilchen in einen gewissen Mindestabstand von den Kernen gelangen, werden die Kernreaktionen eingeleitet. Da aber neben den zu untersuchenden Kernen noch die gesamten Elektronen der Atomhülle vorhanden sind und auch der Wirkungsquerschnitt des Kernes sehr klein ist, wird der größte Teil der geladenen Teilchen von den Hüllenelektronen abgebremst, während nur ein sehr kleiner Teil die gewünschten Kernreaktionen herbeiführt. Erfindungsgemäß wird der Wirkungsgrad der Kernreaktionen dadurch wesentlich erhöht, daß die Reaktion in einem Vakuumgefäß (Reaktionsröhre) durchgeführt wird, in welchem die geladenen Teilchen hoher Geschwindigkeit gegen einen Strahl von den zu untersuchenden und sich entgegengesetzt bewegenden Kernen auf einer sehr langen Strecke laufen müssen. Dies kann in der Weise durchgeführt werden, daß die geladenen Teilchen zum mehrmaligen Umlauf in einer Kreisröhre gezwungen werden, wobei die zu untersuchenden Kerne auf derselben Kreisbahn, aber in entgegengesetzter Richtung umlaufen. Da die geladenen Teilchen dabei nicht von bei der Reaktion unwirksamen Elektronen abgebremst werden und andererseits auf einer sehr langen Wegstrecke gegen die Kerne sich bewegen können, wird die Wahrscheinlichkeit für das Eintreten der Kernreaktionen wesentlich größer und der Wirkungsgrad der Reaktion sehr stark erhöht.

Um die bei der Kreisbewegung entstehenden Zentrifugalkräfte aufzuheben, müssen die umlaufenden Teilchen von nach innen gerichteten Ablenkkräften gesteuert werden, während eine Diffusion der Teile mittels stabilisierender, von allen Seiten auf den Bahnkreis gerichteter Kräfte verhindert wird. Falls die gegen-

Fig. 5.4   At top: Bruno Touschek's February 1960 preparatory notes for construction of a storage ring for electron-positron collisions, later called AdA, from Touschek Papers, courtesy of Sapienza University of Rome–Physics Department Archives, https://archivisapienzasmfn.archiui.com, documents provided for purposes of study and research, all rights reserved. Below, first page of Widerøe's patent for colliding particles, submitted in September 8th, 1943, registered in 1953

content it had in terms of physics discoveries, he would remember it. In February 1960, preparing his proposal to build an electron positron storage ring, Touschek began his notes about what became AdA, acknowledging Widerøe's idea, Fig. 5.4 (top).

Widerøe's mind set was very different from Bruno's. He was an engineer, and through his work at NEBB he was used to submit his ideas as patents rather than scientific articles. Wanting to be assured of his priority, Widerøe went ahead, ignoring Touschek's opinion, and submitted the patent, Fig. 5.4 (bottom). Had Bruno felt less superior to Widerøe on his knowledge of theoretical physics - and humbler in terms of what accelerators could be used for - his name could have been on the patent. It didn't.

In Widerøe's autobiography, there is a curious note about Touschek's reaction to Widerøe's patent submission. Widerøe says that, notwithstanding Bruno's dismissive remarks, he telephoned his friend Ernst Sommerfeld in Berlin, and together they turned the idea into a nice patent.[19] He also notes that, although they played down the center of mass gain in energy, Bruno was 'pretty offended' (Waloschek 1994, 82). The remark is strange, as there was no reason for Bruno to be offended by Widerøe's submission of a patent, nor by Widerøe's disregard of his opinion. So, why was Touschek's offended? Perhaps because, having discussed the idea with Rolf, he was expecting to be included in the patent? Or the person offended had been Widerøe, after Bruno called it a *primary school* idea?

5.4.2   How Reliable is the Date of the Storage Ring Patent?

The storage ring patent carries the officially recorded date of May 11th 1953, and an original date for submission of September 8th, 1943. While there is no doubt about the veracity of the episode, the timing, including the date of September 8th, is inconsistent with the chronology from both Touschek's letters to his parents, and Widerøe's autobiography.

Widerøe says that the episode of watching the clouds, and the idea which followed, "happened during the autumn of 1943, on one of my vacation trips to Norway" (Waloschek 1994, 81). He then goes on saying "After I returned to Hamburg I spoke with Touschek" (Waloschek 1994, 82). Autumn of course could very well be September, so this would not be in real contradiction, but Widerøe elsewhere says that "After he [Touschek] came to Hamburg I made his acquaintance at the house of Professor Lenz where he had taken up lodgings." (Waloschek 1994, 74). If the patent was in fact submitted on September 8th, the whole sequence, with meeting Touschek after he came to Hamburg is strange, since in September Touschek was in Berlin, in fact swimming out of the Landwehr Canal after a heavy bombing, or

---

[19] Widerøe and Ernst Sommerfeld had known each other since they were both students in Aachen. Ernst had specialized in the field of patenting and before the war was in Berlin, working as a patent agent for Telefunken (Waloschek 1994, 27).

looking to saving his note books in the half destroyed Opta building. One can point out as well that, according to Touschek's letter of June 17th, Bruno had presented his criticism to Widerøe's betatron article in June, in Berlin, implying, in the letter, that they had met already in Berlin, not in Hamburg, as Widerøe says. In addition, Touschek, when he moved to Hamburg in 1944, did not live at Lenz's, but close by, in his own place.[20] As it is, Widerøe's tale about the episode and his first meetings with Bruno is inconsistent, or at best confusing. These contradictions are irrelevant as to the priority of the idea, which Touschek always acknowledged, but they are pointed out here to keep in mind that Widerøe's recollections may not be fully reliable.

Such inconsistencies or contradictions in Widerøe's memoirs may impact the appreciation of Touschek's contribution to the success of the 15 MeV betatron. Widerøe acknowledged Bruno to be a 'very important colleague'. In 1979, he wrote to Amaldi: "[Touschek] was of great help to us in understanding and explaining the complications of electron kinetics, especially the problems associated with the injection the electrons from the outside to the stable orbit where they were being accelerated".[21] As such calculations are *de facto* crucial to the working of a high energy accelerator, Touschek's contribution may have made the difference between reaching the goal or not. His studies had equipped him with the theoretical physics knowledge that Widerøe lacked, in particular about the relativistic motion of electrons at the required betatron energies. According to Kaiser, who wrote a detailed report on European induction accelerators, they were very important and valuable (Kaiser 1947). Proof of Bruno's major contribution to Widerøe's betatron operation are the many times he was taken to the UK in 1946, to Wimbledon, for interrogation about his work with Widerøe. In fact Bruno was exceptional, and 'could think the unthinkable', as he was later praised in the docu-film Bruno Touschek and the Art of Physics.

5.5   A Death Sentence, and Fire-Bombing in Berlin: September Through December 1943

While our two heroes of the betatron story exchanged opinions on the interest of center of mass collisions among elementary particles, more and more of the German cities, their people and infrastructures were the target of bombing. Operation Gomorrah had killed more than forty thousand people in Hamburg, soon it would be the turn of Berlin.

At the end of August, on the night between August 31st and September 1st, there was a failed attempt by the Allies to carry through a heavy bombing of Berlin. From the UK national archives, the August 1943 bomber command records show there was clouding in the area planned for the attack, and this, when combined with some

---

[20] See letters to parents from Hamburg in 1944–1945.

[21] See Rolf Widerøe's letter to Amaldi in Edoardo Amaldi Papers in Sapienza University of Rome–Physics Department Archives.

technical problems and 'ferocious' German defense, spared the city population. After this raid, Goebbels ordered the evacuation from Berlin, of children and all adults not engaged in war work, to country areas or to towns in Eastern Germany where air raids were not expected. Bruno remained in Berlin, as his work at Opta was related to the war effort. After the Opta building was heavily damaged during the raid, the work was moved to one of the Flak towers, near the zoo, in the Tier Gardens.[22] Bruno's work involved the development of fluorescent screens for radar tubes, in which, in addition to the instantaneous blue fluorescence, a delayed reddish flash was produced. This work was done for the Strategic Command of Berlin in the building of one of the counterattack towers, which—being situated in the Tiergarten (the Zoological gardens)—was practically unattackable (Amaldi 1981).

Thus, on a dark night, he was a victim of the blackout in front of the Flak towers. Leaving the Flak tower to go home, Bruno almost drowned.

```
I left the flak tower at around 8 p.m. It was very dark
outside and besides, I couldn't see anything at all because
I had looked into bright Wo-light for half an hour.
Beforehand I looked for the way that leads to the Zoo. To do
this, I lit a match. I saw a grey area with leaves on it.
That was obviously the way. It wasn't. Instead it was a
branch of the Landwehr Canal. Of course I swam out, without
getting any pneumonia. Because of this incident, the RLM is
now having a railing built.
That same night E. [Egerer] drove over a wooden bridge in
the Tiergarten - which was not intended for cars, however.
He got off on the right side. One wheel was in the air.
```

In October, the negotiations with RLM to join Widerøe's project came through. Bruno had to sign a very formal declaration of secrecy with a hundred oaths and threats. It was all so serious that Bruno felt as if he had been signing a death warrant.[23] And in a sense it was. If he had hoped to be unknown to the Nazi authorities, his status had now changed. By joining Widerøe's team Bruno was engaging into a secret, war related, project, financed by the RLM, and he had become visible to the Gestapo. If and when he would finish the task assigned to him, he would be exposed to deportation to concentration camps because of his Jewish origin. It almost happened, as it will be seen.

The RLM project implied a move to Hamburg, where the actual work for the beta-tron was starting. But that won't be requested until the next year. In the meanwhile, he had to go through one of the heaviest and deadliest bombing the German capital would be subjected. This was part of the Allies bombing operation which became known as *the Battle of Berlin*. It started on 18/19 November 1943. After a number of smaller attacks to nearby cities, the full force of the Allies hit Berlin, on 22/23 November 1943, with 764 aircrafts. The attack continued on the following night with 383 aircrafts and went on for almost a month.

---

[22] According to wikipedia, Flak came into English as an abbreviation for the German word *Fliegerabwehrkanone*, meaning 'aircraft-defense gun'.

[23] Letter to parents on October 29th, 1943.

On 26/27 night the weather was clear over Berlin. Most of the bombing fell within the city boundaries and particularly on the semi-industrial suburb of Reinickendorf; smaller amounts of bombing fell in the centre and in the Siemensstadt (with many electrical factories) and Tegel districts. The Berlin Zoo was heavily bombed on this night. Many of the animals had been evacuated to zoos in other parts of Germany but the bombing killed most of the remainder. Several large and dangerous animals— leopards, panthers, jaguars, apes—escaped and had to be hunted and shot in the streets.

December brought no relief to Berlin. On 16/17 December, the bombing intensified once more. In the city centre, the National Theatre and the building housing Germany's military and political archives were both destroyed. The damage to the Berlin railway system and the large numbers of people still leaving the city, were having a cumulative effect upon the transportation of supplies to the Russian Front; 1,000 wagon-loads of war material were held up for 6 days. The sustained bombing had now made more than a quarter of Berlin's total living accommodation unusable.

The battle of Berlin, just like the bombing of the other German cities was bringing heavy casualties because the phosphorus bombs dropped to the ground would self-ignite and continue burning after the enemy planes had left. Bruno was right in the middle of the attacks. He lived at the time in Eislebenerstrasse, not far from the Tier Gardens where Egerer had transferred most of the OPTA laboratory in one of the Flak towers.

His description of how he managed to move across the city during two of those November nights, while the aerial battle over Berlin took place, is here.[24]

```
27.11.43
I am writing to you only today because, as a result of
recent events, a letter would hardly have been dispatched
earlier. Well, on the 22nd and 23rd, I was once again in the
center of the attack. Eislebenerstrasse, which has about 20
houses, is 60% pulverized. And the same is true almost for
the whole area.
At the first alarm, the English came very early .... I just
had a visitor. The girl (actually a divorced woman) Monika
and Mr. Fromme. We played tohuwabohu in the air raid
shelter.[25] After the alarm, we found out that the Streisand's
apartment is pretty much [gone to] the devil.[26] Pretty much
means completely up to my room, and then nothing. Even the
windows are whole. Only the ceiling has crumbled
considerably and the room was covered with a thick layer of
dirt. After the alarm we wanted to go to one of the
neighboring houses to extinguish [some fires]. ...The house
in question was finally hit the next evening. There was an
explosive bomb in Eislebenerstrasse. About 3 pieces in the
vicinity.
Monika lives in Friedenau and we wanted to accompany her
there. I then said goodbye to Fromme. He had to save his
```

---

[24] Letters to parents on November 27th, 29th and December 16th, 1943.

[25] *tohuwabohu* is an Austrian-German term derived from a Hebrew word with a complex meaning such as chaos, or darkness.

[26] Streisand was the family Bruno lived with at 4 Eislebenerstrasse.

house, which was in a sea of flames. I still don't know whether he succeeded. The expedition to Friedenau did not succeed because the roads were completely impassable due to smoke and sparks. So we decided to go home again. (The idea to go to Friedenau came from Monika. Since F. was also for it, I was outvoted). We got stuck about 200m from my house. The fire had grown so strong that we could not go back or forward. We went into a cellar and tried to clean our lungs. After 2 hours I made a breakthrough attempt home. It succeeded. I climbed again over the burning rubble of a house in the Fürther-Strasser to Monika to take her with me. She couldn't continue 50m from my house. Nothing helped. I had to smuggle them back - which made me extremely angry. Since she was relatively safe in the basement - [although] she could have slept better with me despite the devastated apartment - I went home a third time and got through [the night].

I was in such agony for my books while I was in the basement that I decided to put them in the flak tower. On the 23rd there was another alarm. Again at the same time. During the bombing, I went into the flak tower. When I came back, the whole Eislebenerstrasse was on fire again. The fires from the previous day had not yet been extinguished. I worked in number 6. Water flowed very sparingly. The fire department had left because there was no water. I crawled around on the various endangered roofs until about 4 a.m. Then I went to sleep. These two alarms were the worst that Berlin has ever experienced. Especially the first. ...The poor defensive successes seem to be largely due to bad weather. In addition, our fighters have a hard time with the heavily armed RAF.

[...] incendiary bombs are mainly used in the attacks [...] The reason is obvious. In the case of incendiary bombs, the energy used for the destruction is drawn from the object, whereas in the case of high-explosive bombs it has to be 'carried along'.
...

Micha and I estimated the building damage in Berlin to be around 20%. That doesn't sound like much, but it's a lot. (Probably a little too high, the estimate). My surroundings are like a heap of rubble. Every third house burned out to the ground. The military authorities are very generous. In the flak tower you get to eat without stamps. Even cigarettes now and then.

I almost had the same trouble with my laundry as on March 1st. The laundry is completely burned out. But today, on the rubble, I find the sign: 'laundry [washed] saved'. A sigh of relief. This morning I was at the flak tower: major attack on Bremen, alarm in Hamburg. Berlin also had a pre-alarm which only the flak tower residents know.

Your electric stove, which I am currently using (burning slippers, which are already badly worn from extinguishing

Fig. 5.5 At left, a sketch of bombed out buildings in Berlin included by Bruno Touschek in his 27th November 1943 letter, Family Documents, © Francis Touschek, all rights reserved. At right, one of the Flak towers in Berlin, |Bundesarchiv_Bild_183-1997-0923-505,_Berlin-Tiergarten,_Flakturm_als_Krankenhaus.jpg

```
the fire in the afternoon – our house had started to burn
for no good reason) is a relief. What would I be without it.
It is embarrassing that we have no water.[...]Of course there
is also no gas. My hosts cook at the stove.
```

Bruno closes this letter with the drawing in Fig. 5.5, which we show together with a photograph of one of the Flak towers in Berlin.

Through all the destruction and danger, Bruno mostly worried to keep his books safe from fire, and went on working. On 29 November, Bruno adds to his letters a very interesting remark, which can only refer to the work on the betatron. He could see his work so far as having been done for others and not for himself, such as had been the one at Studiengesellschaft, or at Opta, with others taking credit for his efforts, but could also see that things had remarkably improved. Until then, Bruno had often obtained results claimed by his office bosses or colleagues, or meant only for his professors. Now, for the first time, he felt that the solution to the problem was in his hands and could be called his own. Making the betatron work at the proposed energy, 15 Million Volt, was not just an engineering feat: theoretical physics calculations, implying the use of advanced mathematical formalism and relativistic corrections, were needed and he was the physicist on the team able to provide them. Touschek saw that the betatron to be constructed had many interesting aspects, and if they succeeded in making it work, it could become a world record. Perhaps, he mused to his parents, something from this time would remain after the war.[27] After

---

[27] Letter to parents on November 29th, 1943.

three years of apparently wasted efforts, it was a comforting thought to cherish, while the cities around him were destroyed, with fires springing out of the burning pavements, increasing shortage of food, lack of minimal heating. The future was to confirm Bruno's vision. Widerøe's betatron as a tool for reaching higher electron accelerations was soon to be taken over by the independent American and Russian discovery of the synchrotron working principle, but Bruno's acquired knowledge in accelerator science stayed with him. Later, in Italy, it all came back when he built the first matter-antimatter storage ring, the collider ring envisioned by Widerøe during his 1943 late summer vacation.

In mid December, more bombing of Berlin took place, once more successfully survived by Bruno:[28]

```
16/12/43
Survived today's serious alarm happily. Five incendiary
bombs in the house. All extinguished.
```

Bruno was in Berlin for Christmas with one of his new friends, a half-Jewish girl like himself, and then traveled with her to the old city of Nordhausen where they spent the end of the year together.[29]

5.6 1944: Academic Lectures and a Work Completed

While bombing of Berlin turned into an almost regular feature and food started to be scarce in the city, everything became uncertain and difficult. Between one alarm and the next, daily life was spent as in a suspended bubble. War was in front and behind. For his father's birthday, which fell in January like his stepmother's, Bruno sent a letter including a 'virtual' gift of sausages and wine, compared with a drawing of his real life, both shown in Fig. 5.6. Bruno wrote that every calculation of time was lost. The normal sense of continuity was gone, and one could expect things both to change as well as remain the same, there being no way to predict what would happen next. Thus, while congratulating his father for having reached his 60th birthday, he also wondered whether there was any sense in keeping up with such congratulations.[30]

---

[28] Postcard to parents.

[29] From https://en.wikipedia.org/wiki/Nordhausen,_Thuringia: Nordhausen, in Thuringia, is located some 60 Km from Göttingen. With a history dating back to the late 900, it was heavily bombed during World War II, with the old city center almost totally in ruins at the end of the war. Nordhausen became a target for Allied bombing, after the construction of rockets was moved there in 1943, following the destruction of the Peenemunde site. On the outskirts of Nordhausen, a concentration camp, Mittelbau-Dora, was established. It provided labor for the Mittelwerk V-2 rocket factory in the nearby Kohnstein hill. Over its period of operation, around 60,000 inmates passed through Dora and its system of subcamps, of whom around 20,000 died from bad working conditions, starvation, and diseases, or were murdered. Around 10,000 forced labourers were deployed in several factories within the city; up to 6,000 of them were interned at Boelcke Kaserne, working for a Junkers factory.

[30] Letter to parents on January 18th, 1944.

Fig. 5.6 Drawings by Bruno Touschek included at top and bottom of his January 18th, 1944 letter to parents. He comments these drawings by writing: '*Ein Traum (Oben links)*', i.e. dream (top left), and '*ein Leben (unten rechts)*', life (bottom right). Family Documents, © Francis Touschek, all rights reserved

Berlin was now the object of frequent aerial attacks. After particularly heavy ones at the end of January, on January 29th and 30th Bruno sent news of his survival by means of already printed cards, which included in the interior the lettering *Leben-szeiche*, meaning they were carrying signs of life from the sender. These express cards, which could reach Vienna within 3-4 days, were much cheaper than telegrams, would not clog transmission lines unlike telephone calls, and could carry the same information.

5.6.1   1944: The Betatron Starts to Be Built in Hamburg and Bruno is Called to Forced Labor in Berlin

In January, returning from his vacation, Bruno had been confident that his new engagement with RLM could turn into a success (making the German betatron a world record). This contributed to lifting his spirits and brought a temporary pause in his secret war with Opta boss Karl Egerer. But it did not last long. Through the previous autumn negotiations with RLM, Bruno had been very annoyed by Egerer's behaviour and, in March, their dealings again took to a negative turn. This is likely to have been motivated by Egerer's unhappiness at Bruno's involvement with Widerøe's project. Egerer considered himself as the initiator of such project and was bitter about not having to gain from it. His fears, of being left out of a possibly advantageous deal, were partly well founded. Indeed, while Bruno continued to work on Opta's projects in Berlin, his move to Hamburg, where the betatron was under construction, was becoming necessary, according to Widerøe. Such move would necessarily reduce Bruno's work for Egerer, and lead to conflicting allegiances.

Around 18th of March, a meeting between Egerer, Widerøe and Bruno took place. Widerøe explicitly told Egerer that, for the construction work to be quickly completed, in the coming months Bruno's presence would be requested in Hamburg, and asked Egerer to give Bruno a leave of absence from work at Opta. Egerer said this was out of the question, Widerøe jokingly said this would have to be seen. At this, Egerer was incensed, as he would not give up Bruno so easily. A discussion ensued, with Egerer claiming priority in having brought Bruno into the betatron's project, and the like. There was an apparent reconciliation between the two of them, but then, later, Bruno and Rolf went to RLM, where the matter was discussed in a manner which appeared to turn out quite negatively for Egerer.[31]

After the confrontation between Egerer and Widerøe, and the visit to the RLM, Bruno started to regularly spend time in Hamburg, working on the construction of the betatron, as requested by Widerøe, commuting to Berlin for his work at Opta. But his work for the RLM was not without danger, as he had understood when, many months before, he had written to his parents of having probably signed his death warrant. Indeed, Bruno was soon to receive his first summon from the Todt Organization (O.T.).

O.T. was an institution founded in 1938 by an engineer, Fritz Todt, to organize industrial laborers. After 1939 it had become the main venue for the forced or slave labor, which sustained to the very end the industrial and military effort of the Nazi regime. Following Todt's death in a plane crash in 1942, Albert Speer became the head of the Organization (Speer 1970). O.T. kept track of all the workers deported from the countries occupied or annexed to Germany, having different rules for prisoners of wars, political dissidents, Jews or mixed race, such as Richard Gans or Bruno Touschek. While a few Jews and a number of half-Jews could be employed in war related activities, they were nonetheless under the ever-watchful eye of the O.T. and could be summoned any time to the labor camps at the Organization usually unappealable decision. Exceptions, to being drafted to the labor camps, were very few and usually required the intervention of very highly placed personalities.

Not long after the discussion with Egerer and the visit to the RLM which had followed it, Bruno received his summon to Berlin from the O.T. When the summon arrived, he was probably in Hamburg.[32] Bruno knew he was too young to be exempted from forced labor and was, at first, quite shaken. He called Egerer in Berlin, to find out what to do. The people from the team in Hamburg got into action, and he learnt that

---

[31] Letters to parents on March 20th and April 4th, 1943.

[32] This can be gathered from a handwritten letter to parents on April 4th 1944, without the sender's address. During the first months of his engagement in Hamburg, Bruno's typewriter stayed in Berlin and he wrote his letters from Hamburg by hand. This letter is difficult to read, but it indicates that the summon may have arrived on Wednesday March 29th.

General Milch was in charge of the case and that he would prepare a 'Reklamation' letter. The letter would state that Bruno was involved in very important work for the Reich and that his presence at the C.H.F. Müller factory was indispensable. The exemption had to be approved by Albert Speer, who was head of the Ministry for Armament and War-production, and in charge of the O.T. The 'Reklamation' was successful and a few days after receiving the order to appear at the O.T., Bruno learnt that his draft had been postponed.

The threat of being drafted to forced labor did not disappear, however, and it surfaced twice before March 1945. Ultimately, when Bruno was arrested on March 17th, 1945, the voluntary summon turned into an order for imprisonment, the last step before concentration camp and probable death.

5.6.1.1 The Last Summer Before the End

The scientific team working in Hamburg on the betatron consisted of Rolf Wideröe, head of the project, Bruno Touschek for theoretical calculations but also for other more practical activities (as when the betatron was finally moved out of the Müller factory), Richard Seifert, and two physicists, Rudolph Kollath, and Gerhard Schumann, who would later write together a report on the 15 MeV betatron, (Kollath and Schumann 1947). Another, external, member of the group was Theodor Hollnack, whom Wideröe describes as the 'head of a relatively small private company who acted as a mediator' between him and the RLM. We shall later see that Hollnack played a role in liberating Bruno from the prison in Altona, on April 30th 1945.[33]

Bruno commuted between Berlin and Hamburg through all of 1944. For a while, he kept his room in Eisenerberstrasse 4 in Berlin, close to the Tier Gardens and the Flak tower, where he worked for Opta. He was keeping a room in Hamburg as well, in a pension called Dammtorpalais. The location was very good, in the University compound, very close (3 minutes) to Wilhelm Lenz, who taught a course on relativity at the University, which Bruno attended, and within walking distance (10 minutes) from Seifert and the Müller factory, where the betatron was under construction.[34] In this period, he became very fond of Seifert, at whose place he would often be invited for dinner, together with Wideröe. Seifert had been Wideröe's first contact in Hamburg and was very supportive of Wideröe's situation, with his family in Norway and his brother in prison in Germany. (Waloschek 1994, 70).[35]

In May, according to Bruno, the scientific situation [was] beginning to become very exciting.[36] The work was also made easier

---

[33] From (Waloschek 2012), it appears that Hollnack may also have been secretly acting for the Allied.

[34] In (Amaldi 1981), Wideröe reportedly wrote that, in Hamburg, Bruno lived in Lenz' house. This is not borne out by Bruno's frequent letters of this or earlier periods in Hamburg, but it could have happened when Bruno was looking for accommodations, before finding the room in Dammtorpalais.

[35] In his biography, Wideröe refers to this situation as a 'strange predicament'.

[36] May 16th 1944 letter to parents.

by his very good relations with Widerøe. This may be surprising, in light of Bruno's negative experience with his bosses, Jobst in Hamburg or Egerer in Berlin. One obvious reason for things to be different with Widerøe is that Bruno and Rolf were never in contact for long periods of time, since every other week or so Bruno would go to Berlin and Widerøe was often in Oslo.[37] Even so, it was quite natural that between Bruno and Rolf there would arise a friendship which lasted beyond the war years. Bruno was impatient with fools, or dishonest people, but capable of admiring and respecting his peers. Mostly, he could recognize scientific brilliance and intellectual capabilities. Both were present in Rolf Widerøe, as is apparent from his work and many accounts from colleagues and friends (Dahl 1992; Voss 1997; Wiik 1998). Given the intellectual stature of Rolf Widerøe, and the human qualities he was also endowed with, as reported through many interviews in (Sørheim 2020), it is obvious that the two would respect each other and become friends. In a November 11th 1979 letter sent to Amaldi, who was preparing Bruno's biography, Rolf Widerøe wrote:[38]

> I liked him very much, we were good friends and understood each other very well. He was so fresh in his thoughts and in his way of behaving, unorthodox but with a sharp critical sense and so lively! In Hamburg I admired his spirit of survival. He never gave up and he had a certain way to get at and concentrate on the essential things.

While providing theoretical calculations for the betatron, it is from Rolf that Bruno learnt the art of making particle accelerators. In the constellation of great minds which forged Bruno into a physicist, Rolf is present together with Arnold Sommerfeld, Werner Heisenberg, Max Born and Wolfgang Pauli, five great scientists who contributed to the flourishing of his genius, each of them in different ways.

The resolution of the O.T. summon, by way of the RLM successful intervention, led Bruno to be commanded to work in Hamburg on the betatron problem. The new formal situation would free Bruno from working at Opta, that would bear the costs of his command with RLM, upon adequate compensation. Egerer was asked to prepare a draft contract as soon as possible, along conditions agreed in advance by Bruno, Hollnack and Widerøe. The main point was that any suggestion and work by Touschek on the subject of the ray-transformer had to be first cleared with Seifert and Widerøe, and that no patent application for Opta should be filed without the consent of Hollnack.

Negotiations followed between Egerer, acting for Opta, Hollnack, acting as intermediary with the Ministry, and Widerøe, who was keen in building the betatron and holding on to any patent claim. Egerer thought he had the upper hand because Touschek was still working for him, and was thus trying to include claims on possible patents arising from the betatron project. Widerøe and Hollnack naturally resisted and the situation dragged on. Hollnack went to Berlin to discuss matters, but Egerer's draft contained a number of provisions implying the diversion to Egerer of much of the insight gained by Bruno in the project. This was unacceptable to Hollnack, and

---

[37] For instance Widerøe was in Oslo, at the time of the May 16th 1944 letter.

[38] See Rolf Widerøe's letter to Amaldi in Edoardo Amaldi Papers, Sapienza University of Rome–Physics Department Archives.

to Bruno as well. Bruno felt the situation to be unpleasant, even throwing a negative light on him, or imperiling his exemption from forced labor.[39] As things kept dragging on, he was distracted and could not work. He moved back and forth from Hamburg to Berlin, carrying on a silent struggle with Egerer, basically ignoring him, while waiting for the negotiations to end.

In June 1944, Bruno saw that the betatron would function, and his interest began waning.[40] This was a mental propensity Bruno always had in his life. Once a problem had been solved, or once he had understood how to go around it, he would lose interest and move on to something else. The challenge of formulating and solving the theoretical problems, and thus make the ray-transformer reach the set record energy, had been met, and the excitement was now over. Bruno felt that the employment in Hamburg was now becoming mostly a financially rewarding affair. It also allowed him to remain in Hamburg, a city he much preferred to Berlin. In March 1942, when he had first arrived in Hamburg, he had thought poorly of the food and was disturbed by the Northern German way of speaking. Since then, he himself had changed. He was not a lonely youth, longing for his family and the comforts he had left in Vienna. Now he had made his contribution to an important scientific project, was sought out by his peers as worth of notice, had made good friends whose interests and intellectual capacities equalled his own, among them a brilliant scientist engineer such as Widerøe, Wilhelm Lenz, a professor of theoretical physics, Richard Seifert, an industrialist bestowed wth honorary degrees after the war. He would often dine at Seifert's, together with Widerøe and Lenz. At Lenz's he enjoyed the food prepared by Miss Wagner, the excellent housekeeper.

In the morning, Bruno would listen to lectures or seminars at the University, which he was often invited to attend, and spent time in the library, where he had permission to access any book, even foreign magazines. In (Amaldi 1981), Bruno's reading of foreign magazines is considered the reason of his arrest in March 1945, but the repeated summons of the O.T. for Bruno to join forced labor tell a different story, as it shall shortly be seen.

As his earning had improved, Bruno could often have lunch at the Curiohaus, a very good restaurant inside the University, with a menu which included trout with cheese, chicken paté, veal stew, etc. During that last summer before the end of the war, bombing had started again to hit Hamburg, but the weather was really excellent and in the afternoon he would go swimming and had meetings with Widerøe. They discussed writing a book about the ray-transformer, as Rolf called the betatron. Occasionally, they planned a Sunday excursions to the Cranz neighbourhood, on the other side of the Elbe. In the evening, Bruno spent time reading English literature, and whenever he could, went fencing.

---

[39] Letter to father 15th June, 1944.

[40] Letter to parents on June 15th, 1944 from Berlin (context).

5.6.2   A Lecture During an Alarm

Much of such idyllic description may have been written by Bruno just to reassure his parents, as in fact, daily life was becoming harder. In late July, he confessed that his shoes were broken, he had no socks and his shirts were all torn. As he had none in good condition, he could not send them to be repaired. Raids over Hamburg had also started again, taking advantage of the good weather and usually clear skies over the Northern city.

On 27–28 July 1944 there was an attack, but without damages comparable to the 1943 bombing of the same period.[41] This time he was in Hamburg and described what happened in one of his usual letters, which he starts by writing: To your reassurance: nothing happened to me in the night from Friday to Saturday [28-29 July].

At the beginning of the week he had asked Lenz to supervise the work in nuclear physics he was doing. Lenz suggested to hold a lecture on this every week, with the first lecture scheduled for Saturday (29th July). The audience was very selected, as he wrote: Prof. Lenz, Prof. Jensen from Hannover, Dr. Artmann - assistant to Prof. Jensen, Dr. Süss - assistant to Prof. Harteck, Dr. Wideroe, including Dr. Kollath (the new experimental assistant to Dr. Wideroe -formerly first assistant to Ramsauer).

Bruno had spent most of the night awake, chasing for a wrong sign and similar things. Then the alarm came, and he had to accompany a lady [from the Pension] to the Klosterstern Station [a metro station in Hamburg since 1929] out of courtesy. Back to his room, he found the mistake in his paper and then it was already half past six in the morning. At 9 o'clock Wideröe telephoned to inquire if he was all right and whether they would meet as agreed. Then the alarm went off again and Lenz postponed the matter. Bruno lived very close to the Physics Institute and went out to find if anything had happened to the institute, which was, indeed, in a very doubtful state. At Planten un Blomen he met Wideröe and Kollath and, having by now reached lunch time, the three of them all went to the Curiohaus.[42] Shortly afterwards, they were joined by Lenz and thus the presentation could start. It was debated through lunch and into the afternoon, with a certain degree of merriment. As Bruno writes: The language that developed could be heard in all bars and restaurants in Hamburg and it never ended.[43]

In late August, Bruno took a vacation to visit his family in Vienna. Leaving from Berlin around 28th August, the train would take him to Breslau in about 3 hours and from there, directly south to Vienna. He had taken this trip a few times in 1943, travelling in rather comfortable conditions, apart from Bruno's complaining against

---

[41] The limited damage was apparently due to clouded skies during the attack.

[42] Planten un Blomen is a park in the center of Hamburg, which contains the Old Botanical Garden of the city.

[43] Letter to parents on July 30th, 1944.

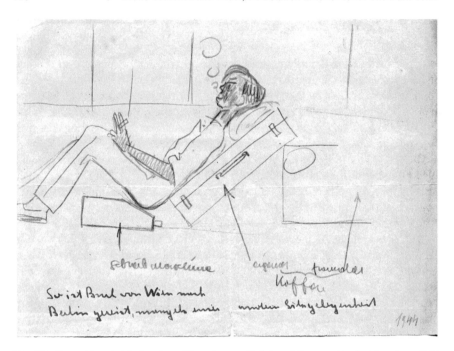

Fig. 5.7 Bruno Touschek's drawing of how he travelled from Vienna to Berlin. The date, 1944, was added by his father, the drawing is included in the September 11th, 1944 letter to his parents, also in (Bonolis and Pancheri 2011), Family Documents, © Francis Touschek, all rights reserved

his fellow second or first class passengers.[44] But transportation between the two capitals, Berlin and Vienna, was now becoming more difficult and less comfortable. There was continuous danger of aerial attacks. Vienna had already been repeatedly bombed, and trains were crowded by people getting out of the cities to find some safety in the country side. Back in Berlin, Bruno made a sketch of himself travelling from Vienna to Berlin and sent it to his parents on the back of his letter. In Fig. 5.7 we reproduce this sketch, where he shows how he had built himself a seating (and sleeping) arrangement out of suitcases and his typewriter.

This was probably Bruno's last visit to Vienna for quite some time. The next such visit we know about, took place well after the end of the war, when he was already in Glasgow studying for his doctorate in physics, in 1947.

5.6.2.1 The Betatron Starts Working

On April 27th, 1944, the commissioning of the first betatron developed by Siemens for medical use had occurred in Erlangen. As the year progressed, Schiebold's project

---

[44] In letter sent on 3rd May1943, he mentions a similar Vienna-Berlin trip, travelling rather comfortably from Vienna to Breslau, in first class, but having to stand from Breslau to Berlin (3 hours).

came under criticism, and by the end of the year was dismissed. At the same time, the Luftwaffe did not abandon its interest in the production of high energy beams regarding Widerøe's project. In a report of August 25th, 1944, Gerlach claimed in fact that the development of betatrons would be of scientific interest to investigations on high-energy electrons and very hard X-rays. Gerlach of course viewed with great favor all project regarding accelerating machines in Germany.[45]

Since April Touschek had moved to Hamburg in order to accelerate the work with the betatron. As he also kept the work at Opta, he commuted between Berlin and Hamburg, maintaining a room in both towns for a while. He was going along very well with Widerøe, who asked Bruno to write a book together on the betatron. He had already collected much material for it, but the project was never realized.[46] By September 1944, according to Touschek's letter to his father, Widerøe's betatron was beginning to work:[47]

```
I work every other day for W. - the thing starts to work -
on the other days I'm preparing for my dissertation. I sleep
very much, eat little [...] [Arbeit jeden zweiten Tag für W. -
die Sache beginnt zu klappen - an den anderen Tagen bereite
ich mich auf meine Dissertation vor. Ich schlafe sehr viel
esse wenig - die Zusatzkarten wurden abgeschaft usw].
```

Work performed by Touschek during the period May–September 1944 is contained in a report preserved in Zurich among Widerøe's papers.[48] Widerøe later told Amaldi that Touschek's theoretical investigations in Hamburg "were of great help to us and I think they might also have stimulated my later work on the betatron in Baden [after the war]".[49]

According to Widerøe's recollections (Waloschek 1994, 85), one day, they were visited in Hamburg by Professor Gentner from Heidelberg and Professor Kulenkampff from Tübingen: "They were full of praise for our results." In autumn 1944 Widerøe was invited to a meeting at the Kaiser Wilhelm Institute in Berlin at which both Heisenberg and Gerlach, as well as other physicists, were present: "We all spoke freely and said exactly what we meant. As there weren't any Gestapo men present, nothing was kept secret. We unanimously agreed that Schiebold's fantasies should be called off as they were so utterly unrealistic. On the other hand, it was decided that the betatron was a very interesting machine, especially with regard to the medicine and nuclear physics of the future."

---

[45] W. Gerlach, Minutes ("Niederschrift") dated August 25, 1944, regarding the session of the Kuratorium Groosthéim on May 6, 1944 in Berlin and the meeting of a sub-committee on August 15, 1944 in Ainring (Waloschek 2012, 208).

[46] Letter on July 8th, 1944.

[47] Septembre 18th 1944 letter.

[48] Bruno Touschek, Bericht: Theoretische Untersuchungen für das MV Verfahren in Hamburg während der Zeit Mai bis September 1944 (ETH-Bibliothek, 175, Rolf Widerøe, "Akten, Korrespondenz und andere Dokumente zu Werk und Leben", section "Korrespondenzen, Protokolle, No. 73–74)".

[49] Widerøe to Amaldi, 10th November 1979, Edoardo Amaldi Papers in Sapienza University of Rome–Physics Department Archives, Box 524, Folder 4, Subfolder 2.

When Schiebold's project was dismissed an organization called 'Megavolt Versuchsanstalt' (Megavolt Research Association, MVRA) was created, which continued the activity on the 15-MeV betatron under the guidance of Widerøe. At the same time, the construction of the 200 MeV betatron was being planned with directors and design engineers of Brown Boveri & Co. (BBC), and Seifert awarded BBC a preliminary order from the Aviation Ministry (RLM) for planning such a machine which, however, was never realized.

5.6.3   O.T. Calls Again: *cui bono?*

As the war raged on two fronts in Europe, the general situation in Germany was becoming very critical. In the fall of 1944, million of forced laborers toiled to support the regime's increasingly desperate attempt to postpone defeat.

Bruno's letters in fall 1944 describe how the situation in Germany, and Austria as well, was rapidly deteriorating. Bombing of Vienna intensified, and Bruno's preoccupation for his parents and the fate of his native city, pervades his letters. Berlin and Hamburg were under continuous attacks, and there were calls in the *Hamburger Fremdblatt*, one of the most important daily newspapers in Hamburg, for a sniper's war in case of emergency. Food cards were abolished, people started getting hungry.[50] His room was also hit, a bomb exploded next to it and the dividing wall collapsed. The clean up work took all-day and he started looking once more for a different accommodation. Still, in those days of early 1944 fall, Bruno did not have yet the sense of what would soon engulf Germany, destruction and desperation, and the unconditional surrender on May 8th, 1945.[51] He expected the war to be near its end, and was making some plans for the future, in terms of asking his parents to eventually move to Hamburg after the war.

Postal communications started being affected and were not as regular as in the past, but there was one type of letters which had no problem reaching Bruno, those from the O.T., the Todt Organization.

Starting in October and continuing through November, O.T. renewed its pressure for Bruno to join mandatory forced labor. There was a first letter on October 20th, which summoned him to report to the O.T. camp in Eichkamp, bringing his clothes as an unskilled worker. After telephoning all the people he knew for their help in avoiding the draft, the matter was put on hold through some high-rank intervention, and adjourned again, as it had happened in the spring. Bruno was only asked to come back two days later to be photographed and examined (Bruno wondered what that could mean).

---

[50] Letter on September 11th, 1944.

[51] Germany's unconditional surrender was staged twice, first in Reims, on May 7th, and then on May 8th, in Berlin, at the request of the USSR, and signed on May 9th. The cease-fire was officially placed on May 8th.

Bruno had been in Berlin when this happened. Before the call from O.T., Egerer had warned Bruno not go to Hamburg, and hinted at some treacherous behaviour on the part of Seifert. According to Egerer, Seifert was plotting against him. The whole thing was very strange, as Bruno had very good relations with Seifert, calling him 'my director' in very favourable terms, or even 'my idol'. Bruno suspected Egerer of duplicity, but could not understand if the O.T.'s renewed draft had been genuine or set in action, and in such case, by whom? By Egerer, who had a vested interest in keeping Bruno away from Hamburg, using him as a pawn to get his part in the RLM bargain, or Seifert? But, if it had been Seifert, as Egerer claimed, what would anyone have gained in sending Bruno away from the betatron project?

On October 24th, Bruno took the usual fast train to Hamburg and had a meeting with Widerøe, returning to Berlin soon after, as the O.T. matter was still pending. Things developed slowly and on November 3rd, Bruno had a final summon for the following week, to appear at the camp with his work clothes (which he did not have), 2 blankets, warm underwear, etc.. On Monday morning, he was informed that his petition with Himmler was still pending, but, in the meantime, he could postpone going to Eichkamp.[52] Once more, the Damocle's sword was lifted and he could temporarily relax.

Apparently, Seifert indeed had some responsibility in this second O.T. call.

In the fall, after Widerøe's betatron started showing its promise, a revision was started, at the RLM, of the scientific and military value of the various ongoing death-ray projects, namely Siemens' betatron, basically useless in terms of war potential, but scientifically significant, as well as Widerøe's betatron and the Schiebold's X-rays project. While Schiebold's project was definitely put out of the picture, in this review of ongoing projects at RLM, some pressure was applied by Seifert about merging Siemens and Widerøe's betatron projects. This was in Seifert's interest because he had some stakes in the Siemens project, but it was opposed by Widerøe and his group, including Bruno, who, as was his habit, was probably very vocal about his opposition. Bruno, like his fellow workers, felt that the two projects had very different merits, and a merge would obscure their success: Siemens betatron had been going on for 3 years reaching only 5 Million Volt electron's energy, while theirs had been in operation 3/4 of a year and had already reached 15 Million Volt. Perhaps on Seifert's initiative, an investigation had been started at the RLM, and the presence of Bruno in Widerøe's team had been put again in discussion. Bruno wrote:[53]

```
As a special delicacy we would like to mention the
following: the RLM research department wanted to combine our
development with Siemens' parallel development. W. should be
sent home (the Moor has done his duty ...) Here are some
figures: we work 3/4 of a year, Siemens 3 years. Our thing
delivers 15 mill. Volt the Siemens 5th, ours is equivalent
to 200g radium, the Siemens 0,2g. Probably my Seifert thing
has its reason in this plan, which S. was very friendly to.
```

---

[52] Letter to parents on November 11th, 1944.

[53] Letter to Parents on 22nd November 1944.

```
This plan was thwarted by the Minister Speer. We are still
autonomous.
```

Once more, O.T. call was put in a limbo. Thus, in November 1944, Bruno was again exempted from being drafted to forced labor, but he remained very worried and upset. Since the autumn of 1944, between 10,000 to 20,000 half-Jews and persons related to Jews by marriage were recruited in fact into special units. The conditions of life and work were so hard that many died even before their deportation to the concentration camps. Such perspective arose a great anxiety in the young Touschek. He understood that his O.T. file was ready to be pulled out anytime, at the will of circumstances out of his control. Seifert, if indeed it had been him, had meant to thwart internal opposition to a merger with Siemens, silencing Bruno. While he failed to obtain the desired result, his move probably cleared the way for future O.T. actions against Bruno. As indeed it happened in March 1945.

5.6.3.1   Back in Hamburg

For now, however, Bruno was safe, although with a suspended sentence. Things seemed to quiet down for a while. The work in Hamburg was going along well, and Widerøe's could plan for a long absence and be with his family in Oslo, at least through Christmas and the New Year. Around November 11, he had visited Bruno in Berlin, and outlined his plans: he would go to Norway at the end of November to write the book on the betatron in three months' peace. In his absence Bruno was to continue the experiment together with Kollath, if the Seifert's affair had been settled by then and O.T. had not yet drafted Bruno. Of these two, the second 'if' was the most doubtful, since exemptions from forced labor in the whole of all the Nazi controlled countries were known to be extremely rare and only in exceptional circumstances. It is indeed indication of the importance attached to Widerøe's project and the relevance of Bruno's work that the O.T. call was still pending.

Previously, when Widerøe in Berlin had let RLM know of his intention to go to Norway for an extended period of time, difficulties had appeared. He had been so far totally free to move back and forth between Oslo and Hamburg or Berlin, travelling by plane and first class trains. But the review of the betatron projects had changed the atmosphere at the RLM. There was some fear that Widerøe may not come back. These fears may have had their justification in light of the envisioned merging with Siemens, implying new contracts to be drafted, probably limiting the freedom of movement he had so far enjoyed, as new conditions could require for instance that he could not leave until the work was really finished. Widerøe's opposition to the merging may have been at the root of such fears. His request to go home to be with his family was met with a counter offer to bring his wife to Germany, instead of him going to Oslo for three months. But the merging plan was put aside by Albert Speer, the Minister in charge of war and armaments, the Hamburg team remained independent, and Widerøe was free to go home as he had planned.

On November 19th, Bruno went back to Hamburg to find that living conditions had changed for the worse: in his building the lift was not working, heat and hot water were either missing or available only a couple of days a week. Outside, it rained in the elevated railway, coal shortage was growing from one day to the next, electricity had to be rationed and even at the C.H.Müller there were problems. Widerøe had to write requests to provide the 5 KWatt needed to switch on the betatron.[54] Life was harder day by day, but invitation at Lenz' place would still get Bruno a proper meal.

Constant in Bruno's mind at the time was also the worry for his parents in Vienna, where heavy bombing had started. There was no more talk of their coming to Germany, but one could envision moving out of the city, to Tyrol, that still appeared an island of tranquillity. Still, it was difficult to imagine leaving Vienna, the former great capital of the Austrian Hungarian empire. On December 2 Touschek wrote a short letter to his parents clearly replying to news about the ongoing dramatic bombing of Vienna:[55] I almost cried when I heard of the Belvedere [being bombed], wrote Bruno, and then: What will remain of Vienna, if this continues? Of a city which actually lives only from its past?[56] In the wake of millions of people dying everywhere on a never occurred scale, many of Europe's monuments, which had taken hundred of years to build and had survived many wars, were following the same destiny of destruction.

After the O.T. affair of October and November, Bruno could make no plans for the month of December. Widerøe was finally going to Norway, and Bruno was glad to have some peace concerning the work with the betatron. At the same time, he worried about missing him, as they had by now become good friends and Widerøe was providing a buffer against Egerer's machinations.

5.7   1945: Towards the End

The war was entering its sixth year, having lasted longer than anyone had expected. The end, feared but also anticipated, was in sight.

January 1945 brought harsher conditions to civilian life. Since Christmas, Bruno's room in Hamburg had no heating and Bruno got used to sitting down with his coat on. The work at the betatron had slowed down and he took advantage of it by doing some physics of his own, writing at least two papers (unpublished) in January and attending classes at the University, whenever possible. Physics was a survival tool, absorbing his attention and providing a mental distraction against the worries about his parents and his own future, immediate and present. O.T. was silent during the

---

[54] Letter to parents on 22nd November 1944.

[55] The first systematic air raids on Vienna had begun at the beginning of September, when air bases had been established in Foggia, a town in a part of the Kingdom of Italy not occupied by the Germans. Heavy bombings had hit Vienna every day from November 12 to November 19.

[56] Letter to parents on 2nd December 1944.

first months of 1945, even though one of his friends, Micha, received the summon. He had been the one with whom they had planned the school for *mischlinge* children in December 1942, when there still was hope something could be changed, or make a dent in the nightmare ahead. Micha's summon was luckily postponed, as it had happened to Bruno. By now, the only hope was that things would end soon, and reconstruction could start.

5.7.1 March 1945: The Betatron Leaves Hamburg

Amid the ruins and desperation which characterized the last months of the war in Germany, as the Allied troops were moving towards Hamburg and Berlin, from East and West, contradictory orders were given from the Ministers of the Reich concerning civil and industrial installations around the country. There were strong differences of opinion in the Nazi government as to whether installations should be destroyed to slow down the enemy's advance through Germany, or preserved for future reconstruction, after the defeat, which many saw to be, by now, inevitable.[57]

Between 8 and 10 March 1945, 312 aircraft attacked Hamburg in order to destroy the new-type XXI U-boats being assembled in the shipyards and considered a great threat owing to their capability of cruising under water for long periods. The British Army was also rapidly approaching the city of Hamburg, so that the German Aviation Ministry ordered Widerøe to move the betatron out of Hamburg. The betatron, which was under the protection of the RLM and officially still considered of military value, was an asset to protect. Seifert suggested to move it to a disused dairy factory, owned by his family in Wrist, a village near Kellinghusen, approximately 40 km North of Hamburg, in Holstein. How this was done and what happened before Touschek's imprisonment on March 17th 1945, is best described through the last letter Bruno sent to his parents before the end of the war, from Kellinghusen, on March 13th, 1945. This letter is long and difficult to understand, as it is written in almost telegraphic style, with many abbreviations, and lacking proper nouns. It is partially translated as follows:[58]

> Kellinghusen.   13.3.45.
> Dear Parents!
> I have not written to you for ages. The main reason for this
> is that, with one brief exception, my room in Hamburg has
> not been heated since Christmas and Berlin's pavement is too
> burning hot to write letters. To give you an overview of the
> situation, a description of the last 4 or 5 days follows:
> Friday [March 9th]: I am insanely cold. Widerøe calls me on
> the phone to persuade me to move to Kellinghusen as soon as
> possible. And for good reasons. Hamburg is once again the

---

[57] See Speer's memoirs about keeping the country's infrastructures safe through the end of the war, in order to be able to rapidly move to reconstruction (Speer 1970).

[58] Free translation by the Author (and Amrit Srivastava).

main target. Saturday the lorry goes to Kellinghusen. The
main concern is a box filled with books, weighing about
80Kg. At 1 o'clock Kissel, a Physics Diploma student whom I
helped to get through the exams in Lenz course and who is
now employed by Lenz and Jensen to do numerical
calculations, comes to me to help.Then comes Wideroe. We
find a car and bring the box (made at Opta) to the elevated
train station at Dammtor. In Ohlsdorf a handcart is waiting
for the stuff to Müller.
Saturday [March 10th]: 8 1/2 meeting place at Müller in
Ohlsdorf. 2 hour drive in an open truck to Wrist (where the
factory is, the village is 3 KM from Kellinghusen). With the
help of some Italians - now my basic Italian comes back to
honour - we unpack the truck. W. [Widerøe], Dr.Kollath and I
rush around all morning. Among other things, a crane is set
up - a job which I do at a lofty height as a result of my
constant agility. The room I had occupied 7 days ago in
Kellinghusen is gone, as the secretary tells me.
Kh.[Kellinghusen] is 2/3 occupied by refugees from the east.
With a cart and a box of books to Kh. first to the Hotel
Stadt Hamburg for dinner. Then room search. 3 hours of
pointless running around the village. In the end, a false
success. But in the room there is still a lady living there,
who is waiting for her transport to Kiel. Together with W.
and Dr. Kollath I carry the damned book box. Dinner at the
Hotel Stadt Hamburg. Back march to Wrist (railway junction)
with some luggage in an excited mood for three. Dinner again
(refugee rest) in Wrist. Alarm.In the evening I am [again]
in Hamburg at 11 1/2. Miss Erhorn tells me that Prof. Lenz
had called me three times and wanted to see me urgently.[59]
Sunday [March 11th]: Prof. Lenz calls at 8 am. I should come
to him immediately. History: Miss Wagner lives with Prof.
Lenz, and takes care of the household for him. As you know,
due to this circumstance L. [Lenz] still has everything.
About 2 months ago Ms Wagner fell ill with a severe
pleurisy. She also has tuberculosis and currently weighs 30
Kg. Since about a week a private nurse is in the house.
Prof. Lenz has always been hypocondriacal. Ghr.Sommerfeld
even told me once that he is crazy. For some time now I have
noticed an increasing aggravation of his condition. He
always has a cloth in front of his nose and doors may only
be opened after lengthy ceremonies. In any case, L. is sick.
I go first to Miss Wagner, who is in bed and who is
obviously not very fond of L. Then to L. The room is - as
the Sister [the nurse] claims - not ventilated for 3 weeks.
Lenz is lying in bed with a winter coat and shoes and - as
it turned out later - three pairs of trousers. Under his
winter coat he has a terribly torn old jacket. Above the
head there is an apparatus that spreads a rather dirty tea
towel over the head and neck like a tent. In front of the

---

[59] Miss Erhorn appears here for the first time. From her later initiatives, *re.* sending a telegram to
Bruno's parents, she was probably Touschek's landlady.

nose there is another bulge of tea towel attached to the
ears with copper wire. Next to the bed is a mobile serving
table with four different thermos flasks attached to its
four legs. There is a chair and a lot of wood shaving lying
around the room. As if from Hades, L.'s voice resounds from
the mummery. He is very ill, the Sister does not care about
him. He didn't know what to do at the alarms - all this
after an introduction that was worthy of a will. The
disorder in his room goes back to his intention to build a
recliner for Miss Wagner to transport him to the cellar.
This takes about 2 hours and involves a lot of talking and
planning. Whenever I come into the flat, Miss Wagner and the
Sister protest vehemently against this enterprise. Miss
Wagner never got involved, etc. Ms Wagner also tells me that
L. is said to have said that he is now the main person in
the house, he is seriously ill and the Sister was only
looking after him. It is clear to me that L. only got ill to
be cared for. According to Freud, such illnesses occur in
hysterics and cannot be distinguished from 'real' illnesses.
At noon I have an appointment with Dr. W. and Dr. Kollath at
Handl in Wellingsbuttel. At Dammtor station I learn that the
S-Bahn does not stop there on Sunday. I was able to get on
Sternschanze. So march to Sternschanze (1/4 hour). Nothing
there either. Back via Dammtor to the main station. Hardly a
station (until Berliner Tor): alarm. I march - the alarm is
always given early in Hamburg - to Lenz (who also lives in
the Dammtor area). That takes 30 minutes. When I arrive,
Lenz is already in the cellar. With all his clothes. It was
a wild attack, but nothing happened in our area. I carry
Lenz's things, except for the unappetizing odds and ends,
about 5 blankets back into the flat. After a meeting with
Miss Wagner I phone Prof. Jensen to talk to him about Lenz.
We arrange another telephone conversation in my flat. At 3 I
go for lunch. Jensen claims that L.'s strange condition
occurs at times, but it is not possible to tell whether it
is a stationary condition or whether the whole thing is
converging towards an institution. In the evening, I have a
whisky with my neighbour Dr. Elbrecht. Then comes the alarm
and the L. theatre repeats itself.
Monday [March 12th]: Small shopping in the city. At 11, I am
at L.'s. Alarm comes prompt. I try to catch the 12 o'clock
train from Altona to Wrist - after I have put Lenz in a
cellar - but the small alarm doesn't stop, the S-Bahn
doesn't run and I take a 2 hour walk with my suitcase. At 2
o'clock Kissel, whom I helped to pass the exams (Dipl.
Physik), and now is assistant to L. and Prof. Jensen, is
coming to me for a kind of follow-up. Then comes W. who
also missed the train to Wrist. He persuades me to go with
him and Dr. Kollath to Wrist at 5 o'clock, although I don't
know yet whether the lady from Kiel has finally moved out.
We arrive in Wrist with the train, of course during an alarm
arrival, at 10 1/2 o'clock, with a heavy luggage. The
electricity is cut off. After a long knocking my landlady

manages to appear. The Kieler lady is not gone yet. Further
march through Kellinghusen, and finally spend the night on a
sofa, in somebody's landlady's house.[60] We write a letter to
the landlady so that she does not be startled. At 5 o'clock
the landlady comes in. (You get up so early in Kh because
there is no more gas later). Didn't I want a room? (Of
course she is afraid of a refugee family). According to a
concurring statement by Dr. Kollath and W. I now have the
best room in Kh.
Tuesday [March 13th]: After 5 [am] I sleep - I'm a bit tired
again. At 8 am opulent 'Norwegian' breakfast with W. Marsch
in the factory. As for lunch, it is resting day in Wrist,
and food on Tuesdays is a disaster because of the rest day
and there is nothing there. Back to to Kh. I still have to
get the 80 Kg box. Organize a ladder cart. Get the box. (All
in all about 6 km). Unpacking the box. Walk into the place
to organize a desk. Failure. Another march to Wrist. Jerk
march. Eat. New suitcase negotiations. At 9, Dr. Kollath. We
sit together until 11 (with W. and create an immediate
program) etc. etc.
Now I am - it has become Wednesday [March 14th] in the
meantime - really exhausted. Mail goes on to Hamburg. Come
safely through the alarms!

What happened next, is known through a letter which reached Bruno's parents only
on October 22nd, 1945. Thanks to this letter, and a second one reaching in November,
we can now continue with Bruno's story.

On the morning of the 16th, Bruno and Rolf were still sitting down in the veranda,
it was a beautiful spring morning and they were reading an issue of *The Physical
Review* in which they had found an article of special interest for the betatron operation,
about radiation damping, *Strahlungsdämpfung* in German, a phenomenon potentially
limiting the energy reached by the electrons. It was an article by two Russians, which
later on would return to Bruno's attention.[61] In the evening, he went back to Hamburg.[62] Even before the letter from Kellinghusen could reach Bruno's parents, they
were alerted that something had happened to him. The chronology of the events to
follow indicates that Bruno was arrested early in the morning on March 17th, after
reaching Hamburg the night before (Bonolis and Pancheri 2011). It is not known how
the news reached the betatron team in Wrist or his parents in Vienna, but Bruno's
father immediate inquiries were sent on March 19th, and were answered by Miss
Erhorn, Bruno's landlady in Hamburg, in a telegram shown in Fig. 5.8. In this telegram, which reached Vienna ten days later, Bruno's housekeeper lets Bruno's father

---

[60] In the letter, it is said this to be the landlady of the dentist's wife, but it is unclear whose dentist.

[61] This is likely to have been a paper by Iwanenko and Pomeranchuk (Iwanenko and Pomeranchuk 1944).

[62] Bruno's parents received a telegram from Bruno asking for his mail to be sent on to Kellinghusen. The telegram is dated March 17th, and it may have been sent by telephone, upon Bruno's arrival in Hamburg, where he reached after midnight, according to his post-war letters. The access to a telephone even in those last days of the war should not surprise, since Touschek was working for a military sponsored project.

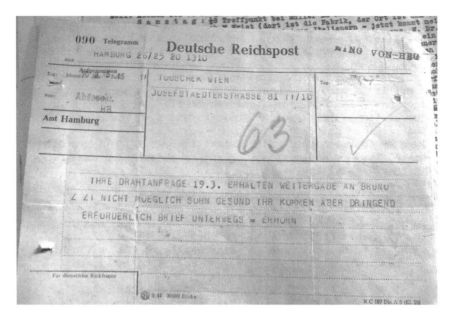

Fig. 5.8 The telegram sent from Hamburg on March 29th 1945, and received on the 30th in Vienna. The text can be translated to say 'YOUR TELEGRAM 19.3 RECEIVED NOT POSSIBLE TO FORWARD BRUNO SAFE LETTER COMING = ERHORN'. Family Documents, © Francis Touschek, all rights reserved

know that he is safe, but cannot be reached. There are many questions: who alerted Bruno's father that Bruno had been arrested? It is very likely that someone from the betatron team had been informed through the RLM, and telephoned Bruno's parents. But why had he been arrested?

5.8   Fuhlsbüttel Prison and a Brush with Death

Bruno's whereabouts after he finished transferring the betatron to Wrist, are described in two letters he sent to his parents, in June and November 1945. In the longer November letter, Bruno describes his whereabouts since March 14th in an almost-day-to day-account (Bonolis and Pancheri 2011). After the arrest, he was brought to the Fuhlsbüttel prison. Two days later, he was brought back to the Gestapo offices, where he could meet with Seifert, Kollath and Widerøe. Bruno had practically had no food, and had been subjected to the usual indignities, such as stinking toilets, no place to sleep, room crowded with sick people. Scared but angry as well, he pleaded with his friends for help, and Widerøe made the Gestapo understand that the future of the Reich for better or worse depended from a research on the influence of the radiation-damping work just started by Mr. Touschek.[63] The trio thus suggested that, as a first measure, Touschek's prisoner conditions be made less harsch, and he be allowed to smoke, read and receive visits, in addition to be alone in a cell. On Friday, nothing having changed, Bruno was ready to hang himself. But on Saturday, March 24th, Widerøe visited him again, bringing cigarettes, schnapps and a book to read, Heitler's *Quantum Theory of Radiation*. His cell conditions also changed, and he could be alone in his cell with a bed to sleep and sit, reading and writing. According to (Waloschek 1994, 91) it was in prison that Bruno wrote an important note about 'radiation damping in betatrons', which he traced in invisible ink in the pages of Heitler's book. Finally, around April 10, Widerøe brought the good news that release papers for Bruno had been signed in Berlin and he would soon be freed.[64]

5.8.1   Bruno's Pardon and the Reich Commissioner for Death-Rays

Date and visits of 'important people', as Bruno wrote in the November letter, tally with a curious story about secret weapons given by then Minister Albert Speer, in his Memoirs (Speer 1970, 464). Speer writes that on the early days of April 1945, he was told that 'death-rays had been invented' and they could be the 'decisive factor' to win the war. He was also told that unfortunately the Ministry had rejected the plea from the inventor. Urged to get the project to go on immediately, Speer humoured the requester, Robert Ley, by nominating him 'Commissioner for death-rays' and charging the head of the entire electrical industry to find the inventor, so as to get the details of the device and start experiments to check its validity. Speer ends the story by mentioning a detail, which would have disqualified the claim. He says that one of the components, on which the inventor insisted, was a 'circuit breaker [which] has not been made for about 40 years', and adds the bitter comment that 'Such wild notions flourished as the enemy approached.'

One is tempted to associate this little story to the pardon promised to Bruno around April 10th. If this anecdote is correctly reported by Speer, whose memoirs were published some 25 years later, the Ministry asked the 'inventor' to provide all information on how to make the device, and had the experts from the Ministry try to make one. Widerøe, who may have written the petition to free Bruno, was certainly not going to give out all his secrets. Always very keen on keeping his patents and their priority, Widerøe may have given some generic information such as to mislead the experts. In any case, the 15 MeV betatron had engaged for almost a year the genius of such as Bruno and Rolf, and could hardly be reproduced, in a few days, by whatever experts Speer could find. Had the experts even tried to do so, they would have failed.

---

[63] Letter to parents on Novemberr 11th, 1945.

[64] June and November 1945 letters to parents. For full text, in English translation, see (Bonolis and Pancheri 2011).

In that tragic last month of war, no expert in Germany would have dared to admit his failure to make a device claimed to win the war. A fib, about it being impossible to find instruments, could have saved somebody's life. As for the curious story of the circuit breaker not having been made 'for about 40 years', it could very well have been part of the betatron construction and referred as such in Widerøe's notes. When inventing something really new and with disrupting technology, scientists often go back to old literature, to pick up threads and ideas, which had been then neglected or overcome by more recent development. Thus, the reference to the old text book could have had such origin.

As for the possibility that the episode refer to *death-ray* projects other than Widerøe's, the list is limited to Schiebold's or Schmelllenmeyer's. But the episode could not have referred to Schmellenmeyer. His betatron had been moved away from Berlin already in autumn 1944, together with Gans, first to the East, near the Czech border, and, as the Soviet troops approached, to the South near Bayreuth (Waloschek 2012, 55–56).[65] The Rheotron laboratory, as it was called, or rather what remained of it, was taken over by the Americans on 14th April 1945, and Richard Gans was released. This leaves Schiebold as the only other possible 'inventor' in Speer's story. It cannot be excluded. But a last minute claim of a 'discovery', after his X-ray device had already been discredited and dismissed in the previous autumn, is unlikely.

Instead, the time coincidence between Speer's story—early days in April—and Touschek's promised release papers, allegedly signed around April 10th according to what Widerøe told him—give credibility to Widerøe's project being the one mentioned in (Speer 1970, 464).

5.8.2   Left for Dead on the Way to Kiel

Bruno was not freed on April 10, as he had been made to hope. Instead, a few days later, a new odissey took place. As the allied troops were approaching Hamburg, Touschek was summarily woken up in the early morning and started on the way to the Kiel concentration camp, together with a group of other 200 prisoners.[66] Luckily for him, and for physics, he did not reach Kiel. Near Longhorn, he fell to the ground, sick and under the weight of his books. A guard pushed Bruno into the ditch and shot him, leaving him for dead. Or, perhaps, the guard missed him on purpose.[67] As recounted in the postwar letters to his parents, Touschek had not been seriously wounded and could walk to a nearby hospital. Confident that his pardon was on the way from Berlin, as Widerøe had assured him, he went to the nearby Langehorn

---

[65] According to Waloschek, Gans had left Berlin with Schmellenmayer through a special permission from the Gestapo.

[66] In (Bonolis and Pancheri 2011, 47), Touschek's letter is translated as to read 'In all 200 of us were deported'. The author is indebted to Klaus Gottstein for pointing out that the prisoners held in Fuhlsbüttel prison were a total of 800 and were divided in groups of 200 prisoners each.

[67] It is rather strange that an SS guard would miss seriously injuring Bruno, when shooting at him twice at short distance. Bruno could have inspired pity or friendship in the guard who shot him.

Hospital to get bandaged. Of course they imprisoned me again and sent me to Hamburg, from one prison to another.[68] But the war was winding down, and the prison was, in Touschek's italian words to Amaldi, "una prigione di pipistrelli".[69]

> All this lasted about 4 weeks. In the meantime, Widerøe had
> gone to Norway and nobody seemed to know where I was. I
> wrote a series of secret messages [to let people know of my
> situation]. Prof. Lenz would smuggle some food into the
> prison for me, and finally my working team at the Ministry
> learned of my existence. A few days before the occupation of
> Hamburg, Hollnack came to get me out of prison, although not
> quite legally. This was timely, since I would have been
> shot, in the best case.[70]

5.9 Why was Touschek Arrested?

The discussion about betatrons and the connection to the proposed secret weapons of the German Reich, presented in the previous chapter, can also suggest why Touschek was allowed, if not downright requested, to work on projects which were classified as secrets, but, then, finally arrested, for espionage, according to what he wrote to his parents. Why espionage? Touschek told Amaldi that it was because his habit to read foreign magazine caught somebody's attention. A different anecdotal reason was given by Touschek to friends in Rome: at a party, Touschek had been listening to a discussion on gyroscopes between some German officers, and, while he was listening, he had started to draw some gyroscopes, thus appearing suspicious, since such devices were used for missile navigation.[71] On the other hand, Widerøe told Amaldi that Touschek was "for no known reason, arrested by the Gestapo."[72] Such statement appears unlikely in view of the fact that Bruno was on the O.T.'s draft list, and that Widerøe's help had been instrumental to get Bruno to avoid the summon. Could he have been protecting Bruno's reputation from a possible charge of collaborating to a war-related project? But Bruno was a first-class *mischling*, and his work on the RLM funded project needed no justification. Widerøe must have known that Jewish

---

[68] Letters sent on June 22nd, 1945, reaching them on October 22nd, and letter on November 17th, 1945.

[69] The expression literally translates as *a prison of bats*, and is typical of Bruno's verbal puns in Italian. He was probably referring to such prison as the one described in Johan Strauss's 'Fledermaus', were people could move in and out. This interpretation was suggested by Rudolf Peierls, while reviewing Amaldi's draft of Bruno's biography, in a 26 March 26th, 1980 letter to Amaldi, after Touschek's death, Amaldi papers, Sapienza University of Rome–Physics Department Archives.

[70] Letter to parents on June 22nd, 1943.

[71] Carlo Bernardini, personal communication.

[72] Widerøe however, is dating such event in November 1944. Widerøe to Amaldi, 10th November 1979, Amaldi papers, Sapienza University of Rome–Physics Department Archives, Box 524, Folder 4, Subfolder 2.

scientists working on classified or war related projects were sent to forced labor when the project was completed, as expected in Gans' case. The most probable explanation to Widerøe's claim of ignorance, is that he preferred to ignore the implications of his own work with nazi Germany during the war, and, mostly, he did not wish to let it become known.

To understand these events in their wider context, let us go back to the often asked question of why Touschek could work on a secret war related project, notwithstanding his being of Jewish origin by mother's side. Certainly the RLM, for whom he ultimately was working, must have known of his Jewish roots. Many concomitant facts suggests such knowledge and tolerance by the authorities, as long as it suited their purposes, as presented in (Waloschek 2012), and confirmed through a correspondence with a contemporary of Touschek, Klaus Gottstein, who at the time was working in Schmellenmeier's team.[73] Since the betatron was (wrongly) considered by the RLM to be or to have the potentiality to become a useful weapon producing *death rays* against enemy airplanes, the project could get the necessary material support in a time of extreme shortages of material and manpower. This is the way in which also Schmellenmeier was able to obtain official support for his rheotron project as a potential weapon and employ Richard Gans who did the theoretical calculations for magnet shape etc. and was irreplaceable in this position. Gans was considered by the Nazis as a full Jew, thus he was scheduled for quite a different work, and to be transported to a concentration camp for elimination as soon as he had finished his calculations. Not unlike what happened with Touschek in 1944 and 1945, Schmellenmeier had to apply every few months for a further extension of Gans exemption from transportation to concentration camp under the pretext of unique calculations that had to be continued for another few months. After the completion of the rheotron Gans would have been arrested immediately. Fortunately, the rheotron was not completed during the war, and Gans survived.

An analogous situation might have existed in the case of Touschek and Widerøe's betatron. He was only a half-Jew, so he was, for the time being, not yet destined for a concentration camp. But as a half-Jew he was certainly not meant to work permanently on secret weapons. In 1944 he should have worked in a mine or dug ditches for the Todt Organization. Only temporarily, as an expert for calculations and measurements that only he could do, he could have been freed for this task from other work. On March 15th, 1945, after the betatron installation at Wrist, his task might have looked completed, at least to the RLM and the Gestapo. Now he could return to the usual fate of half-Jews, i.e., work for the Todt Organization, and from there be deported to concentration camp. To make sure that he did not disappear in the meantime, the simplest way under Nazi conditions was to arrest him, at least until the O.T. destination was established for him. This interpretation coincides with what we know happened to Touschek after his arrest. He was indeed put in prison with other Jews, in the meantime Seifert and Widerøe explained to the Gestapo that "the

---

[73] Klaus Gottstein worked from 1950 to 1970 at the Max Planck Institute of Physics in Munich of which Heisenberg was the director, first as a graduate student and in later years as head of the experimental division.

future of the Reich, for better or worse, depended from a research on the influence of the radiation-damping just started by Mr. Touschek." As he wrote much later to his parents:(Bonolis and Pancheri 2011)

> The first week nothing came from these concessions. I was
> confined without a pencil. Widerøe had put a couple of
> cigarettes in my pockets, but I had no matches. On Friday I
> wanted to hang myself, and on Saturday, Widerøe came. From
> then on the situation got better. I had a 'decent' cell on
> the first floor and Wideröe brought me Heitler's Quantum
> Theory of Radiation and I started research on
> radiation-damping.

5.10   A Note in Invisible Ink and Its Long Life After Fuhlsbüttel Prison

Bruno had been studying Heitler's book for quite some time. The reason for reading Heitler's fundamental volume which had been published in 1936 (Heitler 1984), was certainly that the second edition had just appeared. In Heitler's preface dated Dublin, March 1944, the author explained that war conditions had made impossible to rewrite and amplify it, and that new matter could only be added at the end of the book, in new sections on the cascade theory of showers and on the general theory of damping. This latter issue was fundamental for Touschek, in particular about radiation damping in a betatron. A note with such title (Zur Frage der Strahlendämpfung im Betatron) is actually preserved in the Archives of ETH in Zurich, among Wideröe's documents. The problem was especially related to the design of the 200-MeV machine. The possibility of energy losses due to radiation damping, which would define an upper limit for the energy obtainable with the betatron, were set forth by Iwanenko and Pomeranchuk (Iwanenko and Pomeranchuk 1944), however, as Touschek remarked at the beginning of his note, the "genuine publication which appeared in a Russian periodical was not available according to the war, which made a complete reinvestigation by the author necessary".[74] One can remember that on the day before being arrested in March 1945, Bruno and Wideröe sat on the terrace in Kellinghusen and discussed an article by "two Russians". This paper was one, by Iwanenko and Pomeranchuk,

---

[74] An English version of Touschek's note, "Radiation-damping and the Betatron," is now part of the Air Technical Index collection (ATI No. 71292) consisting of 80,000 documents on microfilm. It was among the thousands of German and Japanese research documents captured by the Allies during and after World War II as part of "Project Lusty," and is available at http://www.dtic.mil/dtic/tr/fulltext/u2/a801166.pdf. The date added by Touschek at the end of the document, before signing it, seems to be July 28, 1945. A typescript with a similar title ("The Effect of Radiation-Damping in the Betatron", with no date but addressed "To the Editor of Physical Review", signed Bruno Touschek, Göttingen, Zeppelinstrasse 2, is kept in Touschek's archive in Rome, Sapienza University, Physics Department, Box 4, Folder 15. Other theoretical notes stored in Wideröe's archive, in Zurich, all dated 1945, are: Zur Theorie des Strahlentransformators, On the Starting of Electrons in the Betatron, Die magnetische Linsenstrasse und ihre Anwendung auf den Strahlen-Transformator.

about which Bruno would later write a note, in July 1945.[75] This is how the young Bruno, while in prison, could take his mind off his present conditions and what could happen if pardon would not arrive in time. He knew that concentration camp and death could be waiting for him.

Radiation damping is related to the well known classical phenomenon of the emission of electromagnetic radiation when a charged article is accelerated. This problem which he started to study during his prison days, became central to Bruno's research during the planning for the construction of the large electron-positron col-lider ADONE. In 1964, after Bruno and the French-Italian team had demonstrated with AdA the feasibility of matter-antimatter storage rings, the construction of a bigger and more powerful ring started in Frascati National Laboratory, in Italy. At ADONE's energy, radiation from the accelerated particle could be so copious as to cloud the experimental results. He thus embarked on a program of what he called "the administration of the radiative corrections to electron-positron experiments". Thus, in 1965, a group of young scientists would be created, which carried on his teaching and ideas, until this day.

---

[75] Iwanenko's work and, in particular, the paper which came to Bruno's attention, were remembered in (Sardanashvily 2016).

Chapter 6
Bruno Touschek in Germany After the War: 1945–46

I want to become a physicist
Ich will ein Physiker werden
Bruno Touschek, Letter to parents, May 9th, 1946

Abstract Through the second half of 1945, in the months immediately following the end of the war, confusion and danger ran in parallel to plans for rebuilding and reconciliation. In 1946, everywhere in Europe, the reconstruction of science started. In this context, Bruno's unique betatron experience made him valuable to the occupying Anglo American force, which questioned him and planned for his future. In Germany, the reconstruction was centered in the University of Göttingen and took place with the return of the German nuclear scientists, held prisoners in UK at Farm Hall, since July 1945. Bruno was moved to Göttingen, where he obtained his Diploma in physics and became Werner Heisenberg's assistant. In the meanwhile, the decisions of the Allied powers, towards restructuring science and technology in the UK after the war effort, determined Touschek's move to the University of Glasgow. In the relatively unexplored period of his life from summer 1945 to the end of 1946, Bruno started his long way to become a theoretical physicist.

On May 8th, 1945, with the official surrender of Germany, Second World War II came to its end in Europe.[1] The immense bloodshed and destruction that had overcome Europe were over.

---

[1] The official date for the end of the war is different from country to country. In Italy for instance, the *Giorno della Liberazione*, the day of Liberation of Italy, is celebrated on April 25, which is the day the freedom fighters, *i partigiani* in Italian, entered Milan, whereas in Paris *la Libération de Paris* falls on 25 August 1944, which is the day the German command in France surrendered. In Asia, the war ended only after the second atomic bomb was dropped on Nagasaki, with Japanese forces surrendering on August 15th 1945. In Germany, the German surrender was staged twice, first in Reims, France, on May 7th, then, in Berlin on May 8th, at the request of the USSR, and signed on May 9th. The cease-fire was officially placed on May 8th.

© The Author(s), under exclusive license to Springer Nature Switzerland AG 2022   137
G. Pancheri, *Bruno Touschek's Extraordinary Journey*, Springer Biographies,
https://doi.org/10.1007/978-3-031-03826-6_6

Amid the million Europeans starting on a new road to peace and collaboration, there are the early protagonists of the story of electron-positron colliders, the Austrian Bruno Touschek and the Norwegian Rolf Widerøe. Bruno and Rolf had come together in 1943, during the darkest times of Second World War II and worked for two years on the 15-MeV German betatron, commissioned to Widerøe by the Reichsluftfahrtministerium (Reich Air Ministry) for alleged war purposes (Amaldi 1981; Waloschek 1994; Bonolis and Pancheri 2011).

Before, during and after the war, many pathways criss-crossed Europe to ultimately lead to the construction of AdA, the first ever electron-positron collider, built in Italy in 1960. When Bruno and Rolf had met, two of these pathways had come together, one from Norway, the other from Austria and Germany. Then, after the war, the destinies of the two scientists took different ways. Between March and April 1945, as described in Widerøe (1984), Brustad (1998), and more in depth in Sørheim (2015, 2020), Rolf Widerøe returned to Norway, where, in May, shortly after the German surrender, he was arrested and accused of having worked on the development of V2 rockets.[2] He wrote an extensive report on his work on the betatron construction in Hamburg and was released in July, but only in February 1946 it was clarified that his work had not been of any military value to Nazi Germany. However, he was burdened with financial penalties and was eventually allowed to move to Switzerland, where he took up a leading position at Brown Boveri & Co and applied his knowledge of accelerator science to medical developments.

As for Touschek, his mind and heart would soon be taken by regaining the lost years and finishing the studies he had started at the University of Rome in Spring 1939, continued at University of Vienna, and which had been interrupted by his expulsion in 1940. In Hamburg and Berlin, from 1942 until 1945, he had attended physics lectures, leading the life of a student without being one, hoping to prepare a secret dissertation, and, mostly, trying to avoid the omnipresent Gestapo and be drafted to forced labor. His escape from death and concentration camp in the last days of the war appear almost miraculous, in view of what was happening around him. A vivid illustration of Bruno's escape from death in the final days of the war, fully reconstructed through the letters he sent home, is found in Fig. 6.1. He was one of the few who survived and, the war over, he could try to fullfil his dream to become a physicist. His year of studies at the University of Göttingen, (Bonolis and Pancheri 2019), became the first step in this direction.

On April 30th, Theodor Hollnack, the administrator of Widerøe's betatron project, eventually had come to free Touschek.[3] Those were the final days of the war in

---

[2] See letter from Widerøe to Ernst Sommerfeld from Baden, dated April 12, 1946. Deutsches Museum Archive, NL 089, box 014: "My question whirled up naturally a lot of dust and quite fantastic things were hypotesized (for example, I was supposed to have invented the V2)" (Meine Sache wirbelte natürlich sehr viel Staub auf und man vermutete ganz phantastische Sachen (beispielweise sollte ich angeblich die V2 erfunden haben)).

[3] In a letter to his friend Ernst Sommerfeld describing all this, Touschek wrote that he was angry because Hollnack had waited too much: "He then explained to me—after having done nothing for three weeks—that without him I would have definitely been shot." Deutsches Museum Archive, NL 089, box 014.

Fig. 6.1 At left, a contemporary photo of the entrance to the Fuhlsbüttel prison in Hamburg where Touschek was held for about 6 weeks in Spring 1945, from http://www.memorialmuseums.org/eng/denkmaeler/view/280/Fuhlsbttel-Concentration-Camp-and-Prison-Memorial-1933-1945. At right, a map of the forced march from Fuhlsbüttel to Kiel, which brought the 200 prisoners from Hamburg to Kiel, between 12–14 April 1945, from Fentsahm (2004), http://www.akens.org/akens/texte/info/44/Todesmarsch_Fentsahm.pdf. At the bottom, within the Hamburg region, one can see Langenhorn, where Touschek was shot and fell to the ground, while the column continued towards Kiel without him. The crosses indicate prisoners' deaths

Europe, indeed the last hours, during which prisoners risked being killed, often to prevent witnesses from surviving. Touschek was lucky, or perhaps, and more likely, the tight grip held by the Nazis was at its end. Two days earlier, the British army had started the final assault on the city of Hamburg, where the German command was holding against the Allied invasion, and the fight moved from block to block through the city. The city surrendered on May 3rd (Fig. 6.1 left).[4]

6.1 German Science and the Mission of the T-Force

The last year of the war saw not only the heavy bombing of German cities and installations, but also planning for the future of the Western world, as it came to be called. The position of eminence of Germany in science and all fields of technology in Europe had been such that, as the various Allied armies progressed through Germany,

---

[4] An eerily silent footage about the entrance of the Allied troops in Hamburg can be found in https://www.youtube.com/watch?v=en3hkuc1QoM. See also details about the Capture of Hamburg in 1945.

Fig. 6.2 At left, an image of the battle for Hamburg, May 1945, from https://en.wikipedia.org/wiki/Capture_of_Hamburg. At right, Farm Hall in Godmanchester, England, where the Uranium scientists were held *incommunicado* for 6 months, from https://en.wikipedia.org/wiki/Operation_Epsilon#/media/File:FarmHallLarge.jpg

they raced to secure what would be the most prized booty, depending on the stage of scientific and technological advancement of the different countries. What the Germans had achieved in science and technology since the late 1930s, would be important to know and to acquire in view of the new world political assessment after the war. To this end, the Western Alliance set in motion a number of different operations which would lead to the capture of a vast amount of German industrial, scientific and technical equipment as well as of the most prominent German scientists, who were quickly transported to the United States and to England (Fig. 6.2).[5]

All along, even before the final surrender of Germany, a special task force under joint American-British command, named the T-Force, had been scouting Germany for its industrial and scientific resources, racing to reach Germany's top scientists before the arrival of the Russian Army (Bernstein 2001; Longden 2009). One of the key actions of T-Force units was the Allied Scientific Intelligence Mission, code-named "Alsos", brain-child of Colonel Leslie Groves, the military head of the Manhattan Project.[6] The Alsos Mission, headed by U.S. Army Lieutenant Colonel Boris T. Pash, was set up to seize key elements of the German nuclear energy project working at Hechingen, in southwest Germany, where the Kaiser Wilhelm Institute for Physics had been evacuated from Berlin-Dahlem (Pash 1969; Goudsmit 1996; Cassidy 2017). Other actions concerned the rocket scientists, as well as biological

---

[5] For a good journalistic overview, see https://www.theguardian.com/science/2007/aug/29/sciencenews.secondworldwar.

[6] "Alsos" is also the Greek word for 'grove'.

and chemical warfare experts (Jacobsen 2014).[7] And, as we shall see in Sect. 6.1.2, particle accelerators as well would be a target of interest.

6.1.1   Operation Epsilon

Operation Epsilon was the codename of the program by which the main protagonists of the German nuclear program—the Uranium Club—were flown to England at the beginning of July 1945 and held in secrecy at the country estate Farm Hall, near Cambridge. A primary aim of the program was to understand how close Nazi Germany had been to building a nuclear bomb by listening to their conversations through hidden microphones (Bernstein 2001; Cassidy 2017; McPartland 2013).[8]

Among the Farm Hall detainees, there was Werner Heisenberg, one of the founders of Quantum Mechanics and one of the most illustrious German scientists, a key theoretical figure and head of the German nuclear project, and, since 1942, official director of the Kaiser Wilhelm Institute for Physics in Berlin-Dahlem (Cassidy 1993).[9] He was taken into custody by Alsos on May 4th, in the little village of Urfeld, where he had arrived after a desperate bicycle ride across war-torn southern Germany to reach his family, as soon as news of the French advance had reached Hechingen, Fig. 6.3. In Heisenberg's own words (Heisenberg 1971, 190–192): "On May 4th, when Colonel Pash, leading a small US detachment, came to take me prisoner, I felt like an utterly exhausted swimmer setting foot on firm land."

---

[7] For the Americans a major target became the rocket scientists, foremost among them Wernher von Braun, who had been in charge of the Nazi V-2 program in Peenemunde, and would later bring the US to land on the Moon.

[8] The 10 leading German nuclear scientists brought to Farm Hall were Erich Bagge, Kurt Diebner, Horst Korsching, Walter Gerlach, Otto Hahn, Paul Harteck, Werner Heisenberg, Max von Laue, Carl F. von Weizsäcker, and Karl Wirtz. The American officers in command wanted to keep them under constant guard, as prisoners, but the British captain in charge explained there would be no need for this, if the scientists could be convinced to give their word of honour not to escape. The transcript of these conversations were held classified until 1992, when they were released following a request addressed to the House of Lords by the President of the Royal Society in London. In the letter, to whose draft contributed Rudolf Peierls, the following 1985 words by the German President Richard von Weizsäcker, are quoted: "We need and we have the strength to look truth straight in the eye without embellishment and without distortions […] anyone who closes his eyes to the past is blind to the present". The British version of the transcripts is now available for free download at http://discovery.nationalarchives.gov.uk/SearchUI/Details?uri=C4414534. For the US copy, see NARA, RG 77.11.1 (Office of the Commanding General), entry 22, Box 163, in Cassidy (2017). A German translation appeared in 1993, as *Operation Epsilon: Die Farm-Hall-Protokolle oder Die Angst der Alliierten vor der deutschen Atombombe,* Wilfried Sczepan, trans. (Berlin: Rowohlt, 1993), Hoffmann, Dieter, ed., but without Heisenberg's important speech on bomb construction on August 14, 1945, https://history.aip.org/web-exhibits/heisenberg/farm-hall.html. An Italian translation was published in 1994 (Frank 1994).

[9] For an extensive bibliography of Heisenberg's work, see http://www.netlib.org/bibnet/authors/h/heisenberg-werner.html. For the collected works see as well (Heisenberg 1984).

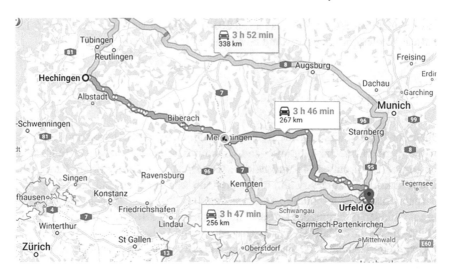

Fig. 6.3 A contemporary map showing the long trip Heisenberg took to reach his family, as the Allied forces were progressing through Germany, from Hechingen to Urfeld, where he was taken in charge by the ALSOS mission

The imprisonment of the German 'atomic scientists'[10] marked the 'zero hour' of the post-war future of German science. During their seclusion in Farm Hall which lasted 6 months, the Western Alliance debated and planned the reconstruction of Europe. In particular, a major interest to the British political establishment was the reconstruction of Germany on its industrial and technological aspects, all of which had to start with rebuilding a strong scientific terrain.

6.1.2   The T-Force and Widerøe's Betatron

Widerøe's betatron was also one of the targets of the T-Force. Particle accelerators had now morphed from the planning and invention stage into the most prominent research tool in atomic and nuclear physics, and would become of future strategic interest, in the mind of politicians and the military. As the war started, a major advancement in the field had taken place in the United States, with the successful operation of a betatron, announced by Kerst in 1941 (Kerst 1940, 1941; Kerst and Serber 1941). This series of articles revived the interest in betatrons, leading to the construction of Widerøe's 15 MeV betatron—the first at this energy in Germany—and a parallel proposal for a much more powerful 100-MeV machine which Widerøe had envisaged,

---

[10] What is now called *nuclear physics*, and which includes studies for nuclear energy uses, both civilian and military, was then called *atomic physics*, hence the still used term *atomic bomb* and, in the context of German scientists at Farm Hall, *atomic scientists*.

but was never built.[11] It is therefore quite understandable that, as the war ended, the German knowledge in accelerator science became of possible interest to the Allied nations, and in particular to the British, less so to the Americans, whose expertise and dominance were not lacking in this field. The German work on betatrons, which had been going on through the war, became part of the British war spoils (Hall 2019).

After Hollnack freed Touschek from prison, they went back to Kellinghusen, near Wrist, where, in mid March, a time which now seemed like centuries ago, Touschek and Rolf Widerøe had brought the 15-MeV betatron. In Kellinghusen, Hollnack had immediately put himself at the disposal of the British authorities, and reorganized activities around the betatron creating a small enterprise called the MegaVolt Forschung Laboratorium, MV-Research Association (MVRA), which gathered all the key members of the betatron group—previously working under the guidance of Widerøe (the Megavolt Versuchsanstalt)—and others.[12] Hollnack asked Touschek to join in. Everybody was trying to survive and, for some, as in Hollnack's case, even to strive. At that moment Touschek had no alternative and accepted the offer; he had been freed by Hollnack, thus avoiding being killed in the last days of fights around the city. While still trying to recover from the painful memories of his losses and the trauma of imprisonment, Bruno immediately made plans for a doctorate, as he told his parents in his first letters written in June, where he recounted the whole story of his arrest. In the meantime, thanks to his knowledge of English,[13] he acted as interpreter and was then able to negotiate with the T-Force and have the MVRA protected by the British troops against looting and other killings, commonly occurring in the first months after the war.

However, he was soon out of empathy with the group. He did not like Hollnack, nor 'his grandiose ambitions', and was also eager to return to the theoretical physics studies he had started during his university time in Vienna, and which had been interrupted by his expulsion. Bruno's dream of becoming a physicist had been reinforced by the correspondence and friendship with Sommerfeld which had followed, and the lectures by prominent scientists he had attended during the war, at Hamburg and Berlin Universities.

By end of June Touschek asked his colleagues in the MVRA to put an end to his collaboration. They agreed that he would have a three-months leave and in late August, as he wrote in a letter to Arnold Sommerfeld dated 28 September, he went to visit some of the scientists whose lectures he had (unofficially) attended during the war: Hans D. Jensen—one of the members of the Uranium Club, now in Hannover—and Hans Süss in Göttingen, who had participated in German nuclear research activities during the war, and whom he knew since his Hamburg days, as well.[14] In Göttingen, Touschek also saw Friederich Georg Houtermans, or Fritz, or

---

[11] For a T-force report on history of betatron development, see B.I.O.S. Report n. 77 in http://www.cdvandt.org/fiat-cios-bios.htm.

[12] November 17th, 1945, letter by Bruno Touschek to his parents in Bonolis and Pancheri (2011).

[13] Touschek had learned English when in Rome, in Spring 1939, when he was applying for a Visa to go to England and study at the University of Manchester.

[14] At that time, Jensen and Süss collaborated on the nuclear shell model.

Fissel, as he was also known, with whom he would remain friends until Houtermans' death in 1966.[15]

Touschek's principal worry now was to formally complete his studies, first by obtaining a degree in Physics, namely the title of Diplom-Physiker, and then continue on with a doctoral thesis. During the visit, Jensen promised Touschek he would arrange for a PhD work and a position as assistant and he also received a similar offer from Hamburg, where Wilhelm Lenz, director of the Institute of Theoretical Physics, who had always protected Bruno during the war, could now openly support him to complete his university studies and continue them towards a doctoral degree.

A way to proceed was in sight, and, returning to Kellinghusen, Touschek was now eager to start writing a dissertation on the betatron. However, this could not happen. As he would write to his parents on November 17th, 1945 (Bonolis and Pancheri 2011):

```
A meeting with the T-Force has decided that things should remain a
state secret so that its use for a thesis is out of the question. I
will be able to leave
Kellinghusen only after an Allied Commission has decided in regard
to the betatron.
```

Writing to Sommerfeld in September, Touschek says that he felt like a T-Force prisoner.[16] Indeed, he was. In this second half of the year 1945, the Allies were making a thorough survey of the scientific achievements of German science and technology, and nothing could really start in Germany until the decisions had been taken as to Germany's future. Not unlike the members of the *Uranverein* (the Uranium Club), who were held in England in Farm Hall, so was Touschek forced to remain in Germany. Unlike them, however, he was free to move within the British zone, still he could not go to Austria or publish anything about the betatron. In the meantime he continued his work on different theoretical topics related to the betatron, in particular on radiation damping, but also on neutrino theory.

In October, following British-American careful investigations held on various German Science and Industrial Institutions, officers from the British Intelligence Objective Sub-Committee (B.I.O.S.) visited the C.H.F. Müller factory in Hamburg, where the 15-MeV betatron had been built, and where Touschek had worked with

---

[15] Fritz Houtermans had arrived in Göttingen in spring 1945, after a tortuous trajectory of persecution by both Nazis and Communists. Fritz Houterman's life was the subject of various books, in particular of an unpublished manuscript on which Edoardo Amaldi was working before his sudden death in 1989. In 2010, the manuscript was donated by Amaldi's family to the Bern University Laboratory for High Energy Physics. It was then edited by S. Braccini, S. Ereditato and P. Scampoli, three researchers from the University of Bern, in recognition of Houtermans' contribution to the development of particle physics in Bern, and in Switzerland (Amaldi 2012). In the Preamble to the unfinished book, Amaldi writes : 'When in 1937 my friend George Placzek arrived in Rome from U.S.S.R, he had mentioned Houtermans as one of the young physicists gone to Karkhov to participate in the construction of a socialist society and recently in serious political troubles. I received a letter from him, from Berlin in 1942, after, as I learned late, he had succeeded in getting out of a prison to which he had been transferred from the Lubyanka in Moscow.'

[16] Touschek to A. Sommerfeld, 28 September 1945 from Kellinghusen, Deutsches Museum Archive, Arnold Sommerfeld papers, folder NL 089,013.

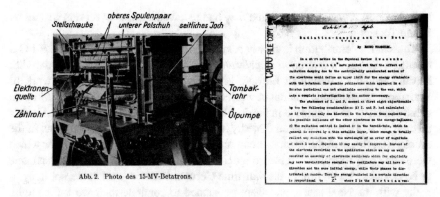

Abb. 2.   Photo des 15-MV-Betatrons.

Fig. 6.4 At left, the 15-MeV betatron mounted on a big table as shown in Kollath and Schumann (1947), 635. At right, the first page of Touschek's 1945 report on Radiation damping in a betatron, unpublished report for the US Armed Forces, part of the Air Technical Index [ATI] collection available at https://apps.dtic.mil/dtic/tr/fulltext/u2/a801166.pdf. Signed by Bruno Touschek at the end of the note, the handwritten date reads as to be 28.9.45

Widerøe and his group. A photo of the betatron, from a postwar publication by Kollath and Schumann, two members of Widerøe's group, is shown on the left panel of Fig. 6.4.[17] The right panel shows the first page of a report on radiation damping in the betatron, where we see what would become Touschek's lifelong interest in the question of how radiation from a moving charge affects the operation of electron accelerators. This report is likely to include the work Touschek was working on during his imprisonment, and which Amaldi mentions as having been written in invisible ink (Amaldi 1981, 5) on Heitler's book on the quantum theory of radiation (Heitler 1984).[18]

The investigation continued with a visit to Wrist, the town near Kellinghusen, where the betatron had been kept since Touschek and Widerøe had brought it in March 1945. The British officers were in the Wrist Laboratory, on October 23rd.[19]

---

[17] In B.I.O.S. Final Report No. 201, Item No. 1,7, 21, dated 8.10.1945, "Visit to C.H.C. Müller, A.G. Röntgenstrasse 24, Bahrenfeld, Hamburg, reported by C.G. LLoyd and G. J. Thiessen, http://www. cdvandt.org/BIOS-201.pdf, on p. 3 it was further specified that "Dr. Fehr [assistant to Manager] stated that the project had been experimented for the Luftwaffe with the hope (?) of obtaining a death ray for anti-aircraft work." These reports covered a wide variety of German scientific and industrial Institutions, and were authored by officers from B.I.O.S., C.I.O.S. (Combined Intelligence Objectives Sub-Committee) and F.I.A.T (Field Information Agency Technical, United States Group Control for Germany).

[18] A possible version of this work, entitled *The effect of Radiation-Damping and the Betatron*, undated but bearing an address in Göttingen—and apparently submitted to *The Physical Review* (according to the first line of the document) where it was never published—is preserved in Bruno Touschek Papers, Sapienza University of Rome–Physics Department Archives.

[19] A.T. Starr, K.J.R. Wilkinson, J.D. Craggs, L.W. Mussel, "German Betatrons", BIOS, Final Report No. 148, Item No. 1, dated 24.10.1945.

In their report it was mentioned that they had received a series of reports written by Bruno Touschek.[20]

Touschek's contribution is clearly acknowledged also in another detailed B.I.O.S report on *European Induction Accelerators* prepared in October 1945 by the U.S. Naval Technical Mission in Europe: "In collaboration with the design work of Widerøe, a considerable amount of work was carried out by Touschek. This is known to have been of invaluable aid in the development of the 15-MeV accelerator. Further theoretical work has also been done by Touschek in the starting of electrons in the accelerator. Some of the work is along the lines initiated by Kerst and Serber which were known to Touschek." It is also specified that "Widerøe and the group that came to be associated with him in the war-time German betatron work were not in sympathy with the Nazi-cause, and were persuaded to continue their work for purely scientific considerations."[21]

The activities of the Megavolt Research Association in Wrist are examined in detail in the second part of report No. 148. The British investigators specified in particular that the experimental work at Wrist should close down at once and that the complete apparatus should be sent to UK. They finally recommended that Touschek be taken to UK for work on theoretical physics. This recommendation sheds light on all of Touschek's whereabouts in the year to follow. As we shall see, starting in the early months of 1946, Touschek would go back and forth between Germany and the UK, Göttingen and London, or Glasgow, until he would finally settle in Glasgow, in spring 1947, remaining there for 5 years.

In November, writing to his parents, Touschek was still hoping he would soon be allowed to visit his family in Vienna, in the Soviet occupied zone. But this could not happen yet. The T-force had other plans for him.

In December,[22] working on his research on radiation damping in the Wrist office, from 10 in the morning until 10 in the evening in the only warm room in the area, the uncertainty of the situation was becoming unbearable. The Western Alliance were making preparations for the reconstruction of Europe, but in the meanwhile, living conditions were dramatic. Winter was coming, there was scarcity of food, hardly any winter clothing, heating was a luxury.

As the year was coming to an end, it was clear that the situation could not go on forever. Exacerbated, Bruno wrote to the officers of the T-Force, but no immediate answer came, as the decisions had yet to be taken. However, a rumor, eventually originated by the T-Force, gave him the hope that he would be brought to England. The probability could be low, but the prospect made the situation more tolerable.

---

[20] Copies of such reports on the theory of the betatron written by Touschek (Zur Theorie d Strahlentransformators. Typoskr.-Kopie, o.D. 10 Bl.+ Beil.; On the Starting of Electrons in the Betatron. Kopie, o.D. 11 Bl.; Die magnetische Linsenstrasse und ihre Anwendung auf den Strahlen-Transformator. Typoskr.kopie, 10 Bl., 1945; Zur Frage der Strahlungsdëpfung im Betatron. Typoskr.- kopie, 7 Bl., 1945) can be found in Rolf Widerøe's papers at the Eidgenossischen technischen Hochscule (ETH) in Zurich (see finding aids at https://www.research-collection.ethz. ch/bitstream/handle/20.500.11850/140811/eth-22301-01.pdf).

[21] B.I.O.S MISC.77, p. 6.

[22] Letter to his parents, December 13th, 1945 from Kellinghusen.

Europe, at that time, pillaged of its scientists and infrastructures, was not appealing for his future as a physicist, and, financially, England would likely be a much better prospect, given that he wanted to help his parents, his father being retired from the Austrian Army, and living under difficult conditions in Vienna, under Soviet occupation. Another reason, the main one perhaps, was that going to England would make a plan he had envisioned before the war come true. In spring 1939, after the Anschluss of Austria shattered the regular course of his life and studies, while in Rome visiting aunt Ada, he had decided to go and study in England. Actual plans had been drafted and he had applied for a Visa to the British consulate in Rome, but these plans never went through.

During these last few months of 1945, the reconstruction of German science was being discussed and planned by the occupying forces: how, how much, where, and under whose direction, these were the questions to pose and solve. Finally, it was decided that the University town of Göttingen, which was in the British zone and had been relatively untouched by Allied bombing, would be the main center for rebuilding German science. This decision had a wide impact: the Farm Hall detainees could be allowed to return home, and Werner Heisenberg would be one of the key figures in the revival of scientific research in Western Germany, and especially in German science policy. Once this path was clear, also other decisions came along and the restrictions imposed by the T-Force on the betatron group were lifted.

In December 1945, work with Wideroøe's 15-MeV betatron had been completed at Wrist and the machine was transported to the Woolwich Arsenal near London, where it was used for some time with the help of Rudolph Kollath.[23] As for Bruno, he could be free to leave Kellinghusen and go back to his studies, the first thing being to obtain his diploma in physics.

Bruno had to now start his life anew. As a 17 year old youth he had gone through the Anschluss, and then the loss of his identity as a rightful Viennese citizen and promising physics student, who had been following courses during Spring 1939 in the University of Rome and, in 1939–1940, at the University of Vienna. He had been living in Germany through four years of semi-hiding, with little food, scarcely any heat, both in Berlin and Hamburg with daily bombs devastating the cities, away from his beloved family, the grandmother, his father, his stepmother, his aunts, and his friends. At some time, he had learnt that his grandmother Josefine Weltmann had never returned from Theresienstadt, the concentration camp circa 60 km from

---

[23] Rudolf Kollath wrote a five-pages long report on their results ("Bericht von Ing. R. Kollath, 11.12.1945, über die Arbeiten am Betatron in Wrist", see copy in Wideroøe's archive in Zurich). Kollath and Schumann, who had operated the betatron up to the end of 1945, wrote together an extensive report on the performance of the betatron and on tests in Wrist which was published only about two years later (Kollath and Schumann 1947). A detailed outline of the 15-MeV betatron and related work carried out by the group, including studies for a large 200-MeV betatron, were reviewed by Herman F. Kaiser in early 1947, also specifying different aspects of Touschek's involvement as a theorist (Kaiser 1947).

Prague.[24] The world of his youth was definitely over. He now had to go on with life. How? Not unlike many scientists in those days, he could do this only by fulfilling his dreams. For Bruno, they were the ones he had pursued through his correspondence with Arnold Sommerfeld in 1941 and which had prompted him to move to Germany in 1942. He had dreamed of studying and becoming a physicist. This is what he was now anxious to do and was the only way he could overcome the grief for the lost past. As we shall see, he was not yet free to decide his destiny, and had only a partial notion of which decisions were taken about the rebuilding of universities in Germany. Likewise, whether he could eventually end up studying in the UK was also rather nebulous. As things unraveled, his first return to normality was to be at the University of Göttingen.

In the section to follow we shall temporarily leave the story of Bruno Touschek, and give a brief overview of what had been happening in Germany and what Touschek found when he joined the University, in spring 1946.

6.2 From Destruction to Reconstruction: Starting Anew in Göttingen

Touschek would not be alone in rebuilding his hopes and dreams. As 1946 started, all around him the titanic effort of the reconstruction of Europe was already taking place, coordinated by the American military, with the UK command on its side.

During the immediate post-war years Germany was facing devastation, poverty, enormous loss of lives, and the collapse of economic and political organization with the country divided into four occupation zones. Among the many challenges, the reconstruction of research in Germany was perhaps the most difficult, as German science had to restart itself practically from the ground up and, at the same time, needed to be reintegrated into the international community. Moreover, throughout the war years, scientific developments had taken a leap into the future and nuclear science was now of political interest, being strongly intertwined with the re-shaping of socio-economic relations in the wider context of Cold War relations between the United States, the Soviet Union, and the countries of war-ravaged Western Europe.

The rebuilding of science in postwar western Germany—and German political revival—is to be framed within the broad contexts of the Allied occupation (Cassidy 1994, 1996; Berghahn 1996; Ash 1996; Gimbel 1990b; Krige 2006). Allied restrictions specifically forbade applied nuclear physics and in particular also commercial production of betatrons, synchrotrons and all particle accelerators over 1 MeV, including many sorts of equipment. The conditions were perhaps most favorable in the British Zone, where the authorities, especially the liaison officer Colonel Bertie Blount, appeared quite open to a dialogue with German scientists. In the British

---

[24] Josefine Weltmann, was deported on 27 August, 1942. According to wikipedia, the first German and Austrian Jews arrived in Theresienstadt in June 1942. At the time of deportation, she was living in Alxingergasse 97, Wien 10. From Shoah database.

Fig. 6.5 A 1944 image of the Auditorium of Göttingen University, courtesy and © SUB GöttingenUniversity /GDZ

zone, the city of Göttingen had survived World War II without major damage, which meant an invaluable starting advantage for the town and the famous university, the oldest in Germany, Fig. 6.5.[25]

With the permission and the encouragement of the British, Göttingen grew into one of the main scientific centers of the Western occupation zones.

The Georgia Augusta was the first German university to resume teaching already in September 1945. In the early 1930s, the University included Institutes for experimental and theoretical physics. The Institute in Experimental Physics was directed by James Franck, the one in Theoretical Physics by Max Born. Both had arrived to Göttingen as Professors in 1921, and, in due time, both were to win the Nobel Prize.[26] In 1933 the National Socialists' rise to power and the contempt for modern "Jewish" physics, which included Quantum mechanics and Einstein's theory of relativity,

---

[25] Named after its founder, George II, King of Great Britain and Elector of Hanover, the Georgia-Augusta University of Göttingen was founded in 1734 with starting classes in 1737.

[26] The Nobel Prize in Physics 1925 was jointly awarded to James Franck and Gustav Ludwig Hertz 'for their discovery of the laws governing the impact of an electron upon an atom'. Born was awarded the 1954 Nobel Prize in Physics 'for his fundamental research in quantum mechanics, especially for his statistical interpretation of the wavefunction', sharing it with Walther Bothe, as from https://www.nobelprize.org/prizes/physics/1954/born/facts/.

forced them to emigrate. James Franck, by then a Nobel laureate, went to the United States. Max Born, one of the founders of quantum mechanics, went to Italy, and then to Cambridge. During his time in Cambridge, he received a letter from Raman, inviting him at the Indian Institute for Science (IISc).[27] After a six month stay in Bangalore, Born moved to Scotland at the University of Edinburgh, where he remained until his retirement in 1953. Born's chair at the Institute for Theoretical Physics in Göttingen was then occupied in 1936 by Richard Becker, who had been transferred from Berlin to Göttingen by order of the Reich Ministry for Education (Hentschel and Rammer 2001). Becker and Born had an influence on Touschek's development as a theoretical physicist, as Becker was Touschek's professor when Bruno studied in Göttingen in 1946. As for Born, as we shall see in Chap. 7, Touschek first met him in Edinburgh, in May 1947, while in Glasgow as a doctoral student. Bruno became a regular attendee of Born's weekly seminars, and, later on, prepared the Appendix on the theory of neutrinos in Born's new edition of his famous *Atomic Physics*.[28]

After the war, the First Institute for Experimental Physics was headed by Robert Wichard Pohl, whose research constituted one of the foundations of solid-state physics. The direction of the Second Physics Institute between 1942 and 1953 was in the hands of a former student of James Franck, Hans Kopfermann (Weisskopf 1964), who initiated and supported nuclear physics in Germany together with his assistant Wolfgang Paul.[29] During the war, Kopfermann and Paul learned of Kerst's success in constructing and operating the first betatron, and decided to build such an accelerator as soon as possible. The project was put aside as they heard that Konrad Gund had built a 6-MeV betatron at Siemens-Reiniger Company in Erlangen (Waloschek 2012) and Paul started taking measurements on the machine in Erlangen. After the war, Paul and Kopfermann, with the help of Ronald Fraser, Scientific Advisor of the Research Branch of the British military, were able to transfer this betatron to Göttingen. Together with Becker, Kopfermann was Touschek's advisor for his 1946 Physik-Diploma dissertation about the betatron, from the University of Göttingen (Amaldi 1981, 7), Fig. 6.6.

In early October 1945, while the University of Göttingen was resuming academic activities, Heisenberg, Hahn and von Laue, still held in Britain, had met their British colleagues at the Royal Society in London to discuss the rebuilding of German

---

[27] Chandrasekhara Venkata Raman was awarded the Nobel Prize in Physics 1930 'for his work on the scattering of light and for the discovery of the effect named after him', an effect discovered with the help of K. S. Krishnan. About Born's visit, see https://connect.iisc.ac.in/2019/12/when-raman-brought-born-to-bangalore/. See also My Life: Recollections of a Nobel Laureate by Max Born, Routledge, https://doi.org/10.4324/9781315779379. In Bangalore Born gave a lecture entitled "The Mysterious Number 137", the inverse of the fine structure constant, the number of Touschek's hospital room in Geneva, before being moved to Austria, where he passed away in 1978. As Touschek would later tell Ugo Amaldi, it was also the number of the room where Pauli died, in 1958 (Amaldi 2004).

[28] This book had several English editions starting from 1935, the last one in 1969.

[29] W. Paul shared with Hans G. Dehmelt one half of the 1989 Nobel Prize in physics "for the development of the ion trap technique", the other half was awarded to Norman F. Ramsey "for the invention of the separated oscillatory fields method and its use in the hydrogen maser and other atomic clocks".

Fig. 6.6 Richard Becker, at left, and Hans Kopfermann in 1949 portraits, courtesy and © SUB GöttingenUniversity /GDZ

science.[30] On 3 January 1946, the ten German nuclear scientists, among them by now three Nobel Prize laureates, were finally released from Farm Hall and brought back to Western Germany by Colonel Bertie Blount, who had studied in Germany.[31] With the British officials Bertie Blount and Ronald G. J. Fraser, himself a physicist, the group of German physicists started to forge working relationships. It was the beginning of a long collaboration which had a great importance for the future of the Federal Republic.

---

[30] In this meeting, two of the German physicists had been Nobel prize winners, Max von Laue in 1914 and Werner Heisenberg in 1932. The third, Otto Hahn, would be awarded the prize a few months later, while still under imprisonment at Farm Hall. On the British side, the meeting included Patrick Blackett, who would be later awarded the 1948 Nobel Prize in physics, for his work on cosmic rays. In early September, Blackett and Heisenberg had held a long conversation in Farm Hall. This encounter had then been followed by a letter addressed to Blackett by Heisenberg, on behalf of the other scientists. In this letter, the conversations and the position of Heisenberg and the other Farm Hall detainees was summarized (Bernstein 2001).

[31] At his arrival, on January 3, 1946, Heisenberg immediately wrote to his wife Elisabeth from the small village Alswede: "My dear Li! This is the first evening back in Germany since the end of the war. This long time of captivity seemed to us only bearable through the scientific work. How it's going to be here, we do not know yet. The purpose of our being here is as follows: The highest authorities have decided that we all should in the future have our workplaces in the British occupation zone." (Heisenberg and Heisenberg 2016).

On January 12 Hahn and Heisenberg, accompanied by Col. Blount, visited Göttingen, where they found Max Planck, who had arrived there as a refugee seeking shelter with relatives.[32] Since 1930, Planck was President of the Kaiser Wilhelm Society, a non-university science organization founded in 1911 to conduct specialized basic research in its own Institutes, predominantly in the natural sciences. The Kaiser Wilhelm Society (Kaiser-Wilhelm-Gesellschaft) had quickly established itself nationally and internationally thanks to its outstanding scientific achievements, but during the 1930s–1940s, the Society's leadership and many of its scientists had become supporters of Hitler's regime, or had been involved in armament research. The Allies were thus urging that the Society should be dissolved. However, with the support of Nobel Prize Laureate Max Planck, who was unanimously regarded as an outstanding scientist with an impeccable international reputation, Otto Hahn's efforts succeeded in gaining British approval for the revival of the Kaiser William Society and on 26 February 1948 the Max Planck Society was eventually founded in Göttingen as successor organisation, initially only for the British zone.[33]

Seven of the ten physicists kept at Farm Hall, were now members of the Kaiser Wilhelm Institute for Physics in Göttingen: Werner Heisenberg, Max von Laue, Carl Friedrich von Weizsäcker, Karl Wirtz, Horst Korsching, Otto Hahn and Erich Bagge. The first four were also given positions as professors at the University, were hosted in the empty rooms of the former Aerodynamics Research Institute (Aerodynamische Versuchsanstalt, AVA) which had been denuded of all its war-related equipment: its large wind tunnels had in fact been partly destroyed and partly dismantled and transported to England.[34] All of the scientific machinery and commodities, with which they had hoped to be able to start anew, had been carried away. Moreover, they had no access to all their instruments and equipment left in Heisenberg's Institute in Hechingen after the scientists fled as the Allied army was proceeding through Germany. The Institute was now located in the French occupation zone, and could not be reached anymore.[35]

In Fig. 6.7, we show a map of how Austria and Germany were divided among the four powers which had won the war, with Göttingen being in the British zone, some 200 kms North of Cologne.[36]

---

[32] Max Planck had been awarded the 1918 Nobel Prize 'in recognition of the services he rendered to the advancement of Physics by his discovery of energy quanta.'

[33] Its first president was the Nobel Prize laureate Otto Hahn. The Max Planck Society was founded in 1949 (Walsh 1968; Dickson 1986) and evolved into one of the mainstays of the science landscape of the Federal Republic of Germany.

[34] On the race to take possession of the German aircrafts as well as research and production facilities see Christopher (2013).

[35] On February 28, Heisenberg wrote his wife: "Well, here in Göttingen things are limping along, more or less. Our rooms in the AVA, at this point, are ugly, some basic office space devoid of any hint of warmth, but useful enough as temporary campsites in the crusade of life." (Heisenberg and Heisenberg 2016).

[36] A day by day account of how the final agreement about the division among the four powers was reached, can be found at https://berlinexperiences.com/potsdam-conference-1945/.

Fig. 6.7 At left, the four zones into which Germany and Austria were divided and became occupied by the winning powers, following the accords which took place in Potsdam, a small city located in occupied Germany, from 17 July to 2 August 1945, map is from https://estonianworld.com/life/remembering-estonias-wwii-refugees/. At right an image of the location of the town of Göttingen in relation to the places in UK through which Touschek would later move

In Göttingen, in the years to follow, Heisenberg devoted himself to two large tasks: the reconstruction of the Kaiser Wilhelm Institut für Physik, as a center for experimental and theoretical research in physics, and the renewal of scientific research in Germany, where, during these early post-war years, research was limited by the directives of the Allied Control Commission.

Heisenberg's efforts took large part of his energies, but he was successful and, during the years of the reconstruction, the Max Planck Institute for Physics gained a growing reputation as a leading representative of German physics in the international arena after so many years of isolation during the Third Reich and World War II. As he wrote to his wife in the early days of his Göttingen's stay, on February 28: "What our future will look like in all its reality, I cannot yet tell at all. In spite of it, I have the clear sense that it will not really be all that bad, if only we are patient." (Heisenberg and Heisenberg 2016).

6.3 1946: Touschek Between Göttingen and Glasgow

As 1946 started, Touschek was anxious to have a clear idea of how and where to continue and complete his studies. As we shall see, this was not so obvious, and through the whole of 1946 he moved back and forth between the United Kingdom, travelling to London and Glasgow, and Germany, following courses in Göttingen.

In 1946, the decisions leading to the reconstruction of Europe were being put in action. Many European scientists, who had left to join the Manhattan project, returned to Europe. Together with those who had remained in Europe, they could resume visiting each other's laboratories and universities and restart pre-war exchanges. The first international conferences since the beginning of the war were held. Everything was slowly starting anew.

In Germany, science was to be rebuilt starting from Göttingen, but all equipment of technical or scientific interest had been taken to England or to the US (as in the case of the already mentioned wind tunnels, (Jacobsen 2014)). Among them, was Widerøe's betatron. As Widerøe says in his biography: "In December 1945, the British authorities decided to take the betatron, as part of the booty of war, from Kellinghusen to the Woolwich Arsenal near London. Apparently, Rudolf Kollath later on took charge of its operation in Woolwich where it was used for non-destructive X-ray inspection of steel plates and such like. The machine has since disappeared without a trace. Many, including myself, later attempted to find it, but with no success. It was most probably scrapped." (Waloschek 1994, 87).

In 1945, to the officers of the T-Force reporting on Widerøe's betatron, Touschek had expressed the desire to go the UK, and such had been the recommendation in the BIOS report. This had been also the plan he had pursued just before the war broke out in Europe. At the same time, he was also part of the war booty, to be interrogated on German science, but in particular on the betatron, and, sometime in January or February 1946, he was taken to England.[37] During this first visit, plans for Touschek's move to study in a university in the UK were put in motion. The English officer in charge of the German scientists in Göttingen was Ronald Fraser, whom Heisenberg remembers as a friendly British officer in his memoirs of the period.[38] Fraser sought to have Touschek to go to the University of Glasgow, where plans for a 300-MeV electron synchrotron, to be built under the direction of Philip I. Dee, were considered, together with a smaller, preliminary 30-MeV machine.[39]

Philip Dee had arrived in Glasgow in 1945 to hold the long established Chair of Natural Philosophy, which had been offered to him already during the war, while he was involved in radar work and other leading war activities (Curran 1984). He had immediately set up major plans for relaunching physics, that included the building of an electron synchrotron, based on the new revolutionary principle of phase stability just discovered, simultaneously, both in the USSR (Veksler 1946) and US (McMillan 1945).[40] At that time, the only European country, whose scientific and technical

---

[37] This early visit to London is glimpsed from a letter he sent to his parents on April 8th 1946, from Glasgow, where he mentions that his entrance to the U.K. had been again refused, an indication that he had already been in the UK, but that refusal of entry did not prevent him to enter the country. The apparent contradiction between immigration authorities and the military, which were accompanying Touschek as an 'enemy alien', is similar to the long drawn fight between the US Immigration and Naturalization Service and military authorities over allowing entry or residence rights to German scientists, who could have been involved in war crimes, as seen in Jacobsen (2014).

[38] Ronald Fraser was a research physical chemist at Cambridge University, where he worked for a few years after having been a lecturer at Aberdeen University, as from footnote (50) in Amaldi (1981).

[39] Fraser knew Dee from the University of Cambridge, where Dee had graduated in 1926, later working at Cavendish Laboratory.

[40] After the war, four types of accelerators were in use: Van de Graaff, Cockcroft-Walton, cyclotron, and betatron. The cyclotrons, which were able to produce the highest energies, had reached their energy limit due to the relativistic mass increase at very high particle velocities, laying at roughly

evolution in nuclear and atomic physics could be compared to the US, was Great Britain, as underlined by John Krige: "[…] as the leading nuclear power in (western) Europe at the time, Britain alone amongst European countries had the human and financial resources, and the political will, to launch a major accelerator construction programme immediately after the war." (Krige 1989, 488).

The foundation of Britain's post-war accelerator construction program were laid out, immediately after Japan's surrender in August 1945, through a government committee (Cabinet Advisory Committee on Atomic Energy) which should advise the new Labour Prime Minister Clement Attlee on general policy for Britain's postwar atomic program. A Nuclear Physics Subcommittee was created on October 4, 1945. It was chaired by James Chadwick and was composed of leading nuclear physicists such as Patrick Blackett, John Cockcroft, Charles G. Darwin, Philip Dee, Norman Feather, Mark Oliphant and George Thomson. One of its first recommendations had been that "immediate support be given to Oliphant's and Dee's proposals to build accelerators at Birmingham and Glasgow universities, respectively." (Krige 1989, 488–490).[41] Dee's plans were thus part of a larger program launched in UK universities between October 1945 and March 1946 which included the building of big accelerators in five universities: a 1.3 GeV proton synchrotron (Birmingham), a 400-MeV synchrocyclotron (Liverpool), the 300-MeV electron synchrotron in Glasgow, as well as two less powerful machines in Oxford and Cambridge.

In this perspective, Touschek's experience with Widerøe's betatron would be an important asset for Dee's department and the foreseen project. As mentioned, there is some evidence that Touschek was brought to the UK in the early months of the

---

25 MeV for protons. The principle of phase stability came as a solution to this problem, making it possible to accelerate particles into the GeV region compensating for the relativistic mass increase either by changing the accelerating high-frequency voltage or the magnetic field strength during the acceleration of the particles. Not only cyclotrons could be operated at higher energies converting them into synchro-cyclotrons, but it was also possible to build a completely new type of accelerator, the *synchrotron*. This new machine could keep the particles on a path of constant radius by varying both the magnetic field strength and the frequency of the accelerating voltage with increasing particle energy. Last but not least, this kind of accelerator could be used for accelerating *either protons or electrons*. Machines based on this principle promised to displace betatrons as accelerators of high-energy electrons: indeed further developments of the betatron mostly took place for medical uses. In US, the leading country in the field, accelerator programs for nuclear physics research were being carried out at Brookhaven and Berkeley, two Laboratories which played a role as models for European physicists. In fall 1946 Lawrence's 184-inch synchro-cyclotron produced its first beam at Berkeley's Radiation Laboratory and new machines were being planned, notably a 10 GeV proton synchrotron.

[41] For a discussion of the British projects on accelerators, see Mersits (1987). As part of the British nuclear-physics program, a variety of different types of accelerators was being also planned at Harwell, the site chosen for the Atomic Energy Research Establishment to cover all aspects of the use of atomic energy, but this program was more oriented towards nuclear physics rather than "meson physics", as nuclear and particle physics was called at the time. When the 400-MeV synchrotron Liverpool machine went into operation in 1954, it was Europe's biggest synchro-cyclotron until 1957, when the CERN 600-MeV synchro-cyclotron was completed. Three of these UK university accelerators under construction would allow to do meson physics. Even if in the meantime higher energies had become available in US, they were a good basis for launching a research program in particle physics.

year 1946, probably to be further interrogated about the German betatron and start negotiations for a move to Glasgow. A suitable salary from the Darwin Panel Scheme, under which German scientists and technicians could be employed in UK, may have been discussed at the time.[42]

To finalize such an appointment, it was necessary to wait for the UK government's final approval of the Committee recommendations about the construction of new accelerators. In the meanwhile, Touschek, still under the 'protection' of the T-Force, was brought back to Germany, firstly to Kellinghusen, to take leave of his apartment and pack his few things. As for the next step, while waiting for the Glasgow situation to become definite, the natural choice was for him to go to Göttingen, where the University was restarting in the British occupied zone. Of interest to Bruno, was also that Wolfgang Paul, Kopfermann's assistant at the Institute for Physics of the University, was working with the betatron built by Konrad Gund for Siemens in Erlangen (Waloschek 2012) and which was later brought to Göttingen. This would give Bruno a good opportunity for discussions with Paul and Kopfermann while he was completing his dissertation to earn his Diploma in physics, the pre-requisite for any further studies.

Sometime in early March, Bruno was moved from Kellinghusen to Göttingen, where the University had been left almost untouched by the war.[43] In Fig. 6.8 the three Nobel Prize laureates from Farm Hall, Werner Heisenberg, Max von Laue and Otto Hahn are seen, in a 1946 photograph. When Bruno reached Göttingen, they were already there, having arrived since January. Bruno knew both Heisenberg and von Laue, whose lectures he had attended in Berlin, and was comforted and awed by their presence. It was a confirmation that the world he had known before the tragic final days of the war had not completely disappeared and could be reconstructed under the guidance of the scientists who had made modern physics before the war.

Touschek also renewed his acquaintance with Fritz Houtermans, who was planning to go to Vienna in the summer, raising Bruno's hopes to go with him and finally be reunited with his family. But he was not yet free to do as he wished. Unbeknownst to him, the British plans for post-war accelerator physics development were being finalized, with approval of the construction of the 300-MeV synchrotron in Glasgow. The plan to bring Touschek to the UK had also been carried through, including arrangements with Philip Dee and a suitable salary higher than could be expected in Germany or Austria. A contract was prepared for a six month position under the Darwin Scheme and on April 1st Touschek was brought to Glasgow, and housed in MacBrayne Hall of the University of Glasgow, Fig. 6.9.[44] The accomodation in the old Scottish University was very different from what he had seen in Göttingen, where

---

[42] Information about the Darwin fellowships, and the scientists whose work in the UK was sponsored through the Scheme, is available at the UK National Archives, https://discovery.nationalarchives.gov.uk/details/r/C258396.

[43] Bruno Touschek's letter, in which he describes his first impressions of Göttingen, is dated 12.4.1946, but the month, as written, is likely to be an error, with Touschek typing a 4 (April) instead of 3 (March). All evidence from the letters of this period, in particular two letters from Glasgow, respectively on April 8th and 12th, points to the date "12.4.1946" to be "12.3.1946".

[44] April 8th, 1946, letter to parents from Glasgow.

Fig. 6.8 Werner Heisenberg, Max von Laue and Otto Hahn in Göttingen, in 1946. Courtesy of AIP Emilio Segrè Visual Archives, Goudsmit Collection, Niels Bohr Library & Archives, American Institute of Physics. One Physics Ellipse, College Park, MD 20740

he had been housed in the buildings of the AVA (the Aviation Institute), which had been deprived of its instrumentation, but had been newly refurbished, all shiny and polished. In McBrayne Hall, his rooms were small and ancient looking, only 1.80 meters high, with beams on the ceiling running as if towards some distant adventure, and crossed by the tubes of all bathrooms, which Bruno decided, right away, to paint pink as soon as possible. He expected to be there for half a year, at least, and had brought his books and few things, which he immediately arranged around the room. Lack of proper clothing was still a worry, but Bruno had an uncle from the maternal side, Alfred Weltmann, who lived in Birmingham, and could help, in case of need.

However, once more, things were not to go on as he had expected. A week after his arrival, a complication arose.[45] As it turned out, this was not a small mishap, instead it was a tough obstacle to overcome. This was so because the Department for Scientific and Industrial Research (D.S.I.R.) had suddenly found out that he was Austrian, whereas the Darwin panel, from which Touschek's salary should come, could only be applied for Germans. Although this had been clearly stated by Bruno in the many questionnaires which he had filled in the intervening months, this 'detail' had obviously escaped the attention of whoever had prepared the contract.

---

[45] April 12th, 1946, letter to parents, from Glasgow.

This was an unexpected drawback. Once more Bruno's path had to be changed. Somewhat used to skirt administrative regulations, Touschek at first thought the objections could be of minor import and underestimated the difficulty of overcoming the D.S.I.R. objections. Since leaving Austria as a twenty one year old, Bruno had either been in semi-hiding or under the control of military authorities, while working in Widerøe's group during the war, or under T-Force authority afterwards. Thus, he did not understand that, outside the control of the military, civilian life was quite differently regulated and administrative obstacles were not so easily overcome. All his life, Touschek would have little tolerance for this type of delays and encumbrances, also a remnant of how, in the war years, he had to find alternative solutions to survive. As ultimate instance, one can remember that, towards the end of his life, he refused to prepare and submit his scientific credentials for promotion to Professor of Physics at the University of Rome (Amaldi 1981). His friends had to do it for him.

6.3.1   Getting a Diploma in Göttingen

The D.S.I.R. proved to be a hard contender and Touschek soon returned to Göttingen, to prepare for the physics examinations and presentations leading to his Physics Diploma, while Dee would look for a way which would allow to have Bruno in Glasgow.

Touschek was not disappointed about having to return to Göttingen. He was very resilient: a life of losses and changes, starting with losing his mother when not yet in his teens, then the expulsion from Vienna University and the years in Germany spent almost in hiding, up to miraculously escaping death during the final days of the war, all this had hardened his resolve to survive and bounce back. He was still young and confident in his future.

Back in Göttingen, he was pleased to be with familiar faces and have nice arrangements for housing in the countryside.[46] He had immersed himself in his studies, to prepare the exams for the Physics Diploma at Göttingen. He passed his pre-diploma exam very well on May 8th. There also appeared a thrilling prospect, namely that, after the exam, he could remain, for a while at least, as a research assistant with Heisenberg's group in particle theory. This was a chance like he had never encountered before and could not be missed, after almost seven years of disrupted life. Thus, the plan to go to Vienna in the summer had to be postponed, notwithstanding his parents' pressing for his return there.

In fact, having seen Bruno return to Germany from Glasgow, his parents had started hoping he would come back to Vienna. Earlier he had also received an offer for a lectureship in Berlin, and similar chances could possibly exist in Vienna, as well, but none of this was in Bruno's plans. He refused the Berlin offer to lecture in electricity and theoretical nuclear physics, because, at that point in his life, the priority for him was to become *a physicist*. A precarious and temporary position in a University, which at the time was completely empty, had no interest for him now that physicists such as von Laue and Heisenberg were no more teaching there.

About his return to Vienna, this had to be postponed at least by another year, as he could not see himself going back without having first gotten his degree and become Heisenberg's assistant in Göttingen, an extraordinary opportunity opening up for him in the coming months.

Now that he was engaged in a clear path for his Diploma, Bruno could enjoy the friendship of other Viennese physicists or professors he had seen in Berlin or Hamburg during the war. One such occasion was on May 10th, when there was an evening at the Houtermans' to celebrate Touschek's pre-diploma exam, with three Generations of Viennese scientists: the 75 year old mathematician Gustav Herglotz, Fritz Houtermans, then 50 year old, and Touschek himself, at 25. During these months in Göttingen Touschek became close to Fritz Houtermans. They had both been born in Vienna, came from similarly assimilated Jewish families, and had both experienced a skirmish with death, from which both had luckily escaped, Fig. 6.10.

---

[46] May 9th, 1946 letter to parents from Göttingen.

F.G. Houtermans mit W. Pauli, P. Jordan und P.G. Bergmann
anlässlich des Einsteinkongresses in Bern 1955

Fig. 6.10 A 1937 photo of Fritz Houtermans while imprisoned by the Soviet Secret Police, when he was held prisoner at the Lubyanka in Moskow in 1937, from Frenkel (2011). At right, from left, Wolfgang Pauli, Pascual Jordan, Fritz Houtermans and Peter G. Bergmann during the conference held in Bern in 1955 to celebrate the 50th anniversary of the formulation of special relativity by Einstein, photo property of the University of Bern collection, courtesy of S. Braccini

This time of Touschek's life reflects a close camaraderie with other German or Viennese physicists, who had all lived through the hardship of war.

On June 9th, 1946, Pentecost, also known as Whitsunday in the English world and an important Christian festivity, took place.[47] The first Pentecost after the war had ended, it held a special importance in Europe. After the carnage, divisions and conflicts of WWII, survivors and warring armies were sharing the hopes and burden of reconstruction, in a kind of suspended peace, which would soon be shattered by new divisions brought about by the *cold* war. But on that Pentecost Sunday, it was a good moment, for all, to celebrate the peace, no matter what their religion was.

Touschek, while studying hard for his exams and presentations, and basically on the eve of his diploma preparations, was one of many other Europeans who shared this holiday with friends, taking a small break from everyday occupations.

The week-end was quite exhausting. On the Friday before Pentecost, Houtermans had to go to the observatory, and asked Touschek to follow him there with Fritz's wife, promising visions of a comet. But, first they failed to find the observatory, wandering around until 11 in the surrounding forest, and then, when they finally got there, there was no sign of the promised comet, and they had to content themselves with a flickering Jupiter. At 2 o'clock they were not yet back home and the night was lost.

On Saturday, Jensen from Hanover appeared and they went together to Houtermans' home, to find Heisenberg, a Military Government official, and Süss, Fig. 6.11.[48] Saturday night was spent at Touschek's place, with a 'little night physics' ('eine

---

[47] June 14th, 1946, letter to parents from Göttingen.

[48] Hans Süss studied physical chemistry at the University of Vienna, receiving his Ph.D. in 1936. He was in Hamburg at the Institute for Physical Chemistry, since 1938. He had a wide range of interests, becoming an expert in heavy water and a scientific advisor to NorskHydro, the Norwegian plant in Vemork. After the war, in 1950 he moved to the US. For details see Waenke (2005).

Fig. 6.11 Left panel: a photograph of Werner Heisenberg, 1947 portrait, courtesy and © SUB GöttingenUniversity /GDZ. Right panel: Hans Süss from the Biographical Memoirs of the US National Academy of Sciences (Waenke 2005)

Kleine Nachtphysik'), in a typical Jensen-Houtermans' meeting.[49] The rest of the night was not much fun, given the rather crampled accomodation in Touschek's place (such as just one bed and both Touschek and Jensen having to share it).

On Sunday morning, they were again all together at Houtermans', and Heisenberg was there as well, but, apart from Heisenberg, they were all quite sleepy. The conversation must have been sleepy as well, given that, for some of them, the hours from 1 to 5 in the morning had gone by with the 'kleine Nachtphysik'.

While all this happened, Touschek was very worried about preparing a lecture he was to hold at Heisenberg's seminar, and, in the same days, completing the submission of his diploma thesis. He was able to bring both to a successful completion, not without obvious effort and strain, and he submitted the thesis on June 14th, as he proudly announced to his parents.[50] The diploma thesis on the theory of the betatron had been done under the joint supervision of Richard Becker and Hans Kopfermann

---

[49] Houtermans was famous for his hospitality. In Amaldi (2012), 27, Houtermans' first wife, Charlotte Riefenstahl, is quoted as remembering that in Berlin, around 1930, their "...small house and the tiny garden were always bursting with guests. It was not unusual to have 35 people dropping in for tea". One evening almost every week, the Houtermans invited their colleagues and friends to what Fissel called 'Eine kleine Nachtphysik' paraphrasing Mozart's 'Eine kleine Nachtmusik'. During these evening get-togethers, discussions around physics often lasted for hours and until late into the night. See also preface in Rößler (2007).

[50] June 14th, 1946, letter to parents from Göttingen.

and was most likely based on the work he had done during the war and the reports he had prepared afterwards for the T-Force. As for the lecture, it was received well and Bruno could rest and relax for a few days.

In the meanwhile, two weeks before, the possibility of going to England had come up again. He had to fill a rather long questionnaire with an English officer (taking him well of two hours), about a still rather uncertain stay for a six month period. Among a number of different opportunities, the UK option was still appealing to him, partly as he felt he owed the British a lot. There were also an invitation from Rolf Widerøe to visit Switzerland for a three week period, and the offer for a lectureship position at the Berlin University, which he had already decided to refuse. In any case, nothing could be decided until his diploma had been granted.

His gymnasium papers, testifying that he had passed the *matura* in Vienna in 1939 at the Staatsgymnasium, were requested and received, and he passed his Colloquium with full honors on June 26th.[51] At this point, after six months of having gone back and forth between London, Göttingen and Glasgow, he started asking what could he do next, or, rather, where would he go. Beyond the six-month position under Heisenberg, the plans for the future included the project in Scotland, the Swiss offer, or remaining in Göttingen and starve. Each of the plans had its own attraction, and staying in Göttingen with Heisenberg was most appealing to him scientifically. Financially, however it was the least secure, because of the lack of research funds available in Germany at the time. Touschek wanted to help his parents in Vienna, where conditions under the Soviet occupation were very harsh, and the possibilities of doing this, as a poorly rewarded Heisenberg's assistant, were scarce. Waiting for the work at the Heisenberg institute to start in August, he envisioned to take a small break, such as driving around the countryside, something he would enjoy, but seemed frivolous. As a matter of fact, the decisions he was agonizing about were not in his power to take.

The British in fact had been preparing his next visits to London, the first of which took place in early July, but in June he would not know about this, and became restless. One night, in late June, after his diploma, he read a book which drove him to reconsider what had happened in Germany. The book was *Darkness at Noon* by Arthur Koestler.[52] It dealt with the fate of a People's Commissar during one of the Russian purges, started in 1933–34, and leads the reader from the arrest to the hanging. Apart from leaving him quite depressed, he was led to consider the difference between what had happened in Germany and the Soviet still ongoing brutality. He could clearly see how things had now changed in Germany, at least in the British zone. He also saw that people around him did not realize this change, neither the British, nor the Germans, who had not seen evil when it was in front

---

[51] June 28th, 1946, letter to parents from Göttingen.

[52] Arthur Koestler (1906–1983) was born in Budapest, from Jewish parents, who left Hungary for Vienna in the 1920s. He became a member of the German Communist party in 1931 and traveled to Russia. He was disillusioned by what he saw, and, after many perilous adventures which included Spanish prisons in 1937 and a stint with the French Foreign Legion, he went to England, and was later naturalized as a British citizen. *Darkness at noon* was published in 1940.

of their eyes during the Nazi regime, and could now hardly wait for it to become history.[53]

6.3.2   Doubts and Uncertainty

Shortly after receiving his Diploma, Bruno was taken to England.[54] In Wimbledon, at Beltane School in Queensmere Road, there was an internment camp, where German scientists and technologists were held in order to obtain information and expertise by interrogating them about techniques in which Germany had been ahead of Britain (Gimbel 1990a).[55] Unlike others, Touschek was actually free to move in and out of the Beltane school and was even financially compensated. By July 19th, he was back in Göttingen, although not for long.[56] The frequent moves between the UK and Germany, which appear to have taken place between July and September, compounded Touschek's feelings of displacement, even affecting the research he was engaging in. For Touschek, the period after his diploma became a period of great uncertainty.

After the diploma, Touschek was offered a six month assistantship in Göttingen and he seems to have entertained various possibilities for his future studies, including to remain in Germany, perhaps doing his doctorate with Heisenberg. Envisaging the possibility of a doctorate under Heisenberg was shaking his original desire to go to England. In any case, he now faced two possible pathways to follow, whether to remain in Germany for his doctorate, either in Göttingen or perhaps in Berlin, or pursue the UK road, to Glasgow, where Philip Dee was continuing his efforts to obtain for him a doctoral stipend. Both personal and financial reasons weighted in, pulling him in one or the other direction, and would make Bruno alternate between different routes.

It was an extremely difficult choice. In Germany, he could have the chance to work with Heisenberg, and be surrounded by some of the most renown German physicists, eager to rebuild the pre-war eminence of German science. From a strictly scientific point of view, however, the occupying forces were strongly restricting the research topics which could be pursued by German physicists. Certainly no accelerator could be built in Germany, for quite some time. And, soon to happen, as we know now, *a posteriori*, the greatest advantages in theoretical physics, the development of relativistic quantum field theories and Quantum Electrodynamics, would in fact

---

[53] Letter to parents, from Göttingen on June 28th, 1946.

[54] July 3rd, 1946 letter to parents from London-Wimbledon.

[55] 'Once the Germans [scientists] had been located by the search teams, escorting officers were detailed to accompany them to London where they were taken to an interrogation center in Wimbledon, based at the premises of the Beltane school.' in Longden (2009). The center was moved to Hampstead in 1947.

[56] July 19th, 1946, letter to parents from Göttingen.

take place away from Europe (Schweber 1994).[57] On a personal basis, while he had known many of the Göttigen professors, who held him in good consideration, he was an Austrian, would still be partly an outsider, and his Jewish heritage clashed with remaining in Germany. He would of course be even more of an outsider in the UK, where he would be an ex-enemy alien, but he also had family in Birmingham, his maternal uncle, Alfred Weltmann, with whom he could relate. Ultimately, Touschek always remained an outsider, and this may have been both the source of his genius, and his demise.

However, it is not clear at this point how and why he followed the original plan and left Germany for Glasgow. As we shall see, at the end, after being literally taken back and forth between the UK and Göttingen from July to December, in April 1947 he was settled in Glasgow. Five years later, from Glasgow he moved to Rome, where he would be protagonist in the early development of particle colliders: *a posteriori* one can say that this turned out to be the right choice.

In the uncertainty, he went back to physics, to a neutrino physics problem he had worked on before. Having lost all his notes because of his many moves, and unable to reconstruct right away the arguments and the calculations, he felt like an old man, losing his capacities. He even doubted of losing his talent.

As for his future, conversations with Ronald Fraser did not help to clarify his mind or what could he expect to happen. Fraser wanted to know about possible work and publications on the betatron, but Touschek was now almost totally disinterested in anything connected with that work. Albeit late at this point, Fraser also gave him a gratifying information, namely that things were no longer secret and that the whole secrecy about the betatron, as it was in the previous November and December, was an invention of subaltern officials.

In the second part of July, while in Göttingen, discussing with Fraser whether Touschek were free to accept a possible offer to go to Glasgow, a British corporal appeared with a telegram from England requesting once more the completion of yet another questionnaire. All this was still non-committal, and Bruno was feeling more and more displaced and without a safe direction to go. Memories of his family were coming back to him more often, and, at times, he dreamed of taking a vacation, three years from now, after his doctorate, and go back to the 'Colle d'oro', the golden hill near Rome, where his aunt Ada had a summer house and where she had taken him, during his visits before the war.[58]

Between August and December, Touschek was in the UK at least one more time.[59] In August he may also have been again in Scotland, where the position in Glasgow University, in the department where Lord Kelvin had held a chair, was appealing and

---

[57] In 1965, the Nobel Prize in physics was assigned to Richard Feynman, Julian Schwinger and Sin-Itiro Tomonaga for 'their fundamental work in quantum electrodynamics, with deep-ploughing consequences for the physics of elementary particles.'

[58] The 'Colle d'oro' is a location near Velletri, one of the many small towns dotting the vulcanic hills South-East of Rome.

[59] August 18th, 1946, letter to parents from Wimbledon.

definite enough that he gave up his room in Göttingen.[60] As it turned out, more time was needed before the offer could be approved by the University administration, and in late September he was back in Göttingen, where the landlady refused to let him back to his room, and he had to sleep on hard floor, until, presumably, rescued by his friends.

During the summer and the months to follow, Bruno worked hard on double β-decay.[61] In those months, traveling between different places and countries did not permit easy concentration on physics. Still, he worked on the problem while in the UK and started writing a paper, which was then submitted for publication. Upon his return to Göttingen, focusing better on his physics, he found there was a mistake in his conclusions, and had to chase the error to correct it before the paper would be published (Touschek 1948b).[62] The anxiety about correcting an error, trace its origin, and rushing to have it corrected while the article was under publication, took most of his energies in October and November. In addition he had to move away from the betatron affairs, where some British officers were still keen on obtaining work or informations from him.

Not receiving news from Glasgow, the uncertainty about where he would be in the next year became a pressing concern and, on November 5th, Touschek solicited Philip Dee for an answer about the Glasgow position.[63] During these months, Philip Dee, keen on having Touschek come to Glasgow, had continued his efforts to have Touschek entering the University Doctoral program.

The solution was near, but it would take another four months before Touschek could take on his research fellowship in Glasgow.

Touschek's anxiety about his future was also entangled with a degree of uncertainty about the direction his research should take. He saw that purely theoretical problems were not interesting him any more, and felt he perhaps lacked the enthusiasm to persevere and solve them. He went to Heisenberg for advice, but could still not see his way out. Various other difficulties piled up, including financial ones. At the end of November, after Dee's letter, the only strategy for Bruno appeared to let the British authorities take care of his next move, although it was clear to him that no solution would be the perfect one. Behind uncertainty and doubts, there looms large the presence of Werner Heisenberg, who befriended Touschek, and may have been one of his inner mentors throughout his life.[64]

Heisenberg was one of the great scientists who constructed the theoretical framework sustaining particle physics, a concerned observer of the influence of science and philosophy, and a controversial protagonist of the debate about the moral imperative of a scientist facing political power. He was also a major influence on Touschek's

---

[60] Bruno Touschek's November 24th, 1946 letter to his parents.

[61] November 24th, 1946, letter to parents from Göttingen.

[62] The paper, submitted to *Zeitschrift für Physik* (now the *European Physical Journal*), on December 2nd, 1946, was published in 1948, and recently cited in Vissani (2021). In this article Touschek thanks Heisenberg for suggesting the problem and for advice.

[63] November 14th, 1946, Philip Dee's letter to Bruno Touschek, Bruno Touschek Archive, Box 1.

[64] The other was Wolfgang Pauli, as shall be seen in Chap. 9.

development as a physicist. Touschek and Heisenberg never collaborated on an actual paper, nor was Touschek to be his doctoral student. However, they often discussed physics together and Bruno Touschek occasionally worked on problems of interest to Heisenberg: when a scientist of Heisenberg's stature makes himself available to intellectual and physics discussions, as in Touschek's case, the effect will last forever.[65] The influence of Werner Heisenberg on Touschek runs deep through Bruno's work with Walter Thirring (Thirring and Touschek 1951) on the Bloch and Nordsieck theorem (Bloch and Nordsieck 1937; Etim et al. 1967a), and in statistical mechanics (Touschek and Rossi 1970). No matter how short, six months or one year, the contact with genius, when the latter allows it, touches one's mind and heart.

By mid December, Dee clarified that the 'unfortunate delay' was that all those involved in the affair had forgotten that difficulties could arise at the University level—not just at the D.S.I.R.[66]

Once this was understood, it had then been necessary to wait for the rectorate decision. This having been favourable to Touschek's hiring, it was now mostly a question for the appointment to go through the usual official channels. This would naturally take some time, but it was now only a matter of few months. This delay would suit Bruno, who was keen on attending a lecture by Heisenberg, to be held in January.

Once the Glasgow position had a definite starting date, April 1st, 1947, Touschek could see a clear way ahead of him, and could make closure with some of his past. In particular, he had to put an end to his parents' pressure to go back to Vienna. He had to definitely let his parents know that he would not look for a position there, as they were rather naturally asking of him. Going to Glasgow was a clean break from the past. The lost time was his to reclaim, he hoped: the five years spent in semi-hiding in Germany, the two years between the Anschluss and the expulsion from the University of Vienna in June 1940, studying at Urban's home with borrowed books in 1941, all that lost time could be retrieved. He was going to begin a new life, and could not afford to make any more mistakes. He would not go back to Vienna, at least not until he had his Doctorate.

He saw that the first mistake had been not to leave Austria in 1938 or 1939, when, from Rome, he had applied for a *visa* to go England. Waiting for it to arrive any day, ultimately he had returned to Vienna. Did he receive the *visa* but lacked the courage to go, or, perhaps, was the family support not forthcoming? It is quite possible that the difficulty may have been on missing family support. In Vienna, they were still hopeful for the worst not to happen. But it was not going to be, as we know. Bruno's grandmother Weltmann, who had moved to Rome to stay with her daughter Ada in 1938, had later returned to Vienna, following the anti-semitic regulations declared

---

by Mussolini's regime. But once in Vienna, in August 1942 she had been taken to Theresienstadt, never to return.[67]

In December 1946, when Bruno's diploma thesis was in print in Göttingen and the Glasgow situation was clearly in sight, Bruno could sever the bond with his native city. He saw that offers for a position in Vienna were not forthcoming: if he were to go back, he would be one of the clamoring many, and would need to enter into the typical academic squabbling and competition, something he did not, and would never have, appetite for. As for his next move, he had no doubt that, from the scientific and intellectual point of view, remaining in Göttingen would be the most favourable way to go ahead towards a doctorate, but this was not to be taken for granted, not to mention the poor financial prospects. In fact the financial situation of Heisenberg's Institute was still a difficult one, with scarce possibilites to support PhD students.

The problem of money was a natural consequence of Touschek's rather desperate economic situation for a number of years. From a rather affluent pre-war, pre-Anschluss life, he had been thrown into the need to support himself when semi-hiding in Germany. The war over, one can see the emergence of a moral imperative to support his parents in Vienna, under Soviet occupation. Touschek's father had been a Major in the Austrian Army and was now retired, and Touschek believed that his father had perhaps left the Army under pressure because of him, his Jewish son, from his first marriage. In later conversations, with Edoardo Amaldi and Carlo Bernardini, his closest friends during the twenty five years he lived in Italy, Touschek let transpire a feeling of guilt in this respect (Amaldi 1981). None of this can obviously be found in Touschek's writings, but the *leitmotiv* of financial concern, and how much he could help his father and his step mother is omnipresent.

Thus, in April 1947, Touschek joined the Physics Department of the second most ancient of Scottish Universities, as a doctoral student in Glasgow. In later years, Touschek regretted not having remained in Germany, but the history of science tells us that this was the right decision. In Glasgow, Touschek would develop into a full fledged theoretical physicist, and establish contact with the young Italian theorist Bruno Ferretti, who would bring him to Rome in December 1952, where one of the great adventures of particle physics was to begin a few years later. There, in the nearby hills overlooking the vast plain where the city spreads down to the Thyrrenian Sea, a new laboratory would be conceived in 1953 and built, and an electron synchrotron constructed and made to operate in 1959. In this laboratory, on February 17th, 1960, Touschek proposed to construct AdA, an electron-positron collider, the first storage ring of matter-antimatter particles in a laboratory. Through AdA's operation and first successes, there came the development of a new type of accelerator, which, in the fifty years to follow, would unravel many of the mysteries of the world of particle physics.

---

[67] Amaldi writes that 'in 1941, on a date which I have not been able to determine, she was arrested by the Nazi and sent to the Theresienstatd concentration camp, where she died shortly after', (Amaldi 1981, 13). The Shoah's archives and information from IKG give exact dates for her deportation, August 27th, 1942 (and death, March 10th, 1943), but she could have been arrested earlier, as one can gather from Bruno's 1942 letters home.

6.4   Who Made the Decision for Touschek's Move to Glasgow in 1947: T-Force, Touschek or Heisenberg Himself?

The sudden change of plans in April 1946, when Touschek was first taken to Glasgow and, one week later, left and went back to Göttingen to get his diploma, can be understood if we place Touschek's personal story in the wider context of how the Allies were planning for the scientific and technological future of the Western world, in a race against the Russians. We have seen the development of Operation Epsilon, through which the German nuclear scientists were chased and brought to England, to be kept for six months, for some time without any contact with family and colleagues. In January 1946 they were released to return to Germany, where they would rebuild German science in its non military aspects, namely no applied nuclear physics, no new accelerators, and other restrictions. Much more sinister, and better known in its general lines, was Operation Paperclip, which brought to the United States many scientists involved in rocket building, and in chemical and biological warfare, (Jacobsen 2014).[68]

In the context of our story, we should recall that the October 1945 B.I.O.S. report about Widerøe's betatron had recommended that Bruno Touschek be brought to the UK.[69] This was also what he mostly wished to happen at the time. As 1946 rolled in, we have also seen that in January a program for constructing new particle accelerators was proposed by the UK Nuclear Physics subcommittee of the Government Advisory Committee on Atomic Energy. This program was then endorsed by the Committee on March 28th, 1946, and, shortly after, approved by the UK government (Krige 1989, 491). This is why Ronald Fraser was able to carry through Touschek's proposed hiring in Glasgow, where Bruno would complete his studies and eventually get his doctorate. No time seems to have been wasted after the UK Government approved the construction of the new accelerators, and in April 1946 Touschek was brought in the UK by the military to start his work in Glasgow. Apparently, the immigration authorities, at Harwich, had some objections and officially refused landing rights.[70] However, this did not stop Touschek from entering the UK, something which had also happened before, during a first visit in January or February, but the military was able to override the civil authorities. However, when the Darwin fellowship could

---

[68] Among the German scientists brought to the US, there was Wernher von Braun. Main scientist of the Nazi rocket program, including the V-2, he became the main artifex of the American space program, as director of NASA's Marshall Space Flight Center and chief architect of the Saturn V launch vehicle which propelled the US to the Moon (see biography of Wernher von Braun at https://history.msfc.nasa.gov/vonbraun/bio.html). Bringing a number of German scientists, some of whom turned out to have been directly involved with slave labor in the concentration camps during the war, often gave rise to contrasts between the military and the US Immigration and Naturalization Service, most of the times resolved in favor of the military by higher political decisions.

[69] B.I.O.S. Miscellaneous Report No. 77, Technical Report No. 331-45, European Electron Induction Accelerators.

[70] See letter April 8th, 1946, from Glasgow.

not be approved because of him being Austrian rather than German, nothing could be done, and he went back to Göttingen, to continue his studies there.

On June 26th Touschek obtained his diploma, which had been a great success as Sommerfeld wrote in a letter to Paul Urban.[71] We can now see various parallel actions being set in motion. While Dee and Fraser were trying to get him to join Glasgow, Touschek, emboldened by his diploma, was now hopeful to remain in Göttigen and do his doctorate with Heisenberg. Other options were also open, but the most coveted would obviously be to remain and work with Heisenberg. It did not happen. He did receive a six month position, but that was all. We have no hints regarding Heisenberg's intentions about keeping Touschek at his institute, but at the end of 1946/early 1947, the Kaiser Wilhelm Institute for Physics was still in a very difficult phase, most probably not yet in the position of funding PhD students. Touschek discussed his prospects at length with Fraser, who would assure him things would be OK if he, Bruno, would remain in Germany. Was this all as straightforward as it appears? Or did the T-Force decision that Bruno was needed in Glasgow influence Heisenberg so that Touschek's only way forward was to go to Scotland? Did Bruno ever have a different choice? We may never know, but the background story is so much larger than what Touschek could see, that various possibilities co-exist. Once the accelerator program was approved by the UK government in March 1946, his move to Glasgow had to take place one way or the other. Dee (and Fraser) could not immediately overcome the obstacles posed by the civil authorities, but eventually they did, and Touschek (and his professors in Göttingen) had no choice. Namely, from the very beginning, it is very likely Touschek was meant to take the Glasgow way, because his expertise was of interest to the British scientists planning for the future of nuclear physics in the UK.

It may appear that we are assigning too much importance to Touschek in this context, but one cannot forget the exceptional intellectual qualities that he possessed and were clearly seen by his peers, Arnold Sommerfeld, or Werner Heisenberg, later Max Born, among them: coupled with the unique experience with Widerøe, this combination is what ultimately led to the success of AdA. In Touschek, one finds the potential for innovation and disruption: he was a theoretical physicist who had learnt the ways of electrons, during the dark days of World War II, under the guidance of Rolf Widerøe, one European authority on electron accelerators at the time. Thanks to such combination, of theory and practical expertise, in due time, Bruno Touschek could envisage and build a new type of accelerator, a matter-antimatter collider.

---

[71] Postcard from Sommerfeld to Urban, in Amaldi Archive, Sapienza University of Rome, Box 524, Folder 4, Subfolder 4. It was probably a document sent by Urban to Amaldi, when the latter was preparing his biography of Touschek (Amaldi 1981).

Chapter 7
Bruno Touschek in Glasgow: The Making of a Theoretical Physicist

...the infrared catastrophe: a problem to solve.
W. Thirring and B. Touschek, A covariant Formulation of the Bloch–Nordsieck Method, *Philosophical Magazine, 42(326):244–249, 1951.*

Abstract This period of Touschek's life covers the five years he spent at University of Glasgow, first to obtain his doctorate in 1949 and then as a lecturer. These years, which have hardly been explored, were important for his formation as a theoretical physicist. They included contacts and correspondence with Werner Heisenberg in Göttingen and Max Born in Edinburgh, as well as Touschek's close involvement with colleagues intent on building modern particle accelerators in Glasgow, Malvern, Manchester and Birmingham. During these years, he lived through the shock of the Fuchs affair, which unraveled in early 1950, and which may have influenced his decision to leave the UK. In 1952, contacts with the Italian theoretical physicist Bruno Ferretti led Touschek to join the Guglielmo Marconi Physics Institute of University of Rome, where he started as a researcher in January 1953.

In the fall of the academic year 1946–1947, the Principal of Glasgow University welcomed a very special student cohort, the one which was returning from the war. As he wrote in the preface to the 1946–1947 Student Handbook, he welcomed not only the new first year students, but equally warmly those who would be resuming the course of studies interrupted by the war. "These are sombre and anxious days" he added, "as they must be after the years of loss and destruction", but the process of recovery had started and, although possibly long and hard and disappointing, he hoped that the new students would set a standard for the future. The Principal's words are poignantly reflected in the front page of the Student Handbook, Fig. 7.1, showing the transition from the war scenario to the purposeful stride of a confident young man under the benign eye of the University tower.

© The Author(s), under exclusive license to Springer Nature Switzerland AG 2022   171
G. Pancheri, *Bruno Touschek's Extraordinary Journey*, Springer Biographies,
https://doi.org/10.1007/978-3-031-03826-6_7

THE UNIVERSITY

Fig. 7.1 The University of Glasgow 1946–1947 Student Handbook, by permission of University of Glasgow Archives & Special Collections, Students' Representative Council collection, GB248 DC157/18/56

Among this new class, joining in the Spring as a graduate student, there would be Bruno Touschek. He had gone through the disruption of his life, suffered loss and fear, and now he was moving ahead to a new country, a country which had won the war. Unlike Germany, still under the weight of restrictions and lacking resources, the UK was ready to pour energy and money into new roads, which scientists, coming back to basic research after their war engagement, were eager to follow.

When the preface to the Student Handbook was written on July 19th, 1946, Bruno was still in Göttingen, had just submitted his Diploma thesis and was worrying about the future directions his studies would take.

The year 1946 had been very hard for Bruno, who was torn between a desire to remain to study in Göttingen, under very dire monetary circumstances, and the worry about his parents' precarious situation in the city of Vienna, under Soviet occupation. His unique expertise in accelerator science, gained working with Rolf Widerøe (Waloschek 1994; Sørheim 2020) during WWII, decided for him. In early January 1946, in fact, the British accelerator program was being defined, and a project for constructing an electron synchrotron at Glasgow University took definite form. Touschek was one of few experts in the field present in Germany at the time, and his knowledge was prized by the Allied T-force, who had advised for him to be moved to the UK.

The uncertainty about a future, which was not in his power to direct, and the desire to continue his physics studies had made him oscillate between hopes of remaining in Göttingen and obtain a doctorate under Heisenberg or moving to Glasgow, where he

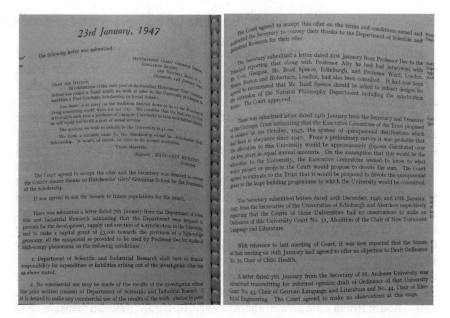

Fig. 7.2 Minutes of University of Glasgow Court 1946–1947, January 23rd 1947, by permission of University of Glasgow Archives & Special Collections, University Court collection, GB248 C1/1/54

would face unknown surroundings and, in some ways, a less prestigious prospect. At the same time, Glasgow would provide a clean break from the past, and be financially more rewarding. It is very likely that, once the decision had been taken (by the T-force), Bruno was happy to leave Germany, which was still under the deprivations and gloom of the war. He had finally made true his old dream of studying in the UK, would have a reasonable salary, and thus could help his family, as he had wanted all along. He would join a great and ancient University, and be part of the British post-war effort focused on bringing to peacetime use the scientific successes and technological innovations which had won the war to the Western Alliance.

In early 1947, events, which had been set in motion in 1946, began to unravel. In January 1947, the University of Glasgow Senate submitted to the University Court a proposal for the construction of a synchrotron in the University, whose development, supply and erection would be provided by the Department of Scientific and Industrial Research (D.S.I.R.), together with a capital grant of 3,000 pounds towards the provision of a high voltage generator. The Minutes of Glasgow University Court, Fig. 7.2, state the decision to build a synchrotron as proposed by the D.S.I.R., and appoint the famous Scottish architect Mr. Basil Spence to design the expansion to house the synchrotron. This was part of the project to build an extension to the original Natural Philosophy Building having the purpose to establish the

Department of Natural Philosophy "as one of the pre-eminent physics departments in the UK."[1]

Following this decision, Philip I. Dee, chair of the Natural Philosophy Department at University of Glasgow, was finally able to carry through the appointment of Bruno, a researcher of exceptional promise, as Dee himself noted in the submission forms a few months later.

Bruno Touschek spent almost 6 years in Glasgow, from April 1947 until December 1952. With the exception of a preliminary posting (Pancheri and Bonolis 2020), little has been published until now about this period of Touschek's life and the crucial influence it had on Touschek's later emerging in Rome as full fledged theoretical physicist, whose exceptional intelligence and sarcastic wit would soon conquer his Roman colleagues, and make him one of the major stars of an Institute bent on taking a central role in the physics scenario of the new Europe. In Edoardo Amaldi 's work (Amaldi 1981, 9–12), only a few pages describe the period Bruno Touschek spent in Glasgow. Amaldi cites the copious scientific output, more than 15 published articles with many different co-workers, and, as a personal note, reproduces a long letter by Philip Dee, which describes Bruno as brilliant but temperamental, able to do clean and tidy scientific work, but occasional unrestrained by normal rules of behaviour. This period of Bruno's life seemed almost wasted, or at least irrelevant to his future as the undisputed genius of later years. His scientific work in the Glasgow period could even appear forgettable, with the sole exception of the article with Walter Thirring on the Bloch and Nordsieck theorem (Thirring and Touschek 1951). And yet, these are the years when Touschek became a theoretical physicist. Indeed, the search for clues to Touschek's genius and tragic life, and to the genesis of the development of particle beam colliders, shows the importance of this period. In particular, inspection of available archival sources highlights the relationship Touschek had with Werner Heisenberg and Max Born during his Glasgow years, and reveals the influence these two giants of 20th century physics had on Touschek's formation as a theorist.

Building on this, and placing Touschek in the context of historical events around him, such as the Fuchs affair, and the effect it had on UK physics, gives new insight on Touschek's life in Glasgow, showing how his destiny became entangled and influenced by his new surroundings. Far from the image of a studious but nervous and eccentric young man, as described by Philip Dee in the 1979 letter to Edoardo Amaldi, Bruno appears as the young protagonist of a future great adventure, well aware of the mainstream of world events and physics developments, bent on his studies and anxious to reclaim the years he had lost since he had been expelled in June 1940 from the University of Vienna.

---

[1] From Minutes of University of Glasgow Court 1946–1947, January 23rd 1947. Courtesy of University of Glasgow Archives & Special Collections, University collection, GB 248 DC 157/18/56Archives. News from the Glasgow University Archive Services with a later photo of the Building can be found at https://www.theglasgowstory.com/image/?inum=TGSD00214. The first phase was completed around 1954 and consisted of a new lecture room and the purpose built accommodation for the synchrotron. A 2007 journalistic assessment of Spence's work can be found at https://www.theguardian.com/society/2007/oct/16/communities.

7.1 Backstage in UK and Crucial Particle Physics Developments Elsewhere

Touschek's move to Glasgow can be seen in the context of the extensive program of particle accelerator construction in which the UK had embarked soon after the end of the war. Initially motivated by the possibility of having more powerful machines for investigations at a nuclear level, preliminary plans were further boosted by the discovery of new elementary particles in cosmic ray studies which definitely set the stage for subnuclear physics.

Up to 1945, particle accelerators in use were the van de Graaff, the Cockcroft–Walton, the cyclotron and the betatron. A new era in the field was inaugurated with the principle of phase stability, proposed in 1945 by Edwin W. McMillan in the US and, independently, by Vladimir Veksler in the USSR (McMillan 1945; Veksler 1946). This idea would allow to accelerate particles to an energy range beyond the reach of cyclotrons, namely thousands of MeV, the GeV region. In cyclotrons the relativistic mass increase with high velocities resulted in an energy limit of about 25 MeV for protons. This obstacle was overcome by the new principle, which opened the way to a completely new type of accelerator, the synchrotron, and also allowed to convert cyclotrons into synchro-cyclotrons, operating them at much higher energies. A remarkable advantage of such device was the possibility of accelerating both protons and electrons, while all the basic knowledge already acquired through betatrons was instrumental for the functioning of electron synchrotrons. The war had shown the terrible power which could be unleashed by the nuclear world, while, at the same time, opening new horizons to research. All this became a powerful trigger for the construction of new accelerators as a tool of investigation of the nuclear world. As such, the new type of machine was immediately included in the British nuclear physics program.

By 1945, eight particles were known: the electron, the positron, the proton, the neutron, the photon, the neutrino, and the so called positive and negative *mesotron*—as it was termed the very penetrating component of local cosmic rays because of its mass (200 electron masses), intermediate between that of electron and proton. At the time, the mesotron was thought to be Hideki Yukawa's meson, the predicted field quantum mediating strong interactions. Such mesons were believed to be associated with the extraordinary attractive forces binding neutrons and protons in the nucleus, and the new challenge was to make mesons in the laboratory and study them in the number and detail otherwise impossible to obtain, when their source was still only the cosmic rays. The needed energy was at least 200 MeV, because of their rest energy, thus a popular target for the builders of machines was 300 MeV. This minimum threshold clarifies the energy choices for the Glasgow and Liverpool synchrotrons, which the post-war UK scientists planned to build.

7.1.1   The UK Accelerator Program in 1945–1946

The UK accelerator program was part of a wider postwar atomic energy program. It was started with the decision in October 1945 to establish the Atomic Energy Research Establishment (AERE), to be built in a site near Harwell, which would become the main center for atomic energy research and development in the UK. At the time, UK was the leading nuclear power in Western Europe, having the financial, industrial and technical resources to launch a large-scale project, whose basic task would be the development of applied nuclear physics at the new center at Harwell, and which would include also a number of accelerators. The new center was to be directed by Sir John Cockcroft.[2] Cockcroft had come back from war work at Chalk River, in Canada, where a heavy-water nuclear reactor to manufacture plutonium and enriched uranium had been built. At AERE, where the first nuclear reactor in Western Europe started up in August 1947, Cockcroft supervised the construction of various reactors. In parallel with nuclear reactors, a variety of accelerators had also been planned to be built there, to be used for such tasks as the production of neutrons by photo-disintegration or the determination of cross-section of neutron induced reactions, of interest also for radiotherapy and for the study of high energy X-rays.

Between October 1945 and March 1946, the decision was made to build accelerators in as many UK universities as could be chosen to be adequately equipped for the task. In early 1946, the Nuclear Physics Committee of the Ministry of Supply sent circulars to thirty universities and other institutions asking about their research programs in nuclear physics (Mersits 1987; Krige 1989). Only five of them replied with requests for large scale equipment, subsequently receiving grants from the D.S.I.R. for the construction and maintenance of the following: a 1.3 GeV proton synchrotron at Birmingham, designed by Mark Oliphant, a 300–400 MeV electron linear accelerator at Cambridge (later abandoned), a 300-MeV electron synchrotron at Glasgow, a 400 MeV proton synchro-cyclotron at Liverpool, and a 150 MeV electron synchrotron at Oxford. Actually, all these machines would go into operation between 1953–1954, when the Bevatron and the Cosmotron, accelerators of much higher energies, were already available in the US.

By the end of 1945 the advantage of a synchrotron for the photo-production of mesons had been recognised, and a research program in synchrotron development had been prepared. Philip Dee's idea of building a 200 MeV betatron at Glasgow had been abandoned in favour of the mentioned 300 MeV synchrotron. A working group, under the leadership of Donal Fry, was created in the government center at Malvern, where a top secret radar group had been hosted during the war.

---

[2] In 1929, John Cockcroft and Ernest Walton, working at Ernest Rutherford's Cavendish Laboratory at Cambridge University, started to build the first 'Cockcroft–Walton accelerator', as it was named since then, a system of capacitors and thermionic rectifiers capable of 600 KV. It was used in 1932 to bombard lithium and beryllium targets with high-energy protons achieving the first artificial disintegration of an atomic nucleus and the first artificial transmutation of an element (lithium) into another (helium) (Cockcroft and Walton 1932). This machine was still used much later to supply voltage in large particle accelerators.

Discussions with industrial groups, such as Metropolitan-Vickers, English Electric and British Thomson-Houston, all of which had participated in the war effort, had started (Lawson 1997).[3]

The decision was also made to build a 30-MeV electron synchrotron, which should serve as a prototype for the 300 MeV machine planned by Dee in Glasgow, and at the same time could be used to study nuclear photo-disintegration and gamma-neutron reactions. Following Frank Goward's idea, the first step in this direction was done converting a betatron at the Woolwich Arsenal Research Laboratory (the first in UK during the war) into the world's first electron synchrotron. This prototype, first operated in October 1946 by Goward and D. E. Barnes (Goward and Barnes 1946), established the practicability of the synchrotron acceleration, together with the machine operated by General Electric in US (Elder et al. 1947). The Woolwich synchrotron was then moved to Malvern where it was modified to be used for more general experiments. In providing the premise for the 30-MeV machine, which would go into operation in October 1947 (Fry et al. 1948), the Malvern synchrotron in turn became the hotbed for analyzing the problems of the 300 MeV Glasgow machine. The magnet for the Glasgow electron synchrotron would be built by Metropolitan-Vickers Electrical Co., who were to be responsible for overall design and construction together with the Malvern group and Glasgow University. The Malvern–Harwell group also gave advice to the building of the Oxford electron synchrotron.[4]

7.1.2 1946: Reaching Out to the Reconstruction of European Science

Alongside the planning for particle accelerators and their transformation as a new tool for exploring particle physics in an unprecedented energy range, the postwar period also brought together many of the prewar European scientists bent on reconstructing their University laboratories and eager to compare each other's ideas about past and present new directions.

The UK was the natural place for such reconstruction to start. English Universities were now seeing many of their scientists coming back from the United States where they had participated in the Manhattan project and the associated Anglo-Canadian Project in Montreal and Chalk River. Among the scientists returning to the UK from the US or from Canada, and of direct interest for Touschek's story, we note Rudolph Peierls, and Hans von Halban. Peierls, who would be Touschek's external Ph.D. examiner in Glasgow, came back from Los Alamos, where he had played a major role in the development of the atomic bomb, and joined the University

---

[3] See also *The CERN Synchrotrons* by Giorgio Brianti in 50 Years of Synchrotrons where Lawson's recollections appear.

[4] See Metropolitan-Vickers Electrical Co 1899–1949 by John Dummelow: 1939–1949, https://www.gracesguide.co.uk/Metropolitan-Vickers_Electrical_Co_1899-1949_by_John_Dumm elow:_1939-1949.

of Birmingham (Lee 2007).[5] Von Halban, returning from Canada, was to become the first director of the Laboratoire de l'Accélérateur Linéaire in Orsay (LAL) and oversaw the construction of the linear accelerator which contributed to AdA's success story (Marin 2009; Pancheri and Bonolis 2018).[6]

There were also visitors from continental Europe, in particular from Italy, where the process of reconstruction of scientific institutions was taking place. In Italy, many of the best scientists had moved to the United States, as it had been the case of Enrico Fermi (Bernardini and Bonolis 2004; Pontecorvo 1993), Bruno Rossi (Bonolis 2011), Emilio Segrè (Segrè 1993), Bruno Pontecorvo (Close 2015), and many others. But some of Fermi's students or young collaborators had remained, notably Edoardo Amaldi (Rubbia 1991).[7] Soon after the war, Italian physicists, in particular from Milan and Rome, in contact with Fermi and Rossi, both in the US, and following their advice, were ready to take up new roads in particle physics.

In 1946, a pivotal moment in the scientific reconstruction of Europe had been the international conference on *Fundamental Particles and Low Temperatures*, held from 22 to 27 July 1946 at the Cavendish Laboratory in Cambridge (The Physical Society and Cavendish Laboratory 1947). As recalled by Amaldi (Amaldi 1979a, 62), "Contacts between physicists in different parts of the world had been impossible for years and this conference provided a welcome opportunity to renew old friendships, and to hear what others had been doing." Many European states—but not Germany where foreign travels were strict restricted—as well as other countries like the US, USSR, China and India were represented at the Conference. All of them presented their latest work on cosmic ray physics. Large attention was also given to the theoretical physics side, in particular the last session was entirely devoted to the S-matrix theory proposed by Werner Heisenberg during the war (Heisenberg 1943a, b, 1944). Focusing on observables such as cross-sections or energy levels, Heisenberg's S-matrix theory aimed on building a theory to calculate only observables, such as scattering cross-sections and energy levels. Heisenberg, although released from enforced stay at Farm Hall (Cassidy 2017), was not present, but talks on the subject were presented by Walter Heitler, Christian Møeller, and Carl G. Stueckelberg (Mersits 1987).

Besides the opportunity of scientific exchanges after the forced isolation that many scientists had experienced during the war, the Cambridge conference was

---

[5] An extensive description of Peierls role in the development of the atomic bomb can be found in Close (2019).

[6] In 1939, at Collège de France in Paris, Hans von Halban had collaborated with Frédéric Joliot and Lew Kowarski in the discovery that several neutrons were emitted in the fission of uranium-235 (von Halban et al. 1939a, b)—a discovery leading to the possibility of a self-sustaining chain reaction. In 1941 von Halban fled to England to subtract the laboratory heavy water equipment from the approaching Germans. He settled in Oxford and then moved to Canada to join the Manhattan project related effort. Back to Oxford, he returned to France in 1954 to become the first director of LAL.

[7] In 1938, Fermi was awarded the Nobel Prize in Physics 'for his demonstrations of the existence of new radioactive elements produced by neutron irradiation, and for his related discovery of nuclear reactions brought about by slow neutrons.' Fermi's group in Rome included Franco Rasetti, Edoardo Amaldi, Ettore Majorana, Bruno Pontecorvo, and Emilio Segrè, who was awarded the 1959 Nobel prize in Physics for the discovery of the anti-proton.

also an important occasion to renew the strong pre-war relationships, in particular between physicists from University of Rome, such as Edoardo Amaldi and Gilberto Bernardini,[8] and Patrick Blackett, then at the University of Manchester, promoting with his group a broadly based cosmic-ray program (Butler 1999; Lovell 1975).

Among the speakers at the Cambridge conference of special interest to Touschek's story, there was Bruno Ferretti, a young theoretician in Amaldi's group, who had given a talk on the absorption of slow mesons by atomic nuclei. He was analyzing the problem of nuclear capture (Ferretti 1947), at stake in cosmic-ray experiments being performed in Rome by Marcello Conversi, Ettore Pancini and Oreste Piccioni, which would soon attract attention on both sides of the Atlantic Ocean (Conversi et al. 1947).

During the Cambridge conference, Amaldi saw the opportunity to renew the pre-war exchanges between Italy and the UK, and was able to secure a British Council fellowship for Ferretti to spend ten months at Manchester in 1946, where Ferretti's theoretical experience in the design and analysis of cosmic ray experiments made him more than welcome. Ferretti was thus in Manchester at the time of major cosmic ray discoveries from UK laboratories, and was invited to teach a course on cosmic radiation at University of Birmingham (Amaldi 1979b, 441). There, he established close contacts with Rudolph Peierls, and wrote with him an article on the quantum theory of radiation damping applied to the problem of propagation of light (Ferretti and Peierls 1947). This problem was related to well known divergences in quantum electrodynamics, a highly debated question at the time and a life-long interest of Touschek's. This common interest and shared acquaintances would be instrumental in Touschek's move to Rome in December 1952.

7.1.3 Post War Revolutions in Particle Physics

The Cambridge conference was a major event which re-established collaborations and communication between the UK and the rest of Europe, but in a few months, the new year 1947 would open up completely new perspectives for elementary particle physics.

Based on Yukawa's meson theory, and related theoretical suggestions, slow positive Yukawa mesons, i.e. positive mesotrons of cosmic rays, traversing matter should strongly prefer to decay rather than be absorbed by a nucleus, because of Coulomb repulsion by protons. Negative Yukawa mesons, on the other hand, should strongly prefer absorption to decay. These predictions were blown to bits by a crucial experiment carried out in Rome during the war by Marcello Conversi, Ettore Pancini and Oreste Piccioni. In four consecutive steps they had directly investigated "the fate of mesons coming to rest in matter" (Conversi 1988, p. 11), disclosing that positive cosmic ray mesons behaved as predicted, but a substantial fraction of negative sea-level

---

[8] An affectionate and humorous reminiscence of Gilberto Bernardini by Leon Lederman can be found in Ledermann (2009). See also Remembering Gilberto Bernardini in Ricci (1995).

cosmic ray mesotrons decayed in a carbon plate (Conversi and Piccioni 1944b, a, 1946; Conversi et al. 1945, 1947). See Piccioni (1988) and Conversi (1988) for a detailed overview of the main steps related to the accomplishment of the four consecutive experiments.[9]

As stressed by Luis W. Alvarez in his Nobel lecture: "[…] modern particle physics started in the last days of World War II, when a group of young Italians, Conversi, Pancini and Piccioni, who were hiding from the German occupying forces, initiated a remarkable experiment…".[10] The surprising results of the Rome experiment showed that negative cosmic-ray mesotrons were almost completely unreactive in a nuclear sense, and provided the first demonstration that this particle was not behaving as it should, if it were the meson predicted by Yukawa as the mediator of nuclear forces (Conversi et al. 1947). The final result of the experiment appeared on February 1st, 1947. Fermi, who had already been informed by Amaldi in winter 1946, immediately reacted writing an article with Edward Teller and Victor Weisskopf, in which a startling conclusion was reached (Fermi et al. 1947, p. 315): "If the experimental results are correct *they would necessitate a very drastic change in the form of mesotron interactions* [emphasis added]."[11]

The question was definitively resolved through a cosmic ray experiment performed by Cecil F. Powell's group in Bristol (Lattes et al. 1947). Improved nuclear emulsions, developed following an idea of the Italian physicist Giuseppe (Beppo) Occhialini, had enabled Powell's group to establish the existence of a new elementary particle in cosmic rays, the π-meson, a strongly interacting particle of short lifetime, which was identified as the Yukawa meson, the accepted quantum of nuclear forces. The mesotron of cosmic rays, now termed the μ-*meson*, was recognized to be the

---

[9] As recalled by Piccioni and Conversi, the first experiment was prepared at the Physics Institute in the Rome University campus, the *Città Universitaria*, close to Termini Rail road station. However, on July 19, 1943, Rome was bombed for the first time by the American Air Force and nearly 80 bombs fell within the perimeter of the University campus, which was located near to the main station, an area especially subject to raids. It was thus decided to transport the experiment to a semi-underground class room of the high-school Liceo Virgilio, transforming the class into a laboratory. The liceo was located in Via Giulia, near the river Tiber, not far from the Vatican City, and hopefully less in danger of being targeted by American bombs. Work was however interrupted by the Italian armistice (September 8, 1943) and, following it, by the occupation of Rome by the German troops. Resumed under very difficult and dramatic conditions, only late in 1943 Conversi and Piccioni finally started to run the first experiment. While they were planning the third experiment, Rome was liberated by the Allied troops (June 5, 1944) and they moved everything back to the Physics Institute at the University. After the liberation of Northern Italy (April 25, 1945), Conversi and Piccioni were joined by Ettore Pancini, a leader in the Partisan movement of the Italian Resistance against German occupation, with whom they performed the third, and then the fourth, decisive, experiment. In this last experiment they replaced the iron with carbon and were able to provide convincing proof that negative mesotrons stopped in carbon did undergo spontaneous decay whereas they did not when stopped in iron. For a reconstruction of these events see also Le particelle elementari by A. Ereditato, one of Pancini's students.

[10] L. W. Alvarez, Recent developments in particle physics, Nobel Lecture, December 11, 1968 http://nobelprize.org/nobel_prizes/physics/laureates/1968/alvarez-lecture.html.

[11] See also Ferretti's publications analyzing results of the Conversi, Pancini and Piccioni experiment (Ferretti 1947, 1948).

product of π-meson decay and was clearly a weakly interacting particle, according to the Rome experiment confirmed by Fermi's interpretation.

Mesotron interactions were still the focus of discussion at the first Shelter Island conference on the Foundations of Quantum Mechanics, held from June 2–4, 1947 near New York. It was the first peace time opportunity for the leaders of the American physics community to meet, exchange new ideas and assess the state of the field, after the Manhattan project (Kaiser 2005). At this time, the May 24 issue of *Nature* with the article by Lattes, Occhialini and Powell had not yet reached the United States, but, not long after, in December of 1947, evidence for the existence of new unstable elementary particles detected in cosmic-ray showers—the so called V-particles because of the characteristic tracks they left in the cloud chamber—was announced by Butler and Rochester, members of Blackett's group in Manchester (Rochester and Butler 1947). Although it took the particle physics community several years to appreciate the importance of the V-particles, the modern day K-mesons, this discovery, together with that of the π-meson, was signaling the existence of a large number of hitherto unsuspected sub-nuclear particles, which would open the new field of particle physics. In parallel with the results of the Conversi, Pancini and Piccioni experiment, these achievements made 1947 a high point in the history of elementary particles, recognized by the Nobel Prize to Blackett, Yukawa and Powell.[12]

This was the scenario which Bruno Touschek stepped into, as he arrived in Glasgow in April 1947. He had lived in Germany the terrible 'Hungerwinter' 1946/1947, noted as the coldest of the 20th century in Europe. The struggle for survival had been extremely painful, and many thousands of people had died because of cold and famine. The winter had severe effects also in the United Kingdom, with fuel and food shortage, and floods caused in March by the thawing snow and heavy rain. Agriculture and breeding were dramatically affected, too, and thousands of British people emigrated, especially to Australia.

At the same time, the initiative of the Marshall Plan, the European Recovery Program, would soon be established by the United States to restart industrial and agricultural production, set up financial stability and expand trade in war-torn Europe.

7.2 1947–1949: Getting a Doctorate in Glasgow

Touschek's time in Glasgow is first presented by Philip Dee in a 1979 letter to Edoardo Amaldi (Amaldi 1981, 9).[13] After Touschek's early death in 1978, Amaldi

---

[12] Blackett was awarded the 1948 Nobel prize in physics 'for his development of the Wilson cloud chamber method, and his discoveries therewith in the fields of nuclear physics and cosmic radiation', https://www.nobelprize.org/prizes/physics/1948/summary/.

[13] One of Glasgow University colleagues, S. Curran, wrote an insightful biography of Philip I. Dee (Curran 1984). Of present note in this biographical sketch are two comments. One, at p. 7, about Dee's publication record, is the fact that machine builders do not have adequate recognition from publication, since during the construction of the machine, their research output in terms of papers is often very poor. This is interesting because the same can be said of Touschek, whose publication

had written to all the people who had been friends with Touschek asking for their recollections and memories.[14] Dee's letter describes Bruno's sparkling intelligence, but also his restlessness and, occasionally, a behaviour outside the accepted norm. On 11th April, 1979, Dee wrote to Amaldi:

> Dear Professor Amaldi,
>
> I did not know about Touschek and I am very sad indeed to have your news.
>
> I will write something for you about his life in Glasgow. For quite a while he lived in our house in the University and became in many ways a member of the family […]
>
> I was quickly impressed by Touschek's obvious ability, his extensive knowledge of physics, and his enthusiasm and I arranged […] for him to have a research appointment in the department, which, at that time, had only one staff member on the theoretical side […]
>
> He was very clever and original, he was also untiringly energetic and extrovert. Bruno led his life to the full extent in all situations and at all times. His enthusiasms were many and, although often brief, were exploited in a manner which most people would have found exhausting.

Dee's high opinion of Bruno is clearly expressed in the application he submitted for a research scholarship for Bruno in June 1947, which is shown in Fig. 7.3. It probably refers to the renewal of the existing scholarships, since at this time Touschek had already been in Glasgow for two months.

7.2.1 Arriving in Glasgow Accompanied by a Guard

When Touschek arrived in Glasgow in April 1947, he was escorted by a guard, as Philip Dee would remember in the letter to Edoardo Amaldi (Amaldi 1981):

> My association with Bruno Touschek began in April 1947 when he was brought to my office under guard (!) for an interview. This had been arranged by Dr. Ronald Fraser (a friend of mine) who had met Bruno when serving on a post-war Allied Commission which was visiting laboratories in Germany and elsewhere. Touschek had expressed the wish to work in a British laboratory and Fraser knew that I had recently come to Glasgow to construct a nuclear physics center in the university here.[15]

Touschek's duties included teaching, research in theoretical physics, and support to the synchrotron program. He had never lectured in English, and this was a challenge. He started preparing a regular course on selected nuclear physics topics, to

---

output dramatically drops after he started working on AdA. Another interesting comment, in p. 5, is that Dee was very kind, and "often he would make ample allowance for illness and the like". This attitude is confirmed in his communications to Amaldi, after Touschek's death, as Dee acknowledges Touschek's original but sometimes unnerving behaviour.

[14] The answers he received are kept in Edoardo Amaldi Papers, Sapienza University of Rome–Physics Department Archives.

[15] From Chap. 6 and the discussion in Bonolis and Pancheri (2019), the examination of Touschek's two letters from Glasgow in April 1946 suggests that Dee may have met Touschek already in April 1946.

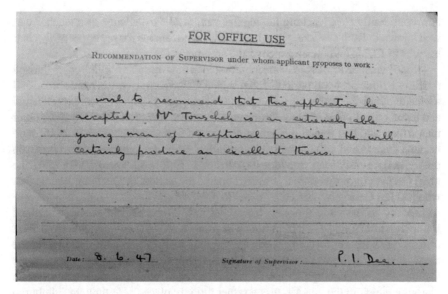

FOR OFFICE USE

RECOMMENDATION OF SUPERVISOR under whom applicant proposes to work:

I wish to recommend that this application be accepted. Mr Touschek is an extremely able young man of exceptional promise. He will certainly produce an excellent thesis.

Date: 8. 6. 47   Signature of Supervisor: P. I. Dee.

Fig. 7.3 Philip Dee's application to the Senate in favour of a research scholarship for Bruno Touschek, by permission of University of Glasgow Archives & Special Collections, University Registry collection, GB248 R18/2/31

be held together with Ian Sneddon, the only other theorist in the department at the time.[16] They outlined the topics, then Sneddon would write down the lectures in "proper english" as Touschek says, and then lectures would be delivered, alternating between the two of them. But his first lecture in English came up as a surprise, as, just on the day before starting his regular course with Sneddon, he was suddenly called to substitute for one of the Professors, who had originally wanted to talk about the synchrotron but had to go to the University Court. Dee asked him to please jump in and Bruno had to rapidly put together a lecture. He picked a subject which he had studied in depth when in Göttingen, so that there would be no problem in preparing the lecture.[17] The well received delivery in English was a confirmation that he could

---

[16] I. N. Sneddon (1919–2000) was born in Glasgow and became a noted mathematician. Following studies in Glasgow and Cambridge, after the war he held a research position in Bristol University, where he worked with N. F. Mott, and authored with him his first book, *Wave Mechanics and its Applications*—which was published in 1948. He returned to Glasgow as a lecturer in physics 1946, or rather natural philosophy as the subject was called in the ancient Scottish Universities at that time. During this period his next major text was the book entitled *Fourier Transforms*, which appeared in 1951, and became such a classic text to be later reissued in 1995. Before the book appeared in print, Sneddon left Glasgow to take up the chair of mathematics at the University College of North Staffordshire (which later became Keele University), and then returned to Glasgow, in 1956 to take up the Simson Chair of Mathematics. Sneddon travelled widely, particularly in North America where he held a number of visiting professorships. From http://www-history.mcs.st-andrews.ac.uk/Biographies/Sneddon.html. See also Chadwick (2002).

[17] Letter to parents, April 23rd, 1947.

now confidently start teaching his regular course. His excellence as a teacher was well known in Rome, where, years later, he became famous for his impeccably clear lectures in Italian and the elegance of the arguments (Margaritondo 2021).[18]

Bruno was soon immersed in the vibrant scientific atmosphere which characterised the early British post-war period, with frequent visits from scientists from abroad and a general atmosphere of expectation and new discoveries, both from an experimental and a theoretical point of view. Bruno and Sneddon, who was only two years older than Bruno and had joined the department just a few months before in 1946, soon became friends and collaborators. They started writing a paper together on meson theory, and, in early May, Sneddon took Bruno to Edinburgh to meet Max Born, the great German born theoretical scientist, one of the fathers of Quantum Mechanics, who had left Germany after Hitler came to power.[19]

Leaving in the early morning from Glasgow, Touschek and Sneddon reached Edinburgh after 1 and 1/2 h train ride. The train crosses the country side from West to East, until the approaching North Sea signals its presence with a change in the clouds and increasing light in the sky. They had time to be tourists before joining the Colloquium. In Edinburgh Bruno discovered a beautiful city, in stark contrast with Glasgow which in the late 40s was a rather gloomy place.[20] He thought Edinburgh most interesting and was enchanted with its hilly setting, the medieval streets and old state prisons, or Princes Street, said to be the most beautiful main street in Europe. In the middle of the city, Bruno and Sneddon visited the ancient castle, from which one can see the Firth of Forth and on clear days the sea, and the city Museum, with paintings by Rembrandt and Brueghel. Bruno's artistic disposition had been nourished through pre-war visits to the great European Museums in Rome and Vienna and fostered by a family immersed in the tradition of the Vienna Secession movement. After the ravages of the war in Germany, where museums had been bombed, masterpieces hidden or expropriated, walking into a museum was like being reborn. His life was picking up again.

In the afternoon they joined the Colloquium. In Edinburgh Born held a bi-weekly seminar, which he had introduced following the German tradition. In the 1920s in

---

[18] See the 2013 interview given by the theoretical physicist Luciano Maiani, 1999–2003 CERN Director General, in the movie Touschek with AdA in Orsay. See also Maiani and Bonolis (2017a).

[19] Max Born (1882–1970) had left Germany in 1933 and in 1936 settled in Edinburgh as Tait professor of Applied Mathematics. The chair had been previously offered to Erwin Schrödinger, but bureaucratic impediments slowed down the hiring process, and, in the meanwhile, a chair in Graz was offered, which he immediately accepted, as it also came with a honorary chair in Vienna (Moore 1989). Born, who had not yet been able to secure a permanent position, and was in India at the time visiting the Indian institute of Science in Bangalore, was glad to accept the offer. In 1954, Born was awarded the Nobel Prize in Physics 'for his fundamental research in quantum mechanics, especially for his statistical interpretation of the wavefunction', sharing it with Walter Bothe "for the coincidence method and his discoveries made therewith". See also Kemmer and Schlapp (1971) in Bibliographical Memoirs of the Fellows of the Royal Society and a translation of the main part of a Memorial address given by Werner Heisenberg on January 12th, 1971 in Göttingen (Heisenberg 1970).

[20] About living conditions in Glasgow, see a recent article in *The Guardian* about the so called 'Glasgow effect'.

Fig. 7.4 Left panel: Max Born and Wolfgang Pauli (at right), in Hamburg, circa 1925, from CERN archives at http://cds.cern.ch/record/42702. This same photograph at https://calisphere.org/item/ark:/28722/bk0016t4k4m/, contributed by UC Berkeley, Bancroft Library, has the two scientists mirror inverted, and is likely to be the original one. Right panel: a photograph of the building housing the Applied Mathematics Department in Drummond Street, in Edinburgh, from Wolf (1995)

the Institute for theoretical physics at the University of Göttingen, one of prominent weekly events had been the Proseminar, conducted by Max Born and James Franck,[21] in which ideas were debated and young graduate students would present their work. It is remembered that in 1926, at one of these seminars, Werner Heisenberg presented a lecture on the development of Quantum Mechanics, and was enthusiastically applauded by the audience (Amaldi 2012, 14). Figure 7.4, shows Max Born with Wolfgang Pauli in 1925, and the Applied Mathematics Department in Drummond Street, in Edinburgh, where Born held the seminar.

Touschek and Born shared many common acquaintances, in particular they both knew Fritz Houtermans, who had graduated in Göttingen with James Frank in the 1920s, when Born was holding the chair of theoretical physics before leaving Germany in 1933. Since these Göttingen times, Houtermans had been imprisoned by the Soviets, divorced and remarried (Amaldi 2012). Fritz Houtermans' witty and adventurous personality was an obvious common argument of conversation. Touschek had been close to Houtermans when in Göttingen in the previous year, in 1946, in particular sharing evening gatherings and occasional visits to the Observatory to watch the stars and had been talking with Houtermans on the last day of his stay in Göttingen, while waiting for the car to take him away, just a few weeks before. This last meeting with Houtermans in Göttingen belonged to another world now. Sharing memories and anecdotes about their common friend soon put Touschek and Born on close grounds. Born, appreciative of Touschek's brilliant intelligence and profound interest in theoretical physics, invited him to the seminars, which were held every week, on Monday and Thursday. Edinburgh, being some 1 and 1/2 hour away by train, biweekly attendance of the Seminar could be difficult, but Bruno decided he

---

[21] The Nobel Prize in Physics 1925 was jointly awarded to James Franck and Gustav Ludwig Hertz "for their discovery of the laws governing the impact of an electron upon an atom".

would do his best to participate at least once a week.[22] Thus they started an acquaintance which later became a true collaboration, with Touschek giving a contribution to one of the Appendices of a new edition of Born's classic *Atomic Physics* (Born 1951), whose first English edition had already appeared in 1935.[23]

In early May, spring was on its way, and the trees were just starting to sprout, but icy winds would whistle down through the fireplaces. It was difficult to believe that summer was just a few weeks away. But it arrived and, early in June, the annual department excursion to the shore took place.[24] Two buses were rented to take the 50 students, the research workers and the lecturers, to the Cobbler, a mountain of 884 m (2,900 ft) height located near the head of Loch Long, a narrow salt water fjord. The trip went past Loch Lomond, where Bruno had already been during a previous outing. They all climbed to the mountain top, including all the girls and Mrs. Dee. Bruno, although he had never climbed in Austria, was confident in his Alpine genes and reached it first. It took more than four hours to be back down. Afterwards, they all lay on the beach along Loch Long, playing ball and throwing stones in the water. Bruno, hot from the climbing and the descent which followed, jumped in the Loch, unknowing of its true nature, getting a good dose of salty water, to everybody's merriment. He could not remember of having had such light hearted fun in a long time. The landscape was astonishingly beautiful, the mood like that of children on some illegal outing, since the trip was not a holiday, but a department yearly date. Bruno could finally appreciate the famed carefree English humour.

In the meantime, Touschek was working hard on topics which would later become the subject of his Ph.D. dissertation. He had also written an article related to the problem of infinities in quantum field theories in which he was commenting on a work by Walter Heitler's assistant, Huanwu Peng (Touschek 1948c).[25] During the war, a radiation damping theory had been proposed by Heitler and Peng, which gave a procedure to calculate scattering amplitudes and extract empirical predictions. The theory could also be used to calculate the nucleon-meson scattering cross sections, or purely electrodynamic scattering processes. Heitler had presented his damping theory at the 1946 Cambridge conference on fundamental particles (Heitler 1947).[26]

---

[22] Letter to parents, May 3rd, 1947.

[23] There is acknowledgment of Touschek's contribution to the volume, but the acknowledgment disappeared from the following editions. This contribution would have remained unknown, had not Touschek included it in his later CV's, when applying for professorship in University of Rome, Bruno Touschek Papers in Sapienza University of Rome–Physics Department Archives. The eighth edition of Born's book was published in 1969, one year before he passed away in Göttingen, on January 5th, 1970. The eighth edition, including revisions by R. J. Blin-Stoyle & J. M. Radcliffe, is available from Dover Publications in paper cover, ISBN 0-486-65984-4.

[24] Letter to parents, June 7th, 1947.

[25] Peng had worked since 1938 with Max Born at University of Edinburgh, then recommended by the latter went to Dublin Institute for Advanced Studies, established by Erwin Schrödinger, Director of the School for Theoretical Physics. He was a post doctoral researcher in Schrödinger's group, from 1941 to 1943, and later an assistant professor until 1947, when he returned to China.

[26] Heitler, who had left Göttingen for the University of Bristol in 1933, had remained there until 1941, when he became a professor at the Dublin Institute for Advanced Studies. In 1946 Heitler became Director of the School for Theoretical Physics after Schrödinger resigned.

However, Heitler's program was now being abandoned after renormalized QED was developed in fundamental papers by Tomonaga, Schwinger, Feynman and Dyson (Schweber 1994).[27] Touschek began also to analyze the use of electrons as particles to produce interactions at a nuclear and subnuclear level (Touschek 1947; Sneddon and Touschek 1948a) and, at the same time, worked on theoretical issues related to the nuclear structure (Sneddon and Touschek 1948c, b), on the eve of the formulation of the nuclear shell model theory.[28]

In August, he developed an idea for the new synchrotron building and started discussing it with colleagues in the synchrotron group. There were trips to Manchester and Edinburgh, and he was busy moving from the University Halls to private lodgings.[29] Not all was work, however. He was also frequently bathing in the Loch Lomond and other nearby places, such as Gerloch, the Clyde and even in the Atlantic, getting tanned as he had never been before.[30] In Fig. 7.5 we indicate some of the location Touschek visited in 1947. Born in a country with no access to the sea, swimming became a long life passion for Touschek. When, in later years, he would be working in Frascati on the electron - positron colliding beam accelerators he proposed and had built, his frequent trips to nearby Lake of Castel Gandolfo were memorable. Aunt Ada had a villa nearby, and bathing and fishing in the lake would be a favourite pastime. He would just leave the laboratories in the early afternoon and escape to the small town of Albano, perched on the rim of the now extint vulcan housing the lake. He also loved the sea. After Aunt Ada passed away in 1959, he started spending some time on the south of Rome, in Sperlonga (Amaldi 1981), or on the Amalfi coast (Pancheri 2004), swimming in the crystal clear waters of Positano, which was chosen for family holidays in September, when the vacation crowd has left, the Mediterranean sea is calm, and the water still warm. But all this was in the future, when he would have settled in Rome and the Glasgow years were a distant past.

The first summer in Glasgow being over, he dutifully kept up the correspondence with his family in Vienna, where life was still difficult, worrying about their well being, proud to help his parents with his earnings, sending both money as well as packages. Then an obstacle appeared in this otherwise almost idyllic picture. Because of administrative requests, some problems had arisen with the D.S.I.R. about his

---

[27] The Nobel Prize in Physics 1965 was awarded jointly to Sin-Itiro Tomonaga, Julian Schwinger and Richard P. Feynman 'for their fundamental work in quantum electrodynamics, with deep-ploughing consequences for the physics of elementary particles', from https://www.nobelprize.org/prizes/physics/1965/summary/.

[28] The nuclear shell model was formulated by Maria Goeppert Mayer and, independently, by Hans Jensen, Otto Haxel and Hans Suess, Touschek's good friends since Hamburg and Göttingen times. Goeppert's first article summarizing the evidence for a shell model of the nucleus appeared in August 1948 (Goeppert Mayer 1948). Her second decisive paper (Goeppert Mayer 1949) was published together with the one by Jensen and colleagues, in June 1949 (Haxel et al. 1949).

[29] In September his address was c/o Fisher, 16 South Park Ave, Glasgow W2. Letter to parents September 2nd, 1947.

[30] Letter to parents, September 2nd, 1947.

Fig. 7.5 A view of places where Touschek went in 1947: at left for the excursion to the Cobbler, with Loch Lomond in the center, Glasgow -not shown- just outside the lower right hand corner. At right a larger view from Scotland to Manchester, down in the middle of England. Both images from Google Maps

contract after the first six months in Glasgow.[31] As the renewal of his fellowship ran into some delay, he started wishing he could leave the UK.[32] He was very annoyed, as he would often be throughout his life when similar impediments forced him to interrupt his work and studies. But the problem was solved, by getting the needed documents from Austria.

All through this first year, Bruno worked hard with the newly established synchrotron group in Glasgow, which included Samuel Curran, in Fig. 7.6, Walter McFarlane, A. C. Robb and, Philip Dee. As one can see from Curran et al. (1948), where Touschek's collaboration is acknowledged, he worked with Samuel Curran on a problem related to measurements and results obtained with the proportional counter, a

---

[31] A copy of his birth certificate, attesting to his Austrian citizenship, was required to renew his contract. The problem was soon solved, though, and the birth certificate received by the D.S.I.R. on September 1st, 1947.

[32] Letter to parents, October 11th, 1947. The letter ends with two verses, paraphrasing a popular song, Sowieso: "Egal was kommt, es wird gut, sowieso/Immer geht 'ne neue Tür auf, irgendwo/Auch wenn's grad nicht so läuft, wie gewohnt/Egal, es wird gut, sowieso [No matter what, it's going to be fine, anyway, a new door always opens, even if it's not going as usual, it's going to be fine, anyway.]" https://lyricstranslate.com/en/mark-forster-sowieso-lyrics.html. The letter implies that happiness may always be elsewhere.

Fig. 7.6 Bruno Touschek with Samuel Curran, at University of Glasgow, around 1948. Family Documents, © Francis Touschek, all rights reserved

new device Curran was developing with his student John Angus (Close 2015, 113–115).[33] This work with Curran, is also described in the December 1947 research report Touschek submitted to the University.[34]

The Glasgow group started preparing plans for the synchrotron and meetings were held with other UK groups working on synchrotrons.[35] The list of attendees of the 8th Meeting of the Glasgow synchrotron group held on November 4th, 1947 in Manchester, at Trafford Park, on the grounds of the Metropolitan-Vickers company, shows that all the major actors in electron synchrotron preparations in the UK were present, among them Frank Goward, who had succeeded in building the world first

---

[33] Before the war, Samuel Curran (1912–1998) had been at the Cavendish Laboratories with Rutherford in Cambridge. In late 1940, he joined Philip Dee, Bernard Lovell, Alan Hodgkin and others at Worth Matravers, in Dorset. A recollection of this period by Bernard Lovell can be found in the 28th October 1982 issue of the *The New Scientist*, p. 246. After the invention of the cavity magnetron, Curran's development of spark gap modulators was critical to the success of the magnetron transmitters. In 1944 he went to the United States with a number of other UK other scientists to work on the Manhattan Project, and in 1946 he went to Glasgow University to join Philip Dee. He assisted Dee and McFarlane in the installation of the 300 MeV Glasgow Synchrotron, which became operational in 1954. Later, as Principal and vice chancellor of the University of Strathclyde, he took the lead in developing Britain's first technological university, as from http://www.purbeckradar.org.uk/biography/curran_sam.htm and https://www.worldchanging.glasgow.ac.uk/article/?id=39. See also Fletcher (1999).

[34] Research Reports (Glasgow), Dic. 1947-Apr. 1948, Bruno Touschek Papers in Sapienza University of Rome–Physics Department Archives.

[35] He writes about his work on the 300 MeV synchrotron in the October 11th, 1947 letter to his parents.

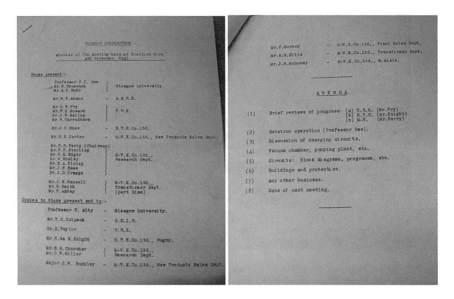

Fig. 7.7 List of attendees and agenda of the 8th meeting of the Glasgow Synchrotron, held in Manchester, at the Metropolitan-Vickers factory in Trafford Park, courtesy of Sapienza University of Rome–Physics Department Archives, https://archivisapienzasmfn.archiui.com, documents provided for purposes of study and research, all rights reserved

electron synchrotron, just over one year before. A copy of the list of participants is shown in Fig. 7.7.[36]

In December Rudolf Kollath and Gerhard Schumann, with whom he had collaborated in the 15 MeV-betatron designed by Widerøe and built in Hamburg, published a review article on the work done during the war and in the early post-war period at Kellinghusen. It included all the important information and many details, making Bruno Touschek's contribution to the project officially known (Kollath and Schumann 1947).[37]

As an expert in the field, Touschek continued to be involved with betatrons, as a consultant for the 20 MeV machine operating at the High Voltage Laboratory of the Metropolitan-Vickers at Manchester.[38] However, his knowledge on accelerators was

---

[36] The agenda for a previous meeting held on September 7th is also available in Sapienza University of Rome–Physics Department Archives, https://archivisapienzasmfn.archiui.com.

[37] Touschek sent his parents a copy of the issue of the Journal of Applied Physics containing the article (letter to parents, October 11th, 1947).

[38] See letter by John D. Craggs of January 27th, 1948: "You may remember that we had a short discussion on the peculiar spectrum we had found for our 20 MeV betatron [...] you may have been able to do some thinking about the problem. We should be most grateful for any light you can throw on the problem." Bruno Touschek Papers, Sapienza University of Rome–Physics Department Archives. After graduating from King's College London in 1938, Craggs had joined the staff in the High Voltage Research Laboratories of Metropolitan Vickers Ltd., where—except for a period in Berkeley, California, in 1944–1945—he stayed until moving to Liverpool University in 1946 as

quickly evolving. From his research report on work carried out during the months December 1947–January 1948,[39] we learn that he was preparing a lecture "on the family tree of accelerators (Cyclotron, Betatron, Synchrotron, Synchro-Cyclotron)" as well as a related paper on the synchrotron for the newly founded *Acta Physica Austriaca*, published by the Austrian Academy of Science, of which Hans Thirring was co-director at the time (Touschek 1949).[40]

Back in Europe, among his mentors and friends, there was appreciation for his achievements, as it appears from a postcard, sent by Arnold Sommerfeld to Paul Urban, on November 2nd: "Touschek has had great success, in Göttingen he passed all his exams one after the other, is now sought after by Hamburg and Hannover and is presently in England [sic!], well paid by the British".[41]

7.2.2 Bruno Touschek and Werner Heisenberg

All along during the doctorate years, Touschek's published theoretical physics output is remarkable. In what follows, we shall highlight contacts and, so far unpublished, correspondence with Werner Heisenberg, which shed light on his formation as a theoretical physicist. Such scientific correspondence was a natural continuation of the relationship established before Touschek's arrival in Glasgow, during his stay in Göttingen, while getting his Diploma and afterwards, when he was, for some time, one of Heisenberg's assistants at the Kaiser Wilhelm Institute (KWI) for Physics.

In un undated manuscript, Touschek remembered having seen Heisenberg for the first time in 1939, giving a public lecture in Vienna, when Heisenberg was already one of the most prominent and influential German physicists. During the war Heisenberg had become director of the Kaiser Wilhelm Institute for Physics in Berlin-Dahlem, leading the German nuclear project and it is in Berlin that Touschek met him for the second time: [...] hatless hurrying to the KWI & I asked him the way because I wanted to visit him & had not re-cognised him. He brought me to his office [...]. After the war, in Göttingen, Touschek attended his lectures on quantum field theory, and later commented on them, in the same undated manuscript: It was not a good course of lectures, but there was one among them, which for me was a complete eye-opener: the harmonic oscilla-tor & its quantization. I had learned Q.T. [Quantum Mechanics] from Sommerfeld's "wellenmechanisches

---

a lecturer in Electrical Engineering. His work in Manchester on the neutron generator resulted in the building of the first full-size Van de Graaff in UK (News from the Archive of the University of Liverpool, http://sca-arch.liv.ac.uk/ead/search?operation=full&recid=gb141unistaffc-d-d835).

[39] Research Reports (Glasgow), Dic. 1947-Apr. 1948, in Bruno Touschek Papers, Sapienza University of Rome–Physics Department Archives.

[40] See also manuscript "Zur Theorie des Synchrotrons", Bruno Touschek Papers, Sapienza University of Rome–Physics Department Archives.

[41] Edoardo Amaldi Papers, Sapienza University of Rome–Physics Department Archives.

Fig. 7.8 The January 12th, 1948 Heisenberg's letter to Touschek, announcing his forthcoming visit to the UK, Family Documents, © Francis Touschek, all rights reserved. At right, Werner Heisenberg at Göttingen, 8.3.1949, from left to right: Otto Hahn, K. F. Bonhoeffer, Werner and Elisabeth Heisenberg, Werner Hoppenstedt, courtesy Archives of the Max Planck Society, Berlin

Ergänzungsband" & I had tried Dirac's famous book both of which lean heavily on wave mechanics. H.'s lecture opened my understanding to the mechanical approach.[42]

Among Touschek's unrecorded work there are his studies of the analyticity properties of Heisenberg's S-matrix.[43] Touschek was bent on understanding Heisenberg's proposed theoretical approach to particle dynamics (Heisenberg 1943b). Heisenberg's seminal work, *Die 'beobachtbaren Größen' in der Theorie der Elementarteilchenphysik*, namely *The 'observable quantities' in the theory of elementary particle physics*, was focused on dealing with observable quantities, rather than order by order perturbation theory calculations, plagued by divergences at a given fixed order.[44] This work had a profound influence on Touschek and is echoed in his later work about infra-red radiative corrections to electron-positron experiments, where Touschek reflected about the incompatibility between "the picture of an experiment [as] drawn by theory and reality" (Etim et al. 1967a).

The analyticity properties of the S-matrix were a strongly debated topic in the theoretical physics community and Touschek had discussed the argument with Heisenberg when he was still in Göttingen and then again after joining Glasgow, Fig. 7.8. In particular, since fall of 1947, Touschek exchanged letters with Werner Heisenberg focused on the S-matrix and other theoretical issues which he was studying at the moment.[45] In January 1948, Heisenberg involved Touschek in a discussion about a work on the S-matrix, which he had just received from Ning Hu, one of Walter Heitler's collaborators in Dublin, but at the moment in Niels Bohr's Institute

---

[42] Quotations are from manuscript in Family Documents, © Francis Touschek, all rights reserved.

[43] Correspondence with Heisenberg cited here is partly kept in BTA, partly in Touschek's Family Documents, © Francis Touschek, all rights reserved.

[44] See Rechenberg (1989) for a historical view of the S-matrix development from 1942 to 1952.

[45] See letters dated October 6th (Touschek to Heisenberg), October 10th, 1947 (Touschek to Heisenberg), Bruno Touschek Papers, Sapienza University of Rome–Physics Department Archives.

in Copenhagen.[46] Heisenberg wanted Touschek to look into Ning Hu's work because he thought it to be in some contradiction with Touschek's unpublished notes.[47] In asking him to check Hu's calculations, Heisenberg suggested him to directly contact Hu. In the same letter, he also announced that he would be in Cambridge for six weeks, starting on January 26th.[48]

The correspondence continued through January, with Heisenberg writing from Cambridge on January 28th, 1948, discussing further points about the S-matrix, Fig. 7.9. Other letters followed.[49]

The exchanges on the S-matrix properties between Touschek and Heisenberg appear in the research report Touschek submitted to the University of Glasgow, for the periods December 1947 & January 1948, Fig. 7.10 (left). In the report, one of the mentioned items is a 'triangular discussion' between Heisenberg, Hu and Touschek, reflecting a correspondence with Hu, mostly lost. A second report was submitted on May 2nd for the period February 1st to April 30th 1948, Fig. 7.10 (right). From this, we learn that Touschek met Heisenberg in Manchester about the matter. In the meantime, on April 15, Hu's article "On the Application of Heisenberg's Theory of S-Matrix to the Problems of Resonance Scattering and Reactions in Nuclear Physics" was submitted to the *Physical Review* and appeared in the July 15 issue of the journal (Hu 1948). On May 25 Touschek thanked Hu for his manuscript, that he was carefully reading, also mentioning previous letters they had exchanged (but which are not present in Touschek's papers in Rome). Finally, on July 27, Touschek wrote to Heisenberg that he had a rather lengthy correspondence with Hu and at the moment

---

[46] After graduation from Tsinghua University in Beijing, Ning Hu had moved to the United States. He had obtained his Ph.D. from Caltech and during 1944–1945 had studied quantum field theory under Wolfgang Pauli at the Institute for Advanced Study in Princeton. During his stay in Europe from 1946 to 1949 he visited Walter Heitler in Dublin and Niels Bohr in Copenhagen. In 1980 he went back to China, and was a long-time professor in the department of physics at the University of Beijing.

[47] The January 12th, 1948 Heisenberg's letter to Touschek, also announcing his forthcoming visit to the UK, is in Touschek's personal papers, courtesy of Mrs. Elspeth Yonge Touschek.

[48] During Heisenberg's visit to Cambridge, Peierls extended an invitation to Heisenberg to come to Birmingham, letter [445] in Lee (2009), and discuss the matter of German scientists responsibility and different attitudes towards the Nazi regime. To this Heisenberg replied in letter [447] in Lee (2009): "Dear Peierls! Thank you very much for your letter. I like to come to Birmingham and I can talk at your seminar about the little thing I think I know about the theory of elementary particles. I also thank you very much for being open to your opinion about a difficult political problem. It is as you suspect: I do not agree with you. But the fact that you wrote to me so openly gives me hope that in a conversation, if not an approximation of the viewpoints so as to understand the other viewpoint, can come. As to the timing: From 10th to 12th I am at Blackett's [in Manchester] in March; I could come to B [Birmingham] on the 12th and stay until the evening of 13. Possibly also March 8 and 9, if from the planned visit to Oxford would come to nothing. Would that be ok for you? Goodbye".

[49] On January 26 and 27 (Touschek to Heisenberg), on January 28 (Heisenberg to Touschek, from Cambridge, UK), on January 31 (Touschek to Heisenberg) on February 29 (Heisenberg to Touschek), on February 23 (Heisenberg to Touschek from Cambridge, Cavendish Lab), on February 29 (Touschek to Heisenberg), on April 20 (Heisenberg to Touschek), on May 2 (Touschek to Heisenberg), Bruno Touschek Papers, Sapienza University of Rome–Physics Department Archives.

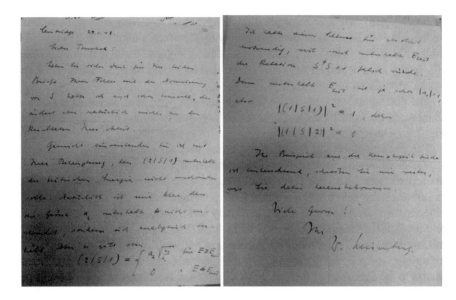

Fig. 7.9 Werner Heisenberg's January 28th, 1948 letter to Bruno Touschek, from Bruno Touschek Papers, courtesy of Sapienza University of Rome–Physics Department Archives, https://archivisapienzasmfn.archiui.com, documents provided for purposes of study and research, all rights reserved

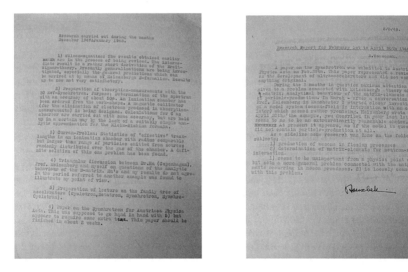

Fig. 7.10 Touschek's research report submitted to University of Glasgow in 1948, Bruno Touschek Papers, courtesy of Sapienza University of Rome–Physics Department Archives, https://archivisapienzasmfn.archiui.com, documents provided for purposes of study and research, all rights reserved

he did not think that Hu's work was correct.[50] From the available documents, we
see that the exchanges on this subject between Touschek and Heisenberg continued
through part of the summer 1948. His closeness to Heisenberg is also seen in a
sequence of letters between Touschek and a colleague, B. J. Warren, about a position
in Vancouver offered to Dr. H. Koppe, a member of Heisenberg's group in Göttingen.
However, after July 27th, 1948, no letters are present in Bruno Touschek Archives
in Rome up to 1958, when correspondence with Heisenberg started again.

The correspondence is proof of the trust held by Heisenberg in the young theoretician, how much Bruno would be learning from this exchange of ideas, and how
much he was deeply involved in strongly debated theoretical questions of his time.
The available documents indicate that the exchanges with Heisenberg during the
Glasgow period lasted about 2 years. There is no doubt these exchanges had a profound influence on Touschek. The direct confrontation with one of the proponents
of Quantum Mechanics, a scientist of great intellectual and scientific stature, did
influence his formation. He could not avoid feeling proud of Heisenberg's clearly
good opinion, and his own self-esteem could now be enforced. He would still make
mistakes, as we shall see from his correspondence with Born, but, encouraged by his
exchanges with Heisenberg, he had no reasons to doubt anymore his capacity to do
physics.

We shall now step back, to see how Bruno's other physics interests and personal
life unravelled during 1948, the first complete year he spent in Glasgow.

7.2.3 1948: Settling in Glasgow

At the end of 1947, Bruno was finally able to travel to Vienna and see his parents,
probably for the first time after the end of the war. He left London on December 16th,
spent Christmas and New Year with them, and was back to Glasgow on January 6th
1948.[51] In between, combining work and family, on his way back from Vienna in
early January, Touschek passed through Malvern, to see the synchrotron with Donald
Fry.[52] The trip to London is also mentioned in letter to parents on January 28th, 1948.

The long sought reunion with his parents fortified his spirits and carried him
through the harshness of the Glasgow winter and his frequent travels to England or
to Edinburgh. His research report for December 1947 and January 1948 describes
many different physics projects, about accelerator physics and possible experiments,
and, of course, theoretical physics. In addition to giving lectures and working with
the Glasgow group on designing and planning the 300 MeV synchrotron, he was

---

[50] BTA, Box 1, Folder 1. Touschek's work remained unpublished, but it is a very interesting result
quoted by the noted theoretical physicist R. J. Eden in a Royal Society communication presented
by Dirac (Eden 1949).

[51] See January 7th, 1948 expense note sent to the D.S.I.R., in BTA, Series 1, folder 1 (boz 1) in
Correspondence 1947–1949.

[52] An expense note to the D.S.I.R. on January 7th, 1948 mentions travelling through to London
onto Malvern.

involved with the operation of the 30 MeV electron synchrotron in Manchester, and giving support to the group taking measurements with the 20 MeV betatron, as also seen by the intense exchange of letters between Touschek and the colleagues in Manchester, which took place through the first months of the year.[53] Both Bosley and Cragg from Metropolitan-Vickers in Manchester came up to Glasgow to discuss with him about their results on X-radiation from the 20 MeV betatron.[54] Touschek's contribution was later acknowledged by the authors as their being "greatly indebted to Mr. B. Touschek" (and F. K. Goward) for "instructive discussions on the work" (Bosley et al. 1948). He also was in correspondence with Goward in Malvern.[55]

But he was keen to continue on with theoretical physics. In January he submitted a paper on the double β-decay to the *Zeitschrift für Physik* for a special issue prepared to celebrate Lenz' 60th birthday (Touschek 1948a).[56] His contribution had been asked by Hans Thirring, the former professor of Theoretical physics at University of Vienna until 1938, reinstated to his position after the war and dean of the Philosophical Faculty at University of Vienna in 1946–1947. During the war years, Touschek had kept in close touch with Hans Thirring, visiting him whenever he could go to Vienna.

---

[53] Letter to F.K. Goward January 8th, 1948, and exchanges with Bosley in February, Bruno Touschek Papers, Sapienza University of Rome–Physics Department Archives.

[54] In Manchester, the Metropolitan-Vickers Electrical Company had participated intensively to the war effort, including having some of its scientists released to work in the United States for the atomic bomb effort. The company was also active in the field of nuclear physics, and a High Voltage (HV) laboratory had been opened in 1930 by Ernest Rutherford, director of Cavendish laboratory at the time, who had been professor in Victoria University in Manchester from 1907 until 1917. At the HV laboratory, cyclotrons had been constructed in 1938, one with Cockcroft for the Cavendish Laboratory, one to be installed at Liverpool for James Chadwick, the discoverer of the neutron. After the war, a research group, finalised to work on accelerators, was started in 1946, and a 20 MeV betatron, the first in the country, was designed and built. A collaboration, also involving the Telecommunication Research Establishment (TRE) in Malvern, was started with Philip Dee in the design of the 300 MeV synchrotron in Glasgow, as described by John Dummelow in the section about *Nuclear Physics* in *Metropolitan-Vickers Electrical Co 1899–1949*, https://www.gracesguide.co.uk/Metropolitan-Vickers_Electrical_Co_1899-1949_by_John_Dummelow:_1939-1949. Metropolitan-Vickers also provided diffusion pumps for the Malvern synchrotron.

[55] The history of the contribution to science and technology from Malvern is kept alive by the Malvern Radar and Technology History Society (MRATHS), a registered charity No 1183001.

[56] Letter to parents, January 28th, 1948. Already in November 1946, Touschek had announced to his parents, that he had come back from one of his preliminary travels to UK with an article on such topic "in his pocket". However he also told having found an error and that it took him a couple of months to set the problem and that it would appear in the *Zeitschrift für Physik*, but further problems must have hindered publication, probably also a normal delay as it happened for all scientific journals at the end of the war. The submission date is marked December 1946, but the article appeared only in early 1948 and was included in the issue dedicated to Lenz. In a later manuscript note, Touschek wrote: "The problem which bothered H. [Heisenberg] & which he asked me to unravel was 'double β-decay'. Haxel felt he could just do it (experimentally) & I ran into the difficulty of distinguishing what was arbitrary & what was sound in Fermi's theory of 'weak interactions.' I saw that clearly in 1947, but what I wrote then was riddled by stupid mistakes, which H. Did not—or did not want to—see" (manuscript, courtesy of Mrs. Elspeth Yonge Touschek). This research argument may also have been suggested by Paul Urban just before Touschek moved from Vienna to Hamburg, (Urban to Amaldi, June 3rd, 1980, Edoardo Amaldi Papers, Sapienza University of Rome–Physics Department Archives.

Fig. 7.11 The letter which Philip Dee wrote to Rosa Touschek reassuring her that Bruno was fine and striving. At right, a snapshot taken on board returning from an excursion to Rothsay, with Walter McFarlane at left and Touschek at right. Samuel Curran was also part of the same excursion, but is hidden behind Touschek, and Miss Meriç is at the center. Family Documents, © Francis Touschek, all rights reserved

The request was an acknowledgment of Bruno's standing as a promising theoretician and of Touschek's closeness to Lenz during the war period, when he was attending his lectures in Hamburg as an unregistered student, and helping him during bombing alarms (Amaldi 1981, 4). [57]

In Glasgow, Bruno was also enjoying the new environment, notwithstanding occasional bouts of annoyance towards a world which appeared to be moving too slowly, as he was anxious to regain the lost years of school.

His involvement with the social life of the department is aptly described by Philip Dee in a letter he sent to Touschek's step mother, Rosa Touschek. Busy between lecturing and doing synchrotron work in Glasgow, traveling to Manchester, Malvern and Birmingham, attending Born's Seminar in Edinburgh, meeting and writing to Heisenberg, it is no wonder that Bruno may have neglected to write to his parents as frequently as had been his custom since his young age, when away from home. Thus, one day, on April 7th, 1948, his step mother took the extraordinary step to write to Dee and inquire about Bruno's well being. Dee replied describing Bruno as "perfectly well and happy", adding how glad he was to have him there, "not only because of his value to the department, but also because his attractive personality makes him a welcome addition to the social activities of our group", Fig. 7.11 (left). This comment by Dee is confirmed by a snapshot taken during a later excursion, shown in the right hand panel of Fig. 7.11. The group is identified by Touschek in

---

[57] Rolf Widerøe mentions to Amaldi after Touschek's death: "Touschek lived in the flat of Professor Lenz in Hamburg [...] and he had considerable difficulty bringing the old and often sick man to the cellar when the bombers came". We notice that Widerøe's description is not completely accurate, since Touschek did not live with him—as it is clear from his March 13th 1945 letter to parents—but Lenz was indeed often sick, as described by Touschek in Chap. 5.

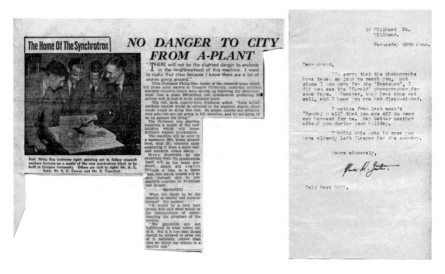

Fig. 7.12 At left, cutting of an article in the local Glasgow daily. The photograph is likely to have been among those included by a journalist from *The Scotsman*, in a June 30th, 1948 letter, right panel, where Touschek's harvesting in Northern Scotland is also mentioned. Family Documents, © Francis Touschek, all rights reserved

the back of the photo, the young woman in the center of the photograph being Miss Merriç, a graduate student from University of Instanbul, who received her Ph.D. in physics in November 1949, in the same session as did Touschek.[58]

As the work on the Glasgow synchrotron progressed, rumors spread, through the town and the countryside, about University professors planning to build some 'atomic' project. To reassure the public of lack of any danger, Philip Dee had to give an interview to the Glasgow Herald, as we can see from a contemporary newspaper cutting shown in Fig. 7.12. The date of this article is not known, but it is likely to have been published in mid 1948, when Touschek was still strongly involved in the synchrotron project.[59] Part of his activity includes exchanges with Emlyn Rhoderick, who was working on the Cavendish cyclotron in Cambridge at the time, and would soon join Glasgow University.[60] Bruno was concerned about the treatment of Coulomb interaction in meson scattering, and the related divergences. This problem was strongly debated, and of interest also to Rudolph Peierls, who visited the

---

[58] Courtesy of University of Glasgow Archives & Special Collections, University collection, GB 248 DC 157/18/56.

[59] Letter to parents, July 3rd, 1948.

[60] Emlyn H. Rhoderick (1920–2007) worked at the Royal Signals and Radar Establishment during the Second World War on coastal defence radar, and studied physics at Trinity College, Cambridge. He then taught at Glasgow University, and went on to become professor of solid-state electronics at Manchester.

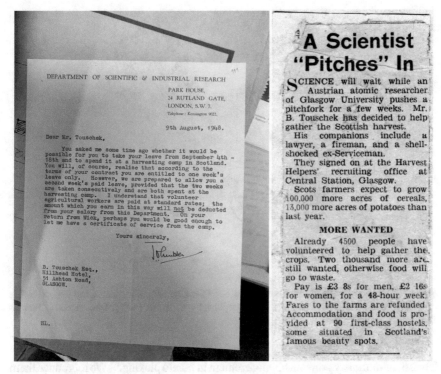

Fig. 7.13 At left, copy of the D.S.I.R. letter of August 9th, 1948, granting Touschek leave to participate in the harvest, Bruno Touschek Papers, courtesy of Sapienza University of Rome–Physics Department Archives, https://archivisapienzasmfn.archiui.com, documents provided for purposes of study and research, all rights reserved. At right a newspaper article about Touschek's participation to the summer harvest in Northern Scotland. Family Documents, © Francis Touschek, all rights reserved

Cambridge group in August, as we learn from Rhoderick's August letter to Touschek.[61] Physics was not all encompassing, however. Always a lover of nature, Bruno joined his colleagues in excursions to the islands, to Rothsay as we have seen, and, in summer 1948, went harvesting to Wick, a town in Caithness county, very far up North. In Fig. 7.13 we show the letter from the D.S.I.R. allowing him to take leave to participate in the harvest in Northern Scotland and, to the right, a notice from the Glasgow Herald, which mentions these activities.

Bruno's spirits in this period were high, and he started including little drawings in the letters to his parents, as he had continuously done through the war years until September 1944. Since then, however, no drawings are present in the home letters until summer 1948, when he related to his parents some adventures of the summer

---

[61] See Bruno Touschek Papers, Sapienza University of Rome–Physics Department Archives, Box 1, Folder 1.

Fig. 7.14 Drawings included by Touschek in one of his letters home, describing his harvesting time in Northern Scotland in summer 1948, graphics by A. Ianiro. Family Documents, © Francis Touschek, all rights reserved

harvest, humorously drawing his engagement in potato picking.[62] We reproduce them in Fig. 7.14.[63] The reappearance of his drawings and the playful nature of their content signal a renewed confidence in his abilities. After the traumas of imprisonment, the tragedy of immediate post-war months in Hamburg, the displacement to Göttingen, and the move to Glasgow, he was relaxing, engaging with fellow harvesters, and enjoying, it seems, the company of a painter, as he was used to have in his family circle in Vienna.

Returning from the harvest, he found an unpleasant surprise. In his recollections to Amaldi, Dee related the incident, writing (Amaldi 1981, 9):

> Naturally, after this early period he gave me many problems! The first was his housing. A small lodging house seemed satisfactory for a while, but after a short 'holiday', which he spent potato picking in the north of Scotland, under spartan conditions, but fortified by the prospect of an early return to his comfortable room in Glasgow, this arrangement came to an abrupt end. On his return he found that the landlady had changed his curtains without prior consultation and, enraged by this destruction of his anticipated homecoming, he immediately returned the curtains to the manageress with a demand for instant restoration of the original ones.

---

[62] One can find some anguished scribbles or doodles on the back of 1947 or 1948 letters he kept in his office at University of Rome and presently in Bruno Touschek Papers, Sapienza University of Rome–Physics Department Archives, *Corrispondenza varia, anni 1947–1949*.

[63] Letter to parents, September 3rd 1948 [our dating], with written date 3/8/48, probably but a typo, in place of 3/9/48.

Fig. 7.15   From left, Paul A.M. Dirac, Wolfgang Pauli and Rudolf E. Peierls, at the 1948 International Conference on Nuclear Physics, in Birmingham: Science Museum London/Science and Society Picture Library, CC BY-SA 2.0, https://creativecommons.org/licenses/by-sa/2.0 [creativecommons.org], via Wikimedia Commons. Right panel: Donald Kerst around 1950 wielding a soldering gun and drill, from https://repository.aip.org/islandora/object/nbla%3A298184 © Department of Physics, University of Illinois at Urbana-Champaign, courtesy AIP Emilio Segré Visual Archives

This request was obviously refused, and one more attempt to find a suitable lodging took Bruno from 51 Ashton Road to Kew terrace.[64]

Shortly after the harvesting holiday, physics was once more at the center of Touschek's life. Between 14th and 18th September, a conference was organized in by Rudolph Peierls, Fig. 7.15.[65] Touschek participated in the conference, where he had the occasion to discuss with some well known scientists, such as Donald Kerst, Wolfgang Pauli, Maurice Pryce, a mathematician and theoretical physicist from Oxford University, and, at the time, Max Born son-in-law (Elliott and Sanders 2005).[66] In Birmingham, he might also have discussed with Rudolph Peierls, about his ongoing work for the Ph.D. dissertation.

In those days, Touschek was working with Sneddon on meson production with electrons and after the conference sent to Donald Kerst a preliminary version of the paper, as possible example of meson (particle) experiments one could do with 300 MeV. At Kerst's Institute at University of Illinois, they worked on the same problem

---

[64] Earlier address given in May 23rd, 1948 letter to parents and c/o Mrs. Boyle in October 29th, 1948 letter.

[65] See Peierls' letter to Hans Bethe [448], in Lee (2007).

[66] Letter to parents on October 10th, 1948 from Birmingham, September 28th, 1948 letter to Kerst, September 27th, 1948 to Pryce and (undated) reply from Pryce, Bruno Touschek Papers in Sapienza University of Rome–Physics Department Archives.

(electron excitation), but Touschek thought they were about one year behind.[67] The paper was later submitted and published. He also had news of his friend Fritz Houtermans coming to England, with whom he had quarrelled for no reason Touschek could remember. It was a silly thing, and Bruno sent him a postcard with a view of the place he had been in Caithness county and a short amicable message. This restored them to the past friendship, and Houtermans, who was at the time in England, proposed to visit Bruno in Glasgow. Bruno was elated at the idea and telegraphed back "By all means come!".[68] But the encounter fell through and they do not seem to have met in this occasion.

In October as the summer fun was over and the days shortened in the approach to winter, Touschek was once again feeling restless and unhappy. He was holding a Nuffield fellowship, his previous contract having expired, and he inquired from Dee what could come next. Dee's frank answer includes the possibility of a professorship in two years after the doctorate, but with some *caveat*: Bruno needed to accept life in the UK, and feel more at ease, as he was discussing and occasionally quarreling with people a bit too often for his comfort. The real point is that Bruno was anxious to progress, to be closer to the places where theoretical physics was taking giant steps forward, such as it was indeed happening in particular in the United States. He feared remaining isolated in Glasgow, without being able to keep the needed intellectual connection to other theorists. He felt that perhaps he should again be moving away, and that he may have been wasting his time. It should be added that he may very well have been going through some exhaustion and its consequent depression state. He was in fact still continuously travelling, like having to be in Malvern on Monday and back to Glasgow on the Tuesday.[69]

In December, painful memories were coming back, as a new movie from Germany was released and shown in Glasgow's Cosmo theatre to a packed audience. The movie, entitled *Die Mörder sind unter uns*, was shown in various countries around the world, and was seen by Touschek together with the whole physics department.[70] The action of the film, by the German director Wolfgang Staudte, took place in allied-occupied Berlin, where Touschek had lived during the war. It was one of the first post WWII German films, the first to use a setting for the story showing the consequences of the bombings, with piles of rubble and destroyed buildings. It was produced by a company, DEFA, established in the Soviet occupied zone. Its aim was

---

[67] Kerst was constructing such a betatron in the United States, which became operational in 1950. See also Kerst memoir in http://www.nasonline.org/publications/biographical-memoirs/memoir-pdfs/sessler-andrew.pdf.

[68] Houtermans' letter to Touschek is kept in Bruno Touschek papers in Rome, dated October 25th, 1948. On the top of this letter Touschek scribbled "By all means come! Wire date!" Then adds his address and the word "sent" pointing to Houtermans' address. The letter and Touschek's added words suggest the text of a telegram he sent to Houtermans.

[69] Letter to parents, October 5th, 1948, also about meeting Kerst in Birmingham.

[70] The title of the movie, *The murderers are among us* in English, recalls the title of the 1921 Fritz Lang's movie *M*, originally *Mörder unter uns*.

to urge the public to see and judge those responsible for the atrocities committed during the war.[71]

None of this could be soothing Bruno's anxiety and possibly incoming depression. In addition, as the year 1948 drove to its end, Touschek went through one more change of lodgings. The occasion amounts to an almost comic story, with a landlady quarrelling with her landlord husband, the husband hitting the wife, Touschek trying to defend the wife and being hit by the husband, who finally called the police. The story is related by Bruno in a letter to his parents, but in later years he narrated it to his friends in Rome, who picked it up to become an often relished anecdote about Touschek's life in the UK. In Carlo Bernardini's version, Touschek described Mrs. Boyle's house in Ashton Road as *una casa piena di generali*, a house full of generals.[72] In his letter to Amaldi in 1981, Dee describes the episode as follows:

> [...] on a Sunday morning [...] during my lunch, I answered the door to find Bruno on the doorsteps, very dishevelled and agitated and exhibiting a severely bruised eye. It transpired that during lunch his host had spoken very rudely to his wife and Bruno's attempts to teach him marital civility had ended in a violent physical encounter.

It is at this point that Professor Dee and his wife, a very affectionate couple, offered Touschek to move into their house. And this is why the end of the year 1948 finds Touschek settled into the top floor of Dee's home, 11 University Square.[73]

Dee and his family lived in one of the 13 townhouses built for the University Professors by the famed architect George Gilbert Scott. Scott built a large number of institutional and domestic buildings in the *gothic revival* style, such as The Midland Hotel near St. Pancras Station in London. Number 11 was especially reserved for the Professors of Natural Philosophy. Its first occupant was William Thomson, Lord Kelvin, who lived in the house from 1870 until 1899, when he retired. It was entirely lit by electricity, probably the first in the world to have such futuristic installation. It still houses a clock, especially designed by Kelvin, which spans two floors.[74] Philip Dee was the fifth resident in the house, from 1943 to 1972, when he retired. In Fig. 7.16 we show a front view of Dee's house and, at right, the plaque commemorating Lord Kelvin's residency.

Touschek lived with the Dees for almost two years in the top floor of their house.[75] Professor and Mrs. Dee were remembered by Bruno as well as by many other colleagues as exceptionally kind, as Curran says in Dee's biography (Curran 1984). As soon as Bruno moved in their house, Mrs. Dee took care of buying proper

---

[71] The movie is available through YouTube. Another, almost contemporary, movie on the same subject is Roberto Rossellini's 1948 movie *Germania Anno Zero*.

[72] Personal communication by Carlo Bernardini (1931–2018), Professor of Physics in Rome Sapienza University, close friend and collaborator of Bruno in the AdA adventure.

[73] Letter to parents from Glasgow, 6th December, 1948.

[74] See https://universitystory.gla.ac.uk/building/?id=85.

[75] Letter to parents, December 6th, 1948: "At the Dees I live now in top-flat [of their house]", letter to Max Born from Oakfield Avenue on October 26th and November 1950 to parents with description of moving and room plan.

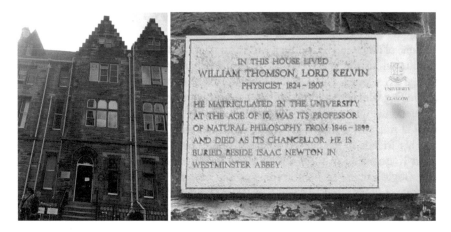

Fig. 7.16 The Square, 11 University: left photo shows a front view of the house, where Bruno Touschek lived in the top floor with Prof. and Mrs. Dee, at right the plaque commemorating Lord Kelvin. © 2019 photograph by the Author

furniture for his room, treating him almost like a son. In his 1979 letter to Amaldi, Dee remembers:

> Our house in the University was an old one on five floors, with rather steep communicating stairways. During the year or two which followed, I never met Bruno on these stairs. His transit time from top to bottom and in reverse were always so short that there was negligible probability for an encounter.

Dee remembers warmly those days and Touschek found a family atmosphere in Mrs. Dee's kind attentions to his needs. Living with the Dees, helped Bruno to reconnect with his past. The more stable situation brought in by comfortable lodgings was more akin to the well-to-do Vienna homes, where he had lived before the war, his parents' or his grandmother's, or in Rome with aunt Ada in the 1930s. The wounds of the war could start to mend. Figure 7.17 shows Philip Dee with Wolfgang Pauli, taken at the 1948 Solvay Conference.[76]

---

[76] The 1948 Solvay Conference took place after a hyatus of 15 years, the longest interval since its beginning in 1911. Through WWI, the conference series had also seen a long interruption, from the second Solvay conference in 1913 until the third in 1921. Thus the eighth Solvay conference, held in Bruxelles on October 28th, saw gathered together all the protagonists of modern physics before the war and some new entries as well. The 1948 conference followed two important meetings held in the United States a few months before in the same year, the Annual APS meeting in New York in January, and the Pocono Manor meeting, April 30–May 2nd, in Pennsylvania. At these two meetings, there are some of the first public appearances of what we now call QED, Quantum Electrodynamics, with Julian Schwinger, at both the APS and the Pocono Manor meeting, giving lectures on the new method to solve and calculate problems in particle scattering (Kaiser 2005).

Fig. 7.17 Wolfgang Pauli, at left, and Philip Dee at the 1948 Solvay Congress in Brussels. Reproduced by courtesy of the Pauli Archive, CERN (PAULI-ARCHIVE-PHO-066), https://cds.cern.ch/record/42755

7.2.4 1949: Getting the Doctorate

1949 was a very important year in Touschek's progression towards becoming a *true blood* theoretical physicist. He started to work with Born on the new (fifth) edition of his *Atomic Physics* book, travelling often to Edinburgh and writing the appendix on β-decay, a subject Touschek knew well from his Göttingen days.

As the year started, Touschek was still involved in collaborations with experimentalists, but was more and more turning to theoretical physics. In addition to the papers published with Sneddon more directly focused on nuclear physics (Sneddon and Touschek 1948b, c), as well as on the "excitation of nuclei by electrons" (Sneddon and Touschek 1948a; Touschek 1950), they had written a preliminary short note on the interaction between electrons and mesons, submitted in October 1947 and published in April 1949 (Sneddon and Touschek 1949), and soon after, on January 20, submitted a more complete paper on the results of their investigations on the "probability of producing mesons by electron bombardment", a relevant question in view of the recent developments in the design of synchrotrons expected to produce high-energy electrons. These results were presented by Bruno in February 1949 at the Annual meeting of the Physical Society, held that year at the University of Edinburgh, where he gave a paper "on electrons as nuclear projectiles".

All this was at the core of Touschek's Ph.D. dissertation, which he submitted in May, and whose title was *Collisions between electrons and nuclei*, and represented "a review of the work on electron excitation carried out in collaboration with Sneddon during the period 1947–1949".[77]

In March he was still living with the Dees, not having yet found suitable accommodations elsewhere, partly because of the cost of a reasonable serviced lodging, but very likely because Dee's hospitality was providing him with a very comfortable home and had a calming effect on him, which he appreciated. He continued traveling to Edinburgh, where he was an active participant to Born's Seminar. Max Born

---

[77] A copy of the dissertation was kindly provided by Prof. D. H. Saxon to L. Bonolis, in 2010.

was fond of Bruno and appreciated his presentations, even when not all of his ideas would turn out to be correct. In one occasion, Bruno gave there a lecture, and two days later, to his great embarrassment, discovered that part of the arguments he had presented was wrong. But Born knew how to encourage students whose capacities he valued (he had been one of Heisenberg's teachers, the other had been Sommerfeld) and kindly told Bruno that he had really liked the other part (which was correct).

As he started feeling more at ease with his research, and confident of being able to keep ahead with the requirements for the Ph.D., he decided to have a real summer vacation. The Dees were going to spend their holidays on the island of Skye and the department would accordingly enter into some lethargic state. Bruno's summer plans were to be with his parents. As a Nuffield research fellow, he could take a full month vacation, quite enough for relaxing in a place nice and warm before returning to Glasgow and the Scottish winter. The plan was to be near some mountain lake in the Austrian Tyrol, where he could find a place to swim, as he loved, and take excursions, walking through the woods and mountains of famous Alpine resorts, such as Kitzbühel, usual vacation site for the Thirring family, or Alpbach. He started proposing the idea to his parents in March, since restrictions still applied for Visas to enter Austria from abroad and travelling documents would take time, not to mention planning for the money to finance the trip.[78]

A month later the question of the summer stay was becoming a full-blown problem. One month vacation in Tyrol would not come cheap of course. His parents in fact, did not think such vacation could be afforded. But Bruno felt confident it could be handled, even recurring to a loan from his bank if needed. He also considered the possibility to go further South, and visit his maternal aunt Ada and her husband, the Italian industrialist Gaetano Vannini. Aunt Ada and her husband were the only ones in the family who were in good financial conditions. They had no children and were very attached to Bruno, who had often spent his school vacation with them, in Rome, before the war. More recently, he had written to aunt Ada from an airplane, probably while going to Vienna a year before at Christmas time, and she had been duly impressed by the fact that he was an air traveller. In a letter to his parents, Bruno, jokingly, muses that his work with pump equipments (for the planned synchrotrons) may have contributed to impress his aunt, who ran a pumping business with her husband. Visiting her, at the house near Lake Albano, would be an attractive, inexpensive prospects and he wrote her a letter.[79]

Summer plans remained undefined until late May, but after debating whether to go to St. Ulrich or to St. Johann or some similar place, and whether to rent a house or stay in a hotel, he finally convinced his parents and decided on Flecken, a small village with the attraction to be near swimming possibilities, such as the Pillersee could offer, in addition to being close to the Thirring family, taking their holidays in near by Kitzbühel, some 20–30 km away.

During this first part of the year, he was also working hard, travelling, occasionally racing all over England, as always. In April he had to go to Oxford and then to

---

[78] Letter to parents, March 9th, 1949.
[79] Letter to parents, April 23rd,1949.

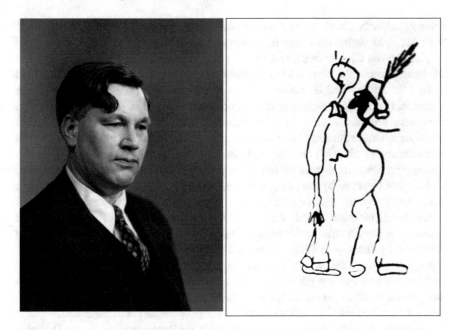

Fig. 7.18 Left panel: Maurice Pryce, with whom Touschek shared a mountain excursion in Tyrol during summer 1949, from Elliott and Sanders (2005). Right panel: One of Touschek's drawings placed together with fall 1948 letters, but possibly referring to 1949 Tyrol vacation with his father, Sapienza University of Rome–Physics Department Archives, graphics by A. Ianiro

Harwell where he met Richard Becker, his professor from Göttingen. Becker was quite optimistic about the situation in Germany and inquired if Bruno would have liked to go back. The idea was appealing, but Bruno did not come to a definite conclusion, perhaps for the poorer financial prospects which Germany still offered, and the occasion to return to Germany slipped away. Later, he regretted not to have come to the opposite decision, but subsequent events in his life may indicate otherwise: Italy, where he settled in January 1953, offered him the way to combine theoretical physics with his knowledge of particle accelerators, and thus build the world's first ever electron-positron collider, AdA, in 1960.

After the intense traveling to all the places in England where synchrotrons were being built, and the hard focusing on his theoretical work for the approaching dissertation, the long sought summer vacation in Tyrol finally arrived. August went by and Bruno enjoyed his parents company. In Fig. 7.18 we show one of Touschek's companions during the Tyrol vacation, the Oxford physicist and mathematician Maurice Pryce, and a drawing with the silhouette of two figures, an older one with a plume in the hat (a characteristic Tyrolean headgear), and a younger slim one, which may be a sketch of Bruno and his father, during the Tyrol vacation. After his parents left, he remained in Tyrol for an extra 5 days in the company of other physicists, among them Pryce, with whom he had a very pleasant mountain tour, and Léon Rosenfeld,[80] with whom he shared his return home through Switzerland.[81]

Back to the UK, there came both the return to physics and the shocking news of devaluation of the pound! On September 19th, the pound sterling was devalued from 4,03 dollars to 2,80 dollars. This was announced by the Chancellor of the Exchequer, Sir Stafford Cripps, in a broadcast to the nation. The devaluation was enormous, almost 30%, resulting in increasing costs for everything imported from abroad. For Bruno this meant that whatever money he could send to his parents would be decreased by almost a third through the new exchange rate. The devaluation was threatening the commitment he felt to his parents financial comfort. He was upset, and kind of betrayed by the country he was now living in. Suddenly insecure about future prospects, thoughts of going abroad, in the United States or Canada, for 1 or 2 years crossed his mind. Preparations for such move could take time, and he wondered about other possibilities in England or, more likely, in Ireland. He had a year-long invitation by Janossy in Dublin, who had graduated from the same high-school in Vienna as Bruno, the Schottengymnasium, but four years before him.[82] Visiting Dublin was an attractive possibility, also because Walter Thirring, Hans Thirring's son, whom Touschek knew well, and considered an outstanding young theorist, was also going to be there in 1950, to study with Schroedinger (Thirring 2008).

All these plans of course would have to wait until Bruno received his Ph.D., but in fact this was going to happen soon. In mid September he was informally told that the dissertation he had submitted in May had come back with a very good rating from Rudolph Peierls.[83] The graduation ceremony was going to be held in early November, and he expected to receive a new contract with an increased salary, which might offset the pound devaluation.

While pleased with looking forward to receive finally his official entry pass into being a theoretical physicist, Bruno could not avoid feeling that Glasgow was as dirty as ever, looking even blander and dirtier after a summer break. The prospect of intensified austerity did not make things any more pleasant, but he hesitated in running away, yet. He had done this all his life, and thus escaped stagnation and probably death. But now, he had gone far, beyond his early dreams, and the next move needed to be carefully thought out. It is quite possible that similar considerations were in his mind when, earlier in the year in Harwell with Becker, he did not follow up with the Göttingen offer.

In October, he received an official letter informing him that his thesis had been approved and no corrections were requested, intimating that he would receive his Doctorate in the coming session, the letter also including the address of the

---

[80] Léon Rosenfeld was a Belgian physicist born into a secular Jewish family. He was a polyglot who knew eight or nine languages and was fluent in at least five of them. Rosenfeld obtained a Ph.D. at the University of Liège in 1926, and was a close collaborator of Niels Bohr.

[81] Letter to parents on September 20th, 1949.

[82] Lajos Janossy (1912–1978) had joined the Institute for Advanced Studies in Dublin upon invitation from Heitler and Schroedinger, but, not long after, he left and joined the Central Research Institute for Physics in Budapest, returning to the country where he was born.

[83] Letters to parents on September 20th, 1949.

Fig. 7.19 Touschek's drawing about leaving the garage where he had just wanted to buy a car, and almost crushed the attendant seller. Family Documents, © Francis Touschek, all rights reserved

robe-makers, in case he planned to be present at the official Graduation Cerimony (which he did). He also received the assurance that he could continue staying in Glasgow as "Official Lecturer in Natural Philosophy" (Amaldi 1981, 12). With the salary coming with his prospect of improved status as a Lecturer, he could feel financially more secure. Emboldened by the official confirmation that his application for doctorate had been approved, he sought to buy a car, even though he still had neither a driving license, nor the money. Indeed his impatience with the slow process of learning to drive, led him to potential troubles, as he related his first driving attempt, with the car to be purchased, in a humorous letter to his parents. Figure 7.19 reproduces a small drawing included in the letter, showing the effect of his driving the B.S.A.—sport—(170 pounds), as he was rushing out of the shop and trying it.[84]

Touschek's doctoral thesis was on the interaction of electrons with mesons, a topic on which he wrote several papers with Sneddon. John C. Gunn, who held the newly established Chair of theoretical physics, was his internal examiner, Fig. 7.20.[85]

Touschek received his degree on November 5th, 1949, as in Fig. 7.21 showing the certificate by the Academic Senate and Bruno's official Ph.D. photograph. As Amaldi writes, immediately after his doctorate, Touschek was appointed 'Official Lecturer in Natural Philosophy' at University of Glasgow (Amaldi 1981, 12), a position he held for three years.

Shortly after Touschek's doctorate cerimony, a conference was organized by Max Born at the University of Edinburgh on November 14–17 (Anonymous 1950). The

---

[84] Letter to parents, October 24th, 1949.

[85] John C. Gunn was a Professor of Natural Philosophy at University of Glasgow. He was born in Glasgow, and studied at St. John's College in Cambridge. After the war he was lecturer in Applied Mathematics at the University of Manchester and then started research in nuclear and particle physics at University College in London, as from https://www.universitystory.gla.ac.uk/biography/?id=WH1433&type=P. In London, Gunn published a paper on "Interaction of Mesons with a potential field" (Gunn and Massey 1948). The paper, published by the Royal Society, was presented by Harrie Massey, one of the active supporters of CERN from the UK side. Gunn was appointed to the Chair of Theoretical Physics in University of Glasgow in 1949. See also https://www.universitystory.gla.ac.uk/biography/?id=WH1433&type=P.

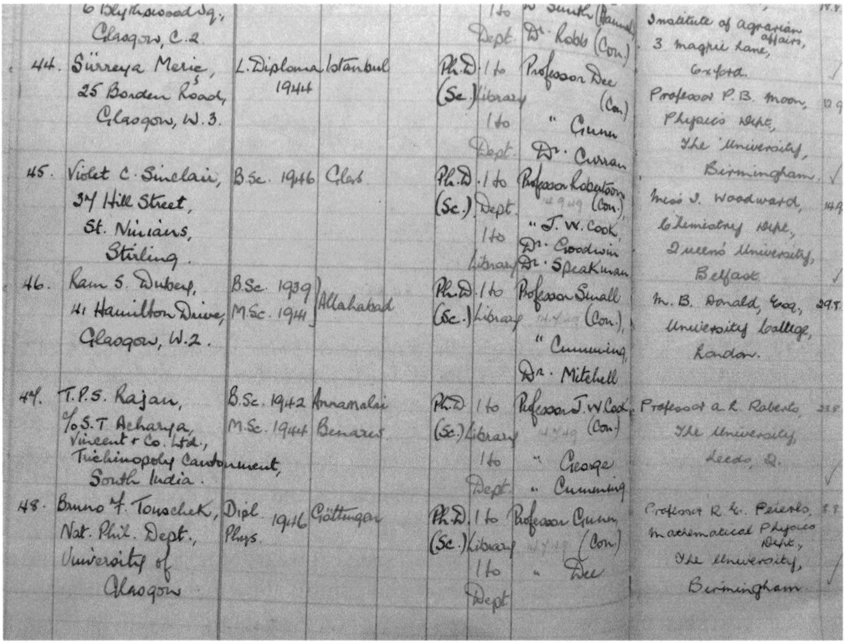

6 Blythswood Sq., Glasgow, C.2.				1 to Dept.	Dr. Robb (Con.)	Institute of agrarian affairs, 3 Magpie Lane, Oxford.
44. Süreyya Meriç, 25 Borden Road, Glasgow, W.3.	L. Diploma 1944	Istanbul	Ph.D. (Sc.)	1 to Library, 1 to Dept.	Professor Dee (Con.), " Gunn, Dr. Curran	Professor P. B. Moon, Physics Dept., The University, Birmingham.
45. Violet C. Sinclair, 34 Hill Street, St. Ninians, Stirling.	B.Sc. 1946	Glas.	Ph.D. (Sc.)	1 to Dept., 1 to Library	Professor Robertson (Con.), " J. W. Cook, Dr. Goodwin, Dr. Speakman	Miss J. Woodward, Chemistry Dept., Queen's University, Belfast.
46. Ram S. Dubey, 41 Hamilton Drive, Glasgow, W.2.	B.Sc. 1939, M.Sc. 1941	Allahabad	Ph.D. (Sc.)	1 to Library	Professor Small (Con.), " Cumming, Dr. Mitchell	M. B. Donald, Esq., University College, London.
47. T.P.S. Rajan, c/o S.T. Acharya, Vincent & Co. Ltd., Trichinopoly Cantonment, South India.	B.Sc. 1942, M.Sc. 1944	Annamalai, Benares	Ph.D. (Sc.)	1 to Library, 1 to Dept.	Professor J. W. Cook (Con.), " George Cumming	Professor A. R. Roberts, The University, Leeds, 2.
48. Bruno J. Touschek, Nat. Phil. Dept., University of Glasgow	Dipl. Phys. 1946	Göttingen	Ph.D. (Sc.)	1 to Library, 1 to Dept.	Professor Gunn (Con.), " Dee	Professor R. E. Peierls, Mathematical Physics Dept., The University, Birmingham.

Fig. 7.20 Pages from Touschek's official Ph.D. record from Glasgow University Record of Higher Degrees, courtesy of University of Glasgow Archives & Special Collections, University Registry collection, GB248 R14/2/1

subject of particle decays was at the center of debates, as one can see from the Conference preliminary program, Fig. 7.22, with cosmic rays results still the winner of the day in terms of experimental high energy particle physics. Giving a talk on *The decay products of the μ-Meson*, there was Bruno Pontecorvo, who had spent the war years mostly in Canada, and was then based at AERE, the Atomic Energy Research Establishment Harwell.[86]

The conference had been organized by University of Edinburgh to take advantage of the presence in Scotland of Niels Bohr, who gave that year's Gifford Lectures, held bi-annually in Scottish Universities (Anonymous 1950).[87] The Glasgow particle physics group, including Dee, Touschek and students, drove everyday to Edinburgh, with Bruno holding the wheel of Dee's car, and the students huddled in the back of the car, frozen with fright at every turn of the road, since Bruno was still a rather

[86] Within less than a year, in summer 1950, and following Klaus Fuchs' February 1950 sentence to 14 years in prison for passing atomic secrets in favour of the Soviet Union, Bruno Pontecorvo would secretly emigrate to Soviet Union.

[87] Niels Bohr gave the 1949 Gifford Lectures at the University of Edinburgh under the title Causality and Complementarity. The lectures remain unpublished, but the audio recordings of 9 of the 10 lectures [lecture 2 is unfortunately missing] are maintained at the Niels Bohr Archive (http://nbarchive.dk). A manuscript entitled "Summary of Gifford Lectures is reproduced in Niels Bohr Collected Works, vol. 10, *Complementarity beyond Physics (1928–1962)*; edited by David Favrholdt.

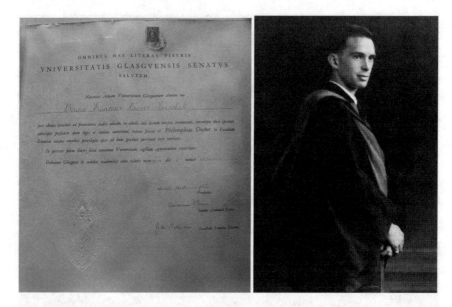

Fig. 7.21 Touschek's certificate of award of Doctor of Philosophy from University of Glasgow, Bruno Touschek Papers in Sapienza University of Rome–Physics Department Archives, and his official Graduation photograph (Amaldi 1981, 10)

inexperienced driver.[88] After the Conference, Bohr visited Glasgow as part of his Gifford Lecture tour. For the Glasgow Physics Department, this was the big event of the year, taking place on Thursday and Friday, 17 and 18th November. Bruno wrote to his parents a rather hilarious description of this visit. The highlight was an inaudible lecture at the physics institute during which Bohr moved from the lectern to the blackboard, to clarify a point which nobody could follow anyway.[89] The talk was given to a packed audience of students and professors alike, who had all come to listen and see *der Physik papst*, the pope of physics, as he was called. A visit from the King of England would not have given half as much trouble. Bohr was accompanied by

[88] The list of participants does not include Bruno, who must have been considered in the Glasgow contingent, as 'Gunn and seven other' from University of Glasgow, as from list of participants.

[89] Bohr's lectures were famously long and often not understandable. One such reaction is reported from the Pocono Manor meeting, March 30–April 2nd, 1948, in Pennsylvania. At this conference, Feynman gave the first public introduction to his method, since then referred to as *Feynman diagrams* and universally used to calculate particle interactions. An interesting article related to The unveiling of Feynman diagrams at the Pocono Manor Conference tells how Feynman's talk, presented at the end of the day, was poorly received. In particular, 'Bohr leapt to the mistaken conclusion that they [Feynman diagrams] represented a violation of Pauli's exclusion principle.' and after more questions were asked which Feynman appeared unprepared to answer, 'Bohr rose and approached the blackboard where he delivered a long speech on the Pauli exclusion principle'. At the end of the session, it seems that almost nobody had understood what Feynman's method could do. See also a 2018 article by Ashutosh Jogalekar on The Birth of a New Theory: Richard Feynman and His Adversaries, in *3 Quarks Daily on-line magazine*.

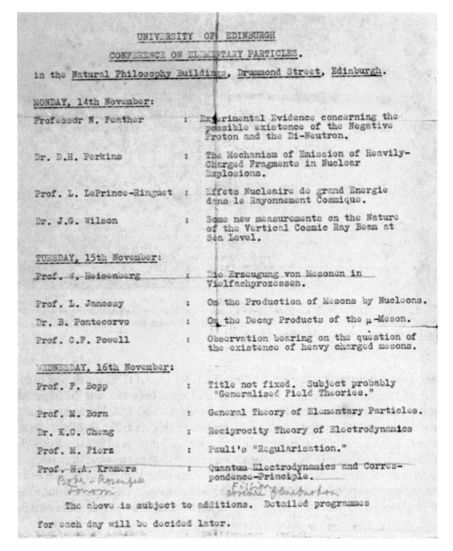

UNIVERSITY OF EDINBURGH

CONFERENCE ON ELEMENTARY PARTICLES.

in the Natural Philosophy Buildings, Drummond Street, Edinburgh.

MONDAY, 14th November:

Professor N. Feather   :   Experimental Evidence concerning the
                                   possible existence of the Negative
                                   Proton and the Di-Neutron.

Dr. D.H. Perkins   :   The Mechanism of Emission of Heavily-
                                   Charged Fragments in Nuclear
                                   Explosions.

Prof. L. LePrince-Ringuet   :   Effets Nucleaire de grand Energie
                                   dans le Rayonnement Cosmique.

Dr. J.G. Wilson   :   Some new measurements on the Nature
                                   of the Vertical Cosmic Ray Beam at
                                   Sea Level.

TUESDAY, 15th November:

Prof. W. Heisenberg   :   Die Erzeugung von Mesonen in
                                   Vielfachprozessen.

Prof. L. Janossy   :   On the Production of Mesons by Nucleons.

Dr. B. Pontecorvo   :   On the Decay Products of the μ-Meson.

Prof. C.F. Powell   :   Observation bearing on the question of
                                   the existence of heavy charged mesons.

WEDNESDAY, 16th November:

Prof. F. Bopp   :   Title not fixed. Subject probably
                                   "Generalised Field Theories."

Prof. M. Born   :   General Theory of Elementary Particles.

Dr. K.C. Cheng   :   Reciprocity Theory of Electrodynamics

Prof. M. Fierz   :   Pauli's "Regularisation."

Prof. H.A. Kramers   :   Quantum Electrodynamics and Corres-
                                   pondence-Principle.

          The above is subject to additions. Detailed programmes

for each day will be decided later.

Fig. 7.22 Preliminary program of the 1949 Conference on Elementary Particles, held in Edinburgh 14–16 November 1949, from Bruno Pontecorvo papers at Churchill Archives, Cambridge University

his assistant Stefan Rozenthal and his long-time collaborator Léon Rosenfeld, whom Touschek knew from Birmingham and Alpbach. Touschek drily reflected that Bohr's esthetically supreme choice in his adepts' name allowed no choice for someone with a non descript name such as his own. But while Bohr could neither catch nor pay attention to Touschek's name, he must have been sufficiently impressed by the young man's intelligence that he invited him to come to Copenhagen on his next trip. This

Fig. 7.23 From left: Werner Heisenberg and Niels Bohr, https://www.flickr.com/photos/193673378@N02/51376340345/, licensed under the Creative Commons Attribution-Share Alike 2.0 Generic. At right, Walter Heitler and Léon Rosenfeld, https://cds.cern.ch/record/42896. For other photos of interest see https://cds.cern.ch/search?cc=Pauli+Archive+Photos&ln=en. Photograph at right is also to be found in https://calisphere.org/item/ark:/28722/bk0016t4h8s/ where it is indicated as around 1934, Copenhagen conference

was an invitation which Touschek was glad to accept for the next fall. Heisenberg, Bohr, Heitler and Rosenfeld are shown in Fig. 7.23, from around 1934.

Bohr and his wife arrived with something like 71 pieces of baggage which Dee personally brought down from the third floor into his car, when they left. At the train station, Dee was having great difficulty keeping track of all the porters running in different directions with that much luggage, not to mention the enormous tip to give them after the Bohrs left. All through this, Bohr very quietly and steadily was clarifying the subtleties of the uncertainty principle.[90]

For Christmas, Touschek decided he had enough of Glasgow and took a break to the North of Scotland, planning to go to Glencoe and Fort William in Invernessshire, climb Ben Nevis and visit the island of Skye, enjoying the North in the heart of winter, Fig. 7.24.[91]

As the year was coming to an end, from the wilderness of the North, he was thinking of his future, looking for possible new research directions or places to go. His thoughts wandered to Rome and aunt Ada. The idea of a summer vacation with a trip to Italy and to the South began once more to be appealing. He thought of writing to her, but did not have her address and postponed.

---

[90] Letter to parents, November 21st, 1949.

[91] Letter to parents, December 25th, 1949.

Fig. 7.24 The places Bruno Touschek toured during his Christmas vacation in 1949

7.3   Lecturer in Glasgow: 1950–1952

While Bruno in Glasgow had been working on his doctorate and, afterwards, continuing as "Official Lecturer in Natural Philosophy", in continental Europe the massive process of reconstructing European science was undertaken. In France, in Italy, in Germany, in all the universities which the war and fascism had decimated of their scientists and laboratory equipment, younger scientists joined forces with the few who had remained and had started to rebuild Europe. Such massive effort, which culminated with the foundation of CERN, but also with the launch of national laboratories and research institutions such as the Laboratoire de l'Accélérateur Linéaire in Orsay or the National Institute for Nuclear Physics (INFN) in Italy, led to the construction of new powerful particle accelerators, which, unexpectedly, could compete with the American ones. Indeed at the end of the 1950s, a proton synchrotron was working at CERN, a linear electron accelerator in Orsay, an electron synchrotron in Frascati. While all this was in the future, the year 1950 saw Europe on the stepping stone of an epochal change, which was only partly reflected in the UK. Britain had won the war, and throughout it had maintained—fostered and increased—its intellectual and technological power, but the drive to start anew in a joint effort, as the continental nations were doing, was not as strong. One recalls that Britain itself criticized the project of a European laboratory, even considering it a crazy idea (Hermann et al. 1987, 114), and hesitated in joining CERN at the beginning. Only in 1954 did Britain become one of CERN member states, after its scientists fully endorsed the idea and lobbied the government (Amaldi 1979a).

Only part of this was resonating in Glasgow, as the world entered into the 1950s. At this time, Glasgow appeared stagnant to the young Touschek and he began to actively look for a place to go and continue his path in theoretical physics. Then, as 1949 ended and 1950 took over, there unfolded events which would affect nuclear physics in the UK, and, with it, Touschek's life. These events were the consequence of actions which had begun 8 or 10 years before, and had been known only to a handful of people, from both sides of the Atlantic. In February 1950 they became public, after having been brought to the attention of the US and UK highest political and scientific authorities in January. The knowledge of what had happened shook the world and changed the course of many people's lives, as well as the direction of nuclear physics research in UK. We refer here to so called *Fuchs affair*. Rudolph Peierls was most directly affected, but Bruno's life as well was indirectly influenced.

7.3.1 Fallout from the Fuchs Affair

In early 1950, the British scientific community was shaken by the discovery that the theoretical physicist in charge of the Nuclear Physics program at Harwell, Klaus Fuchs, was, and had been, a spy, who had passed crucial information about the making of the atomic bomb to the Soviet Union (Close 2019). Fuchs was German born, and had left Germany for the UK before the war. After getting his Ph.D. at the University of Bristol in 1937, he went to Edinburgh, where he worked with Max Born and was awarded a Doctor of Science degree. He was interned as an enemy alien at the start of World War II, but was soon released as his known enmity to the nazi regime cleared his political allegiance. His importance as a theoretical physicists was underlined by Max Born, who wrote that Fuchs "was the soul of my research group [...] He is in the small group of theoretical physicists in this country" (Close 2019, 45). He was also highly considered by Peierls and in May 1941 he became Rudolph Peierl's assistant in Birmingham, working with him on "Tube Alloys", the British atomic bomb project, beginning to pass information to the Soviets. In 1942 he became a British citizen, and could thus move with Peierls to work on the Manhattan Project, later joining the Los Alamos top secret laboratory in New Mexico.[92] After the war, Fuchs had returned to England, going to the nuclear research center at Harwell.

When Fuch's betrayal become public, suddenly, in the UK, the climate for foreign born nuclear physicists changed dramatically. As Richard Wilson remembers in his memoirs: "The problem began in December 1949 when the nuclear physicist Alan Nunn May was arrested as a spy at McGill University in Canada just after a lecture. He had been working on the atomic bomb project and was accused of giving information to the USSR some five years before. The USA panicked" (Wilson 2011, 111).[93] Fuchs' espionage activities were detected, and he was arrested on February 2nd 1950, upon which he admitted passing information to the Soviet Union since 1943, Fig. 7.25.[94]

Rudolf Peierls was particularly shaken by the uncover of Fuchs as a Soviet spy. Peierls, also German born, had supported Fuchs' eligibility for the Manhattan project. The two scientists were very close friends, having shared the experience of leaving Germany as Hitler came to power, and gone to the UK to continue their work. Peierls may have felt not only betrayed, but also himself in danger.[95] Peierls' wife was Russian, and he had frequently visited the USSR before the war. He had collaborated with prominent Russian scientists such as Lev Landau. In the witch hunt atmosphere which would soon engulf the United States with the rise of McCarthysm, German scientists in the UK started feeling insecure. Peierls himself was the target of suspicions and criticism (Close 2019, 392–401).[96]

Everywhere in Europe and the US, the fear of a communist threat to Western society led to major changes to the life of some prominent scientists. One can remember that on April 26th, 1950, Frédéric Joliot (Pinault 2000) was made to resign from the Chairmanship of the French Atomic Energy Commission for his sympathies for the Communist Party and activities in favour of an international ban on nuclear weapons

---

[92] From https://www.britannica.com/biography/Klaus-Fuchs: "Klaus Fuchs, in full Emil Klaus Julius Fuchs, (born December 29, 1911, Rüsselsheim, Germany, died January 28, 1988, East Germany), German-born physicist and spy who was arrested and convicted (1950) for giving vital American and British atomic-research secrets to the Soviet Union."

[93] Richard Wilson (1926–2017) was an English born experimental physicist, who was Professor of Physics at Harvard University, USA, and designed, constructed, and used the Cambridge Electron Accelerator 6 GeV synchrotron, which, from 1962 on, further probed nucleonic structure. Later on, after ADONE's successful operation, the CEA morphed into an electron-positron collider, confirming early ADONE results on multi-hadron production, Chap. 13.

[94] He was sentenced to 14 years in prison. After his release in 1959 for good behaviour, he went to East Germany, where he was granted citizenship and was appointed deputy director of the Central Institute for Nuclear Reactions.

[95] In his memoirs *Birds of Passage* (Peierls 1985, 223) Peierls recalls: "Our most dramatic experience was the Fuchs case. […] On the day I heard of his imprisonment under the spy charges I went on to Brixton prison to see Fuchs. We had a long talk. Yes, he had given secret information to Soviet contacts."

[96] As Close writes: "With his phone and mail continually monitored, Rudolph Peierls became part of the communist witch-hunt until 1954" when the British security closed their file on him. But the United States did not relent and in 1957 asked the British Department of Atomic Energy that Peierls be given no access to American secret documents. At which point, Peierls decided to resign from his consultancy at Harwell.

Fig. 7.25 Left panel: Klaus Fuchs, at the time of his arrest for espionage in favour of the Soviet Union, © ImperialwarMuseum, from https://history.blog.gov.uk/2020/03/02/whats-the-context-sentencing-of-atomic-spy-klaus-fuchs-1-march-1950/. Right panel: Bruno Pontecorvo with Enrico Fermi, shown at right, in a photograph taken on occasion of the International Conference on Nuclear Physics, Quantum Electrodynamics and Cosmic Rays held in Basel, CH, and Como, Italy, in September 1949, from a contemporary newspaper, courtesy of Giovanni Battimelli, Sapienza University of Rome–Physics Department Archives, all rights reserved

(Close 2015, 157). And, in early September 1950, the Italian physicist Bruno Pontecorvo, one of Fermi's collaborators before the war, seen in Fig. 7.25, suddenly disappeared from Italy with his wife and children, joining the Soviet Union, as it became known only five years later (Turchetti 2012; Close 2015).[97]

The *cold* war had started. This atmosphere weighted heavily on the morale of UK scientists, and had strong impact on career prospects for non-British citizens working in the UK.

The impression from the Fuchs affair is present in Touschek's letters in early 1950. Fuchs' trial started on February 12th, very soon after he had been arrested, only one week after he had confessed (Close 2019, 334). Touschek mused on the prosecutor's description of Klaus Fuchs as one with Dr. Jekyll and Mr. Hyde's personality, and drily observed that a double personality has nothing to do with nuclear secrets. He did not sympathize with the Soviet Union, but felt that the prosecutor's hype was excessive.[98]

Unfortunately, the Fuchs affair was not a passing episode.

---

[97] Pontecorvo was born in Pisa in 1913 from a prominent Jewish family. He moved to the USSR in 1950, returning to Italy for the first time only 28 years later, in 1978, on the occasion of Edoardo Amaldi's seventieth birthday celebrations. He died in Dubna in 1993. See also Mafai (1992), Turchetti (2007).

[98] Letter to parents, February 14th, 1950.

7.3.2   Bruno Touschek and Max Born—1950–1952

During his stay in Scotland, Bruno exchanged letters with Max Born in Edinburgh and frequently visited him. From this correspondence, one can see that Touschek's formation into a theoretical physicist owes much to the relationship with the great scientist. The letters often include questions of theoretical physics, articles to be discussed, even some fatherly advice, such as we glimpse from a letter where Touschek acknowledges some errors he had made and accepts Born's suggestion that `a little Puritan classical electrostatics would do me no harm.`[99]

Most of Born's correspondence with Touschek is kept in the Churchill Archives Centre, in Churchill College, Cambridge, UK.[100] The sequence of letters kept in Churchill Archives Center starts with May 18th, 1950, with Born informing Touschek that he had just come back from Cairo finding the galley proofs of his book *Atomic Physics*. He could not yet work on them, because of a commitment to write an article on the physics of the last 50 years for the *Scientific American*, and asked Bruno to help him with the proofs, a copy of which he should have received as well, and invited him to come over for a day to Edinburgh.

Bruno replied ten days later, sending the corrected proofs. After mentioning an interesting effect observed by his experimentalist colleagues and for which he was trying to find a theoretical explanation, Bruno inquired about a position in Cairo, from Born's letter: `On Monday you mentioned a possible vacancy in Cairo and I could not help thinking about it.` Since passing his doctorate in November, Touschek had been looking for a way out of Glasgow, and Born's letter started some fantasy about such position. Uncertain about writing directly to the Physics Department in Cairo, he mused whether this would be a form of 'escapism', born from growing doubts about his future in the UK. He had loved Scotland, but not anymore now that he had gotten used to it. This is an interesting observation, since it reflects an internal pervasive restlessness, which may have had a remote origin, probably when his mother died and, as a young boy before the war, he would leave Vienna for periods of time to stay with aunt Ada, in Rome.

In any case, he reassured Born that he won't want to leave for another year, because of work he had started and wanted to finish. He closed the letter, asking Born if he would be angry at him if he applied for the position in Cairo. Shortly after, on May 31st, Born thanked for the proofs, hoping to see him before Bruno would leave for Germany, for the summer, and promising to inquire about the position in Egypt.

On September 29th, after visits to Göttingen, Copenhagen and Austria (vacationing in Tyrol), Touschek picked up again the correspondence, apologizing for some

---

[99] Touschek to Born, September 25th, 1950, Churchill Archives Center.

[100] This author was made aware of this correspondence by Ms. Antonella Cotugno, from Rome University Library, and the quoted text comes from 23 letters she copied with the kind permission of Churchill Archives Centre, Churchill College, Cambridge University, UK, where the original letters are kept and are available in the Cambridge University, Churchill Archives Centre, The Papers of Professor Max Born. These copies are kept in Bruno Touschek papers at Sapienza University of Rome–Physics Department Archives.

mistake in a paper he had been working on, and informed Born of a pending visit by Walter Thirring in October–November. During the summer Touschek had attended a small private conference organized in Kitzbühel, at Hans Thirring's Tyrol retreat. This may have prompted plans for a visit to Glasgow by Thirring's son, Walter, in the coming fall. In Thirring's autobiography, the tour which he took in fall 1950, first to Dublin invited by Schrödinger and then to Glasgow, is seen as an apprenticeship and travel tour (Thirring 2008), before Thirring would go back to Göttingen for the rest of the academic year. In this letter to Born, Touschek proposes to take Thirring with him to Edinburgh, upon his arrival in Glasgow.

When Touschek wrote next, on October 26th, he apologized for the silence, partly due to having been very busy, having left Dee's house to move to his own place in Oakfield Avenue. He also mentioned the work he had started with Thirring. At the time of this visit, Walter Thirring was interested in going through electrodynamics with the covariant formalism of Schwinger and Feynman. From this letter, the why and when of Touschek and Thirring's paper (Thirring and Touschek 1951) on the Bloch and Nordsieck (BN) problem (Bloch and Nordsieck 1937), first appear. Born welcomed the idea of Thirring coming to Edinburgh and on November 10th invited both Touschek and Thirring to come on the following Thursday for discussions, 'high tea', or possibly for lunch. During November, Touschek and Thirring wrote the paper on the covariant formalism of the BN problem and then, after having submitted it to the *Philosophical Magazine* Touschek sent a copy of the manuscript to Born on December 18th. By that time, in early December, Thirring had gone back to Göttingen. There is no mention of this paper in Thirring's memoirs, whereas for Touschek it would later become a milestone in his formulation of infra-red radiative corrections to electron positron experiments (Etim et al. 1967a).

The next letter in the Churchill Centre Archives is dated April 3rd, 1951. Touschek apologizes for the silence, and, implicitly, for lack of visits to Edinburgh during these 4 months, informs Born of his recent theoretical physics output and asks him for a referral for a position in Oxford as Senior Lecturer. He had written to Maurice Pryce, Born's former student in 1930s and son-in-law, having married one of Born's daughters, but apparently Pryce was in Princeton, according to Sneddon, and there was no reaction. Born replied immediately, after a couple of days, that he was happy to give a referral and thought that Pryce, who should come back from Princeton next July, should be delighted to have Touschek in Oxford. He was happy to have heard from Touschek, inviting him to come again over to Edinburgh.

As of May 28th, Touschek had not received any reply from Oxford, neither from Pryce nor from the University Registrar, and became worried that there would not be enough time to give notice to Glasgow University, in case of a positive answer from there. One reason for this could be that Pryce, after a one-year sabbatical leave at Princeton, on his return would become head of the theoretical physics division at the Atomic Energy Research Establishment at Harwell (Elliott and Sanders 2005), replacing Klaus Fuchs (arrested on February 2nd 1950 and convicted, on March 1, on the charge of spying) (Close 2019). In the same letter, Touschek mentions his latest physics paper, a work which he calls 'no more than a patent application', meaning that he had an idea and wanted to have it down in print, to establish his priority. We know from correspondence with his family that he felt he often worked out results which were developed around the same time by other physicists, but which did not get attention because he had not published them, or, more likely, did not propagate outside his own restricted circle. In a later CV, Touschek mentions one such case concerning the appendix about meson theory he wrote for Born's book. In the appendix, he anticipated the universality of weak interactions, but had not published his intuition about the subject, having discovered that it had been developed elsewhere, in particular by Puppi (1948).[101]

The preoccupation to establish a priority explains the term 'a patent application' for the paper he submitted on May 3rd, 1951. This paper (Touschek 1951) is of note because it includes citation of an article by the Italian theorist Bruno Ferretti, submitted from Rome and published in Ferretti (1951), which may have attracted Bruno's attention to the Physics Institute in the University of Rome.

In the May 28th, 1951 letter, Touschek also informed Born of his summer plans, which included driving the motor cycle he had just bought, and travelling with a friend, Dr. Rae, through the continent. But Born had also already left for the Continent, and would only be returning at the beginning of August. The department, in Edinburgh, offered to forward Touschek's letter to Born in Göttingen.

No more letters are recorded until the next year. What prompted such long silence? Letters between September 1951 and January 1952, when Touschek was back in Glasgow after the summer leave, may have been lost. But it is just as likely that Touschek, after unsuccessful attempts to find an alternative position in Oxford, or Germany, turned his attention to Italy, starting his contacts during summer 1951, as shall be seen. This preoccupation would engulf him. Back to Glasgow, he was also very busy with teaching, working with his research students, and worrying about a (never published) book with Ian Sneddon.

The next group of letters starts in January 1952, but they are less frequent. From January through March, there are various exchanges about slides and photographs of notable scientists in Born's possession which Touschek wished to use for 'a chatty talk' on the historical development of quantum mechanics he had been asked to give at the Glasgow Physics Institute. In Fig. 7.26 we show the scheme he sent to Born, together with a Glasgow Herald cutting about the talk. A long silence follows after this, as no other letter appears to have been exchanged until November.

On November 6th, Touschek writes that he is moving to Rome shortly, and would like to visit Born and his wife, once more. To which Born replies on November 10th, in a somewhat reproachful way, that he had heard of Touschek's move, and congratulates him for the new position. Adding his hope that Touschek will be happy there, he also comments that 'to live in Rome, alone, is a great privilege'. He will try to find a time to see him, among all his engagements, such as the Gifford lectures taking place in Edinburgh. Born's letter ends by saying that they (himself and Mrs.

---

[101] Puppi is credited for having been the first to discuss the muon-electron universality, which actually had been already mentioned by Pontecorvo (1947), but seldom recognized. Oskar Klein, too, had realized that all weak processes investigated thus far seemed to be due to the same universal Fermi interaction (Klein 1948), as did Tiomno and Wheeler (1949).

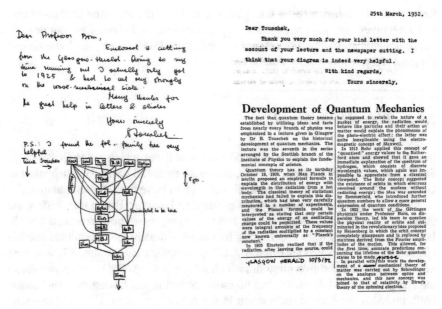

Fig. 7.26 Touschek's handwritten letter to Max Born, and a Glasgow Herald article about the lecture he held at the Glasgow Physics Institute. Courtesy of Churchill Archives Center of Churchill College, Cambridge University

Born) are 'quite well, apart from getting old and always being tired'. Then, in a last letter written just on the following day, concerned, perhaps, that his previous message had not been sufficiently welcoming, he urges Touschek to come, adding, as an inducement, that he would like Touschek to explain him a recently published paper by Heisenberg (1952). Born was going to give a lecture on Heisenberg's work in the coming month of December, in London, and since Heisenberg referred to Born's old non-linear electrodynamics, he wished to mention something about it. Born confesses that despite his efforts he has not been able to understand Heisenberg's work, but hopes Touschek knows something about the matter. It is amusing to compare Born's comments with the interest these papers have generated through the years: Heisenberg's paper has received continuous attention since the time of his publication, and Born's non linear electrodynamics work (Born and Infeld 1934, 1935), has recently been subject of growing interest, as one can see from the 2020 citation results for Heisenberg and Born's papers (Pancheri and Bonolis 2020).

It is quite possible that the long intervals in the Born–Touschek correspondence both in the summer of 1951 and more so in summer 1952, may have been related to Touschek reluctance to tell his friend and mentor that he was going away, abandoning Scotland, the frequent visits to Edinburgh, the warm friendship between them, the train rides through the countryside from West to East and back, afternoon 'high tea' at the Borns' home, the intellectual stimulus of Born's genius and his Seminar. He was focusing on finding a position in Italy and leave Glasgow. Perhaps he could not face to tell his mentor that he was planning to leave the UK. The Oxford option had not worked out, and possibilities to go to Germany had either dried out or were not sufficiently appealing. For reasons which included both his physics interests and personal story, it was to Italy that he chose to go.

7.3.3 Planning Betrayal

In order to unravel the motivations and the possibilities which made Bruno to go South, it is necessary to return to 1950, and to the three full years Touschek spent in Glasgow after his doctorate and how his decision to leave the UK was matured.

Indeed, only a few months into his position as Lecturer, Bruno was finding the atmosphere in Glasgow rather stifling and was trying to see how to escape from it. His physics interests were also leading him more and more into theoretical particle physics, and Glasgow was not offering sufficient stimulus. He had slowly lost interest in the Glasgow synchrotron program, which was stagnating. The building was grow-ing and growing, but Touschek had no more interest in just having the synchrotron built, probably for lack of enthusiasm in the physics program. He could envisage the synchrotron to be finished in one or, rather, two more years, and then what? What to look for? Which new programs could be seen as natural follow-up to the 300 MeV synchrotron? In Glasgow, Touschek mused, only Dee was still interested in the project, everybody else was reading jobs openings in *Nature*, and looking for positions elsewhere.

Touschek was right in feeling that the UK was not offering him much of a way forward. In two years, by 1952, when the Glasgow synchrotron could be close to start operating, European physicists in the continent would be ready for the challenge of building a large international laboratory, such that it could possibly house much more powerful accelerators. In the department in Glasgow, the atmosphere was changing as well. Touschek's friend and collaborator Ian Sneddon was leaving, first for the United States, and later moving to a new position in Staffordshire. Mostly because of the Fuchs' affair, this road was hardly open to Bruno. He regretted to have refused a position in Göttingen the previous year, but he now decided he did want to leave and would use the summer to look for a job in the Continent. The winter climate did not help, of course, nor did the town, which did not offer much entertainment, the only exception being the occasional evening at the cinema, and this as well was often depressing if not making him dizzy with boredom.[102]

To get over his low spirits, Bruno started a detailed planning for the summer, something which always helped him to overcome the winter blues. In 1947, he had stayed around Glasgow for his first summer (occasionally swimming in Loch Lomond and nearby places), in 1948 he had gone harvesting to the Northernmost

---

[102] Letter to parents March 21st, 1950. He also mentions having been rather annoyed by the movie *Gioventù perduta*, by Pietro Germi (1947). The movie depicts young bourgeois Italian youth, whose life leads to ennui and self destruction, and it is rather depressing.

part of Scotland, while in 1949, as travel restrictions to Austria were eased, he spent a full month in Tyrol with his parents. Now it was the time for his first vacation as a Lecturer, and not as a research student. The plan was to be again in Tyrol, even to organize a stay for Professor and Mrs. Gunn, but not only. As he had written to his parents, he was really planning betrayal.[103] Soon after obtaining his doctorate he had already started thinking about looking for a position in the continent, now he was going to work on it. In June 1950, the summer program was completed: 20.6–9.7 in Göttingen; 10.7–18.8: Tyrol and surroundings; 18.8–25.8: Hamburg and surroundings [Umbegeung]; 26.8–18.9: Copenhagen.[104]

The summer plans included, for the first time, a very long absence from Glasgow, a three months leave, of which almost one month to be spent in Göttingen, and then visiting Hamburg and Denmark, following Bohr's invitation in November 1949. In Tyrol he was with his family, but was also part to the already mentioned small, private conference in Hans Thirring's house in Kitzbühel.

The Gunns' planned vacation in Tyrol took place during this extended visit as well.[105] M. J. Gunn, Gunn's son born in 1954, remembers his father often talking about Touschek: "—it was clear he liked Bruno and he had been a lively figure in the department in Glasgow. I heard the story about his escape from the Gestapo (or SS?) during that period. I [also] remember hearing about the Tyrol trip. It was my mother's first trip abroad. My father was more adventurous than sensible—so one of their walks was on their hands and knees across a glacier".[106]

Bruno arrived in Copenhagen on August 31.[107] But he was unwell, having caught a very annoying ear infection, and could hardly enjoy his visit. Notwithstanding the rain and himself being sick, he liked very much the city and was anxious for a bit of good weather to tour around. Unfortunately the language was a hard barrier to overcome, as neither his native German nor his quite good English could help.[108] During this time, Touschek was also keeping a close correspondence with Arnold Sommerfeld. He had planned to visit him in Munich, on his way back from the Austrian vacation, but it did not happen, and Touschek apologized to him in a letter sent on October 5th.[109]

While planning betrayal, Touschek had also decided it was time to move out of Dee's house and find his own home. Since he was leaving for the entire summer, this was particularly easy. On September 29th, back from his extended leave, he was still writing to Max Born from 11, The University, but, on October 26, 1950, we find

---

[103] Letter to parents on March 21st, 1950.

[104] Handwritten letter to parents June 23rd, 1950.

[105] In November 1950 letter to his parents, Touschek mentions that the Gunns had very much enjoyed the visit.

[106] Private e-mail communication to the author from Prof. M.J. Gunn, February 17th, 2020.

[107] See letter to parents from Copenhagen, with handwritten date hard to read, 1950, September or October 6th, likely to be September 6th, since he writes to Born from Glasgow at the end of September.

[108] Letter to parents from Copenhagen, September 6th, 1950, as discussed in previous note.

[109] Bruno Touschek to Arnold Sommerfeld, October 5th, 1950 Deutsches Museum Archiv, NL 089, 013.

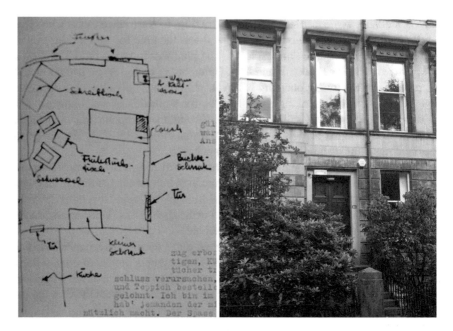

Fig. 7.27 At left, a drawing by Bruno Touschek which describes the layout of his new lodgings at 61 Oakfield Avenue, in Glasgow, extracted from a November 1950 letter to his parents. Family Documents, © Francis Touschek, all rights reserved. The right panel shows a present-day view of 61 Oakfield Avenue, photograph by the Author on August 8th, 2019, all rights reserved

him writing from 61 Oakfield Avenue, announcing Thirring's presence and his house move. A detailed drawing of his new room is shown in Fig. 7.27, together with a view of his new residence, also listed in the University records for the 1950–1951 academic year.

7.3.4   A Visit from Walter Thirring

In Fall 1950, an important development in Touschek's theoretical works, had taken place. Such development was stimulated by a visit by the young theorist, Walter Thirring.

Since his arrival, Touschek's scientific interests had shifted from accelerator physics towards particle physics, where a new world was opening through the experiments made possible by post-war accelerators, such as synchrotrons. The new experiments allowed to observe particle interactions in laboratory settings at higher energies than ever before. Unlike the case of cosmic rays physics, where particles come from the sky and their origin is outside the experimenter's control, accelerators allowed a choice of projectile and target—electrons or protons—, and control of their energy and momentum. The initial particle state being known, the observation of the scattered particles (the final state), could give precise information on what happened during the transition, and the interactions between elementary particles in the newly opened high energy regime could be studied with greater accuracies. New theoretical formulations and techniques were developed, and took central stage, as the infinities plaguing the calculation of scattering processes were being cured by including virtual processes, in a combination of relativistic field theory and perturbation theory.[110] The covariant formalism, in which relativity is formulated, made calculations more transparent and became the modern language of particle physics, where the new techniques by Richard Feynman, and Julian Schwinger for the calculation of scattering processes were opening a whole new world.

It is in this scenario, that Touschek started looking again at an old problem, the so called "infrared catastrophe", discussed in a seminal paper by Felix Bloch and Arnold Nordsieck, in 1937. This problem needed to be reformulated in modern language, namely through the new covariant formulation, and the occasion to do so was the visit of Walter Thirring, Hans Thirring's son, with a Nuffield Fellowship. [111] During the summer, Bruno had learnt of the young man's travels from Vienna, first to Dublin visiting Schrödinger (Moore 1989), and then to Glasgow. He held him in great consideration, as the most brilliant young theoretician in all of Europe, and had been looking forward to welcome him in Glasgow.[112]

Walter Thirring arrived in October 1950.[113] It was an extremely busy time for Bruno. He had never before taken full responsibility for his lodgings, the furnishing, the cleaning, or procurement of food. The many house caring activities engulfed him during October and November. All this had to be done in addition to looking after his research student, going to Edinburgh with Thirring, and, most of all, working with Thirring on the problem of the Bloch and Nordsieck method, writing the paper. They submitted it on November 29th to the *Philosophical Magazine*, where it was published in March 1951 (Thirring and Touschek 1951). This paper is very important in Touschek's future thinking about electron accelerators: after AdA's proof of feasibility in 1961, Touschek planned the construction of a much more powerful electron positron accelerator, ADONE. Because of the much higher energy range of the new machine, the extraction of physics results was connected to the Bloch and Nordsieck theorem, and, in 1964, Touschek would start on the problem of summing the infinite number of soft photons emitted before and after the collisions.[114]

---

[110] See also David Kaiser's *Drawing Theories Apart: The Dispersion of Feynman Diagrams in Postwar Physics* (Kaiser 2005, 53).

[111] Walter Thirring (1927–2014) became Professor of Theoretical Physics at University of Vienna in 1959. For Walter Thirring's comments about Touschek, see Thirring's autobiography (Thirring 2008).

[112] Letter to parents, November 1950.

[113] See Born–Touschek correspondence, where the date is October, versus September–October, according to Amaldi (1981, 12).

[114] This problem led Touschek to train a group of young theoretical physicists recently graduated from University of Rome, as described in the Italian article entitled *Frascati e la fisica teorica* in http://www.analysis-online.net/wp-content/uploads/2013/03/greco_pancheri.pdf.

Thirring left in early December 1950, barely after only 2 months, and Touschek started again feeling that Glasgow could not offer the type of environment he sought. The joys and pains of teaching were also hitting him. Students sometimes complained about difficulties in his course, or he felt that his research students were too slow and lazy to carry through some new research lead he gave them. In February, he observed with some amusement the frolics on the occasion of the 500th anniversary of the University of Glasgow foundation, when the students behaved 'pleasantly ghastly' and threw eggs and rolls of toilet paper at the Senate.[115]

7.3.5   The Southern Way

After Thirring left, Touschek started looking for a position elsewhere, considering Switzerland, where he heard rumors of a new institution to be created by the U.N.— probably what became CERN—and Germany, in Munich, if Heisenberg were to move there. He expressed his hopes and plans in his letters home.

The summer vacations in Europe made winters in Glasgow harder to bear. Max Born was going to retire in one or two years and planned to go back to Germany, so did many others. As seen from the correspondence with Born, in spring 1951 Touschek aimed for a Senior Lectureship position in Oxford, but nothing materialized. The solution was to come from a different direction, through the Italian theorist Bruno Ferretti, who, at the turn of the 1950s, was already deeply involved together with Edoardo Amaldi in discussions about the foundation of CERN (Hermann et al. 1987, 67).

The association between Touschek and Ferretti follows rather standard research channels. It is possible that the two of them met in 1947 in Birmingham, where Ferretti was giving a course and working with Peierls. Bruno spoke Italian, a fact which made for an easy way to approach Ferretti, and the occasion could have been the fact that Ferretti and Peierls were working on radiation damping (Ferretti and Peierls 1947), a problem which had interested Touschek since the war days. Bruno may have read and studied Ferretti and Peierls' paper and wished to talk with them about their work. However in summer 1947, when the article had appeared, Touschek may not have been thinking much about Italy, and he was still a first year research student, hardly in confidence with a well known senior scientist such as Rudolf Peierls, or his guest Bruno Ferretti, Fig. 7.28.

The connection between Touschek and Ferretti was certainly established or enforced in spring 1951, when Touschek saw one of Ferretti's papers in *The Nuovo Cimento* (Ferretti 1951), which particularly interested him. Soon after, he wrote, and submitted for publication, his own article citing Ferretti's (Touschek 1951). From this to start writing to Ferretti about physics is just an obvious step, and either before or after Touschek's article, the possibility of meeting Ferretti in Rome during the

---

[115] The letter to parents is dated February 12th, without the year, but mentions the 500 year anniversary of the university, which was founded in 1451.

Fig. 7.28 From left, Felix Bloch, Bruno Ferretti, Homi J. Bhabha, Wolfgang Pauli, at the 1948 Solvay Congress. Reproduced by courtesy of the Pauli Archive, CERN (PAULI-ARCHIVE-PHO-063)

upcoming summer vacation for a scientific discussion, arose. A visit to University of Rome was likely included in his summer plans. At the same time, independently of physics interests, the visit to Rome would allow to resume contact with his aunt Ada, as he had been thinking of Italy and the holidays spent in Rome before the war, at her beautiful apartment in the Parioli neighbourhood.

The visit to Rome took place in mid July 1951,[116] and it is conceivable that a meeting between Touschek and Ferretti in the Physics Institute in Rome had been foreseen to discuss physics questions. A description of their first meeting in Rome is given in Amaldi (1981, 13): "[…] Touschek went to visit Ferretti at University of Rome. A few hours after their first meeting, spent discussing mutual scientific questions, they established such marked professional respect and personal attachment for each other that Touschek decided to remain permanently in Rome." These lines give a very vivid picture of Touschek and Ferretti's encounter, except that Amaldi places it in September 1952. However, from Touschek's accurate descriptions of the 1951 and 1952 summer travels in letters to his parents, a visit to Rome in 1952 is unlikely, whereas he was certainly planning to be in Rome in July 1951. One possibility is that Touschek and Ferretti met in Rome in July 1951, discussed physics together and the idea of applying for a position in Rome came up. As recalled by

---

[116] Letter to parents, June 25th, 1951.

Amaldi, "It was mainly Ferretti's personality that, in 1952, attracted and permanently fixed in Rome an occasional visitor, Bruno Touschek […]" (Amaldi 1979b, 441). Following the extended European tour in summer 1951 and 'an occasional visit', a correspondence took place between the two of them about such possibility, but the thing needed time to be perfected.[117] There was in fact the question of giving notice to the University of Glasgow, and properly tie the end of Touschek's commitment to Glasgow with the new position he was hoping to have in Rome. The details were probably worked out through such correspondence, mentioned in Touschek's letters home, but with no other records.

Christmas and New Year came, and brought very cold days. In January 1952 it was so cold that the water in the bathroom would freeze.

His unhappiness about the weather was not made better by his getting a cold around Christmas time. It forced him to stay home, and not be able to attend the Dees' party. Christmas and New Year Day were thus terribly boring, and many small problems aggravated his spirit. Taxes to prepare, rations to obtain, writing a chapter of a book with Gunn he could not really enjoy doing, and again the cold, made him unhappy and discontent. In between, always a keen sportsman, he would not neglect some winter outing, such as going to ski with Mrs. Gunn, Fig. 7.29.[118]

He went on with his teaching, and caring for his students, but as the spring came he knew he would be going away, either to Italy or somewhere else. In late August, possibly after the usual vacation in Tyrol, Touschek went down to Milan from Austria, through the Brenner pass. The program, included a visit to University of Milan. No explanation is given as to the reason for such visit, however, some physicists of that University had established a laboratory devoted to the development of applied nuclear physics, CISE (Centro Informazioni, Studi ed Esperienze), in which Ferretti and other physicists from University of Rome, where he was a Professor, were also involved (Amaldi 1979a, 58). Thus, Touschek's visit appears to have been planned to take advantage of Ferretti's presence in Milan at this time, and to discuss the details of the planned engagement with the University of Rome. Touschek had not yet decided whether to get one year unpaid leave from Glasgow, or resign from his Lecturer position and he needed to clarify salary questions related to either option. After a couple of days in Milan, he crossed into Switzerland through the Gotthard Pass, and reached Bern, where he received a letter from Ferretti who wrote he would agree to the postponement demanded by Glasgow until the first of January.[119] In the meanwhile, Touschek was also seeking some other possibility, like a fellowship with UNESCO, which Heisenberg could try to get for him in case of need, or a Lectureship in Bonn, which Wolfgang Paul could support.

On September 15th, the situation cleared up when Amaldi, then head of the Rome branch of the National Institute for Nuclear Physics (INFN), wrote officially to Bruno offering a one year (renewable) position, adding that in case of need, one month

---

[117] Letter to parents, November 8th, 1951.

[118] Letter to parents, January 27th, 1952.

[119] Letter to parents, September 10th, 1952.

salary could be advanced to him.[120] It is of interest to see Amaldi's list of Touschek's
expected duties, namely: (a) research work in theoretical physics, (b) discussions and
advice to experimental physicists working on cosmic rays and possibly accelerators;
(c) not more than two hours per week of teaching to advanced students in theoretical
physics.

Touschek in the meanwhile had decided to resign from Glasgow, thus solving the
problem of missing salary and, on September 23rd, 1952, replied to Amaldi accepting
the offer.

The final exchange of letters put Bruno on his way to Rome, sealing his future and
leading him to conceive AdA less than 10 years later. Bruno's future was now traced.
He would be immersed in a unique moment of hopes and dreams, when all seemed
possible, and when much was indeed accomplished. In the 1950s, Italy, which had
been unified less than one hundred years before, and had given universal voting rights

---

[120] The newly founded institute was bypassing some of the burocracy inherent at the time in University administration. Such possibilities, one month advanced salary, would in fact be unheard of
in the regular university administration. For the history of INFN see Battimelli et al. (2001).

Fig. 7.30 The physics faculty at University of Glasgow in 1952, with Bruno Touschek seated at right, from https://universitystory.gla.ac.uk/images/UGSP01004.jpg, credit Professor David Saxon

to every citizen only in 1946, entered into a period of great cultural, economic and political development, which led to what was later known as the *Miracolo italiano*.[121]

In 1952, Bruno was ready to leave the UK road, one of the many roads ultimately leading to particle colliders. A 1952 photo of the Physics Faculty from the Department of Natural Philosophy of the University of Glasgow, Fig. 7.30, shows Bruno on the verge of flight, with his friends and faculty colleagues, Emlyn Rhoderick, Walter McFarlane, Philip Dee, John Gunn and Samuel Curran. A few months later, Bruno, seen seated in the front row, would be in Italy.

When Bruno Touschek arrived in Rome, it was the end of the year. He had left Glasgow where winter brought early darkness and freezing cold, and after a brief stay in Bern, crossed the Alps through Switzerland, and was charmed by warmer weather and the Roman food. As it often happens to travellers from Northern lands, he found he did not need to walk fast to keep warm, rather he needed to slow down his nordic walking, and must have welcomed the lighter days of the Rome winter. On his arrival he dutifully visited his aunt Ada, and was housed in a nice family pension near the University, booked for him through the Physics Institute. He found the Institute

---

[121] Italy had started its unification process in 1861, making Rome the capital city of the new Italian state on February 3rd 1871. The full unification took place in 1918, when the two regions south of the Brenner pass, Trentino and South Tyrol, became part of Italy.

really excellent, with no feeling of estrangement, as he met there Patrick Blackett whom he knew from Glasgow and Wolfgang Pauli. He also found the colleagues from the Institute very interesting and welcoming.[122]

Bruno's life had been broken, but it was now being pieced together again. Physics in Italy was fully flourishing and the future there looked really exciting. He felt ready for a great adventure.

---

[122] Letter to parents, December 30th, 1952 from Rome.

Part II
Towards Ada through Italy and France

Chapter 8
A Laboratory on the Hills: Frascati and the Italian Road to Particle Accelerators

By October 1954, we knew what we wanted, and that we really wanted it.
Ormai (ottobre 1954) si sapeva quel che si voleva, e bisogna dire che lo si voleva in modo piuttosto deciso. *In* 'L'Elettrosincrotrone e i Laboratori di Frascati', pag. 51, Zanichelli 1962.

Abstract As Bruno arrived in Rome, the construction of the Italian road to high energy particle accelerators was under way. In time, these beginnings can be traced to Enrico Fermi in Italy, before the war, and before he were to leave for the USA. What led the Italian physics community to build a laboratory and an electron synchrotron in Frascati, was part of Edoardo Amaldi's vision for reconstructing Italian physics. This is behind his decision to call Bruno Touschek to Rome, and ask the 33 years old Giorgio Salvini to build the laboratory where AdA would be built in 1960. Salvini and Touschek were about the same age, and their lives ran along parallel lines. Salvini in 1941, in Milan, during the war, was reading the same Kerst's article which put in motion Widerøe's war betatron, in Germany. Afterwards, during most of the 1950s, the construction of the laboratory in Frascati and Touschek's life in Rome developed rather independently. While Bruno followed the novel theoretical physics ideas in particle physics which would bring him to propose AdA, Salvini assembled a team of scientists, technicians and engineers and succeeded in having a powerful accelerator springing to life in April 1959, in a new laboratory built in Frascati, near Rome. At this point, Bruno's interest shifted to the synchrotron and to the laboratories on the Tusculum hills.

When Bruno arrived in Rome at the end of December 1952, Italy was laying the grounds for the cultural and industrial expansion which became known as the *Miracolo Italiano*, the Italian miracle of the 1960s, which saw Italy developing powerful infrastructures, highways and bridges, linking the industrial North to a less developed South. By the end of the 1950s, the national diffusion of television brought language unification between social classes, some of which had only spoken ancestral dialects. While Italy as a whole was modernized, the country started a cultural and scientific

© The Author(s), under exclusive license to Springer Nature Switzerland AG 2022   235
G. Pancheri, *Bruno Touschek's Extraordinary Journey*, Springer Biographies,
https://doi.org/10.1007/978-3-031-03826-6_8

revolution which would bring it on a par with the other European countries. Visitors and tourists from abroad contributed to bring new ideas and knowledge, which fostered artistic and literary flowering.

This is the country which Bruno joined and to which he left his scientific legacy. At the end of December 1952, Italy was moving its first steps along the way to become a particle accelerator builder, a member of the very selected club of countries which can design and construct new accelerators. The plan to build a large facility where a modern accelerator could be designed and constructed was under way. This was going to be the Frascati National Laboratories, which are part of the journey taken by Italian nuclear and particle physics between 1926, the year Fermi arrived in Rome to the Chair of theoretical physics, and 1953, when a project for the construction of a modern particle accelerator was approved by the National Institute for Nuclear Physics, Istituto Nazionale di Fisica nucleare (INFN) (Battimelli et al. 2001). It had been a long road, which had started with Enrico Fermi and which the war had interrupted (Battimelli and Gambaro 1997).

8.1   Italy Before the War

The road which led to the official participation of Italy to nuclear physics research is best illustrated by the comparison between two photographs. The first is the famous official photograph of the 1927 Solvay Conference, which includes all the major physicists of the time, from all the great laboratories or institutes in Europe, but not a single Italian scientist. Italy however was soon to step in. Four years later, an International Conference of Nuclear Physics took place in Rome, sponsored by the Alessandro Volta Foundation of the Royal Academy of Italy.[1] The Academy's President, Guglielmo Marconi, had supported the idea of such conference, strongly advocated by Enrico Fermi. A photograph of the 1931 Rome Conference, Fig. 8.1, has Marconi at the center, many of the scientists from the Solvay conference and a number of Italian physicists, among them Bruno Rossi, from University of Bologna, working on pioneering cosmic rays research, and Orso Mario Corbino, the director of the Rome Physics Institute, in addition to Enrico Ferrmi.[2]

The Conference took place in Rome, on 11–18 October 1931 at the Institute for Physics in Via Panisperna. It included the participation of all the major scientists of the age, except for Albert Einstein, "because he did not want to enter Fascist Italy", as recalled by Emilio Segrè, (Segrè 1993, 46). In Fig. 8.1 one can distinguish Marie Curie, partly hidden behind Marconi at the center, and Jean Perrin from France, Patrick Blackett from the Cavendish Laboratory in Cambridge, Arnold Sommerfeld,

---

[1] The history and connection between the Royal Academy of Lincei (1870–1923), the Academy of Italy (1926–44), the Academy founded in 1603 by Federico Cesi, and Accademia dei Lincei in the period 1926–44, up to present days is reconstructed at the Academy site http://www.lincei-celebrazioni.it/iacca_lincei.html.

[2] Orso Mario Corbino (1876–1937) was a noted Italian scientist and held political influence, as an Italian Senator and former Minister.

Figura 3 - Nell'ottobre del 1931, per impulso di Guglielmo Marconi incomincia a Roma l'*era nucleare*, con il famoso convegno che vide riuniti, intorno al Presidente dell'Accademia d'Italia, ben otto premi Nobel e i maggiori scienziati del mondo. La fotografia ritrae gli intervenuti al convegno.

1. Richardson; 2. Millikan; 3. Marconi; 4. Bothe; 5. Rossi; 6. Lise Meitner; 7. Goudsmit; 8. Stern; 9. Debye; 10. Compton; 11. Curie; 12. Bohr; 13. Aston; 14. Ellis; 15. Sommerfeld; 16. Wataghin; 17. Perrin; 18. Corbino; 19. Trabacchi; 20. Cantone; 21. Parravano; 22. Rasetti; 23. Heisenberg; 24. Brillouin; 25. Townsend; 26. Ehrenfest; 27. Fermi; 28. Beck; 29. Persico; 30. Vallauri; 31. Giordani; 32. Bonino; 33. Mott; 34. Rupp; 35. Quirino Majorana; 36. Garbasso; 37. Lo Surdo; 38. Carrelli.

Fig. 8.1 A group-shot of the International Conference of Nuclear Physics held in Rome in 1931, from Grandolfo et al. (2017), https://www.iss.it/documents/20126/0/Quaderno_12.pdf/5b73b8e9-6203-3782-2d43-0dab5a56fa19?t=1593710943681. Other similar photos are available in Sapienza University of Rome–Physics Department Archives

Walther Bothe, Werner Heisenberg, Lise Meitner (hard to be seen, but listed in the photo caption) and Otto Stern from Germany, Robert Millikan and Arthur Compton from US, and Niels Bohr, with whom strong relationships continued to exist through the years.[3] Many of these scientists were already Nobel Laureates or would be awarded the Nobel Prize in Physics in the near future or after the war.

---

[3] Fermi and Bohr would later work together in the USA, on the Manhattan project. Their joint communication in 1939 about latest developments in nuclear fission is graphically remembered in https://www.osti.gov/opennet/manhattan-project-history/Events/1890s-1939/fission_america.htm.

The Conference took place only a few months before James Chadwick announced of having detected the neutron—the first uncharged subatomic particle to be identified, a discovery inaugurating the nuclear era (Chadwick 1932). This and the year to follow were extraordinary years for nuclear and particle physics: after the discovery of the neutron, the first artificial disintegration of the atomic nucleus was realized with the accelerator built by John Cockcroft and Ernst Walton (Cockcroft and Walton 1932), and Lawrence built the first cyclotron (Lawrence and Livingston 1931). Similarly notable advances took place in cosmic ray physics, where the first antiparticle to be ever detected—the positron—was discovered by Carl Anderson among those of other particles from the sky, analyzed in his photographic emulsion plates.[4]

After his arrival in Rome, Fermi had gathered around him an exceptional group of young scientists, and had them visit the other important nuclear research groups in Europe. Frequent contacts between the various groups working in Europe on nuclear physics, around Rutherford in Cambridge, Marie Curie in Paris, Fermi in Rome, Kurchatov in Leningrad, were fostering the scientific interactions which led to the great results of 1934. This is the year when major steps towards modern physics were taken in Paris, at Collège de France by Frédéric Joliot and Irène Joliot-Curie with the discovery of artificial radioactivity (Curie and Joliot 1934; Joliot and Curie 1934), and in Rome, in October, when Fermi's group discovered that neutrons could produce artificial radioactivity (Fermi et al. 1934).

Thus, in the mid '30s the leading groups in nuclear physics research could be said to be based in Cambridge, Rome, Paris, Berlin and Leningrad, (Close 2015, 25), with Rome even ranked first in Fraser (1997, 18), as the place where one of the decisive experiments on the role of neutrons and nuclear fission took place.[5] In Fig. 8.2, a well known photograph of Fermi's group in the 1930s is shown.

Fermi's group was initially based in the old physics institute in Via Panisperna, in a part of the city center near the *Suburra* of ancient Rome. Around 1936, the Institute of Physics was moved to the new university campus, outside the Aurelian walls, between the San Lorenzo freight station and the Verano Monumental Cemetery. Even in the new spacious grounds, Fermi soon realized that space and financial limitations prevented Italy to adequately progress and compete in the new area of nuclear physics, and in January 1937 presented a request to the National Research

---

[4] The Nobel Prize in Physics 1936 was divided equally between Victor Franz Hess 'for his discovery of cosmic radiation' and Carl David Anderson 'for his discovery of the positron', from https://www.nobelprize.org/prizes/physics/1936/summary. For a detailed story of the discovery see Close (2018).

[5] In Close (2015, 25) the list does not include Berlin. Close quotes this from the interview given by Maurice Goldhaber, to Charles Weiner and Gloria Lubkin, 10 January 1967, Brookhaven National Laboratory, Niels Bohr Library & Archives, American Institute of Physics, College Park, MD USA, https://www.aip.org/history-programs/niels-bohr-library/oral-histories/4632. Goldhaber did not mention Walther Bothe, a renown experimental physicists whose work with Herbert Becker was at the root of the discovery of the neutron. In 1934 Bothe became the director of the Physics Institute at the Kaiser Wilhelm Institute for Medical Research in Heidelberg, where he built, with his assistant Wolfgang Gentner, a van de Graaff accelerator and the first operational cyclotron in Germany. Since 1939, Bothe was a protagonist in the German nuclear project, the *Uranverein*, the Uranium Club.

Fig. 8.2 Fermi's group: Ettore Majorana, Emilio Segrè, Edoardo Amaldi, Franco Rasetti and Enrico Fermi. Photograph taken by Bruno Pontecorvo, reproduced courtesy of Sapienza University of Rome–Physics Department Archives, https:// archivisapienzasmfn.archiui. com, documents provided for purposes of study and research, all rights reserved

Council (CNR) for the creation of a National Institute for Radioactivity. The decision took more than one year to arrive.

By this time, menacing clouds had been gathering over Europe. Less than 4 years after the discovery of slow neutrons in 1934, Fermi's group had begun to scatter away. Emilio Segré had gone to Palermo to the Chair of Experimental Physics, and, during a visit to Berkeley in 1938, because of Mussolini's anti-semitic laws, the so called *racial laws*, left Italy to spend his scientific life in the United States. In 1936, Bruno Pontecorvo, from a prominent Jewish family as Emilio Segré, had joined the Joliots' group in Paris.[6] In 1938, the political situation in Italy made his return dangerous and he remained in Paris, first with French fellowships, then with

---

[6] See letter from Fermi to Frédéric Joliot-Curie recommending Pontecorvo's joining his group in (Vergara Caffarelli 2004, 14).

a permanent position.[7] Pontecorvo remained in Paris until the occupation of France by the Germans. As the Germans reached Paris, he fled to the South by bicycle, and, from there, through Lisbon, left for the United States and Canada, never to return to work in Italy (Close 2015; Guerra and Robotti 2014; Turchetti 2012). One more blow to the group's research activities was Ettore Majorana's tragically mysterious disappearance early in 1938.[8]

By 1938, only Fermi and Edoardo Amaldi, the youngest of his group after Pontecorvo's departure for France, would still be systematically working together in Rome. Another sudden stop to further research in nuclear physics in Italy came from the national government advisory body in 1938. In 1937 Enrico Fermi had asked the Italian government for extraordinary funds to build a national institute for radioactivity. It is not clear if Fermi's request in January 1937 already foresaw the construction of a cyclotron, however a summer correspondence from University of Stanford between Fermi and the director of the Physics Institute in Rome, Orso Maria Corbino, indicates that a conversation with Ernest Lawrence convinced Fermi of the economic and technological feasibility of a 5–6 MeV cyclotron (Battimelli and Gambaro 1997).[9] The answer to Fermi's request was disappointing. In June 1938, the major Italian research institution, Consiglio Nazionale delle Ricerche (CNR), did not back Fermi's request for the creation of a national institute. The CNR's presidency argued that the type of state support obtained in France by Frédéric Joliot for his laboratory (and the construction of a cyclotron), would require much larger expenditures than what Fermi himself had approximately asked (Russo 1986). Thus, while extra fundings had been approved for continuation of Fermi's research, they were insufficient for an enterprise at the level of what was being built in France. Shortly after, in October, Fermi was awarded the 1938 Nobel prize in Physics.[10] By that time, Italy had passed the anti-semitic laws which would decimate the Italian scientific community. Fermi worried for his family, his wife Laura Capon being of Jewish ancestry.[11] In December, he left Italy with his wife Laura Capon and their children to attend the official Nobel ceremony in Stockolm. From there, they traveled to England, and then, on December 24th, 1938, Fermi and his family boarded the *Franconia* at Southhampton, bound for the New World. Two days before, Otto Hahn and Fritz Strassmann had sent to *Naturwissenschaften* their paper announcing

---

[7] Pontecorvo's brother Gillo, the future film-maker, was also living in France at the time. The eldest brother Guido Pontecorvo had left Italy for Scotland, where he obtained his PhD at University of Edinburgh and established himself as a leading genetist at the University of Glasgow.

[8] There exists a vast literature about Majorana's life and many conjectures about the reasons for his disappearance, as well as doubts about a possible suicide. He disappeared in spring 1938, during a boat trip from Naples to Palermo (Esposito 2017).

[9] As in footnote (9) of Battimelli and Gambaro (1997).

[10] The prize was awarded for 'his demonstrations of the existence of new radioactive elements produced by neutron irradiation, and for his related discovery of nuclear reactions brought about by slow neutrons', https://www.nobelprize.org/prizes/physics/1938/summary/.

[11] Laura Capon's father Naval Admiral Augusto Capon, a hero of World War I, was arrested in October 1943 during an infamous German raid against Jewish citizens of Rome. He was deported to Auschwitz, where he was killed shortly after arrival.

the surprising results of their experiments which would be immediately interpreted by Lise Meitner and Otto Frisch as nuclear fission. A chain of extraordinary events would soon begin. The world was on the verge of terrible changes. In less than four years, in Chicago, Fermi would realize the first self-sustaining nuclear chain reaction.

Thus, in 1938–1939, the internal politics of the Fascist government and, especially, the laws against Italian citizens of Jewish origin, drove away from the country some of its best scientists. The departure to South America of Giuseppe Occhialini, a prominent cosmic rays experimentalist, the emigration of Bruno Rossi (Bonolis 2014), and the death in 1942 of the young theoretical physicist Giovanni Gentile Jr., further depleted the ranks of active physicists (Bonolis, Gentile Jr in Milan).

8.2   Edoardo Amaldi: Creator of CERN and the Friend Who Would Write Bruno's Biography

In his first letter from Rome on December 30th, Bruno appears pleasantly surprised in finding an international atmosphere with two Nobel Prize winners visiting the institute, and comforted by a positive and friendly attitude pervading it. He was now returning to a city he loved, with a climate almost unparalleled in Europe, where months stretch from October to March like a mild autumn, and spring follows almost without solution of continuity. His typical northern brisk walk could slow down, being unnecessary in a city where he hardly needed to wear an overcoat in January, used as he was to the Glasgow winters. When leaving the house in Rome, he could just put on a jacket and go to the physics institute without catching cold or having to run against icy winds.

Bruno was housed in a Pension, not far from the physics institute. He had been worried about taking the step of coming to Italy as his scientific destination. He had liked Bruno Ferretti, his first contact with Italian physics, and appreciated his work, but did not know much about the other colleagues. He was soon very favourably impressed and some vague fears of having taken the wrong step in his scientific career, were allayed.

What he did not know, but would soon find out, was that the Rome physicist who had hired him, Edoardo Amaldi, was one of the artifices of the scientific resurgence of Europe, a protagonist of the creation of CERN, and the one who helped Italy to reconnect to its past scientific eminence.

8.2.1   Through the War Towards the Reconstruction

Amaldi had visited the United States before the war and in 1939 had been encouraged by Fermi to remain and work there. Instead, he had decided to return to Italy and keep alive whatever would remain of the physics institute, even trying, until 1942, to have a cyclotron built in Rome for the planned exhibition of 1942. Before the war, Mussolini had planned a Universal Exhibition, E42, to take place in Rome, and be a show-case for the fascist regime (Battimelli and Gambaro 1997). The Exhibition major location was to be a new neighbourhood built on the outskirts of the city, on its west side, on the road which takes the Roman citizens to the sea, towards the ancient port of Ostia. Some of the best architects in Mussolini's time, Adalberto Libera and Marcello Piacentini among them, had been engaged to build a congress hall, an Exhibition palace tall and square, dotted by porticuses and large inscriptions, a Museum of Archeology, and more. E42 never took place, but EUR, an acronym which stands for Esposizione Universale di Roma, is now a municipality of the metropolitan city, and the magnificent buildings, created or designed before the war, now impressively welcome the visitor approaching from the nearby Fiumicino airport or from the many localities on the shore. E42 was cancelled by the war, and no cyclotron was built, but the dream of having an accelerator stayed with Amaldi and the other physicists in Rome, and surged again after the war.

In the absence of an accelerator, there was always the possibility of using what the sky would provide free of cost: particles from the cosmic rays which constantly bombard the earth. In 1943, three young physicists, Conversi, Pancini and Piccioni (CPP) had built a detector in the laboratory of the Rome physics institute to study the properties of one of the copiously produced cosmic ray particles, the so-called *mesotron*. This particle was thought to be responsible for the forces which kept the nucleus together. It wasn't. The experiment, carried out through the bombardments of Rome in 1943–44 (Conversi et al. 1947), showed that the mesotron was not what everyone thought it to be, rather it was a particle which carried a different type of force.[12] A major discovery in particle physics had come from the sky, under the least favourable circumstances. The instrumentation for this experiment had been assembled in the machine shop of the Rome physics institute near the San Lorenzo freight station, outside the city walls, and had been transported to relative safety with a hand held cart, to a high-school in the center of Rome, while aerial attacks were taking place.[13] Important as the CPP experiment had been, this mode of research belonged to the past. As the war ended, new avenues of research were needed to keep up with American science, in particular with particle accelerator science.

In those years, it was difficult to obtain the scientific literature from the United States, but as soon as Rome and the South of Italy were liberated from the German occupation and the Allied troops took over control, the American scientific literature returned to the Physics Institute, and with it, there came the articles by Kerst and Serber about the possibilities to build a new type of particle accelerator, one which could accelerate electrons to higher energies, not attained before. These articles announced the successful operation of a betatron and were the same which had put in motion Widerøe's project in Germany and Bruno's work with it, during the war.

---

[12] The full nature of this particle, was clarified in a subsequent experiment by Lattes, Occhialini and Powell, who gave the name μ-meson to this particle (Lattes et al. 1947).

[13] The high-school was in via Giulia, nearer to the Vatican City, and supposedly safer from bombing by the Allied forces. The center of the city of Rome was spared heavy damage both because of its incalculable archeological value, and by the presence of the Vatican.

The war ended and Amaldi went into action and proposed to the Italian Nuclear Research Council, CNR, that a research center for modern nuclear physics be founded and an accelerator be built. The new center would have to address the nascent science of particle physics, which was still called *nuclear* physics, but was now directed to interactions between the components of the nuclei. For the instrumentation to install, Amaldi's 1945 proposal was not for a cyclotron, as in the original pre-war plan, but for a new machine, a betatron (Battimelli and Gambaro 1997), which in 1945 was the accelerator of the future. Amaldi's request did not immediately go through, but in a few years led to the creation of the National Institute for Nuclear Physics, which would allow Amaldi to hire Bruno, in 1952, and bring the construction of the Frascati synchrotron.[14]

Thus Amaldi began placing the foundations for the reconstruction of nuclear physics studies in Italy, and in 1946 returned to the United States to visit his old friend Enrico Fermi in Chicago. Fermi first urged Amaldi to remain and work in the United States, where funding resources were almost unlimited, after the atomic bomb success. As Fermi wrote to Amaldi: "…everybody talks about million dollars …the main problem will be to device enough things to spend the money for …." (Amaldi 1997). Amaldi resisted the temptation, and, once more, returned to Rome, to Italy, and with it to Europe.[15]

In 1947, Amaldi, looking ahead to reinforce research in particle and nuclear physics, called Gilberto Bernardini from the University of Bologna to Rome, and together they took upon themselves the burden of postwar reconstruction.[16] Their joint efforts would allow Italy to join modern physics as equal partner to the other European countries, which had won the war.

---

[14] Italian law at the time did not permit a foreigner to participate in the national competition for University positions, but this was possible under INFN regulations, and this allowed hiring Bruno as a level-2 researcher, in short as an 'R-2'.

[15] Letter from Fermi to Amaldi and Giancarlo Wick, January 24th, 1946: "… Anche in America la situazione della fisica ha subito cambiamenti molto profondi per effetto della guerra. Alcuni sono per il meglio: ora che la gente si è convinta che con la fisica si possono fare le bombe atomiche tutti parlano con apparente indifferenza di cifre di vari milioni di dollari. Fa l'impressione che dal lato finanziario la maggiore difficolt consisterá nell'immaginare abbastanza cose con cui spendere...", in Edoardo Amaldi Papers in Sapienza University of Rome–Physics Department Archives, courtesy of Giovanni Battimelli. A new, enlarged edition of Amaldi (1997) is under preparation by Battimelli, De Maria and Adele La Rana.

[16] See https://www.aif.it/fisico/biografia-gilberto-bernardini/.

Gilberto Bernardini (1906–1995) was one of the founders of the Testa Grigia cosmic ray laboratory, on the Plateau Rosa, in the Italian Alps. In 1947 Gilberto Bernardini had been called from Bologna to the chair of Spectroscopy in Rome, but starting from 1948 onward he began to spend long periods in the United States, first at Columbia University (New York), where he gave a course on cosmic radiation and where he made Leon Lederman, then a PhD student, discover the excitement deriving from experimental physics with his "remarkable sense of wonder about simple things". Lederman's description of their joint work constructing a kind of Geiger counter is exhilarating: "We machined, soldered, polished, flushed with clean argon gas and watched the oscilloscope. Soon we had tracks. Bernardini went nuts. 'Izza counting!' he screamed. Half of my height and weight, he lifted me and danced me round the lab to the music of Bernardini's sense of wonder. He explained: 'Dese particles, cosmic ray, come from billions of miles away to say buongiorno to us on de tenth floor of Pupin Physics Building. Izza beautiful! So little particle, so long da trip.' So, through Bernardini, I began to recover my love of physics, of searching for simplicity and elegance of how the world works." Lederman had recently been discharged from the US army at the end of WWII and was discouraged after discovering he had forgotten even simple equations and "how to study", from Leon Lederman, in *Life in physics and the crucial sense of wonder*, CERN Courier, 30 September 2009, 22–23 https://cds.cern.ch/record/1734431/files/vol49-issue8-p022-e.pdf. Later, in 1951, Bernardini moved to the University of Illinois (Urbana), where Donald Kerst had built betatrons of increasing energies since 1940. A 80 MeV machine was built in 1948 and a 300 MeV very competitive machine, used in an extensive series of experiments in meson physics and in photo-disintegration of nuclei, was completed in 1950. Bernardini could thus be among the new staff members recruited to exploit this new facility. See list on pages 21–22, in Gerald M. Almy's talk, *A Century of Physics at the University of Illinois, 1868–1968*, before the History of Science Society in December, 1967, about the Centennial Year of the University of Illinois, http://hdl.handle.net/2142/48722, edited in 1973 with minor changes and additions https://www.ideals.illinois.edu/handle/2142/48722.

Welcome in the United States by their friends who had left Italy for political or racial reasons, both Amaldi and Bernardini visited or spent time in the US. While Amaldi's visits were comparatively short, Bernardini spent extended research periods at Columbia University in New York, and in Illinois, exchanging a correspondence with Edoardo Amaldi about the type of accelerator to build in Italy, following US developments on this front. In the United States the Brookhaven National Laboratory had been formally established in 1947 by the US Department of Energy. Located in Upton, Long Island as a federally funded facility equally accessible to American scientists from the different States, early in its history, in April 1948, it approved a plan for the construction of a large accelerator, the Cosmotron, a synchrotron. It soon became evident that new accelerators, able to compete with what was being planned in the US, needed resources not available to a single country in Europe. Thus the first germ of the idea of a supernational European Laboratory, to host large particle accelerators, began to take form.

8.2.2 *A Laboratory for All European Scientists*

Following the example of international organizations, a handful of visionary scientists imagined creating a European atomic physics laboratory. Raoul Dautry, Pierre Auger and Lew Kowarski in France, Edoardo Amaldi in Italy, in addition to Niels Bohr in Denmark and Werner Heisenberg from Germany, were among these pioneers. Such a laboratory would not only unite European scientists but also allow them to share the increasing costs of nuclear physics facilities. French physicist Louis de Broglie put forward the first official proposal for the creation of a European laboratory at the European Cultural Conference, which opened in Lausanne on 9 December 1949. A further push came at the fifth UNESCO General Conference, held in Florence in June 1950, where American physicist and Nobel laureate Isidor Rabi tabled a resolution authorizing UNESCO to "assist and encourage the formation of regional research laboratories in order to increase international scientific collaboration". Finally, at an intergovernmental meeting of UNESCO in Paris in December 1951, the first resolution concerning the establishment of a European Council for Nuclear Research was adopted. Two months later, 11 countries signed an agreement establishing the provisional European Council for Nuclear Research, the *Conseil Européen pour la Recherche Nucléaire*, the acronym CERN was born.[17]

The new laboratory would be supported financially by its founding countries, in a proportion to their national product, and would be housing powerful particle accelerators, meant to compete and even surpass similar machines being planned in the United States for protons, both based on the synchrotron principle, discovered by Veksler and McMillan (McMillan 1945; Veksler 1946) and already successfully applied in the UK, US, USSR, the nations which had won the war.[18] Its first Secretary General was designed to be Edoardo Amaldi.

Even before the official foundation of CERN in February 1952, it became clear to some of its founders that to sustain both politically and scientifically such a large and ambitious organization, national laboratories with their own accelerators needed to coexist in Europe. Scientifically, such laboratories and their parent universities could be the fertile ground where to train and prepare the scientists needed to make CERN the excellent scientific institution which was in its founders' mind. Politically, a national laboratory would prevent the dispersion of major resources, and have a technological return to the country of origin. Among the largest European countries, Germany, having started and lost the war, was barred from building any type of 'nuclear' machine, and had to wait until the second half of the 1950s to start constructing particle accelerators (Heinze et al. 2015). Of the others, in continental Europe, two were the nations with the potential, and the political backing, to pick up the idea to build national laboratories, endowed with their own accelerators: France,

---

[17] From https://timeline.web.cern.ch/origins.

[18] About the synchrotron principle, Markus Oliphant's contribution and Veksler's first presentation of the idea at the USSR Academy in 1943, see *Fifty years of synchrotrons* by E. J. N. Wilson at https://accelconf.web.cern.ch/e96/papers/orals/frx04a.pdf.

and Italy. They both succeeded in doing so, thus laying the foundation for the birth in the 1960s of a completely new type of accelerators, the particle colliders.

The UK was a special case. They had built the first synchrotron (Goward and Barnes 1946), and had both synchrotrons and betatrons under construction or already built, but hesitated to join CERN. British scientists, who had been among its initial proposers, and the most influential of them, strongly supported the idea. But if CERN was to be an International organization, it would cost money, and government backing was to be sought. To convince the UK government, Amaldi was diverted from going to a conference in Brookhaven, and instead asked to go to London. There, John Cockroft and Ben Lockspeiser took him to meet Lord Cherwell, a member of the Churchill government. Amaldi's plea for CERN was snubbed by Cherwell who labelled CERN as being "one more of many international bodies consuming money and producing papers of no practical use". Amaldi was not prepared for such snub and answered sharply that the efforts would bring success, then forcibly argued for the merits of the project. The meeting ended with Cherwell apparently unmoved, but saying that the government would look into the matter. Amaldi left the meeting feeling that he had failed (Amaldi 1979a, b). His British colleagues were not as downcast, since they understood that bringing the matter to the government level was a step to move forward. And so it was. The CERN convention was then signed in 1953 by the 12 founding states Belgium, Denmark, France, the Federal Republic of Germany, Greece, Italy, the Netherlands, Norway, Sweden, Switzerland, the United Kingdom and Yugoslavia, and entered into force on 29 September 1954.

8.2.3 Meanwhile, in Italy

While CERN was on its way to become reality, Amaldi wanted Italy to be a full partner in the European road to scientific recovery. The plan for a national organization, and ultimately an accelerator accessible to the many Italian universities, which had failed to materialize before, was now on its way. INFN was founded in 1951 and its first president was Gilberto Bernardini. Plans for INFN to be funded by the government to build a particle accelerator, and a laboratory to house it, were soon put in motion. It had to be a national effort, and its realization needed agreement among many partners. Not unlike CERN and its founding countries, Italian universities, where nuclear physics had developed more or less in parallel before the war, were now called to join a common effort.

Amaldi's strategy was based on two major allies, and collaborators, Gilberto Bernardini, the experimentalist, expert both in cosmic ray physics and accelerators, and Bruno Ferretti, who had taught Fermi's course on theoretical physics the year after Fermi's departure to the United States. Soon after the war, Amaldi had sent Ferretti to attend the first international conference on *Fundamental Particles and Low Temperatures* held in Cambridge in 1946, and obtained for him to spend one year in the UK, first at Manchester and then in Birmingham. Returned to Rome to fill the chair in theoretical physics, Ferretti played an important role in rebuilding

Fig. 8.3 From the right, Bruno Ferretti, Edoardo Amaldi and Giorgio Salvini (in front), at the 1949 Como-Basel International Conference on Cosmic Ray Physics, reproduced courtesy of Sapienza University of Rome–Physics Department Archives, https:// archivisapienzasmfn.archiui. com, documents provided for purposes of study and research, all rights reserved

the reputation of the Rome physics department. In 1952, when Ferretti approached him about Bruno Touschek, Amaldi immediately foresaw that Bruno's experience on up-to-date accelerator physics could be very valuable for Rome. Thus it is no wonder that the contract finally offered to Bruno included 'help to experimenters at accelerators', even though no modern particle accelerator had yet been operational in Rome.[19]

The decision to build in Italy an accelerator, which could be competitive or at the least complementary with what was planned at CERN, was taken on 19 January 1953 by the Council of INFN. It was not yet made clear which type of accelerator it would be, whether the machine should accelerate electrons or protons, be of a linear of a circular type, and, if a synchrotron, whether based on the weak or the strong focusing principle. Gilberto Bernardini had presented a proposal without further specifications. This strategy probably led to a speedier approval, and things moved rapidly afterwards. In the same INFN meeting which approved the construction of a national laboratory and of a particle accelerator, it was also decided that Giorgio Salvini (1920–2015), then professor of experimental physics at the University of Pisa, would be given the task of creating the new laboratory. In Fig. 8.3, Amaldi, Ferretti and Salvini are shown in a 1949 photograph.

Although the accelerator to build had not been officially chosen, the possible options were rather restricted. In Geneva, CERN had pronounced in favour of building a 30,000 MeV proton synchrotron, a circular machine. Scientific policy suggested it

---

[19] Before the war, a Cockcroft-Walton electrostatic accelerator had been built in Rome, by the Istituto Superiore di Sanità, but it only reached an energy of 1 MeV and, while very useful for a number of nuclear physics experiments, it would be insufficient to produce and study the 'new' physics brought on by cosmic rays experiments, that of particles such as pions.

was unwise to build something similar to what was already planned at the European level, with international backing and funding. In France, approval of the CERN project had been accompanied by the decision to build both a proton synchrotron and a linear electron accelerator. France was among the countries which had won the war, and its scientists had participated to the war effort working on the UK and US projects. It had thus financial, industrial and human resources. Last but not the least, France had already developed the needed know-how thanks to the cyclotron built at Collège de France, through the early efforts by Frédéric Joliot and Irène Curie. Even though no such scenario was present in Italy, Italian physicists now wanted to join the field of modern particle accelerator builders, prerequisite for participation to the nascent world of nuclear and particle physics. To reach this aim, they had to concentrate all their efforts, starting from scratch to build a machine of comparably high international caliber. This machine should bring complementary results to CERN's, where proton machines were planned, and be feasible in the context of Italy's resources. All this was sufficient reason for choosing to build an electron synchroton, a circular electron accelerator.[20] According to Maurice Lévy, a well known theoretical physicists from l'École Normale Supérieure in Paris, it was the American physicist Bob Wilson, in visit to Europe from University of Cornell, who convinced the Italian scientists to build a synchrotron, namely a circular machine. Wilson, in Europe for his sabbatical year, had first visited Paris and tried to convince the French to build a synchrotron, like the one they had built in Cornell, and whose electronic components could be sold in Europe for the construction of similar accelerators. In France, like in Italy, a national laboratory was planned, but Yves Rocard, in charge of the decision, was favouring the construction of a linear accelerator, whose main component, the klystron, was available from the national industry (Marin 2009). Not having succeeded to convince the French, Wilson came to Italy and convinced the Italians. At least this is the story told by Maurice Lévy, in an interview in 2013, and also on the occasion of 50 years of Laboratoire de l'Accélérateur Linéaire.[21]

In time, the debate about electron versus proton machines never ceased. Even now, 70 years later, the choice of which type of machine to build for the 2050s is still raging, just like the discussion of a linear versus a circular collider. The one aspect however on which there is general consensus, is that the future particle accelerator, if there will be one, will still be a collider, one in which particles would collide, in motion one against the other. The road will still be the one which Touschek had started in Frascati in 1960.

---

[20] Gilberto Bernardini had at least two good reasons for proposing electrons instead of protons for the Italian effort. The first belonged to scientific policy, namely it was unwise to build a machine similar to what was already planned at the European level, and, because of limited Italian resources, less powerful. A second reason, was based on scientific considerations, which later Touschek summarized in an unpublished draft: 'A beam of protons loses its identity in the mass of nucleons, when striking [other] nucleons which constitute the atomic nucleus [...] An experiment with protons is like asking a father what he thinks of his sons', wrote Bruno, draft for a book by Bruno Touschek, File 1-AT_11_92(8) from Sapienza University of Rome–Physics Department Archives.

[21] Maurtice Lévy was interviewed by the Author, in his home in Paris in May 2013, for the docufilm *Touschek with AdA in Orsay*, by E. Agapito, L. Bonolis and G. Pancheri.

8.3   How Italy Entered the Road to Become a Particle Accelerator Builder

One of the projects, for which Amaldi had engaged Bruno, physics with accelerators, would soon occupy center stage at the University of Rome. When, in January 1953, INFN had approved the construction of an accelerator, the decision was also taken to appoint Giorgio Salvini to be director of the future laboratory, where the accelerator would be constructed. The events leading to the final decision of building an electron synchrotron unravelled through 1953 and 1954. By the end of the decade its construction was completed and some experiments were running, without Bruno playing particular attention to them. But when he finally did it, in a few months Bruno would make the scientific world turn its attention to Italy. Later in his too short a life, Touschek started writing the story of how AdA and ADONE came to be. These stories have been told many times, not least by Bruno himself, and since this book tells Touschek's story, we shall follow his viewpoint, transcribing from a draft he prepared around 1974.[22]

> How it all started.   …The history of the LNF (National
> laboratories of Frascati) started in 1953. Gilberto
> Bernardini then suggested in a meeting of the I.N.F.N.
> (National Institute for Nuclear Physics, [at the time still]
> a loose association of physics professors from all over
> Italy) that in order to keep up with the international trend
> in high energy physics one should try to construct an
> electron synchrotron of no less that 500 MeV. This, it was
> thought, would be a national effort. That the machine, which
> would bring Italy to a level with international and, in
> particular, U.S. high energy physics, should be an electron
> accelerator was a courageous choice, if confronted with a
> general tendency of physicists who at the time were bent on
> producing proton accelerators.

8.3.1   The Director Who Built the Frascati National Laboratories

It is no wonder that Giorgio Salvini would be the scientist who oversaw the construction of the first Italian particle accelerator. Just as in the case of Touschek, Salvini's choice appears to be part of a design, in which Amaldi searched for, and engaged, two physicists with previous knowledge in particle accelerators: Touschek who had worked on Widerøe's betatron and Salvini, who had graduated with a thesis on Kerst's betatron. To have them to come to Italy and to Rome, in particular, allowed Amaldi

---

[22] 1_AT_11_92(8), date unknown, probably 1974, after a talk at Academia dei Lincei, draft for a book by Bruno Touschek about Storage Rings, reproduced courtesy of Sapienza University of Rome–Physics Department Archives.

to lay the foundations for Italy's scientific renaissance in the new science of particle physics.

After graduating from Milan in 1943, Giorgio Salvini had spent the post-war years in the United States, and come back to Italy in April 1952, having been appointed to the chair of Professor of Experimental Physics at University of Cagliari, in Sardinia. Called to professorship in Pisa in November, just a few months later (in January 1953) he was asked to become director of the future laboratory.

This is how Fermi's vision of a national institute, and a large laboratory equipped with a modern day particle accelerator, became reality after the war. The story of how this happened comes alive through the words of the first director of the Italian National Laboratories in Frascati, Giorgio Salvini, interviewed in 2013 for the docu-film Touschek with AdA in Orsay.[23] At the time, Giorgio Salvini was 93 years old. He had been President of INFN, Minister for University and Research, President of the Accademia dei Lincei, but first, and foremost perhaps, professor at the University of Rome for over fifty years, where his course to the first year physics students was the hardest to pass.

8.3.2   How Giorgio First Learnt of Betatrons

Salvini gave many interviews throughout his life, mostly in national television, as on the occasion of the first operation of the Frascati synchrotron or about ADONE and its major discoveries. The transcript to follow was his last interview, taken at his own home, in 2013.[24] The formidable memory of people and places, which had carried him through his scientific and professional life, was occasionally in need of some help from his wife Costanza, companion of over 70 years, whom he had met in the Physics Institute of the University of Milan during the last dark years of World War II, then became his wife and mother of their five children.

> *Giorgio Salvini:* This is a story which needs to be told.
>
> It must start from Milan. In 1942, when I graduated in physics, I was a young officer in the Alpine infantry of the Italian Army, in the Julia Division, and I was given a furlough to go and take my examination, the *Esame di laurea* in Italian.
>
> I have to say that in the previous two years I had studied continuously, non stop, and had reached a good preparation in all physics subjects. The subject of my thesis had been assigned by two professors of the university of Milan, Polvani and Bolla, who had been fascinated, *incantati*, with a 1941 work by Kerst and Serber, where they described the first betatron, the machine they had built and which was the first electron accelerator working with a variable magnetic field (Kerst 1940).

The article, which enchanted Salvini's professors in 1942 was the same which, in faraway Trondheim, in Norway, had caught Roald Tangen's attention in September

---

[23] Docu-film prepared to by E. Agapito, L. Bonolis and G. Pancheri, ©INFN 2013.

[24] Author's translations from Italian into English.

1941, and led Widerøe in 1942 to submit his proposal for the 15 MeV betatron (Waloschek 1994).

> *Giorgio Salvini:* Polvani and Bolla had found this machine enchanting, and thus procured a photostatic copy of the two articles, through some acrobatic efforts, but they were used to deal such acrobatics towards our "enemy" [the USA, in those days]. Thus Bolla said to me: "Look at this article and see what you can do with it, as it could be a good topic for your thesis". I looked at the paper. Of course I still had quite some difficulty with the English of the text, but once started, I could push myself to understand anything. I understood this work was an important one. Thus the articles became the subject of my thesis. In a sense, Kerst and Serber became my guardian angels, as my scientific origins lay with them.

In 1942, Italy was still in war with the United states and one can wonder how Kerst's articles had reached Milan. According to Carlo Salvetti, one of Salvini's colleagues, it seems that while American journals did not arrive in Rome in 1942, enterprising booksellers in Milan would get them from Switzerland, bartering for them on the [black] market.[25]

> *Giorgio Salvini:* I thus graduated, with honors, of course, and, with my degree, I returned to the war, where I was stationed, north of Gorizia. I was to go to Russia, right away to the battle front. I had thought the train would take us to the front, but, as I arrived to the station with my soldiers, the trains came in, filled with the ragged men returning from the Russian battlefront, the few survivors from the Julia Division, which the Russian counteroffensive had practically decimated. I understood that we would not be leaving so soon.
>
> Those were difficult years. We all kept a set of civilian clothes in our closet, and we understood that sooner or later the Italian army would collapse. In those days, hard as they were, I was lucky. I could live and see again my Milan, unlike many of my colleagues, which ended up in the trains to Poland's concentration camps and did not return.
>
> We stayed in Gorizia for a few weeks, not knowing what to do, until the news came of the heavy bombing of Milan and that the city center had been reduced to ruins. The house where my family lived had been destroyed, the family had moved elsewhere. It was my mother who probably saved me [from joining the war front in Russia or being deported to Poland], as she applied for me to receive a 7-day permit to go back to Milan, which was offered by the authorities to those army soldiers or officers whose family home had been severely damaged by the bombing. I was waiting for the permit, when on September 8th, 1943, my commander came with the permit, urging me to go to Milan.

Salvini's commander probably heard that the Armistice had been signed, and understood that the Italian Army, or some part of it, was likely to go under German control, with those refusing to accept it facing deportation and death. He thus sent Salvini to Milan, having the good excuse of the previously applied furlough.

On September 8th 1943, Italy, its population and its army, were divided in two, the South with the King of Italy and in an armistice with the the Allied Forces, while the rest of Italy was occupied by the Germans Army. Some other memories of Salvini highlight how ordinary Italians handled their loyalties in a divided country.

> *Giorgio Salvini:* [In Milan] I deposited the uniform, told my mother that the war was over and that we had to think of the future. [But shortly after] a fascist government was reinstalled. I

---

[25] Carlo Salvetti (1918–2005), Professor of Physics at University of Rome Sapienza, was interviewed by L. Bonolis, Rome, 18 July 2002, from L. Bonolis, *Fisici italiani del tempo presente*.

did not know what I should do. I had some colleagues, among them Carlo Salvetti, a physicist and an army officer, who had been my military instructor in Pavia. It became peremptory to go to the German [military] headquarters to declare if we wanted to associate ourselves to the Germans and the Salò Republic.

Salvetti and myself decided this would have violated the loyalty oath we had made to the King of Italy, when we had joined the army. Thus, we went to the fascist headquarters and said clearly that we did not wish to join the Republic. They started menacing us, but after our declaration, well, we ran rapidly away, and hided in the nearby hills of Erba [where my family had moved after the Milan house had been destroyed by the 1943 heavy bombing]. Thus my military life came to an end.

What could I do, after this? I was rather lost. In the midst of doubt and confusion, I picked up again my thesis, and, by then, I was in love with physics. Time went by, and I sought to contact again the physics institute in Milan. Although we were two clandestines, we would go to the institute, and use the library to continue studying. There were two physics students, two girls, who would cover our presence.[26] Thus we developed a safety system: I could study in the Institute, using the library, and when either the fascists or the Germans came to inspect looking for young men to draft to their Army or to deport them, we would hide and the two girls would say that no one was there.

My knowledge of physics increased in these months, I could study 8 hours a day, mostly theoretical physics, a work by Bethe about mesonic fields, and I decided I liked physics, but in the experimental field. I also succeeded in having some contact with the German headquarters, who understood that we were *transfughi*, though. We even tried to get funding for doing some research. [Polvani helping] The idea was to make a particle accelerator, a low energy one. And this made time go by. And I arrived to the year 1944, hidden, but having learnt so much, since I could do nothing else [but studying]. I even published my first work.

Then 1945 came. …

In Italy, the war ended on April 25th, 1945, when the partisans entered Milan. Those last few days before the end are vividly recalled in a 2013 interview with Costanza Salvini.

Costanza Salvini: People would suddenly disappear in those days. Nobody knew where they had gone. One day, the local priest from my neighborhood asked if we, myself and my friend, were willing to do some work for the partisans, helping to bring letters from those in the city prison to their families, to let them know they were alive. Nobody should know of this of course, neither parents, nor siblings or friends. Returning messages, from the prisoners, were given to the Sisters during Mass in the prison, brought by them to the local church, and then exchanged between the Sisters and the faithfuls, at Communion times.

It was dangerous of course. If discovered, we would have been deported to Germany right away.

And then it ended. On that day, [April 24th 1945], the priest called me and told me to go immediately to Archbishop Schuster and tell him that tomorrow at 12 [noon] the Allied forces will enter Milan, and the bells of all the churches in Milan must be heard.

In Fig. 8.4 we show Giorgio Salvini with his young wife Costanza, during a visit to the United States.

Giorgio Salvini:[As the war ended] we returned to work [officially]. I built my first Geiger counter, and with Cocconi decided to build a telescope and study cosmic rays. Now, it looks

---

[26] One of them was Costanza Catenacci, who married Giorgio Salvini in 1951, the other one was Piera Pinto, who married Carlo Salvetti.

Fig. 8.4 Giorgio and Costanza Salvini, early 1950s. Courtesy of Luisa Bonolis, who received it from Giorgio Salvini, (Salvini 2010), with kind permission from Giorgio Salvini's family

preposterous that we should think of reconstructing science in those conditions, but that's how we felt, that we had to do it.[27]

Salvini rapidly showed his capacities as a good experimenter in cosmic rays research and to possess courage and ambition. In 1946, the 26 year old Giorgio Salvini, had gone to Rome to meet Amaldi and share his dreams and ambitions with the man who was, at the time, the most famous physicist in Italy, the only one from Fermi's legendary group to have remained in Rome. Amaldi saw the potential of a future leader in the young man, and, when the time came, played his hand and made Salvini director of the future laboratory.

> *Giorgio Salvini:* In 1950 I was still working in cosmic rays and acquired some international reputation. And I went to the US. My work immediately went very well. I was given funds for my own laboratory, and did some good work on liquid scintillators. In Princeton, I met Pauli, and I started thinking of emigrating and living in the United States.
>
> At the same time, I was writing regularly to Costanza [who was in Italy, not yet his wife], and applied for a university professorship in Italy, following advice from my Italian peers, among them Polvani and Amaldi. In 1951 I won the [university competition for the physics] chair in Cagliari, and not long after I was called to Pisa.
>
> …
>
> [Soon after] Amaldi, Bernardini, Ippolito and others were able to secure the funds to create a new laboratory, which they would ask me to build. From where? From nothing. We did not know where, [thus] Amaldi, Bernardini and myself, we toured Italy [in search of an adequate location].

---

[27] Giuseppe Cocconi (1914–2008) was remembered by Ugo Amaldi, Guido Barbiellini, Maria Fidecaro and Giorgio Matthiae, in CERN Courier, July 15th 2009 in the article Giuseppe Cocconi and his love of the cosmos.

8.3.3   The People Who Built the Machine

The first step was to assemble the team which would build it. Salvini had already given proofs of his capacity as a leader, both in his cosmic rays work and in founding an inter-university center in Milan, and knew that to make a modern particle accelerator of international caliber, money and political will are not enough, one needs technicians, scientists, engineers.

Salvini started his search immediately. He toured Italy's excellent higher learning institutions, going directly to ask their presidents to find for him the best new graduates in physics or engineering. He did not worry about them having previous knowledge in accelerator physics, that would come later. In his search he applied a simple driving principle: get the top students from the best schools, and let them study the subject. Since they were the best, they could learn enough to start working in six months. This was the principle he had used for himself and always did, as when he was given Kerst and Serber's article for a subject of his thesis: still an undergraduate, he knew very little English, if any at all, but he studied it and soon made it his own.

Although no official decision had been taken about the type of accelerator, it was clear to Salvini that it would be a synchrotron, and for this he prepared his future team. The basic elements on which a synchrotron is built are the vacuum chamber in which electrons to be accelerated will circulate, the radiofrequency cavities which will accelerate the electrons, the magnets which bend their path and direct them around. In Fig. 8.5 a schematic view of the Frascati synchrotron. Thus he hired Mario Puglisi for the radiofrequency cavities, Gianfranco Corazza for the vacuum, Giorgio Ghigo for the magnets. He also needed a head engineer for the machine. His choice fell on Fernando Amman, who had graduated in 1952 from Politecnico of Milan in electrotechnical engineering.[28] Amman would direct the construction of the synchrotron, and then, starting from 1961, design and build ADONE, the bigger and better AdA, of which more shall be learnt later. He also counted on the expertise and collaboration from a number of colleagues from Italian institutions, such as the theorist Enrico Persico, from University of Rome, and Ruggero Querzoli from the Istituto Superiore di Sanità (ISS), where invaluable experience with the Cockcroft-Walton accelerator had already been gained. The youngest scientist in the team was Carlo Bernardini, newly graduated from Rome, with a thesis on theoretical physics.

As for the technicians, Salvini's work with cosmic rays had taught him their importance as the other building block in a high level scientific and novel enterprise, and he applied the usual principle of searching for the top students. Italy's excellent secondary schools for technical training were just graduating an extraordinary batch of skilled students, and Salvini could chose their best graduates as well. Thus assembled, the team started working in Pisa in 1953. Here Salvini added another ace to his team, Icilio Agostini, a law graduate, who typed physics students' theses to increment his income. Agostini would provide the administrative backbone for the

---

[28] Fernando Amman (1929–2022) would build ADONE and become its director.

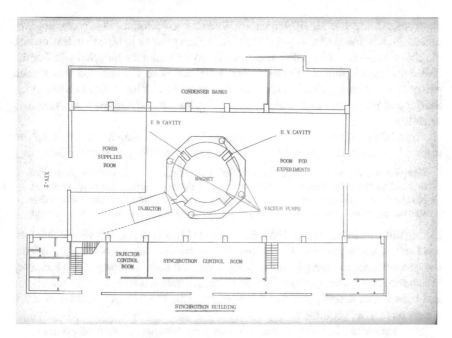

Fig. 8.5   The layout of the Frascati synchrotron plan, G. Salvini, in Proceedings of the 1957 Padova-Venezia International Conference on "Mesons and Recently Discovered Particles. Padua-Venice 1957 Conference, © Societá Italiana di Fisica

newly founded National Institute for Nuclear Physics and later became its Director General.

8.3.4   How to Build Things from Nothing (e.g. a National Laboratory and a Particle Accelerator)

Looking back to his life, sixty years later, Salvini reflected on having been charged of getting Italy into modern particle science. This was to be accomplished by building from nothing a particle accelerator for a laboratory which did not yet exist, in a place which had not yet been found.

As Salvini accepted this challenge, he laid down his priorities: the kind of accelerator and the people to build it. One also needed a site to host the yet undefined machine, but that could come later, together with construction of the facilities. While the site was going to be decided, the team assembled in Pisa, and start learning about accelerators.

The next problem to solve was which accelerator to build. The choice was between accelerating protons or electrons was easily solved, as already said, because CERN had decided for protons. To avoid duplication, Italy was compelled to choose electrons. Although in his 1974 memoirs, Touschek gives a somewhat poetic description of the advantage of using electrons rather than protons to probe nuclear matter, the choice was more of a political nature. There remained the question whether the Italian accelerator should be of the linear or circular type. The decision took a good part of 1953. In June, Salvini and Amman visited Widerøe's laboratory in Baden, then Salvini and Bernardini took a trip to the US, looking for advice from the American scientists.[29]

From 10 July 20 to 10 September 1953, Gilberto Bernardini and Giorgio Salvini made a big tour in the United States visiting several accelerator centers.[30]

> *Giorgio Salvini:* It was an unbelievable trip. Being short of money, with Gilberto we decided to travel by bus. At a certain point we had to go on by foot, carrying our heavy suitcases. Now and then, walking in the sun, we would stop under some shadow. Gilberto would then start saying, 'About the injector …', and then, after some rest, we would return to the road.

During this trip, in summer of 1953, Salvini and Gilberto Bernardini visited Stanford and Cornell, discussing the different merits of possible accelerators to build, with the top American accelerator physicists. They consulted with Wolfgang (Pief) Panofsky at Stanford and Robert (Bob) Wilson in Cornell. Upon returning to Italy, the choice had been made, it was going to be a weak focusing electron synchrotron of beam energy as high as 1000 MeV.

The American physicist Matthew Sands gave his own version of events in an interview to the American Institute of Physics (AIP)[31]:

> *Matthew Sands:* Bruno Rossi, my PhD professor, was an Italian […] I had known Fermi at Los Alamos and liked him a lot […] I had also gotten to know [Gilberto] Bernardini when he first came from Italy to the US and was at Columbia, and we were both working in cosmic rays. Then he switched to accelerators at Illinois, and we had been in contact as people do in the same field. Then I had admired a number of Italians, Rasetti and others. So I knew about the work they had done in cosmic rays. They had been very powerful. Rossi had started his cosmic ray work there [in Italy]. And so it seemed like a logical place to go, maybe. It was 1952, and this was the time when things were going to change. CERN was just getting organized. The government was proposing that it could find money to support research. So it was a period when Italy was clearly going to take off. In some aspects, it was quite an exciting time. They were very hungry for news about the wartime developments, and so I

---

[29] In Baden Widerøe was constructing a 300 MeV synchrotron for the University of Turin.

[30] "Un viaggio incredibile. Ricordo che avevamo preso un autobus di linea—perche' dovevamo limitare anche le spese—e che poi eravamo costretti ad andare a piedi con delle grosse valigie. Facemmo un bel tratto, sotto il sole, e ogni tanto ci fermavamo e Gilberto diceva: "Pero' Giorgio, io penso che per l'iniettore potremmo forse fare cosi' …", allora all'ombra parlavamo un po' e poi ripigliavamo il cammino. Io avevo trent'anni: ma lui, del 1905, ne aveva quarantotto con un vigore e uno spirito generoso e ispiratore. Un gran Gilberto, veramente …Ci furono dei dubbi tra il fare un acceleratore lineare o un elettrosincrotrone, e Pief Panofsky suggeriva un acceleratore lineare. Ma intanto eravamo diventati piu' consapevoli sulle possibilita' italiane dell'industria, sicche' la scelta, nel 1953, fu per il sincrotrone." (Salvini, interview in Bonolis and Melchionni 2003, 390), Author's translation.

[31] Interview of Matthew Sands by Finn Aaserud on 1987 May 4 and 5, courtesy of Niels Bohr Library & Archives, American Institute of Physics, College Park, MD USA, www.aip.org/history-programs/niels-bohr-library/oral-histories/5052. Sands would later be co-author of the famous book Feynman Lectures on Physics (Feynman et al. 1964–1966).

gave lectures in Italian on electronic circuits, and that was published in a little book by a couple of my students [A. Alberigi and B. Rispoli]. And then I gave lectures on accelerator theory to a group that was thinking of building an accelerator. They had originally thought of making a linear accelerator, and they asked me to bring information on that from Stanford, but by the time I got there, they said, "Well, while you're here, let's think about a round accelerator." So, I don't know if that's the main reason they changed their mind. I think I tried to convince them that it was much more expensive to build a linear one than a round one [...] I teamed up with Touschek who was a theoretician, and he showed how we could use the matrix algebra to solve the problem quantitatively in the more general case. So we published a note on this, which hurt a lot of people's feelings, it turned out, because some of the people working at CERN and Brookhaven had just done this, but they hadn't published it [...] The Fulbright administration used to like to exchange people, and so I was asked to come to Paris to the Sorbonne in the springtime, in April, and give lectures, so I went there. I accepted and went for a ten day visit, sponsored by the Fulbright, to give some lectures. I lectured on accelerators, on electronics, and on particle physics [...] It turned out, when I got back –because of my European experience both in teaching the Italians about accelerators and collaborating with the French and writing the paper on the integral questions and so on — I was suddenly considered the expert on accelerators.

8.3.5   The Site

Soon after the approval of the project in its general lines, i.e. a laboratory and a particle accelerator, a competition had started between various Italian universities as to where the new laboratory would be housed. The project was to be national, on a scale yet to be attempted in Italy, and, in the ambition of its proposers, to be, if not competitive, at least on the level of similar science efforts beyond the borders, in Europe or even further away in the U.S. Wherever such laboratory would be built, construction jobs and industrial contracts would arrive, and benefit the development of local enterprises. For the university, which would win the competition to sponsor the project, having the laboratory built nearby would mean positions for the researchers, additional funds for travel and research, international prestige. On the paper, many were the universities able to compete on the basis of their academic standing and international contacts, Milan, Bologna, Pisa, just to name the three front runners. And, of course, there was Rome, where Fermi had taught and started Italy on the road to excellence in particle physics.

The geographic distribution of Italian academic institutions is spread through all of Italy. From the northern plains to Sicily, a large number of prestigious institutions, many of them hundred of years old and none accepting to be second to the other, compete and collaborate. This distribution of learning sites comes directly from the political history of Italy, reflecting the co-existence, for centuries, of a number of independent states, in the North, the Center and the South of Italy, with the two main islands belonging to different kingdoms, Sardinia to a northern one, Sicily to a southern one. The diversity and the competition between the many small states, unified under a single state in 1861, have given Italy uncounted numbers of artistic masterpieces, and are reflected in the constellation of great universities from South to North.

Originally, in order to avoid having the project stalled by crossed vetoes, Gilberto Bernardini had not specified in his proposal neither the site nor the type of accelerator to be built. But, not to waste time, once the director had been identified, the basic personnel had been chosen and the team had started assembling in the University of Pisa, which Salvini had joined as professor. In the meanwhile, the decision about the site was maturing.

As the competition grew for housing the new laboratory in one or another of Italian universities, Amaldi waited. The construction of the laboratory was to be funded from the Ministry of Industry, headed by Pietro Campilli, whose electoral college was the hillside town of Velletri, one of the many old towns dotting the slopes of hills of vulcanic origin, overlooking Rome from the south-east. Among them, there are Albano, Marino, Grottaferrata, Frascati, closest to Rome and reachable by train since 1856, through a 20 km ride. This area if generally known as that of the *Roman Castles*, Castelli Romani in Italian. Aristocratic villas and Roman ruins made the area part of the *Grand Tour* of eminent European artists and writers. In 1935, an astronomical observatory had been constructed in the papal residence of Castelgandolfo, on the rim of the crater housing the Albano lake, with a view of the city of Rome from the Tyrrhenian sea in the west to the Appennine mountains in the East. The observatory belonged to the Catholic Church, which made the decision to build it after electrical illumination in the city hampered celestial observations from the other astronomical laboratories situated in the Vatican city (Consolmagno 2009).[32] For almost 20 years it was the first and only scientific installation in the whole area.

In 1954, when a large estate on the slopes just below the town of Frascati, less than 30 Km south-east of Rome, was offered by the municipality, Amaldi knew that Fermi's dream of a national center for nuclear research would come true in Rome and his legacy would not be dispersed. Thus, at the end, the choice came to be Frascati, a sleepy town with a nice wine and an ancient history. The choice appeared to be almost like a gamble to some of Salvini's friends, who tried to discourage him from accepting to be buried in the middle of nowhere. But money and political will were behind this choice and Salvini's main character trait would carry him to win an almost impossible bet: he was stubborn. Once he would have accepted a task, he would do it. That did not come without giving up something. Giorgio was in no way a 'one dimensional' man, devoid of interests and passions. He was a gifted painter, as some family portraits he kept at home were witness to, but he knew what many scientists learn early in their life: one cannot have more than one passion at a time, and to be a physicist you need to make a choice, if not right away, pretty soon. He was not alone. Giorgio, like many other scientists of his generation, could also count on the full support of a wife, who had herself chosen to study and graduate in physics, but then decided to give up research to be his companion, and the mother of their five children. Thus Giorgio went ahead and built the laboratory.

Giorgio Salvini: I was told: here is the place, here is the money - few millions [Lire], and thus the laboratory was born.

---

[32] Before the unification of Italy, the Church maintained a number of observatories in various locations in the city of Rome, subsequently moved inside the Vatical city.

In the proffered land, nothing was yet built, not even a real road, but the location was excellent. It was close to Rome, some 1/2 h by car—at the time—and on reasonable walking distance (40 min) from the railroad station in the town of Frascati. At the same time, it was sufficiently distant from the city not to appear to be an extension of the University of Rome to the other competing universities. The laboratory was conceived to be for Italy what CERN would be for Europe: a national, interuniversitary center with autonomous facilities, independent from one given university with full and equal access to all the collaborating national institutions.

8.4 To Higher Energies Through the Kinematical Advantage of Head-on-Collisions

While the Frascati laboratory and its synchrotron were being built, the international particle physics community was looking for ways to reach particle energies higher than those attainable with synchrotrons.

In 1953, unbeknownst to the particle physics community, Rolf Widerøe had been able to officially register his 1943 patent for a colliding beam accelerator (Waloschek 1994). He was now living in Switzerland and actively working on betatrons for medical research, keeping contact with current literature on particle accelerators and even participating to topical conferences and workshops. His old idea of exploiting the kinematical advantage of head-on-collisions, was now coming of age. After all, as Touschek had remarked when Widerøe first mentioned it to him in Hamburg in fall 1943, it was something every physicists could know, even from high-school books. Sooner or later, as it often happens in science, what's in the books becomes reality. The time for Widerøe's idea to be taken up by the accelerator community had now come.

At CERN, as soon as the proton synchrotron was on its way to be built, a major conference was launched to investigate how to overcome the limitations in reachable particle energy. This was the 1956 CERN Symposium on High-Energy Accelerators and Pion Physics, the First International Conference on High-Energy Accelerators (Regenstreif 1956), a turning point in accelerator physics. Giving the opening address at the 1956 Conference on Accelerators held at CERN on 11th to 23rd June (Regenstreif 1956) was the president of CERN's governing body, the Council. Things had changed since Edoardo Amaldi had gone to London, where, together with Cockroft and Lockspeiser, he had unsuccessfully tried to convince Lord Cherwell to have Britain joining the nascent effort for the construction of a joint European Laboratory. Now the UK was a full-fledged CERN Member State, and Sir Ben Lockspeiser was its President.

At this conference, the idea of colliding beam accelerators made its appearance, by way of two contributions to the first session on *New ideas for accelerating machines*, one by Donald Kerst (Regenstreif 1956; Kerst et al. 1956, 36) and one by Gerard O'Neill (Regenstreif 1956; O'Neill 1956, 64). The idea had been in the air for some time, and was also considered by Krobeck from Berkeley (Regenstreif 1956, 60).

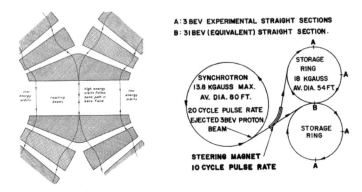

Kerst focused on the problem of storing enough protons in two accelerators side-by-side as in the left panel of Fig. 8.6. O'Neill presented the project worked out at Princeton of two tangent beams of identical particles, accelerated by the same external magnet, as in the right panel of same figure. O'Neill discussed both the case of protons, possibly injected from the existing Princeton-Pennsylvania synchrotron, and that of electrons, if an electron synchrotron in the GeV range could be available.

Following this session, a discussion started. Salvini had discussed problems for a 1000 MeV electron synchrotron (Regenstreif 1956, 40), which, by the way, turned out to be lower than the energy finally obtained by the Frascati machine. During the discussion, intrigued by the idea of colliding beams, and keeping in mind the goal of reaching such energy for the machine under construction at Frascati, he asked about the possibility of storing oppositely charged particles:

> *Salvini :* When we have two beams, one of positive particles and one of negative particles (travelling in opposite directions), can we expect extra-focusing by the magnetic field of one beam acting on the other, or will the particles simply collapse?

Kerst could only answer that the extent of such effect was unknown. Salvini's suggestion was not unlike what Widerøe had proposed in his patent. What would however make the idea interesting could be the possibility of using particles and antiparticles for the collisions. But it did not come up in the discussion. Salvini's question did not go further. Four years later, in Italy he would obtains the funds to build in Frascati the electron-positron collider proposed by Touschek in February 1960, and which would be the first in the world to register collisions in 1964.

Attending the conference, there was also the scientist who, 13 years before, had stressed to Bruno Touschek the kinematical advantage of an accelerator for colliding particles, Rolf Widerøe. During a discussion time, Widerøe questioned about the priority of the colliding beam idea. This discussion is not recorded in the Proceedings, but remembered by other participants in the *Panel on accelerators and detectors in the 1950s* from Brown et al. (1989) or by Per F. Dahl in Dahl (1992). Quoting Larry Jones (Jon 2022, 199), Dahl mentions that Widerøe was picked at not being given sufficient credit for his original idea, and gave an 'improptu polemic' on his storage ring scheme, whose patent had been registered three years before. See also Telegdi in *Pions to quarks in the 1950s*.

The idea of colliding particles to reach higher energies was now taken in some serious consideration. Following the Geneva 1956 conference, it travelled from scientific articles and physicists meetings into laboratory planning and construction. In the US, between 1957 and 1959, an electron–electron machine with tangent rings as envisioned by O'Neill, started to be built. Like-wise in the USSR (Baier 2006).

Meanwhile, in 1957, the construction of the Italian National laboratories in Frascati was sufficiently advanced to deserve a postal address, *Strada del Sincrotrone Km. 12*, opening the way to Italy's future as a particle accelerator builder (Valente 2007).

8.5 Strada del Sincrotrone Km. 12

From 1955 onwards, three different roads started to converge toward the construction of the first electron-positron collider in Frascati. One was coming through new discoveries, such as that of the anti-proton (and the anti-neutron), and the non-conservation of parity in weak interactions in 1957, which would revamp Touschek's interest in particle symmetries and neutrino physics. The second road came through exploration of the kinematical advantage of head-on-collisions between accelerated particles, whose discussion had started at the 1956 Geneva conference. Last, but hardly the least crucial element in AdA's conception and building in Frascati, was the commissioning of the Frascati electron synchrotron, which began working in late December 1958.

It had been an arduous road, exemplary in its reflecting the changes which took place in Italy during the same years, from 1955 to 1959. The construction of the laboratories can be recalled using Touschek's own description in excerpts from a talk he delivered at the Accademia dei Lincei on 24 May 1974.[33] Digging and clearing the land had started in 1955 near Frascati, in a place locally known as *Macchia dello Sterparo*, meaning a thicket, where there lives the *sterparo*, a labourer who pulls twigs to clear the soil.

```
Frascati is a small town, about 20 Km from Rome, one of the
"Castelli Romani", the Roman castles, situated on the north
slopes of the "Tuscolo", a small hill, prelude to the
impressive mountains which form the spine of Italy.
```

---

[33] *A Brief Outline of the Story of AdA*, reproduced courtesy of Sapienza University of Rome–Physics Department Archives, https://archivisapienzasmfn.archiui.com, documents provided for purposes of study and research, all rights reserved.

As the slopes start flattening out to give way to the flat land where the city of Rome spreads out, there the Laboratory was built.

```
Until 1955 there was nothing except dark shrubs and
scattered vineyards which produced a sentimental wine dear
to the Romans. There was a dust road leading up to the place
and there was no water. The nascent laboratories had to wait
till 1957 when a well desperately driven to near the
antipodes started to promise a sufficient yield.

Already in 1953 a manhunt had begun, which by 1955 had
yielded about 25 between scientists and engineers in charge
of designing the various parts of the synchrotron: the
magnet, the RF Radio-frequency, the vacuum chamber, the
injection mechanisms as well as a liquid Helium (He)
laboratory. There was also a theoretical [physics] group
staffed by 4 and a building and service commission by 7.
```

As buildings were getting ready in Frascati, in 1955 the center of activity was moved from Pisa to Rome, closer to the designated site, with people and equipment transferred in three trucks, Fig. 8.7. The assembled group in Pisa would finally start the great adventure of constructing the first Italian particle accelerator.

The transfer to Frascati took place in 1957. The team of technicians, scientists, and support staff was photographed shortly after their arrival at the new laboratory, Fig. 8.8. Many of the future protagonists of the AdA adventure are appearing: Mario Puglisi, in charge of the radio-frequency equipment, Italo Federico Quercia, laboratory director during AdA's construction, Giorgio Ghigo and Giancarlo Sacerdoti, who will construct AdA's magnet, Gianfranco Corazza, whose extraordinary vacuum expertise will be at the core of AdA's operational success, Carlo Bernardini, the young theorist who became Touschek's main ally and friend. There were no women in the scientific or machine shop team, but they were integral part of the technical staff, in the drafting group and the library, and appear in their white laboratory uniforms.

A somewhat similar enterprise was started in France, with the construction of the linear accelerator at Orsay, but many of the scientists and technicians came from the UK or even from the US, where they had participated to the war effort, for instance working on radars, and even on atomic bomb related efforts. In Italy, the situation was different. The personnel of the nascent laboratory came from all over Italy, with different dialects and school formation. Bruno wrote[34]:

```
[In the nascent laboratory] physicists and engineers as well
as administrators all talked a different languages and the
noise must have been Babelian, when the project started.
Also a good part of the mechanics [from the technical staff]
had to learn their skills on the spot and it must have been
clear that if anything touchable and successful came out of
this initial chaos it has something of the miraculous. To
make it even more so, add the geographical diversities of
```

---

[34] *ibidem* Lincei talk.

Fig. 8.7 The electron synchrotron group preparing to leave Pisa for Rome in 1955 from Valente (2007), © INFN–LNF, http://w3.lnf.infn.it/multimedia/

Fig. 8.8 The Frascati laboratory staff in the synchrotron hall, on October 4th, 1957. In the front row, Giorgio Salvini is seated at the center, with Carlo Bernardini and Giorgio Ghigo, respectively second and third from the right, Mario Puglisi and Italo Federico Quercia second and third from the left. Stanislao Stipcich, head of information services, is in the third row, just above Puglisi, Giancarlo Sacerdoti is seen standing up third, on the right side of the photo. Icilio Agostini, head of administrative services, in a light grey jacket, is standing next to Salvini, © INFN–LNF, http://w3.lnf.infn.it/multimedia/, all rights reserved

```
the staff, a fact that in a state united less than 100 years
ago [at the time] still carries considerable weight.
```

8.6   July 1957 to April 1959: Assembling the 1100 MeV Frascati Electro Synchrotron

In July 1957, the final transfer of people and material from Rome to Frascati had taken place. The various parts of the machine started to be assembled, and, at the end of the year, one could count on a number of remarkable accomplishments: the magnet and its power source had been mounted, the Cockcroft-Walton (C-W) electrostatic accelerator to inject the electrons in the synchrotron had arrived from Rome, the vacuum chamber was almost complete, likewise the radio-frequency cavities to pump energy and accelerate the electrons (Salvini 1962, 57). On the construction side, everything was also in place by the end of 1957: the synchrotron building, the power station, the machine shops, the criogenic laboratory.[35]

The assembling of the various parts of the synchrotron began in January 1958, soon after the New Year. Pieces were put together, the ring was completed, the magnet was in place. A careful and slow process was going to start, as everything had to be checked, step by step. At this time, all the doubts, which had been present during the construction, had to be put aside. As Salvini wrote, the time for alternative ideas, with unnecessary calculations, and comparisons with what others were doing, was gone. All these things did not exist anymore, the only thought was of making things work.

The magnetic field, the grand protagonists of particle accelerators, was first very carefully tested, while the injection system, with the C-W electrostatic generator, was prepared in Rome at the Istituto Superiore di Sanità. Preparations were going ahead, when it became clear that a higher injection energy would provide a more stable machine operation. These requests were higher than what the C-W could provide, and the order was placed for a Van de Graaf, with the needed performance. In the meanwhile, one could start testing the C-W. In Summer 1958, the ring was assembled and the system sealed to inject the electrons and search for the beam. The search moved step by step by checkpoints, to make sure that the electrons circulate through the ring, where the magnetic forces make them pass. First one quadrant, then two, three, one circle, two circles. No more, at that point, since the system had been checked and worked correctly. Now, one could start preparing for the Van de Graaf, which would make the synchrotron work at the desired intensity and energy.

When, in November, the Van de Graaf arrived, the system was opened and the new injection system put in place. On December 1st 1958, the first electromagnetic cavity, only one, was inserted in the ring. With the new accelerator now providing electrons with higher energy and intensity, the search for the beam started anew. After twenty turns, which the electrons took without problems, the electromagnetic cavity was turned on, and the electrons, which had entered the ring with an energy of 3 MeV, were brought to an energy, first of 47, and then 300 MeV. The second cavity could be inserted. This second phase took a bit longer, but after a number of trials and errors, a final acceleration to an energy of 1100 MeV was reached on February 9th, 1959. A dinner in a Frascati restaurant, such as the one in Fig. 8.9, sealed the day.[36]

---

[35] The engineer in charge of the design and construction of the buildings was Giovanni Scaccia Scarafoni, the same old friend who had warned Salvini about the danger of remaining buried in Frascati, in the middle of nowhere. Scaccia Scarafoni came from the Istituto Superiore di Sanità, in Rome, where the Cockcroft-Walton had been constructed and in operation since 1939, see *Il deposito del radio e l'acceleratore Walton-Cockfroft*, by S. Risica and G. Grisanti in https://publ.iss.it/Items/GetPDF?uuid=3fe2bd76-92c6-451c-8f67-2d47179f76db, (Grandolfo et al. 2017, 19).

[36] The precise dating of this photograph is unclear, since it says 'March 8th, 1956' but another copy, courtesy of Mrs. Angela Turrin, carries the date '1957' in the back. It shows a typical dining occasion in one of the many restaurants and 'trattorie' in the Frascati area, and could have been taken any time during the construction of the synchrotron.

Fig. 8.9. Dinner with visitors in Frascati. Counteclockwise from the right one distinguishes: G. Diambrini-Palazzi, A. Turrin, E. Amaldi, E. Persico, G. Salvini, C. Bernardini, G. Sacerdoti (head of the table), F. Amman, B. Woodward, M. Ageno, I. F. Quercia, M.Puglisi. Photo Courtesy of A. Mantella -Turrin, also © INFN–LNF, http://w3.lnf.infn.it/multimedia/, all rights reserved

As Touschek would write years later, one day the machine, which had appeared to belong to a distant future, was suddenly there.[37]

> The machine finally burst into action on the 19th December 1958. ... The completion of the Frascati synchrotron gave a tremendous boost of morale and self confidence to Italian physicists. There were only two machines of comparable kind and size in the world – one at Cornell and the other one at Caltech. This in itself would be a valuable achievement, but it assumes gigantic dimensions if one remembers that it was achieved by a staff of scientists and engineers most of which had no experience of the characteristic mixture of technological know-how, industrial and administrative organization and the art of confident improvisation, which makes these big machines.
>
> The completion of the machine came as a surprise to some of the experimenters. The synchrotron was something in the distant future and then one day it was there and working at full efficiency.

Salvini's friend, the American physicist Al Silverman, visiting Rome from Cornell in January 1959, had warned him that one day, 'at breakfast', the Frascati scientists would wake up and find that the machine was there, without the experiments ready to

---

[37] *ibidem* Lincei talk.

Fig. 8.10 At left, the miniature model of the synchrotron at the 1956 Milan Exhibition and, at right, the actual accelerator, surrounded by concrete blocks, three years later, 04/04/1959 (Vasari Studio photograph), both © INFN–LNF, http://w3.lnf.infn.it/multimedia/, all rights reserved. For the scale, the concrete blocks, surrounding the synchrotron at right, were 2.0 m high

use it. This advice had underestimated both Salvini and the Italian tradition in experimental physics, which had been nurtured, with success, through experimentation with cosmic rays, since the 1930s. Salvini had been planning for the experiments since 1956, but still detectors and plans were only half ready when the electron synchrotron started working. However, preparations had been effective and realistic. The first measurements could be put in place already in May, and, in July, the first physics results could be presented at 9th International Conference on High Energy Physics, the familiar 'Rochester Conference, held in Kiev (Academy of Science USSR and IUPAP 1960). As Touschek wrote:[38]

> After a slow start in the first half of 1959, an interval of
> time in which some of the teething troubles of the machine
> and of the personnel servicing the machine were cured,
> experiments began to take up the major part of running time
> of the synchrotron. A full regime of experimentation was
> reached in the 2nd half of '59.

Figure 8.10 shows how the synchrotron became reality, from the model miniature at the major Italian industrial exhibition, the 1956 *Fiera di Milano*, to the actual machine in the new national laboratories. The electrostatic accelerator which injects the electrons in the synchrotron, at right in the scale model, appears at the upper left in the actual machine, surrounded on all the other sides by concrete blocks to shield dangerous radiation and protect from accidents.

When the electron synchrotron started operating in Frascati, the pride and enthusiasm spread beyond the Laboratories staff, who had succeeded, in just a few years, in the Herculean effort of reaching the level of the most technologically and scientifically advanced countries in the world: the country as a whole new modern nation had put its hand on the future. Local and national magazines and newspapers gave the news, national television crews came to film the official inauguration of the machine

---

[38] *ibidem* Lincei talk.

as the President of the Italian Republic was immortalized cutting an inaugural ribbon. The popularity of physics in Italy, already fed by the launch of the Russian satellite Sputnik and popular stories about the US atomic bomb, grew.

During the inauguration of the Laboratories, Italy's President suggested to Salvini to give lectures in physics to be broadcast through the national television.[39] Salvini followed the suggestion and his TV lectures, broadcast throughout Italy, led to a large increase in the number of first year physics students in the academic year 1959–60. The brightest students from the classical high-schools of Rome enrolled in physics, fascinated by Salvini's clear, concise, enlightening lectures. He was a man of principles and ambition, and, as his story tells, an extremely inspiring figures as well. In 1959, this author was one of the students conquered to physics by his TV lectures. The 1960–61 class, in addition to the brightest young men, also had a relatively large percentage of very bright women students, 30 out of an initial enrolment of 330. The attrition rate at the time was high, but less so among the women, many of whom graduated with honors and had successful physics careers at international level.[40]

A spirit of pride and playfulness pervaded the laboratory, while visitors came from all over the world. And from Rome, Bruno Touschek, who had been often visiting his aunt Ada in nearby Albano and whose interest in accelerators was slowly returning, started stopping by to see what was happening.

The set was staged for the Ninth Rochester Conference, to be held in Kiev, in the USSR, in the coming month of July. From its very beginning in 1950, attendance to the conference series was by invitation only. This time it was going to be attended by 60 Americans and 60 Europeans, in addition of selected scientists from Australia, India and Japan. From the Institute in Rome, Nicola Cabibbo, Marcello Cini, Marcello Conversi, Raoul Gatto, Giorgio Salvini, and Bruno Touschek, were invited, a remarkable acknowledgment of the work being carried out in the previous years at University of Rome. At this conference the three roads leading to the construction of particle colliders came together, as the next chapters will make clear.

---

[39] Personal communication to the author by G. Salvini.

[40] Personal reminiscence from Galileo Violini (b.1942), enrolled in the 1960–61 freshman year, Professor of Physics at University of Cosenza, founder and Director Emeritus of the Centro International de Fisica, Bogotá.

Chapter 9
Touschek in Rome in the 1950s

The area around Rome is a fairy tale, there is still snow on the mountains and nothing has changed in the last two thousand years.
Die Umgebung von Rom ist ein Märchen, auf den Bergen liegt noch Schnee und in den letzten zweitausend Jahren hat sich nichts geändert.
Letter to parents, April 16th, 1953.

Abstract While the Frascati electron synchrotron was being built, Bruno's intellectual and emotional potential opened to new life. Between 1953, when he joined the Physics Institute in University of Rome, and 1959, when he started thinking about colliding beams, he made new friends, married, wrote interesting papers, had his first child. He went skiing at 3500 m above sea level, but also swimming and fishing in Italian waters, North and South. He attended conferences on both sides of the Atlantic, and some of the most brilliant students graduated under his supervision, Nicola Cabibbo and Francesco Calogero, among them. His new friends included Nobel Prize winners such as Wolfgang Pauli, whose death he mourned as loss of a mentor, and T.D. Lee. He was universally recognized as a brilliant theoretical physicist, when his life changed direction following the Frascati electron synchrotron coming to life in spring 1959.

While Enrico Fermi, Edoardo Amaldi and Gilberto Bernardini, on the two sides of the Atlantic ocean, from Chicago to Rome, were joining efforts to make Italy a full fledged member of the nascent European accelerator club, Bruno arrived in Rome, where he would join his destiny with the Italian road to particle science (Amaldi 1979a, b).

Bruno had arrived in Rome at the end of 1952. The Physics Institute immediately struck him as really excellent, with two Nobel prize winners visiting the Physics Department, Patrick Blackett and Wolfgang Pauli, the first from the UK, the other from Zurich.[1] American scientists were also visiting Europe, often including Paris

---

[1] The Nobel Prize in Physics 1945 was awarded to Wolfgang Pauli 'for the discovery of the Exclusion Principle, also called the Pauli Principle.'

© The Author(s), under exclusive license to Springer Nature Switzerland AG 2022
G. Pancheri, *Bruno Touschek's Extraordinary Journey*, Springer Biographies,
https://doi.org/10.1007/978-3-031-03826-6_9

and Rome in their post-war *grand tour*. Among them, there was Matthew Sands, an accelerator physicist from the Synchrotron Laboratory in Caltech, the California Institute of Technology.

9.1   First Papers

Sands had come to Rome with a Fullbright scholarship for the year 1952–1953. On his arrival to Rome, he heard from Amaldi about the evolving creation of CERN and the plan to build a large proton-synchrotron. He began looking into some of the proposed solutions for the planned CERN machine, and had strong doubts about some of them (Sands 1989). He shared his doubts with Bruno, whose experience on Widerøe's 15 MeV betatron in Hamburg in 1944 made him one of few physicists in Rome sufficiently expert on such problems. They immediately started working on the problem which worried Sands and wrote together the first of Bruno's paper after his arrival to Italy (Sands and Touschek 1953). The paper, sent out for publication to the *Nuovo Cimento* in February, put Bruno in a very favourable light with his Institute colleagues. It was a small success, but everyone was struck by Bruno's capacity to work right away with a well known accelerator scientist such as Sands.[2]

This first paper restored Bruno's confidence in not having wasted his time in accepting (even urging) Amaldi's offer of a position in Rome. Even though he was not particularly interested in building accelerators at that time, his knowledge and experience were a unique scientific experience and he felt that it could come of use, perhaps not now, later may be. There were also other reasons for not indulging in his fear of having moved his life in the wrong direction. He liked Rome, the city and the countryside. Lying down in the sun, in February, was an unexpected luxury. The lovely weather enjoyed in Rome, even in winter, was well worth withdrawing from the northern chimney atmosphere, even if losses would come with it.

9.1.1   Good Times

Amaldi was immediately taken by Bruno's sparkling intelligence and wit. On his part, Bruno found a mentor and a friend. Soon after arriving in Rome, they started playing tennis together, a habit they kept on for many years. Touschek was taken aback by the hard pavement of the University tennis court, calling it a 'lunar landscape' in obvious comparison with the grass courts in Scotland.[2] But, always ready to adapt and take on challenges, he soon mastered the game, as Amaldi remembers (Amaldi 1981, 15). In Rome, Bruno's drawing virtuosity developed in frequent sketches, which illustrated his leisure times or university life, Fig. 9.1. Bruno was soon part of Amaldi's circle, and joined in excursions to the nearby Tusculum hills, where Amaldi was eyeing a possible site for the future national laboratories. We show Bruno with Edoardo and

---

[2] Letter to parents on February 24th, 1953.

Fig. 9.1 Left panel: one of Touschek's drawings, Family Documents, © Francis Touschek, all rights reserved. The photo shows Bruno Touschek, with Edoardo and Ginestra Amaldi at his right, in the Tusculum hills, in 1953, Joseph Astbury first from left, Georg Placzek with dark glasses, from Edoardo Amaldi Papers, courtesy of Sapienza University of Rome–Physics Department Archives, https://archivisapienzasmfn.archiui.com, documents provided for purposes of study and research, all rights reserved

his wife Ginestra in the right panel of Fig. 9.1, together with other friends and visitors of the Rome institute, such as Joseph Astbury and Georg Placzeck.[3]

Not long after his arrival, he left the pension, where the Institute had initially found him a room, and went to live with one of his colleagues, Alessandro Alberigi Quaranta, and his family, in Via Nomentana, not far from the University. Alberigi Quaranta, who would join the synchrotron effort, (Alberigi et al. 1959), came from an old ancestral family. Much to Bruno's amusement, he was the first in his long family line to have to earn his living.[4] Bruno was older than Alberigi, but of similar academic standing, and they shared many interests. Most importantly, they both loved touring the countryside with their motorcycle.

On Sundays, with Alberigi, or some other motorcycle enthusiast such as Aunt Ada's engineer,[5] Bruno would drive out of Rome along one of the seven ancient Consular roads, *strade consolari* in Italian, so named as they were constructed under a Roman Republic Consul's term, sometimes taking the Consul's name, as in the

[3] Astbury and Plazeck in this photograph were identified by Giovanni Battimelli from correspondence in Edoardo Amaldi Papers.

[4] Alessandro Alberigi Quaranta (1927–2012) had come to Rome from the native Reggio Emilia, to study and graduate in physics from the University of Rome, when he was only 21 years old. He was born in a noble family, Quaranta being a Piedmont land property of the Gonzaga Dukes of Mantova, as from https://www.sif.it/riviste/sif/sag/ricordo/alberigi#Note1. Before returning Norh, to Modena and Bologna, where he had a very distinguished career, Quaranta became assistant professor at University of Rome and was at the Frascati National Laboratories from 1956 until 1960, participating to the construction of the synchrotron, and early experiments.

[5] This engineer is likely to have been the one who had directed the construction of Bruno's aunt Ada's villa in Genzano.

case of the Via Appia, from the Consul Appius Claudius. Like rays out of a wheel, Aurelia, Cassia, Flaminia, Salaria, Tiburtina, Casilina and Appia had been constructed to receive goods and move troops in and out of Rome in all directions, towards all of Italy, and the rest of Europe.

Sunday drives would thus take Bruno to Terni along the Via Flaminia, back to Rome from Rieti by the Via Salaria, to the Etruscan town of Veio along the Via Cassia, to the Sabine hills by the Casilina, or to Aunt's AdA in Genzano, by the Via Appia. In Bruno's words:[6]

```
Easter Saturday: Rome, Monte Circeo (on the litorale, a
dilapidated street, that consists almost entirely of sand
pits and, in places, leads through the sea), and back to
Rome. Montecirceo - the Mountain (with cave) of the
temptress sorcerer Circe - is halfway between Rome and
Naples. Sunday: with Barbieri (Aunt Ada's engineer) after a
first class Easter lunch at T.A. [Tante Ada] to Veii (an
Etruscan site) etc. Last Sunday a desert trip to Terni over
the Sabine Mountains, lunch at the lake (Pie' di Luco) back
via Rieti. The area around Rome is a fairy tale, there is
still snow on the mountains and nothing has changed in the
last two thousand years. On the drive to Terni we drank wine
in an old osteria (800m). The osteria still has room for
horses, and the traveler finds oats and a bed made entirely
of straw. The wine was something very special. There was no
telling whether it was sweet or dry and it was the color of
wormwood.
```

Bruno brought to Rome his own version of MittelEuropean charm and taste for sometimes outrageous jokes. On April 15th, his Italian was good enough and his work on the theory of mesons, as it was then called, advanced enough, that he gave one of his first seminars. Amaldi and Bruno Ferretti, who had been instrumental in Bruno's move to Rome, had clearly advertised it and it was much attended by the institute researchers. It was something of a mixed success. He tells the story in one of his letters home[6].

```
... Yesterday I gave a long lecture in the most terrible
Italian on meson theory. The listeners (I do not know whose
propaganda Amaldi and Ferretti together brought into this
lecture) just twisted themselves like that. Everyone was
very nice and didn't laugh at the worst things I said - but
they probably came to have fun, as in one of my last
lectures I was said to have brought up things like
Hamiltoniana masturbativa (instead of perturbativa) and
similar cabaret jokes. The lecture was not over until 8
o'clock in the evening [...] Afterwards I drove through the
Via Nomentana with great energy and pouring rain, using my
beautiful voice instead of the horn. Alberigi, who drove
away with me, arrived at the garage 5 minutes later because
he was probably embarrassed.
```

---

[6] Letter to parents on April 16th, 1946.

9.1.2 Stuck in Italy?

Enchanted as he was by the beauty of the countryside, the excellent food and ubiquitous quality of wine, Bruno was however worried about his professional future. He was now 32 years olds, and could see that his contemporaries were on their way to become professors in their universities. For many of his friends and colleagues, this was an already traced path. Others, already professors, were dislocated in provincial universities, waiting for their turn to be called to the more prestigious places, such as Rome, among others. Bruno's situation was atypical. His father was still hoping for him to come and work in Vienna, and, quite possibly, Bruno also was hoping it could happen. It didn't. Bruno lacked the type of local support in University of Vienna, where many scientists, of previous Nazi sympathies, were still obtaining positions or were even called back to the post held during the Nazi regime (Moore 1989, 456).[7] Nor would or could Hans Thirring help.[8] The road to professorship in Italy also appeared complicated by him not having Italian citizenship.

To allay his worries, as spring moved on toward summer, Bruno started to plan for an extended vacation to the North of Italy, Tyrol and Switzerland. He thought to reinforce his ties to Fritz Houtermans and Wolfgang Pauli, and sound job possibilities outside Italy, combining planned holidays in the Dolomites with visits to Bern and Zurich. The summer went by, but new prospects for a position abroad did not materialize. He seems to have even considered a position in Australia, but this possibility also vanished. In the meanwhile he spent some time at CERN and participated to a conference, in Cagliari, Sardinia, where his new collaborator Giacomo Morpurgo, nicknamed 'Pimpi', presented the work they had done together (Morpurgo and Touschek 1953).[9] One of Bruno Ferretti's assistants in Rome, Raoul Gatto, Fig. 9.2, was also at the conference. Gatto, Touschek's ally in the AdA adventure through his work on electron-positron physics, future mentor of a cohort of renowned particle theorists (Casalbuoni and Dominici 2018; Giovanni Battimelli et al. 2019), remembers (Gatto 2004):

> *Raoul Gatto:* My most intense memories go back mainly to the years from 1953 to 1956, when I met Bruno almost every day and I talked with him and learned so much from him. [...] Most important were his friendliness and consideration.
>
> At that time, especially at the beginning, I felt rather lost and insecure, in a career which seemed to be very competitive and where some people occasionally exhibited an intense pride of hierarchies. Bruno, on the other hand, was friendly cordial, encouraging. I remember I was 22, at a Conference in Cagliari. He was sitting at a café with Pauli, who participated to the meeting, and I was passing on the side walk, rather trying to get

---

[7] Such was the case of Georg Stetter, director of the Second Institute for Experimental Physics in Vienna and a member of the *Uranverein*, who had been urging his colleagues to stop helping students not having pure Aryan blood, as Urban wrote to Amaldi in 1980 (Amaldi 1981, 8).

[8] In a letter to his parents on November 3 1953, Bruno comments that Thirring prefers to help former Nazis in his efforts at reconciliation, as also mentioned in (Moore 1989).

[9] Morpurgo (b.1927), member of the Academia dei Lincei, is a former Professor of Physics at University of Genova.

Fig. 9.2 At left a young Raoul Gatto, in a 1954 personal photograph given to Luisa Bonolis and Carlo Bernardini for a conference, courtesy of Luisa Bonolis. At right Bruno Touschek at Lake Albano, in the 1950's, Family Documents, © Francis Touschek, all rights reserved

unnoticed.[10] He called me and and wanted me to sit down with him and Pauli and take part of the discussion. Similar things happened many times. [...]

As the fall came, Bruno's situation was still fluid. A professorship was not a clear prospect. Bruno had followed a very unusual road in order to reach his remarkable scientific and professional status, most of it completely on his own. He had overcome losses, discrimination and incredible hardships during the first part of his life, but also enjoyed the friendship and companionship of some of the most famous physicists of his time, such as Sommerfeld, Born and Heisenberg. Now, they were retired or too old or too distant to help him. While still in Glasgow, after his graduation, he had inquired with Philip Dee about his chances for a professorship in the UK, such as his collaborator Sneddon had obtained, for instance. Dee had replied that it could be possible—not at Glasgow, but perhaps elsewhere—but that he needed to be more at ease with the country he lived in. Dee probably meant to say that Bruno still had to adapt and integrate in the British way of living, but Bruno either didn't not want or would not be able to follow that path. This was a path which could also have meant to change citizenship. And Bruno never did it, feeling that his father would have minded.[11]

Once in Italy, the problem of Bruno's Austrian citizenship came up again, since Italian law did not allow foreigners to enter into the mandatory national competitions for university professorship until 1974.[12] For a while, he thought the problem could

---

[10] Raoul Gatto (1930–2017) was proverbially shy, as all his students and colleagues well remember.

[11] See Nicola Cabibbo's interview in Bruno Touschek and the art of physics, 2004, and also (Amaldi 1981).

[12] Changes in University law allowed Bruno to become Professore Aggregato in 1969, and later, in 1974, full professor (Amaldi 1981, 25), see also footnote n. 90), *ibid.*.

be solved, hoping that he could have both Austrian and Italian citizenship, but, at the time, this was not possible. On the other hand, Amaldi was keen to keep Bruno in Rome. He valued his extraordinary and unique combination of theoretical strength and accelerator experience. He would not want to lose Bruno, who was certainly part of his plans for the future of accelerator physics in Italy. And, as the future would show, Amaldi was right. Ten years later, Bruno gave to Italy the glory of having built the first ever matter-antimatter collider, and his success gave him a place in the history of 20th century physics.

In the fall of 1953, apart for the professorship question, Bruno was also reflecting on the direction his research should take. Amaldi suggested him to look into cosmic ray physics and, to start with, visit the Testa Grigia Laboratory, established by the University of Milan, in 1948 on the Monte Rosa, in Piedmont, at over 3000 meters altitude above sea level.

In Milan, a remarkable tradition in cosmic ray studies had been established since the late 1930s by Giuseppe Cocconi and his wife Vanna Tongiorgi, but these activities were nearly stopped during the war, with few noticeable exceptions. They started again in 1945 on the mountains near Bergamo and later at the high-altitude Laboratorio della Testa Grigia, at 3500 m above sea level, whose construction had been promoted by Gilberto Bernardini and which was officially inaugurated in January 1948. It played the important role of a national laboratory, and provided Italian physicists the possibility of systematic research work and in particular for training a new generation of students and for establishing personal and scientific relationships among physicists from different Universities in Italy, hence preparing the stage for the foundation of the National Institute for Nuclear Physics in 1951 (Battimelli et al. 2001). The Testa Grigia Laboratory had frequent visitors from Rome, where experiments in cosmic rays had continued under the Allied bombing, resulting in the well known Pancini-Piccioni-Conversi experiment (Conversi et al. 1947). In Fig. 9.3 we

Fig. 9.3 From left: Marcello Conversi and Oreste Piccioni in 1943, Edoardo Amaldi, Gilberto Bernardini and Ettore Pancini in 1947 at the Testa Grigia cosmic ray Observatory, in the Italian Alps, at the southeast limit of the Plateau Rosà glacier. Reproduced from Edoardo Amaldi Papers, courtesy of Sapienza University of Rome–Physics Department Archives, https://archivisapienzasmfn.archiui.com, documents provided for purposes of study and research, all rights reserved

show a snapshot of Marcello Conversi and Oreste Piccioni in Rome, around the time of their first experiment, and a later 1947 photograph of Ettore Pancini, with Amaldi and G. Bernardini, at the Testa Grigia Laboratory. Together with other mountain laboratories such as the Jungfraujoch in Switzerland and the Pic du Midi in France, the Testa Grigia also became an attraction for the European scientific community, stimulating coordination of experiments planned by different groups and paving the way to wider and more ambitious collaborations. The INFN contract offered to Bruno by Amaldi had indeed also included research in cosmic ray physics. A winter trip to the Testa Grigia Laboratory would provide Bruno with a taste of what cosmic ray research meant. At the same time, on the side, it also offered the opportunity for some fantastic skiing, with lodging free of cost. Thus a plan for being in the Alps, for the end of the year, between Val d'Aosta and Switzerland, was drawn up.[13]

The prospect of spending Christmas and the New year in a high altitude physics station in the highest part of the Alpine regions, was very appealing to young Bruno. It was going to be a daring adventure to be at the Laboratory and ski at such altitude. This is exactly what Bruno was looking for. He was young and, mostly, loved to take over any challenge he met. In summer 1947, in Glasgow, during the yearly physics department excursion, he had been first of his Scottish university companions to reach the top of Ben Nevis, the highest peak in the U.K. rising above sea level to 1345 m.h. at 79,6° North. At the time, he had never gone climbing before, not even in his native Austria, but he had felt he had to be true to his upbringing in an Alpine country. In Scotland he never missed bathing in the icy waters of the Lochs, and he had also gone swimming in Fort William at Christmas time. Skiing at the Testa Grigia was a new interesting challenge, one which his peers, such as Amaldi and other physicists, took regularly. He would of course go. For most of his life, personally and professionally Bruno took risks, with a courage partly innate, most probably grown out of his many personal losses and bad turns. These risks and the damage to his physical and mental strength took a toll on him, but Bruno's propensity and courage in taking risks is what kept him alive, saving him for a great scientific adventure, later in his life. When AdA had to be built and show its feasibility before the Americans would work out how to do it, he won. He thought of the unthinkable, and carried it through disappointments and failure. He would not win the hoped for original prize with ADONE, but, in the long range perspective, AdA was more important that ADONE.

Thus, on December 22nd 1953, Bruno started the long trip to reach the Testa Grigia Laboratory. He left Rome at 3 a.m., reaching Turin at midnight. Then a slow train took him to Châtillon, where he arrived at 6 a.m. From there one would reach Cervinia, which is some 2000 meters higher, with a 3 hours bus drive. From Cervinia a cable car led to the plateau itself in 3 stages, Plan Maison, Cime Bianche, Plan Rosà. The south side of the plateau ends in a wall about 300 m high and on the edge there was the laboratory for cosmic rays. The last stop of the cable car was on the Plateau, where there was also a meteorological station (with a television system, a small suite of rooms for the 3 carabinieri of the border control). Then there was a

---

[13] Letter to parents on December 21st, 1953.

restaurant closed in winter and a house where the Swiss customs officers lived. The border between Italy and Switzerland runs across the plateau.[14]

```
The laboratory consists of the actual laboratory, 3 bedrooms
with 4 beds each and a living room-cum kitchen - there is
also a toilet with a bidet - which is used to repair the
toilet tank without a ladder. There's no water, the cable
car brings the drinking water and the usable water has to be
warmed up by shovelling snow. After the innkeeper left, I
took over this work and it takes almost an hour and then you
have to lie down.
I got there with a terrible cold (from the train ride).
After dinner, I wanted to slide down a bit - and that was a
mistake. I went down to about 3100 m (to the so-called
Teodul). It took me over an hour to get back up, the wind
(100km) kept blowing in my face and I had to lie down to
breathe. Sleep was also very difficult because of the lack
of air. In addition, the air is still incredibly dry and if
you blow your nose - which I unfortunately had to do very
often, blood comes out. But after two days I got used to it.
```

At the station there were only three physicists on site for this festive period. At the time, cosmic ray stations needed to be manned, but there was not much to do, apart from keeping a check on the installations which record the arrival of the cosmic rays which hit and impress the photographic plates. Bruno skied most of the time, often in the company of a friendly *carabiniere* from the Italian border patrol. On the Sunday after New Year's, Bruno was joined by Gleb Wataghin, a theorist from Turin, returning with him down into the plains and to civilization. In Turin, they had dinner together with the theorist Marcello Cini. Cini belonged to a very distinguished family in Turin, and had joined the University of Rome in 1953. In the years to come, he became a leading theorist and, from 1965 until well in the 70's, a leading political figure during the years of the student movement at the University of Rome.[15] With Cini, Touschek wrote two papers, the second of which, in 1958, remains one of his most influential theoretical physics papers (Cini et al. 1954; Cini and Touschek 1958).

9.2   When Bruno Met Elspeth

By April 1954, all the decisions about a new national laboratory, and the future accelerator were taken (Salvini 1962, 54). The laboratory would be built in Frascati

---

[14] Letter to parents on January 20th, 1954. Notice that television public broadcasting started in Italy officially only in 1954—January 3rd, 1954. The station at the Plateau was obviously a military installation.

[15] See also the book he wrote with Giovanni Ciccotti, Michelangelo De Maria and Giovanni Jona-Lasinio *L'ape e l'architetto*, Franco Angeli Chimera Ed., 1976, which had a lasting influence towards the formation of the environmentalist movement in Italy.

and be a national center which would carry Italy into the modern world of elementary particle physics.

As these plans started to consolidate, so did Bruno's life and the year 1954, which had started on the high slopes of the plateau Rosa, marked a new beginning. The distant future seemed to be slowly taking shape. Amaldi, having now secured to Rome the electron synchrotron project, was able to reassure Bruno about his future and told him to consider Rome as the base from where he could eventually move out spending time at CERN, as many Italian physicists would soon be doing. Such possibility for Bruno was aired. Salaries at CERN were paid at the level of other international organizations and could integrate the low Italian university wages.

Bruno was relieved. His fears that the decision of coming to Italy would turn out to be a mistake, were partly allayed. The problem of not having access to a faculty position was not solved, but his research was proceeding well, he was finding interesting and stimulating collaborators, in particular Giacomo Morpurgo, who was in Rome at the time (Morpurgo and Touschek 1953; Cini et al. 1954). He was invited to workshops and conferences, one in Padova later in April, one in Glasgow in July. On April 2nd, he was going to Naples to start some work, to proceed in collaboration with Morpurgo and Luigi Radicati di Brozolo, a theorist, who had just come back from Birmingham, after two years as a post-doctoral researcher with Rudolph Peierls, (Morpurgo et al. 1954).[16]

Luigi Radicati (Salvini 2004, 67):
I have had the great fortune to be close to Bruno Touschek for more than twenty years. I knew him as a physicist and I have learnt from him a great deal: I would even say that Bruno has been one of the persons who had the greatest influence on my scientific life. His great strength was his extraordinary originality of thought, based on a deep knowledge of physics. He clearly was coming from a physics tradition much richer than the one I had been brought up in. For Bruno physics was something to be born with, something which had its origins in the thinking of Boltzmann, Gibbs, Sommerfeld, Heisenberg. For him statistical mechanics was something intrinsic to his being. (I had to painfully learn it and the little I know came from him). But I remember Bruno not only as a physicist : for me he has been mostly a very dear friend with whom I spent unforgettable days and hours, with whom I discussed about everything, history, art, literature, philosophy. He had an extremely rich and multi-faced personality: we came from very different cultural backgrounds, but this diversity, instead of dividing us, was the greatest bound.With him I have lost one of my dearest friends and to him, after these many years, I still think with immense longing.

---

[16] Letter to parents on April 1st, 1954. About Radicati, it is not clear whether Bruno could have met him earlier in the UK. Check https://discovery.nationalarchives.gov.uk/details/a/A13530333 at Oxford: 'Radicati was a post-doctoral research student at Birmingham 1951–1953, later returning to the Scuola Normale Superiore at Pisa where Peierls was a regular visitor. (Bird of Passage 237, 333).'

9.2.1 Elspeth Yonge

On April 2nd, 1954, Bruno went to Naples with his motorcycle, for a few days, to work with Radicati and his life changed.[16] He met Elspeth.

In the year 2009, this author, and Luisa Bonolis, went to visit Touschek's widow, Elspeth Yonge Touschek. We asked how had she met Bruno, and what had mostly impressed her at that time. We knew Elspeth was from Scotland, and that her father, Sir Maurice C. Yonge, was a very distinguished marine zoologist, who had been Professor at University of Glasgow.[17] We had thus naturally assumed that Bruno and Elspeth had known each other from Glasgow or Edinburgh. To our surprise, Elspeth said: "It was in Naples".

Naples houses the famous Dohrn Zoological Station, which Elspeth's father had been visiting, and she must have accompanied him. Elspeth used to help her father by drawing the delicate shapes of marine life, which grace some of his books. She told us that drawing dead sea-urchins or star-fish was rather disgusting, but none of this transpires from the lovely illustrations in the book she showed to us, on that day in 2009, in her villa near Rome.

We then asked Elspeth where in Naples did she meet Bruno and what did she think of him. "It was at a party", she told us.

Naples has a long tradition of great intellectuals and scientists, also fed by exchanges with Rome to the North and Palermo to the South. The 'party' could have been one of the typical Neapolitan evenings, where Naples intellectuals used to gather to discuss everything, from politics to philosophy, and science. One avenue for such parties was the Dohrn Zoological Station, in Via Caracciolo, looking out to the sea. Named after Anton Dohrn, founder of the Station, contemporary and a friend of Charles Darwin, the Station was a major world center for marine science studies. It was directed by Pietro Dohrn, Anton's descendant. In the 1950's Naples was an important cross-point of cultural and political events and the Dohrn Station used to organize parties where physicist were also invited.[18] Elspeth's father was a good friend of Pietro Dohrn and would naturally take along his daughter to the Station's social occasions.

What was young Elspeth's first impression of Bruno? She answered us, very simply, with a smile: 'He was the only person alive in the room'.

Things appear to have moved rapidly after April. In July Bruno was in Scotland to attend the Glasgow Conference on Nuclear and Meson Physics, which took place between 13 and 17 July, 1954, where he presented his work with Morpurgo and Radicati (Morpurgo et al. 1955). When he came back, he wrote to his parents that Professor Yonge would like to meet them for Christmas in South Tyrol.[19] Then,

---

[17] For M.C. Yonge's life and accomplishments, see Charles Maurice Yonge (1999–1986) in Memoirs of the Royal Society.

[18] Courtesy of Guido Cosenza, particle physicist, retired Associate Professor of Physics at University of Naples, where he moved from University of Rome. In recent years Cosenza has authored books on economics and social impact.

[19] Undated letter to parents, file no. P1030694, clearly written after his trip to Glasgow.

after the summer and Bruno's trip to Glasgow in July, Professor Yonge had become *Papa Yonge*, pointing to a clear indication of closeness between the two families.[20] Less than a year after their first encounter, Bruno and Elspeth sent a telegram from Glasgow, announcing they were married, in March 1955.

After the days in Naples, Bruno went North to Padua, where an 'International Congress on Heavy Unstable Particles and High Energy Events in Cosmic Rays' had been organized to discuss the latest cosmic rays results (Ceolin 2002). At this conference, the main object of interest was the so-called $\tau - \theta$ puzzle, which was attracting everybody's attention, and had Bruno engaged. A committee which included Edoardo Amaldi and Bruno Touschek was put in charge of preparing an overall report on τ-meson data (Amaldi et al. 1954). Bruno was clearly a nascent star in the Italian theoretical physics panorama, and Edoardo Amaldi was his friend and sponsor.

9.2.2   What Motorcycle Accidents are Good for

The year 1954 was a special one for Bruno. Not only Bruno met Elspeth, he also started two life-long friendships, one with Luigi Radicati, the other with a neurobiologist, Valentin Braitenberg from Meran. Both had a deep impact on Bruno's life.

Braitenberg was not a physicist, but they bonded through a common love for the Dolomites, where Valentin came from. Their origin had its roots in a similar cultural background, which is still reflected in the population of the former Austro-Hungarian Empire. It is a Mitteleuropean sense of belonging, which, until not long ago, was strongly present not just in the German speaking countries in and around the Alpine range, but also down from South Tyrol into the Italian speaking valleys of Trentino. Bruno and Valentin spoke the same language, really and metaphorically.

The encounter between Bruno and Valentin has become part of Bruno's legend of his 1950's Rome days. There exist at least three written versions of it, one told by Valentin himself in a letter to Amaldi after Bruno's death (Amaldi 1981, 14), one told many times by another of Bruno's close friends, Carlo Bernardini (Bernardini et al. 2015), and, perhaps closer to the truth, there is what Bruno wrote to his parents, soon after things happened, long before his friends retold the story.

Bruno's original version places his first encounter with Braitenberg sometime in May 1954.[21] At that time, Bruno did not yet have a car and was moving around with his motorcycle. One day, Bruno had what he called a small accident, colliding with an American car. The motorcycle, which he called Marie Luise, was damaged, Bruno's head was hit, and he lost consciousness. He was taken to the psychiatric ward of the Polyclinic of Rome University and hospitalized for the night. In the morning, he vainly tried to be released. In Bruno's words:

---

[20] Letter to parents, October 27th, 1954.

[21] Letter to parents on June 15th, 1954. In his 1980 letter to Amaldi, Braitenberg wrongly places the incident in 1964.

```
The next morning they didn't want to let me leave because
they didn't believe I was me - [they said that] 'everyone
says that here'.
```

Bruno had declared himself to be a theoretical physicists and Vice-director of the University Physics Graduate School. His heavy Austrian accent and the effects of the concussion were not helping either to get his freedom. Finally a doctor came who spoke German. He declared Bruno to be sane and finally had him freed. He was Valentin Braitenberg, a German speaking native of South Tyrol, a region in the north of Italy, where the population is bilingual. They immediately became friends for life. Breitenberg was one of the friends who were the last to see Bruno in 1978, shortly before Bruno passed away, in Innsbruck, in his native Austria. After Bruno's death, on Amaldi's request, Valentin Breitenberg related the accident in his own words, as follows (Amaldi 1981, 14):[22]

> *Valentin Braitenberg:* One morning of the year 1964 [sic!], my colleagues from the psychiatric ward asked to come down from the laboratory where I was working to fill the hospital sheet of a person who has been admitted the night before. [dots] The injured man had declared he was a theoretical physicist, Vice-director of the Scuola di Perfezionamento in Fisica Nucleare, and a specialist in time reversal. …We immediately became friends. His story was convincing and not at all psychotic, his German was delicious, rich and precise, his humour was uncontrollable even in those embarrassing circumstances. I wrote his clinical story with his help. …Bruno [said] that if I wanted to associate myself with his habits, he would be glad to take me to Nemi on the following Sunday. I accepted and after that for many months we spent almost all week-ends together. …We started to talk about cybernetics …

The story was later a bit embellished by Carlo Bernardini or by Bruno himself in his telling or retelling the story. Carlo writes that Breitenberg not only freed Bruno from the psychiatric ward, but, most crucially, saved him from being subjected to electroshock, also giving some details not found neither in Bruno's letter nor in Braitenberg's. Here is Carlo Bernardini's version:[23]

> *Carlo Bernardini:* He used to take his favourite motorcycle and come to Frascati. Arriving in Frascati, near the road bent for Vermicino, he got his motorcycle under a truck. I mean, to bump into a truck, that's not trivial. He was brought to the neurological clinic of Professor Cerletti, who examined him, noticed he was somewhat incoherent and kept him for a while under observation. Coming back to check on him, Cerletti heard Touschek explain how the accident had happened. He heard him saying: 'I was standing still when a truck backtracked and hit me.' That didn't sound plausible to Cerletti. When he told so to Bruno, Bruno replied: 'If you make a time transformation, you will see this is possible.' At which point, Cerletti cut short the visit, saying: 'Enough! He is delirious: Electroshock!'.

---

[22] See also letter from Breitenberg to Amaldi, in Edoardo Amaldi Papers, Sapienza University of Rome–Physics Department Archives, https://archivisapienzasmfn.archiui.com.

[23] From docu-film Bruno Touschek and the Art of Physics, at 28':03". Carlo Bernardini in this interview taken in 2005 by L. Bonolis, implies that Bruno was already visiting the Frascati laboratories at the time of the accident. This being impossible, since the Laboratories did not yet exist in May 1954, when Bruno told the story to his parents, other details may also be wrong. In particular, neither Breitenberg, in 1980 to Amaldi, nor Bruno in 1954, mention electroshock. The story may have arisen because Ugo Cerletti (1867–1963), the physician who first visited Bruno, is known to have built the first electroshock device (Aruta 2011).

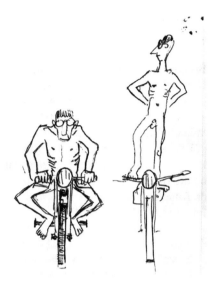

In that instant, one of Cerletti's assistants, Valentino Braitenberg happened to come by, and,
stopping short Cerletti, exclaimed: 'He may not be delirious. He could be a physicist!'

This brief comparison, between three different ways to tell the same story, shows
how difficult it has been to separate facts from fiction about Bruno Touschek. This
may have had the effect of relegating Touschek into an anecdoctical view, which
obscured, in the long run, his greatness.[24] The morale of the story, however, is what
Bruno wrote to his parents:'…it shows what motorbike accidents are good for',
namely to forge a life-lasting friendship.

Soon after the accident, Valentin and Bruno started taking rides together to Lake
Albano and to Aunt Ada's place. They went swimming, drinking and eating together.
At Lago Albano they discovered a piece of rocky coast from which one could jump
into unbelievably clear deep (but still warm) water, as he wrote to his parents.

Bruno and Valentin spent hours together discussing philosophy, psychoanalysis
and cybernetics. Together they studied Jung's Correspondence Principle in psychol-
ogy. Braitenberg was a brain specialist and, after a long time, this was the first
occasion when Bruno would learn of things that had nothing to do with physics. In
Fig. 9.4 we show a well known self portrait of Bruno on his motorcycle.

This was not going to be the last accident Bruno had with Marie Luise. There is
a hint, in letters to his parents, that he may have had an accident in which Elspeth
was also injured. What is certain is that Professor Yonge prohibited Elspeth to ride
with Bruno in a motorcycle. This rule gave him the motivation to buy his first car, an
MG, an English sports two-seater from 1952, very well preserved, which he named

---

[24] Other versions circulating in the physics department included Bruno saying to Cerletti that a
truck backtracked while he, Bruno, was standing still in his reference frame, from G.Violini. In
Braitenberg's account, it is mentioned that the only abnormal detail was Bruno's daily wine quantity.

Carla.[25] Driving a car rather than a motorcycle did not make Bruno less of a daredevil. He taught himself to drive through the Rome traffic, which was at the time rather tame compared to the present days, but still quite chaotic.[26] He even tried to go as fast as 130 Km per hour, not quite a city speed, but satisfactorily tuned to Bruno's love for trying his limits. This car is fondly remember by Raoul Gatto in (Gatto 2004, 70):

> *Raoul Gatto:* When a foreign visitor arrived, we often went with the visitor to Albano or Nemi, two small towns here in the neighborhood, for a walk and a glass of wine. He had bought at that time a strange sport car, I think it was a Triumph, an extremely uncomfortable convertible. He used to drive in full winter with the windshield lowered so that all the air would blow directly into our faces. Before returning Rome, in the not very dense, but totally disordered traffic of the Roman fifties, he would not separate from the colleagues without pronouncing the historical sentence that the fighters in the Coliseum would tell Cesar in the old Rome: "morituri te salutant", in his wonderful precise latin. He was referring to the precarious conditions of his car. I think he lived in that period a rather adventurous life, but the friendliness and generosity of Bruno were an incomparable and unforgettable compensation.

9.2.3   1955: Marriage and Other Things

The year 1955 represents a landmark and a watershed point in Bruno Touschek's life. On the personal front, his marriage to Elspeth was obviously the great change in his emotional life. But, while he was totally absorbed in the new attachments and duties his marriage expected of him, he was also inwardly absorbing and elaborating the novel ideas pouring into elementary particle physics, as new experiments were bringing unexpected results from cosmic rays and accelerators.

In June 1955 Bruno attended the Pisa International Conference on Elementary Particles, which, according to Milla Baldo-Ceolin (Ceolin 2002, 15), was the last conference where cosmic rays results were still dominating the field.[27] Afterwards, results from accelerators burst onto stage and the vision of the particle world took a turn towards modern physics. The great protagonist of the Conference was Giuseppe Occhialini, who had understood that things were changing, when he said that '...cosmic rays would forever remain a source of information and study of unknown phenomena, as coming from outside the Earth, but the time had come to acknowledge that a competing discovery tool had arrived on the scene'. Bruno soon became a good friend of the Occhialinis, Giuseppe, nicknamed Beppo, and his physicist wife, Constance (Connie) Dilworth (Gariboldi 2004). In Fig. 9.5 we show two photographs of Bruno with the Occhialinis at the conference.

---

[25] Letter to parents on November 26th, 1954.

[26] Rome traffic is immortalised in the 1953 American movie *Roman Holidays*, directed by William Wyler.

[27] Proc. Int. Conf. Elementary Particles, Pisa 1955. Nuovo Cim. 4, Suppl. 2 (1956).

Fig. 9.5 Bruno Touschek, Giuseppe (Beppo) Occhialini and Connie Dilworth at the 1955 Pisa International Conference on Elementary Particles, courtesy of L. Bonolis and M. Baldo Ceolin

Since some time, Bruno had been working on time reversal (Morpurgo et al. 1954). The problem was so much on his mind that, when the motorcycle accident had taken place, in May 1954, Bruno is said to have invoked a time inversion to explain his accident. He had begun this work with fellow theorists Giacomo Morpurgo and Luigi Radicati, in a scientific collaboration which started Bruno into the road played by symmetries in particle physics, a time honoured tool to understand observed physical phenomena and predict new ones. On his arrival in Rome, he had also been impressed by the presence of Wolfgang Pauli, and they had become friends, through visits and physics discussions. In those years, Pauli had been concerned about Reflections in space-time and Charge Conjugation properties, and in 1955 had given an explicit formulation of what is now known as the CPT theorem (Pauli 1955). This and the discovery of non-conservation of Parity in weak interactions became a major influence in Bruno's progress in theoretical physics in the years to follow.

9.3   How Different Roads Converged to Make AdA

Starting from 1955, three different roads began converging toward the construction of the first electron-positron collider in Frascati. One was coming through new discoveries in particle physics, such as that of the anti-proton (and the anti-neutron), and the non-conservation of Parity in weak interactions in 1957, which revamped Touschek's interest in neutrino physics. With his work on Time reversal, he had already taken the way of approaching particle physics from the point of view of the symmetries they satisfied. The second road came through the accelerator (particle physics) community interested to reach higher particle energies. The possibility of exploiting the kinematical advantage of head-on-collisions between accelerated particles was aired in 1956, picking up momentum until, three years later, in Kiev, an American project for an electron-electron collider was presented. Undisclosed until 1963, Russian scientists had also been working on a similar project of electrons against electrons.

The third crucial element in AdA's conception and building, was the commissioning of the Frascati electron synchrotron, which started working in late December 1958. This is how it could happened that everything was in place in Frascati in 1960 to accept Touschek's proposal to build an electron-positron collider. The machine shops, the technical know-how, a well tuned team of scientists and engineers, were ready and expectantly looking for further challenges.

9.3.1 New Particles and the Non Conservation of Parity

While the Frascati laboratory and the synchrotron were built, particle physics was entering an era in which new discoveries made their world appear like a puzzle without a guiding picture.

In 1956, the physics of elementary particles took a leap into the future. Two international conferences set the pace, as accelerator physics experiments were overtaking cosmic rays as the source of data and theoretical inspiration. First came the 6th Annual Conference on Elementary Particles, in short the *Rochester Conference*, which was part of an already well known series, and took place in Brookhaven, in the USA, in April 1956 (Peierls et al. 1956). It was devoted to developments in theoretical and experimental physics. The other one was in Geneva, in September, about new ideas and advancements in accelerator physics (Regenstreif 1956), see Chap. 8.

The 1956 Rochester Conference represents a landmark in particle physics. antimatter made its official appearance in a sessions fully dedicated to anti-nucleons on the Friday afternoon. On the following day, non conservation of Parity was put in doubt.

For what concerns ordinary matter, which consists of electrons, protons and neutrons, the first antiparticle, the positron, had been discovered in cosmic rays in 1932, but the next antiparticle had to wait for more than twenty years and the development of particle accelerators. In 1955, Emilio Segrè and Owen Chamberlain, from University of Berkeley in California, announced having found anti-protons in a laboratory experiment.[28] This discovery was followed one year later by that of the anti-neutron, (Cork et al. 1956), fully confirming experimentally Dirac's idea.[29]

---

[28] Segrè and Chamberlain received the 1959 Nobel Prize for their discovery. For Edoardo Amaldi's contribution, see G. Battimelli, D. Falciai, "Dai raggi cosmici agli acceleratori: il caso dell'antiprotone", XIV and XV Congresso Nazionale di Storia della Fisica (Udine 1993 - Lecce 1994), Eds. A. Rossi, Ed. Conte, Lecce 1995, pp. 375–386.

[29] The existence of antimatter was fully established only after the antiparticles of the nucleus components were discovered, and had a large impact on public imagination. *The New Yorker* December 1956 issue had a letter by E.Teller, from University of Berkeley and father of the hydrogen bomb, and a poem about antimatter, which shows the popularity acquired by the idea of the existence of antimatter and the annihilation when matter meets antimatter. Teller's letter came as a comment to one of the usual New Yorker citations of strange events in the regular press. The title of the poem

The variety of new particles which had been discovered, either analyzing cosmic rays emulsions or laboratory produced, constituted a *zoo* with unknown classification and properties. To make order among them, theoretical physics turned to an old ally, mathematics and group theory. During the war, Touschek had studied Wigner's book (Wigner 1931), the formation book of any group theorist, which he had found in Karl Egerer's personal library, at Löwe Opta, in December 1942, in Berlin. He had found the subject so deeply fascinating to promise himself to become an expert in the field. After the disruptions of the war and his Glasgow work on mesons, Bruno's attention was coming back to his old interest.

Long the subject of speculation in literature and science fiction, the existence of differences between mirror images in the chemical and biological world was a long accepted knowledge, as was the unfortunate fact that macroscopic entities cannot reverse their path in time.[30] Symmetry under space reflection was however generally believed to be true in the elementary particle world, as for what concerned electromagnetic and strong interactions. However, as particles with strange properties started to be discovered, doubts arose.

Bruno had been invited to attend the Rochester Conference, not a trivial honour, stemming from his work on time-reversal with Morpurgo and Radicati, which had attracted some attention, as was some work he had done with E. Fabri (Fabri and Touschek 1954). He travelled to New York, and then to Long Island, where the Conference was held, together with Edoardo Amaldi, who was presenting results about identifying anti-protons from cosmic rays emulsion data. At this conference, a question was asked, which set in motion a revolution in contemporary thinking about elementary particle symmetries and invariance principles.

As related both in the Conference proceedings as well in the memoirs of some physicists attending the conference, the session where the crucial question was asked, took place on Saturday Morning, and was chaired by J. Robert Oppenheimer. The session was devoted to *Theoretical interpretation of new particles*.

9.3.1.1   A Crucial Question was Asked in Brookhaven

On the evening before the Saturday morning session, two physicists, sharing the same accommodation, held a discussion about the $\tau - \theta$ puzzle, the nature of what we now call the K-meson: one was Richard Feynman, the great theoretical physicist, the other Martin Block, a brilliant young experimentalist, then at Duke University. At a certain point, Block asked his room mate (Feynman and Leighton 1985, 247): "Why are you

---

was *The danger of modern living* and it told how Teller and his anti-Teller meet 'well beyond the troposphere', and shake hands, annihilating each other with the result that 'the rest is gamma rays'. This magazine page was found among Touschek's personal papers at home, and testifies to his interest in annihilation of matter into antimatter since early 1957.

[30] The most famous example of mirror adventures and the possible differences is Lewis Carroll's *Through the looking glass*. See also Dorothy L. Sayers's 1930 novel *The Documents in the Case*, a 1950 short novel by Arthur C. Clarke, and more in https://en.wikipedia.org/wiki/Chemical_chirality_in_popular_culture.

guys so insistent on this parity rule? May be the τ and the θ are the same particle. What would be the consequences if the parity rule were wrong?". Block was not only a high class experimenter but also a phenomenologist with the gift to directly attack a problem at its root, and was addressing the question of whether symmetry under parity transformations was conserved in weak interactions, recognized to be at the heart of the puzzle. In fact Feynman replied: "I don't know that's so terrible." In (Ceolin 2002, 15) the conversation is slightly different, namely 'Feynman, after considerable thought, concluded that actually there was no compelling evidence.' Urged by Feynman to ask his question during the theoretical session on the following day, Martin Block shied away: "No, they won't listen to me. *You* ask". Martin meant that theorists would not take his question seriously, him being an experimenter, while the same question from Feynman would receive serious consideration.[31] And so it happened.

On the Saturday morning, following the evening exchange between Block and Feynman, C.N. Yang, from Brookhaven, started the session (Oppenheimer et al. 1956, 1–2). In Yang's words, the conference attendees had been introduced to the sub-nuclear *zoo* (in a public lecture by Oppenheimer) and had taken excursions through it during the last two days. Before leaving the *zoo*, it was thus time to ask what had been learned. In his talk Yang pointed to the major puzzle facing current theories of elementary particles, the fact that there were 'strange' particles, copiously produced, but very slowly decaying. This enormous time scale difference, of order 10^{-12}, was difficult to reconcile with other experimental evidence and the known and accepted rule of conservation of space-time symmetry in particle interactions, what was already known as the CPT theorem. CPT stands for Charge Parity and Time transformations of a system of elementary particles and had been first stated by (Schwinger 1951) and Lüders (Luders 1954), followed by Wolfgang Pauli. The three transformations can be applied separately, or in separate combinations. Touschek had worked on Time reversal, and was obviously very interested in the question.

After the individual talks, an extensive discussion involved all the major theorists of the time, such as Richard Feynman and Murray Gell-Mann. Oppenheimer wrote: 'In connection with Gell-Mann's idea assuming that CP commutes with the electromagnetic field, Yang felt that so long as we understand as little as we do about the $\tau - \theta$ degeneracy, it may perhaps be best to keep an open mind on the subject'. Pursuing the open mind approach, Feynman brought up Block's question: could it be the θ and τ are different parity states of the same particle which has no definite parity, i.e. that parity is not conserved? Yang answered Feynman's question, by saying that he and T.D. Lee, had looked in this matter without arriving at a definite conclusion (Oppenheimer et al. 1956, 27). Lee, also present at the session, was a young theoret-

---

[31] The episode is narrated in Feynman's book *Surely You're Joking, Mr. Feynman!*, (Feynman and Leighton 1985). A slightly different version by M. BaldoCeolin, in https://physicstoday.scitation. org/doi/full/10.1063/PT.3.3336 points out that Martin had the gift of asking the right people the right questions at the right time. The same episode was also told to Y. Srivastava and through him to the Author, who has been Martin Block's close friend and collaborator in a numer of scientific papers.

ical physicist, who had graduated with Fermi in Chicago, and had joined Columbia University in 1953.

It was a lively discussion, as it still transpires from the Proceedings almost 70 years later. 'It is now time to close our minds...', said Oppenheimer to put an end to the discussion and move on to the other talks of the session. Lee and Yang had been thinking about the problem, and did obviously not follow the chairman's advice. Two months later they submitted an article to *The Physical Review*, where they put forward the possibility that some invariance properties of elementary particle systems should actually be checked. They envisaged that parity transformations— mirror images—may not leave invariant a system of weakly interacting particles and suggested a number of experiments to prove (or disprove) this hypothesis. The paper was published in December (Lee and Yang 1956). During these months, following intense exchanges with the two theorists, a group of experimenters at Columbia University, led by C.S. Wu, designed and performed the experiment which provided 'unequivocal proof that parity is not conserved in β-decays' (Wu et al. 1957).[32]

Weak interactions became the new frontier in particle physics. In the trinity of particle interactions, meson physics was exciting but like a puzzle yet to be put together and offered too many clues, electrodynamics was triumphantly used to calculate all kinds of reactions and ideas floated around to prove its limits, weak interactions were both mysterious and simpler to analyse. At the heart of weak interactions, the absolute protagonists are the neutrinos. Lonely stars in the particle world which partake of neither strong nor electromagnetic exchanges, their world may appear easier to explain. Bruno had already looked into the $\tau - \theta$ puzzle without any remarkable insight, but was present at the discussion in Brookhaven, and the discovery of non-conservation of parity in weak interaction represented a turning point in his physics interests. One of his earliest works in particle physics had been on double β-decay, (Touschek 1948b), a subject suggested by Heisenberg, and, before him, by Paul Urban. Later, a Nuffield lecturer in Glasgow, he had prepared the appendix about weak-decays for Born's *Atomic Physics* book, and had been recently preoccupied with time reversal properties. Thus, during the months after the Conference and through early 1957, Bruno's attention turned to the properties of neutrinos.

9.3.1.2 Bruno looks at neutrinos

Returning to Rome after the conference, Bruno began thinking along new lines. By the time he saw Wu's experimental results announced in the January issue of *Time* magazine, even before the *Physical Review* with Lee and Yang's article reached Rome, he had already began working on parity conservation and the mass of the

---

[32] Lee and Yang shared the 1957 Nobel prize 'for their work on the violation of the parity law in weak interactions'. The experiment became universally known as "M.me Wu's experiment".

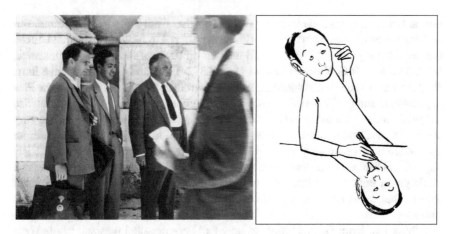

Fig. 9.6 Bruno Touschek with T.D.Lee, Wolfgang Pauli and Robert Marshak (in profile) at the September 1957 Padova-Venice International Conference on "Mesons and Recently Discovered Particles" from (Ceolin 2002). At right, a cartoon mimicking the non-conservation of parity under mirror reflection, in a portrait of T.D. Lee, by Bruno Touschek, Family Documents, © Francis Touschek, all rights reserved

neutrino.[33] By April 1957, he had written three papers, which he deemed 'very important', as indeed they were (Touschek 1957a, b; Radicati and Touschek 1957).[34]

In September, he attended the 1957 Padova-Venice International Conference on "Mesons and Recently Discovered Particles", where Bruno started a life-long friendship with T. D. Lee, humorously depicted in a cartoon, in a not-quite mirror image, Fig. 9.6.

9.3.1.3 A position Bruno refused: why?

While Bruno theoretical thinking had taken a new road, where particle, antiparticles, symmetry properties and invariance laws were dominating his attention, his professional future was still quite uncertain. His friends and collaborators, such as Morpurgo or Radicati, were appointed professors in various Italian universities, but this road was closed to Bruno because of his Austrian citizenship. Ferretti had left (for Bologna) and Marcello Cini was rumoured to be soon called to fill his chair in Rome. Bruno felt his own position was becoming of lower rank that all of them, particularly in Cini's case, whom he liked and had known since his first trip to the Testa

---

[33] In (Touschek 1957a) Touschek cites the January issue of Time magazine where Lee and Yang's work is mentioned, an indication that his paper was prepared before Wu's experiment was published. As for Lee and Yang's article, he had probably seen its preprint, distributed, as was the custom, to interested groups in major international laboratories and universities.

[34] Letter to his parents on Holy Saturday, April 7th, 1957.

Grigia laboratory, when they have travelled together from Turin back to Rome. But nothing could be done in Italy. A position was offered to him in Geneva, at CERN, but he did not pursue it. As it would happen other times in his life, for instance, later, with a chair in Vienna (Amaldi 1981, 23–24), Bruno hesitated to leave Italy. He gave various reasons to such decisions. He wrote to his parents that with the Pisa appointment and the Rome salary he would earn just as much, and it would be unwise to spend money on a move to Geneva, a city that neither him nor Elspeth liked very much. But one can wonder as to the real reasons not to accept the CERN offer he mentioned to his parents.[35] It could not have been because of money, as a CERN stipend and future prospects were well known to be highly superior to whatever one could get in Italy. Was it that he may have wanted a faculty position? Possibly, but the fact that, years later, he refused a chair in Vienna, makes this explanation just as unlikely.

I think that Bruno's attachment to Italy had very personal reasons, whose roots were laying in his rarely mentioned past. As a child, he may have travelled to Rome with his mother to visit Aunt Ada, his mother's sister. Later, he spent there after-school vacations. The deepest attachment we carry through our lives had been severed by his mother's death when he was just 9 years old, and those early memories of travelling to Rome deeply influenced him. He may have felt that, with Elspeth, his life was now being put together again, in a country which is not difficult to love. The beauty of the country side, shaped by centuries of civilization, and a peninsular climate which allows preservation of memories and artistic wonders, help hiding the negative sides which one can encounter in abundance. While, at the end, Bruno chose to die in his native country, where his parents and family had spent their lives, in 1957 he was still young and sufficiently confident in his possibilities. What may have tipped the balance is that Italy is beautiful. He decided to stay in Italy.

A posteriori, he may have made the right decision. In a few years, he would build something for which he will be remembered in the history of 20th century particle physics. The final place to rest, was far away in the future.

9.4 Physics with a Baby in the House

Starting from 1957, after Bruno Ferretti had left and Marcello Cini had just arrived in Rome to occupy his chair in theoretical physics, Bruno had been asked to be thesis advisor for a new crop of physics students. The first three were Francesco Calogero, Nicola Cabibbo and Paolo Guidoni.

Bruno's intellectual and training capacities were becoming well known and appreciated outside Rome. Bruno was giving lectures in Pisa, and asked to organize the 1958 Varenna Summer School on Pion Physics, sponsored by the Italian Physical Society, held at the beautiful Villa Monastero, overlooking Lake Como.

---

[35] Letter to his parents on Holy Saturday, 1957, April 7th.

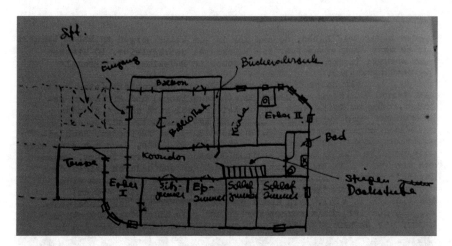

Fig. 9.7   Floor map of the apartment where the Touschek family moved in Spring 1958, from June 5th 1958 letter to parents, Family Documents, © Francis Touschek, all rights reserved

His personal life was on the happy side. In January 1958, his first son Francis was born. His lectures at University of Pisa and the INFN salary, implemented by visits to CERN, allowed for easier living and, in May, the little family, Bruno, Elspeth and the baby moved from the small two room apartment near Piazza Vescovio to a beautiful new place, spacious and nearer to the University. By June 5th, the main works to make the apartment ready for the family to move in, were behind. He enthusiastically described it to his parents, sketching its layout, Fig. 9.7. The house where they moved is on the corner of Viale Regina Margherita and Via Morgagni. A 10 minutes walk would take Bruno to the Physics Institute, and a long tram running around the city center allowed reaching almost any place in Rome without changing. At the time, there were two versions of this tram which ran external to the ancient city walls: they were called *circolare nera*, the inner one, and *circolare rossa* the outer one, going clockwise or anti-clockwise around the city center.[36] Because of the tram, the traffic was a bit loud, but the convenience, the space in the apartment, the terrace, the pleasant and welcoming neighbours made it a minor obstacle. The location and the space made it easier to invite friends and visitors to share dinners and hold scientific or literary discussions, on the terrace, during the hot summer days.

Shortly after Touschek and his family had settled in the new apartment, the time came to make plans for the summer. The major event ahead was the organization and direction of the School in Varenna at the end of July. Letters had been written to invite some 10 lecturers, 20 listeners and 50 students, a rather unusual organizational task for Bruno. Travel and an extended period away from home with Elspeth and the baby had to be arranged. In the Varenna School picture, Fig. 9.8, one can see

---

[36] 'Una interna, denominata circolare nera, ed una esterna (circolare rossa, completata nel 1931) concentriche tra loro, che, nei due sensi di marcia, circoscrivevano l'ambito dei rioni e quello dei primi quartieri', from https://it.wikipedia.org/wiki/Rete_tranviaria_di_Roma.

SOCIETÀ ITALIANA DI FISICA
SCUOLA INTERNAZIONALE DI FISICA "E. FERMI"
IX CORSO - VARENNA SUL LAGO DI COMO - VILLA MONASTERO - 18 - 30 Agosto 1958

Bruno Touschek, director of the course "Fisica dei Pioni", indicated by the circle in the first row.

Fig. 9.8 Course group-shot from the Proceedings of the International School of Physics "Enrico Fermi", Course IX, "Fisica dei Pioni", edited by B. Touschek (1958). Elspeth is seen at right, with Francis in baby carriage. Pedro Waloschek is in the last but one row, right of the tree. © Societá Italiana di Fisica, reproduced with permission

Bruno at the center, and Elspeth with baby Francis in an open carriage at the far right. Next to Elspeth one sees Walter Thirring and his family. Nicola Cabibbo, Bruno's first student, with dark jacket, is in second row to Bruno's right. This photograph includes Pedro Waloschek, the experimental physicist who would later become Rolf Widerøe's biographer.[37] In (Waloschek 1994, 3), Pedro remembers having first heard of Widerøe through Bruno Touschek in Varenna in 1958, although in (Waloschek 2012) he placed the conversation in 1959. They certainly met in 1958, perhaps the conversations continued in 1959.[38] Waloschek (1929–2012) had lived in Argentina since 1937, as his father, the German architect Hans Waloschek (1899–1985), had to flee from the Hitler regime via Vienna at the end of 1933. In 1957 Pedro was a research assistant in Bologna, and later worked for a number of years at the Frascati National Laboratories.

It was the first time Bruno had participated to the Varenna schools. Bruno, Elspeth, the 5 month old Francis and the babysitter, arrived on July 29th, and returned to Rome almost in mid September, most of the expenses paid by the School. His commitment was first to attend the Course on Mathematical Methods, and give a lecture, invited

---

[37] In the Conference Proceedings, Pedro Waloschek is wrongly identified as being directly above Touschek, standing on fourth row, but contemporary photographs of Rolf Widerøe, courtesy of his daughter Karen Waloschek, identify him as indicated in the figure caption.

[38] Memories may not always be truthful. Mention of the 1959 School appears in (Waloschek 2012), about death-rays, published posthumously, in 2012. Waloschek may have been confused about the exact year the conversation took place.

by Pauli, and then to be the director in charge of the Course which followed it. The Villa Monastero, where the School courses were held, was located on Lake Como, one of the lakes of glacier origin in the Lombardy side of the Alps. The beauty of the place was overwhelming, with a terrace with elegant flowering plants alongside the waters, and vines covering the walls in a garden about a kilometre long (but only about 20–30m wide) with cypresses, pines, Greek fountains, 'not so Greek fountains', bay windows looking out over the water, Fauns, Dianas, 'aloes that poke you in the knee (some even in the eye) at night, spiders and even a scorpion in the office: quite fairytale-like'.[39] During the intervals between morning and afternoon lectures, Bruno would swim in the lake, trying to catch the local fish lurking in the lake deep waters.

The return to Rome saw visits from Bruno's parents as well as from Elspeth's father, on his way to some conference in Nairobi. The new apartment was quite larger and it made for pleasant and easy hospitality. In addition, there was even a study room, where Bruno could work undisturbed, while Elspeth used the new living room instead of sitting on the bed when Bruno wanted to work. Money was much less of a problem, and the apartment could slowly be furnished, allowing for an intense social live, with friends for dinner, such as the Amaldi's and Bruno's student Cabibbo, and overnight visitors, such as Bruno's friend and collaborator, Luigi Radicati. Radicati had a teaching assignment in Rome, just like Bruno had in Pisa, where he had to give three lectures per week, every two weeks. When in Pisa, he would stay with the Radicati's, in a very practical reciprocal arrangement.

9.5 Pauli is Gone

December brought sad news to Bruno. On the 16th he received a telegram that his friend and mentor Wolfgang Pauli has died from of a very bad pancreas cancer.[40] Pauli was really sick for just about a week, and had been active and interested in physics until the last moment. Bruno felt this to be the worst loss physics had had in a long time. In a letter home, Fig. 9.9, he remembered that Weisskopf used to call him the conscience of theoretical physics. Bruno attended the funeral in Zurich on December 19th, a very painful affair, with friends like Houtermans, Thirring, father and son, attending and in tears after the funeral. Without him, physics for Bruno suddenly felt to be only half as interesting. As it often happens with one's mentor, Bruno realized that sometimes he only did it so that he could talk to Pauli about it, behind a glass of wine in the summer at a lake or somewhere in Italy.

Bruno had become friends with Pauli during his first winter in Rome, as he had arrived from Glasgow. And then, not long after, in Cagliari (Sardinia). They met at all the congresses of the Italian Physical Society (SIF) of which he was a member,

---

[39] Letter to parents from Villa Monastero, September 2nd, 1958.

[40] Pauli died on December 15th 1958, although, in a letter to his parents on December 24th, Touschek gives the date as December 14th.

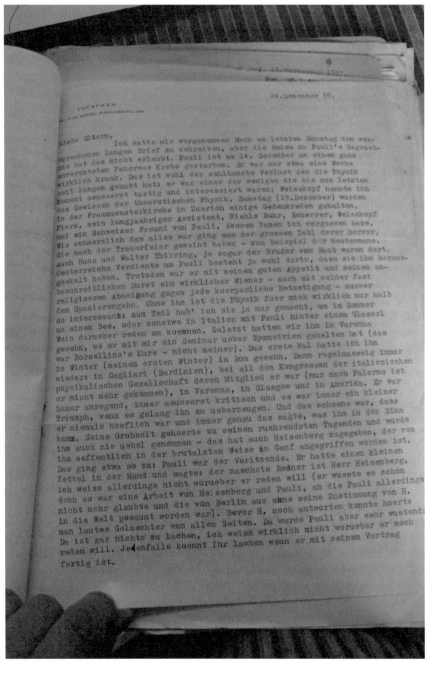

Fig. 9.9 Bruno Touschek's letter to his parents from Rome, announcing Wolfgang Pauli's death.
Family Documents, © Francis Touschek, all rights reserved

in Venice in 1957, Fig. 9.6, in Glasgow and in America, in 1956 at the Brookhaven conference. Bruno saw him as always stimulating, always extremely critical and it was always a small triumph when one succeeded in convincing him. He had recently seen him in Varenna, in July. A nice thing which Bruno appreciated was that Pauli was never polite and always said exactly what came into his mind. His rudeness was one of his most touching virtues and was never resented. Writing to his parents,[41] Bruno recalls an anecdote about Pauli in Geneva, in July 1958 (Cassidy 1993, 543):[42]

```
Pauli was the chairman. He had a small piece of paper in his
hand and said 'the next speaker is Mr Heisenberg, but I
don't know what he wants to talk about'. He actually knew it
because it was a work by Heisenberg and Pauli, in which
Pauli no longer believed and which H. had announced to the
world from Berlin without his consent. Before H. could
answer, loud laughter could be heard from all sides. But
Pauli, obviously, got very angry: there is nothing to laugh,
I really don't know what he wants to talk about. In any
case, you can laugh when he's finished with his lecture.
```

When Pauli died, Bruno's physics interests started changing. The scientist who fuelled so much of Bruno's involvement in theoretical physics since he had arrived in Rome, even in those many evenings or days when Pauli was not really with him, his close friend, was gone, and so did some of Bruno's passion for theoretical physics. He was now alone: without Pauli, he saw the need to move on to something different, which he had known and also understood well, and could feel to be his own creation. As Bruno later wrote:[43]

```
At the time [when the synchrotron started working] I felt
rather exhausted by an overdose of work which I had been
trying to perform in the most abstract of field of
theoretical research: the discussion of symmetries which had
been opened up by the discovery of the breakdown of one of
them, parity, by Lee and Yang. I therefore wanted to get my
feet out of the clouds and onto the ground again, touch
things (provided there was no high tension on them) and take
them apart and get back to what I thought I really
understood: elementary physics.
```

As one reads these later day notes, one is struck by Touschek mentioning 'elementary physics', which sounds like a reminiscence of his criticism of Rolf Widerøe's colliding particles idea: had he not said, according to Widerøe, that the idea was trivial, one that every school child would know about? Now perhaps it was the time to make reality of that old idea.

---

[41] Letter to parents on December 24th, 1958.

[42] The episode and the scientific exchanges between Pauli and Heisenberg which led to it, are described in Cassidy's work. Touschek's description is contemporary and points to Pauli being deeply offended by the public discussion.

[43] Undated typewritten notes, probably preparation of the talk delivered at Accademia dei Lincei, on 24.5.1974. Reproduced courtesy of Edoardo Amaldi Papers in Sapienza University of Rome–Physics Department Archives, https://archivisapienzasmfn.archiui.com.

Fig. 9.10 Elspeth Yonge Touschek with son Francis during the 1987 Bruno Touschek Memorial Lectures, held at the CNEN site of the Frascati National Laboratories. Photograph by the Author, all rights reserved

His imagination went back to particle accelerators, when, on his way to aunt Ada's house near the Tusculum hills and swimming expeditions to the Albano lake, he would occasionally drop by the nascent Frascati Laboratories and wonder about all the excitement. At the end of 1958, when the beam was finally detected, running a full circle through the vacuum chambers, Bruno's interest in the new machine grew. At the same time, Bruno was also still looking for some better position, now that he was a husband and a father, Fig. 9.10. There were exchanges with old Thirring in Vienna, but efforts from both sides appear half-hearted, reading the letters Bruno was sending to his father. Then, sometime in Spring 1959, the possibility that Bruno took charge of a theoretical physics group in the new Italian national laboratories, in Frascati, was vented.

Chapter 10
1959–1961: The Making of AdA

Ada the storage ring will be waiting for me on Monday when I return to Frascati. But then the peace is gone. In November, December we have to measure the magnet and in January it gets really serious. From the reports of my spies I learn that we are really the first in this area: the American competition won't bother us for a year.

Ada der Speicherring wird mich am Montag bei meiner Rückkehr in Frascati erwarten. Dann ist allerdings die Ruhe hin. November, Dezember müssen wir den Magneten ausmessen und im Jänner wird es dann wirklich ernst. Von den Berichten meiner Spione erfahre ich, dass wir auf diesem Gebiet wirklich die ersten sind: die amerikanische Konkurrenz wird uns ein Jahr lang nicht zu schaffen machen.

Letter from Bruno Touschek to parents, Rome, November 6th, 1960.

Abstract How AdA arrived to represent worldwide the prototype of future accelerators in the quest for higher energies in particle collisions, is a story which can be divided into three acts and a preamble. The preamble was staged in places which were thousands of kilometer apart, the Universities of Stanford and Princeton in the USA, the University of Rome and the Frascati Laboratories in Italy, the Institute of Atomic Energy in Moscow and the Institute of Nuclear Physics in Novosibirsk. Developing between July and October 1959, the story picked up momentum in the last months of 1959, when Bruno Touschek and other theorists in Rome began thinking of what could one learn from the physics of colliding beams. Then, on February 17th, 1960, the first act starts, when Touschek proposes the construction of an electron-positron ring in Frascati. From this date in February 1960 until the summer of 1961, AdA takes central place in a stage now set in the Frascati Laboratories, in Italy.

In the two years 1959 and 1960, an epochal change in accelerator physics took place. Nothing would be the same any more. In a few years, particle colliders would become a major discovery tool in particle physics, allowing for the experimental confirmation of the Standard Model of elementary particles in the second half of the 20th century.

© The Author(s), under exclusive license to Springer Nature Switzerland AG 2022   297
G. Pancheri, *Bruno Touschek's Extraordinary Journey*, Springer Biographies,
https://doi.org/10.1007/978-3-031-03826-6_10

Such change was initiated by Bruno Touschek, when he conceived and drove to success the construction of the first ever electron-positron collider AdA.

As always, one can wonder about alternative scenarios: one could ask how would have physics evolved if electron-positron colliders had not been proposed in Italy in February 1960. Or in which direction electron-electron colliders, the way chosen by the Americans, would have gone. What would have happened in particle physics if the search for possible breakdown of Quantum Electrodynamics (QED) had continued for years, in a sort of search for the Holy Grail? More realistically, what if the Soviet Union had been able to build and operate electron-positron colliders before the Western countries? This hypotheses cannot be discarded. According to the Russian physicist V.N. Baier (Baier 2006), in October 1959 a visit to Novosibirsk in Siberia by Isaak Pomeranchuk had the effect to start electron-positron accelerator studies in the USSR. Baier clearly says that hearing of the Frascati results in 1961 strengthened the resolve of the Russian physicists to work in that direction, so that in 1963 an electron-positron collider, VEPP-2, in advanced stage of construction was shown in Novosibirsk to a selected group of Western scientists (Marin 2009). Things went otherwise, as we know, and when VEPP-2 started working in 1967 (Skrinsky 1995; Auslander et al. 1967), its initial successes were soon shadowed by the more powerful colliders built in Europe and the United States, encouraged by AdA's proof-of-principle of electron-positron storage rings feasibility. The ifs and buts could make for a science fiction story, but the questions imply that the construction of AdA was a pivotal moment of historical importance. There was a combination of facts, a joining of roads, which had AdA conceived and built in Frascati. Among them, perhaps foremost, is that the creation of AdA could only happen through Bruno Touschek. In Bruno, one can see the perfect synthesis between hands-on experimentation and theoretical knowledge, youthful experiences and their transformation into a world breaking record. The special momentum in Italian scientific recovery from the disasters of the war, did the rest.

Whereas AdA's early construction days and first successes are well known, both from (Amaldi 1981, 27) and (Bernardini 2003; Bonolis and Melchionni 2003; Bernardini 2004), the creation of AdA has often been presented in the literature as something suddenly happening between February and March 1960, and not much has been published about the year in Touschek's life which preceded AdA's proposal in 1960. However, AdA's story really starts in 1959, and AdA's success can only be gauged by highlighting its non casual genesis through different roads. Among those which led Touschek to propose the construction of AdA, there are encounters and conferences Bruno attended in 1959 and his intellectual evolution after emerging from two personal losses. This alone however would not have been sufficient without other favourable circumstances. Of particular importance was the presence in Rome and Frascati of young theorists and courageous experimenters ready to follow Bruno through the most creative period in his scientific life.

In addition to known sources, the outline of this process is based on Touschek's letters to his family, complemented by later writings about this period and memories of some protagonists of the story, laced together here for the first time. In this narration, the aim is to highlight that AdA was not just a 'curiosity' (as Burton Richter later called it) born out of a lucky idea which also led to ADONE, but represented the confluence of many events and personalities, of which Touschek was the catalysing center. Of special interest is also to see how a group of scientists in a large institution took a crucial decision such as the construction of a totally new type of accelerator, dedicating funds and human resources to it. This decision time will be highlighted from minutes of the meetings where AdA's construction was decided.

10.1   Aunt Ada

January 1959 was a month of multiple birthdays with Bruno's parents, father and step-mother, both celebrating it. In 1959, there was one more family birthday, as Bruno and Elspeth's son Francis turned one year old and took a central place in their life.

Bruno was very busy, teaching both in Rome and Pisa, and was often invited to give talks about his work on neutrinos. In April he had to go to Naples to lecture at the Spring School in theoretical physics, organised by Edoardo Caianiello, Fig 10.1.[1] This School was part of a Graduate Course in Theoretical and Nuclear Physics, which had been established in Naples upon Caianiello's initiative in 1956–57. Caianiello had also solicited the opening of an INFN branch, which would acknowledge the existence of a strong nuclear physics research activity south of Rome and further reinforce it (Battimelli et al. 2001, 132).

Bruno's summer plans included lecturing in the Varenna School on 'Weak Interactions'. In the previous summer, Bruno had been Director of one of these schools. In 1958, the organization of the School had meant a heavy toll of administrative work for Bruno, but this year Varenna just entailed physics. The invitation was both prestigious and convenient. It allowed Bruno to escape the summer heat in Rome, and cover part of travelling costs back and forth to the North of Italy, where summer vacations always included a stay in the Dolomites or trips to see his parents in Tyrol, or Vienna. With a growing family, money matters had to be kept in mind.

He was also busy with INFN meetings, where scientific policy matters were discussed. Spring 1959 was to see a major INFN effort come to fruition, since, in Frascati, the synchrotron was coming to life. In addition to organizing experiments and official inauguration events, discussions were starting about the future of the laboratories. INFN president Gilberto Bernardini and Edoardo Amaldi, who had set in motion the great adventure of building an Italian accelerator and a national laboratory, were keen to plan ahead and maintain the momentum to Italian physics. Bruno as a major ranking INFN researcher and a rising star in theoretical physics was called to take part in these discussions. When INFN meetings were called on

---

[1] Eduardo R. Caianiello (1921–1993) was a mathematical physicist, who introduced the field of cybernetics in Italy. He was Professor of Theoretical Physics in Naples and was instrumental in establishing the Institute for Theoretical Physics in 1957 at the site Padiglione n.19, Mostra d'Oltremare.

Fig. 10.1 Eduardo R. Caianiello from www.treccani.it/enciclopedia/eduardo-renato-caianiello_ (Dizionario-Biografico)/, and the Institute for Theoretical and Nuclear Physics at the Padiglione n. 19 of Mostra d'Oltremare in Naples, in 1962, from https://biblioteca.fisica.unina.it/biblio/index. php?it/134/memorie-fotografiche., © 2010 Biblioteca "Roberto Stroffolini" del Dipartimento di Fisica "Ettore Pancini", Università degli studi di Napoli "Federico II"

a Sunday, Bruno thought it was getting too much, in his frustration calling them a 'Sunday School'.[2]

In March, Bruno was informed of an important invitation to attend the *Rochester Conference*, taking place in Kiev, in July. This series of conferences had designated locations agreed to be in the USA, in the New York area in particular, at CERN and in Moscow in the USSR.[3] In 1959 the conference site was moved from Moscow to Kiev, in Ukraine, at the time part of the USSR.[4] Participation was restricted by national quotas, in an even share between Americans and Europeans, 60 each. Bruno was very proud of the invitation to participate and give a talk on his work on neutrinos, looking forward to the trip, which would bring to Kiev many Italians, with as many as seven participants only from the University of Rome.

His travel to Kiev was put in doubt in May, by Aunt Ada's death. She was 84 years old. Bruno's visits during his first years in Rome had naturally become less frequent as his marriage and family obligations grew, a fact compounded by travels for teaching between Rome and Pisa. On March 26th, he had gone to see her after she had a stroke, on May 27th she passed away.[5]

---

[2] Letter to parents, March 28th, 1959.

[3] The Rochester conference, nicknamed as such from the location of its first occurrence, is now called 'International Conference in High Energy Physics', ICHEP for short. Initially held annually, it became biannual in 1960. Its location rotates through many different world centers, including China, India, Japan, etc.

[4] Not long after the Kiev conference, the USSR announced the launch of Lunik II, on September 12, 1959. The rocket reached the Moon 36 hours later. The event was announced during the 2nd International Conference on High-Energy Accelerators and Instrumentation, HEACC 1959, taking place at CERN, 14 - 19 Sep 1959.

[5] Letter to parents, May 28th, 1959.

In the brief span of five months, from December to May, Bruno had suffered two crucial losses in his life. Of Pauli, we have said. Bruno mourned for him internally, wondering for his interest in theoretical physics to return or other directions to follow. As for Aunt Ada, the loss could be expected given her age, but it ran deep into Bruno's mind.

Ada Weltmann Vannini had been a constant presence in Bruno's life. Bruno's closeness to her, during his early years, is likely to have been one of the reasons why he had moved to Rome in 1952. The memory of those distant days had comforted him after the war, through the Glasgow winters, when he would dream of spending time in the Rome country side, where, with her husband, she would build a villa. Now the villa was going to be sold, and a burden of legal and administrative work fell on Bruno's shoulders. The disentangling of inheritance proceedings and taxes to be paid took a good part of the year, as, in addition to the villa, there was the question of the firm *Garvens*, the pump business which had belonged to Aunt Ada and her husband. After Ada's death, Bruno and Elspeth took over its running. When later, in 1974, Bruno published his last physics article, he gave his affiliation as Garvens s.p.a., Piazza Indipendenza, Rome, the address of the hydraulic pump business owned by Aunt Ada and her husband (Touschek 1974). It has been said that this affiliation was a protest against the disruptions taking place at the University after 1968.[6] This address, considered by Bruno's friends as one more of his puns, may have instead been a personal, last tribute to the aunt who had been a refuge to the young orphan.

10.2   Summer 1959: Evenings in Varenna

In those days without general air-conditioning, the heat in Rome made many families leave the city by the end of June. For Bruno, the invitation to lecture at the SIF School in Varenna was the ideal solution, fitting well with his summer plans. From Varenna he would then go to Vienna and Kiev, where he had now confirmed his presence. He had to return to the University because of the summer examination session, the *Sessione estiva*, and then he could go back up North for the usual vacation in the Dolomites in August. His favourite place was Oberbozen, a location in a wide plateau on a mountain ridge which rises above the town of Bolzano (Bozen), up to over 1000 meter altitude, and then, climbing higher up, offers spectacular views of the Dolomites. One of Bruno's early vacation spots after he moved to Rome, Oberbozen is easy to reach from Bolzano by cable way, with trails across meadows and woods, such as *Il sentiero di Sigmund Freud*, in memory of its most famous visitor.

The 1959 Varenna Summer School on 'Weak Interactions' took place at the Villa Monastero on the shores of Lake Como from June to July (Bhattacharjee 1960). Bruno gave lectures on the theory of the neutrino he had developed together with

---

[6] Starting with the first occupation of University of Rome in 1966, student protests turned violent in 1968, when students and police clashed. A period of disruptions and general political fighting followed, amid political killings and terrorist bombings, including, in 1978, Prime Minister Aldo Moro's death by the extremist movement of the *Brigate Rosse*, the Red Brigades.

Fig. 10.2 Pedro Waloschek in the 1950s, courtesy of Karen Waloschek, all rights reserved. At right, a view of Villa Monastero in Varenna, the site of the Italian Physical Society Summer School, where Bruno lectured in 1958 and 1959, and where Bruno and Pedro met and had occasion to talk about Rolf Widerøe, https://www.primapagina.sif.it/article/885/online-i-nuovi-corsi-di-varenna-gennaio#.YYuHU73MJE4

Morpurgo and Radicati.[7] The beauty of the School location was a major attraction for Bruno, and he was looking forward to escape the summer heat, which grips Rome from June to September.

In 1958, the future biographer of Rolf Widerøe, Pedro Waloschek, at the time a young particle physics experimenter, had been among the participants, Fig. 10.2. He was of mixed parentage from both his maternal and paternal side. From Waloschek, one has many memories of Bruno (Waloschek 2004). In 1987, he contributed a story about Bruno's Hamburg days during the war:

> *Waloschek:* …Hamburg, I think, is the prettiest city of Germany. The first good thing I want to remember of Bruno is that the Professors at the University of Hamburg gave him the opportunity to attend lessons, an opportunity he did not have in Vienna. …he gave lots of trouble to his professor there: one of the things I was told in Hamburg is that when he came in the classroom the first thing he did was to turn head down the picture of Hitler. So that made life not easy with Bruno. He was not only original, he also had an enormous courage because at that time it was in the middle of the War. It was really risking your head.

A second memory comes from Waloschek's last book (Waloschek 2012, 106), and sheds light on Bruno's path to AdA. Waloschek remembers meeting Bruno in Varenna in 1959:[8]

> *Waloschek* …] I knew quite well the temperamental and sympathetic Bruno Touschek (1921-1978) and therefore I would like to report a bit more about him. We met several times in physics conferences and summer schools. Being both originally from Vienna we used to have longish evening chats (in our dialect) in general with some good food. It was

---

[7] Schools with different topics are organized each summer by SIF. Luigi Radicati was the designated Director for this particular one.

[8] Waloschek is not listed as one of the participants to the 1959 Varenna School (Bhattacharjee 1960), whereas he was a participant in 1958. In (Waloschek 1994, 3), he remembers the conversation as having taken place in 1958.

in 1959, in a restaurant at the Lake Como, when Touschek told me about a strange Norwegian engineer who wanted to publish a scientific report on an idea which he (Touschek) considered as absolutely obvious and which could be concluded from data you find in any normal physics school-book. The engineer was not happy and decided to apply for a patent on his idea.[9]

Touschek was very disappointed but he could not avoid it. As Touschek told me in our meeting, the strange engineer was Rolf Widerøe and the discussion took place in Hamburg during World War II.

In Varenna Bruno Touschek and Waloschek had continued to speak at length about Widerøe's genius and Touschek's mind went back to those days and to the idea of center of mass collisions. Thus Widerøe's ideas about colliding particles may have come back to Bruno, while in Varenna he was looking ahead to the next big trip, to the USSR, where these ideas and concepts were going to be reinforced.

10.3 Conferences and New Entries: Positrons and Electrons

After the days in Varenna, Touschek left for Kiev, to attend the Ninth International Conference on High Energy Physics, to be held between July 15th and 25th, and where advanced American plans for electron-electron center of mass collisions were going to be presented by Wolfgang Panofsky, from Stanford.

Since the presentations by Gerard O'Neill and Donald Kerst at the 1956 CERN Conference (Regenstreif 1956), center of mass collisions, as a possible road to reach higher acceleration energies, had gained some momentum. Proposals had been developing in parallel to the more conventional approaches, which in summer of 1959 could boast a remarkable string of successes (Sands 1989). Just coming to life in Europe, there were two proton synchrotrons (PS), one at CERN, and one in Saclay, and two electron accelerators, the LINAC (linear accelerator) in Orsay and the (circular) synchrotron in Frascati. Since 1957, in the USSR, a proton synchrotron, the Synchrophasotron in Dubna, had been in operation, designed and constructed under the supervision of Vladimir Veksler (Baldin and Semenyushkin 1977). The range of reachable energies and the diversification in machine types was a promise of success for the experiments to come. It was now the time for the accelerator builders to look ahead and plan for new machines.

Thus, as summer 1959 moved on, different roads started converging to make colliding beams the future discovery tool in particle physics.[10] In Varenna, where he had shared war time memories with Waloschek, Touschek may have been reminiscing of his days with Rolf Widerøe, his thoughts wandering back to the idea about colliding oppositely charged particles. Immediately following Varenna, the conference in Kiev brought up a different thread, with the presentation of an apparently minor experiment, performed at the Stanford Linear Collider, where the contribution of the

---

[9] This was in fact done by Rolf Widerøe, as seen in Chap. 5.

[10] See details in Chap. 8 about the CERN 1956 Conference.

annihilation of electrons and positrons was detected for the first time in a scattering experiment.

10.3.1   July 1959: To Kiev

Touschek's contribution to the Conference in Kiev was to be a talk on the neutrino gauge group (Touschek 1960). From Varenna, he went directly to Vienna, and from there, early in the morning of July 15th, flew to Kiev through Budapest and Lviv.[11] The trip was typical of those days. Some twenty people were seated in the small plane, among them Edoardo Amaldi, Giacomo Morpurgo, Richard Dalitz, Antonio Rostagni, from University of Padua and Italian representative on the Atomic Energy Commission in Vienna. It was a terribly noisy twin-engine plane, whose inside looked to Bruno like a second class train compartment. For reasons unknown to Bruno, but perhaps limited food supply, breakfast was served only to a fraction of the passengers.The plane was flying at an altitude of 1000 m, it was cloudy and there was nothing to see. After a brief stop in Budapest, they entered the USSR airspace: the next stop was in Lviv, Ukraine, where custom formalities naturally took an endlessly long time.

On the 16th morning, the actual conference started. The list of participants shows an exceptional roster. It included many of the protagonists of the new field of particle physics, as it would soon be called, among them present and future Nobel prize winners: Luis Alvarez, Lev Landau, Leon Lederman, Murray Gell-Mann, Werner Heisenberg, Robert Hofstadter, Yoichiro Nambu, Abdus Salam, Julian Schwinger, Emilio Segrè, Jack Steinberger. But the list of future protagonists in the new field of particle physics, is much longer: it includes Gersh Budker, Geoffrey Chew, Richard Dalitz, Sidney Drell, Isaak Pomeranchuk, to name just a few (Academy of Science USSR and IUPAP 1960).

From France there was Francis Perrin, from the French Commissariat pour l'Énérgie Atomique, in addition to George Bishop, Jeanne Laberrigue and Jean Teillac from the Orsay laboratory, where a linear electron accelerator, of future interest to AdA's story, had just been commissioned.

The conference saw the appearance of Bruno Pontecorvo, the Italian theorist, one of Enrico Fermi's collaborators, who in 1950, at the pinnacle of the Cold War, had fled from Italy to the Soviet Union. One of the last western conferences attended by Pontecorvo before disappearing to the USSR had been the 1949 Edinburgh Conference on Elementary Particles, which Bruno had attended and where Pontecorvo had given a talk about 'The decay products of the μ-mesons', (Pontecorvo 1947), possibly influencing Touschek's later work for the 1951 edition of Born's *Atomic Physics* book.[12]

---

[11] Letter to parents, July 29th, 1959.

[12] See correspondence with Born in Chap. 7.

Fig. 10.3 Present-day plan of the city of Kiev, with the river Dnieper, where Bruno went swimming during the Kiev conference in July 1959. At right, Wolfgang K. H. Panofsky, from Stanford, (left side with glasses) holding court with other lab directors during an excursion on the Dnieper, Kiev 'Rochester' Conference, Photograph by J.D. Jackson, courtesy of AIP Emilio Segrè Visual Archives, Jackson Collection, Catalog ID: Panofsky Wolfgang D19

From Italy, the participants included Gilberto Bernardini, Bruno Ferretti, Nicola Dallaporta, Carlo Franzinetti, Giacomo Morpurgo, Giampiero Puppi, Gleb Wataghin. In addition, from University of Rome, there were Edoardo Amaldi, Marcello Conversi and Giorgio Salvini, under whose direction the Frascati electron synchrotron had just started operating, and the theorists Marcello Cini, Raoul Gatto and Bruno Touschek.[13]

As it would often happen in the conferences organized in the USSR in those days, the participant had to suffer through non air-conditioned halls and incomprehensible simultaneous English translations. Sometimes the translations were not even properly synchronized, and the inconveniences hampered the listeners' concentration. July weather in Central Europe, in Kiev in particular, is hot and humid, making for uncomfortable, sleepless nights. Commonplace activities encountered many obstacles: one could not visit the countryside, as visas were only valid for the city of Kiev, and booking return flights implied long hours at the state agency. On Sunday, Bruno took to one of his favourite pastimes: swimming in the Dnieper, the city river, that runs from North to South and whose abundant waters allowed for Kiev to be called a city of gardens, Fig. 10.3. Always irked by bureaucratic restrictions to his freedom, Bruno was frustrated by the many technical or organizational difficulties, such as the non availability of simultaneous translation during the discussion periods, so that one had to ask for repeats, which took life out of the debate. When he left, directly for Vienna on a special plane provided by the Ukranian Academy, he thought that nothing particularly exciting in physics had been reported.[11] *A posteriori*, a reading of the Conference proceedings tells otherwise (Academy of Science USSR and IUPAP 1960). Two talks, one by Wolfgang Panofsky, Fig. 10.3 (right), and one by Hofstadter were to have a strong impact on his future.

---

[13] Amaldi does not appear in the list of participants, although Bruno mentions traveling with him to Kiev, see [11].

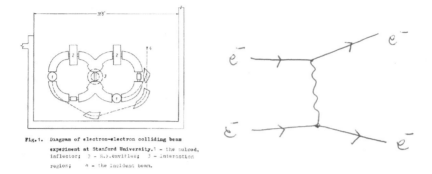

Fig.1. Diagram of electron-electron colliding beam
experiment at Stanford University.1 - the pulsed,
inflector; 2 - H.F.cavities; 3 - interaction
region; 4 - the incident beam.

Fig. 10.4 At left, the schematic description of the experimental set-up for the Stanford-Princeton project for electron-electron scattering from the Proceedings of the 1959 Kiev Conference on High Energy Physics (Academy of Science USSR and IUPAP 1960), presented by Wolfgang Panofsky. The interaction between the two electron beams is made to occur at the tangential point between the two rings. At right, the theoretical description from a Feynman diagram representation of the process, with the direction of time from left to right, drawing by the Author

10.3.2 What was New in Kiev

The Conference had opened with a mournful observation:[14] since the previous venue of the Conference, three grave losses had taken place, most recently that of Wolfgang Pauli, and, before him Frédéric Joliot's and Ernest Lawrence's.[15] A theorist, an experimenter and an accelerator physicist, who had been among the most prominent actors in the new field of nuclear science, were gone. An epoch was closing and a new era was ushered in.

The electron-electron project, which had been proposed by Gerard O'Neill during the 1956 Conference in Brookhaven (O'Neill 1956) had, since then, gained grounds. In Kiev, a progress report on the design of the Stanford-Princeton experiment by the method of colliding electron beams from two separate storage rings was presented (Panofsky 1960), Fig. 10.4, left panel.[16]

This part of Panofsky's presentation may not have surprised Bruno, who had been remembering Widerøe's old suggestion for colliding particles, just the week before in Varenna. The Stanford-Princeton experiment would be measuring the photon propagator in the space-like region, its ultimate goal being to check the validity of QED at the highest possible energy and also to probe the electron size, a question often debated among theoretical physicists, but not much news to Bruno.

---

[14] See Opening Session in (Academy of Science USSR and IUPAP 1960).

[15] They had passed away respectively on December 15th, August 14th and August 27th, 1958.

[16] The experiment was built at Stanford, with some parts prepared in Princeton, where O'Neill was based.

More interesting, both to him and to Raoul Gatto, may have been the second experiment presented by Panofsky, when he illustrated how the positron beam from the Stanford linear accelerator was made to scatter from electrons. The data from this Stanford experiment showed the interesting possibility of measuring positron-electron scattering in an accelerator set-up, and detecting the distinct contribution from the annihilation process, in which positrons and electrons exchange a virtual photon (in the time-like region, see box). The process had been calculated by Homi Bhabha and is known as *Bhabha scattering*, (Bhabha and Fowler 1936). The experimental results had just been submitted to *The Physical Review* (Poirier et al. 1960). For theoreticians, such as Bruno and Gatto, there was nothing new in the presence of annihilation in electron–positron scattering and this might also explain why Touschek felt that there was nothing particularly exciting reported at the Conference. However, this Stanford experiment showed the possibility to create a positron beam and scatter it against electrons: the annihilation graph had made its first appearance in a laboratory experiment! Bruno may have begun thinking of what one could discover in a set-up where electron and positrons were made to collide. In the intense atmosphere of a dedicated conference such as the one in Kiev, such thoughts may surge to later become reality (Fig. 10.5).

Another talk of interest for the three theorists from Rome was presented by Robert Hofstadter about studies of the electromagnetic nucleon form factor (Hofstadter 1960).[17] The probing went through a virtual photon emitted by the scattered electron, as shown in the left panel in Fig. 10.6.

Hofstadter's experiment showed discovery possibilities quite different from those accessible through electron-electron scattering, which would probe limits and possible breakdown of electrodynamics, but not the so called 'strong' interactions, which regulate the world of newly discovered heavier particles, such as pions and kaons, and their interactions with the nucleons. On the contrary, this was the world which Hofstaedter's experiment was accessing. But it was still framed in the context of fixed target accelerators, with their limitations about the range of energy available to the projectile. In a few months, Bruno would know how to solve this limitation. Once more, while Bruno did write to his parents that nothing exciting had happened in Kiev, the germ of how to go ahead was there, in devising a type of experiment where both electrodynamics and strong interactions would be accessed.

---

[17] For his experiments, Hofstadter was awarded the 1961 Nobel prize 'for his pioneering studies of electron scattering in atomic nuclei and for his consequent discoveries concerning the structure of nucleons'.

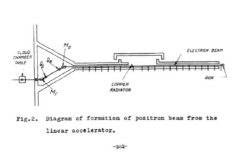

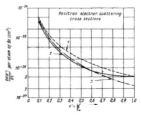

Fig.3.   Positron-electron scattering cross-sections.

Theoretical curves shown are:

1- Rutherford scattering; 2-Bhaba formula;

3 Bhaba formula without annihilation term.

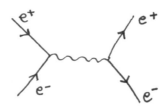

Fig. 10.5   The scheme of the formation of a positron beam from the linear accelerator in Stanford, and the cross-section plot for the positron-electron scattering, respectively Fig.(2) and Fig.(3) of Panofsky's talk (Panofsky 1960) at the Ninth International Conference on High Energy Physics in Kiev, 1959 (Academy of Science USSR and IUPAP 1960). The cartoon shows the Feynman diagram representation of the annihilation term contributing to the cross-section, time running from left to right, drawing by the Author

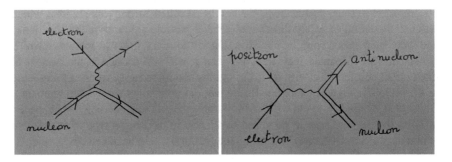

Fig. 10.6   At left, Hofstadter's experiment in its Feynman diagram representation of electron scattering with a nucleon, a proton or a neutron, with the time flowing from left to right. At right, a rotated version of the process, with the outgoing electron rotated counterclockwise to appear in the initial state: by virtue of a quantum mechanical property called 'crossing symmetry', the outgoing electron in the left panel becomes an incoming anti-electron, namely a positron, like-wise the initial state nucleon is rotated into an outgoing anti-nucleon, drawings by the Author

In the right panel of Fig. 10.4, the interaction between the two electrons is represented by a virtual photon (the wavy line), whose properties could be precisely investigated in the Stanford-Princeton experiment in the search for a possible breakdown of Quantum ElectroDynamics (QED), the relativistic quantum field theory for electron-photon interactions. The relativistic kinematic description of the scattering attributes an imaginary mass to the photon, which is then defined as being *space-like*. A similar description is given for Hofstadter's experiments in the left panel of Fig. 10.6, where the scattering particle is indicated by a double line, indicating some internal structure, As in the case of electron-electron scattering, the interaction takes place *via* a *space-like* photon. At right, one can see a counterclockwise rotated version of the process. By virtue of the quantum mechanical property called 'crossing symmetry' (Schwinger 1951), the outgoing electron in the left panel becomes an incoming anti-electron, namely a positron (e^+): likewise the initial state nucleon is rotated into an outgoing anti-nucleon. The interaction takes place *via* a virtual photon as well, but the different kinematics describing the annihilation between electrons and positrons leads to a positive mass for the virtual photon. Virtual photons with positive mass are called *time-like*.

Did Bruno pay attention to the different physics in the three experiments reported at Kiev? For a theorist, the three Feynman diagrams shown in Figs. 10.4–10.6 are very similar. In particular, Bruno's profound knowledge of particle symmetries, coupled to an extraordinary capacity to draw and make graphical puns, would make rotation of one diagram into the other, and envisaging the different physics opportunities, a very natural thing to do. In any case, if not at the time, what he heard in Kiev certainly came to his mind a few months later, when Panofsky came to Rome, visiting Frascati and giving talks on the Stanford two-mile Accelerator. The date of the Rome seminar is not known, but the visit to Frascati is proved by the list of seminars held during the academic year 1959–1960, with one talk by Panofsky taking place on October 26th, 1959, Fig. 10.7. This list of seminars had been kept by Vincenzo Valente, an INFN Frascati National Laboratories researcher, who extensively wrote about the history of the laboratories and was in charge of the Frascati Laboratories publication services for a number of years.

Through 1959, Bruno's involvement with Frascati had grown, both on professional and personal grounds. The electron synchrotron had started functioning on April 4th, 1959, Fig. 10.8, and was at the center of attention of national and international *media*, as well as of the scientific community. As trusted advisor for INFN scientific policy, he participated to the Institute planning and there was pressure on him to be involved in the future of the new Frascati National Laboratories.[18] In October, his increasing attention to what was taking place in Frascati was sanctioned by the official

---

[18] Bruno's participation to the INFN scientific advisory board, is mentioned in a January 16th 1960 letter to his father.

ELENCO DEI SEMINARI TENUTI PRESSO I LABORATORI NAZIONALI DI FRASCATI
DEL C.N.E.N. DAL 1°/7/1959 AL 30/6/1960. –

31/7/1959 – Prof. John DE WIRE   – "LAVORI IN CORSO A CORNELL CON L'E
                                          LETTROSINCROTRONE".

3/9/1959   – Prof. WIEGAND   – "ALCUNI RECENTI SVILUPPI DELLA SPE
                                          RIMENTAZIONE IN BERKELEY".

16/10/1959 – Proff. QUERZOLI,   – "POLARIZZAZIONE DEL PROTONE DI RIN
              SALVINI   CULO NELLA FOTOPRODUZIONE DEL π^o".

22/10/1959 – Dr. DIAMBRINI   – "SULLA ESPERIENZA SPETTRO   ".

26/10/1959 – Prof. W. PANOFSKY   – "SULL'ACCELERATORE LINEARE DA 2 MI
                                          GLIA".

Fig. 10.7 Detail of list of Frascati seminars for the year 1959–1960, courtesy of Vincenzo Valente
to the Author

Riproduzione del permesso che
consentiva a Touschek l'ingresso
nei laboratori

Fig. 10.8 An April 4th, 1959 photograph of the road leading to the synchrotron building, visible
at the end of the road. © INFN–LNF, http://w3.lnf.infn.it/multimedia/picture.php?/278/category/
8. At right, the formal permission granted to Bruno Touschek to enter the Frascati Laboratories,
dated October 30th, 1959, from (Valente 2007), © INFN 2007

permission to regularly visit the laboratories, Fig. 10.8. Bruno was also intrigued by
the Frascati synchrotron, the accelerator, from a more personal and scientific interest.
The construction of the synchrotron, now successfully accomplished from scratch,
may have reminded him of the betatron, the much smaller electron accelerator he
had collaborated to build during the War, in Hamburg. The synchrotron was a much
bigger machine, for the times almost an industrial enterprise, but its scale was still
one to give the possibility of being involved in something tangible, to touch and
manoeuvre.

10.3.3 A Seminar in October 1959, a Question from Touschek and How Theorists in Rome Got on the Action

Either before or after October 26th, date of his seminar in Frascati, Panofsky gave also a seminar in Rome. This seminar was remembered by Cabibbo in a Conference organized for the closing of ADONE, the successor to AdA, (Valente 1997, 219): "I still recall vividly the seminar given in Rome in late '59 by R. [sic!] Panofsky, where he presented the current activities at Stanford, speaking in particular of the Princeton-Stanford e^+e^- [sic!] ring then under construction. It was after the seminar that Bruno came up with the remark that an e^+e^- machine could be realized in a single ring, 'because of the CPT' theorem." Mention of a Stanford e^+e^- ring under construction is probably a typo, since the machine under construction in Stanford was for electrons against electrons. However, it cannot be excluded that Panofsky mentioned some interest or plan for an electron-positron machine, since the notion of the American competition was mentioned later by Touschek, in the February 17th 1960 meeting, where Touschek suggested to the Frascati scientists to make an 'electron-positron experiment', Sect. 10.4.

There also exists an oral testimony about Panofsky's seminar, given by Nicola Cabibbo, in a 2003 interview, partly included in the docu-film Bruno Touschek and the art of Physics.

> *Cabibbo:* I remember he [Bruno] had this brilliant idea, I was present, but this is a story that surely will be told better by others,[19] at a seminar by Panofsky, on electron-electron machines, for collisions between electrons and ...almost as a comment, Bruno said: 'Ah, but why not electron-positron, surely it is more interesting, because electron and positron can annihilate', and then, he said: 'Ah, but then... it's even more practical because electron and positron can circulate in the same ring, there's no need of two rings', and he pulled out, in that moment, let's say, a part of his knowledge of physics. He immediately sensed that it would be better to have collisions between particles in which all the energy had become available to create new particles: therefore, theoretical physics on one side, [for what concerned] the particles, but, on the other, his knowledge of machines. Bruno, during the war, had been working - and, even immediately after the war - on the construction of accelerator machines. Therefore, he had this double ability and passed from one to the other with the same mathematics. I remember a work that impressed me a lot , which he had written with Caianiello, another theorist, and in which they had established the theory of the stability of the beams of an accelerating machine, themes that were a kind of theory of groups, therefore expressed in terms of high mathematics, but taken in an absolutely simple way, analyzed to the end, and then remained, let's say, those formulas you will surely find even now in the books in some place, for those who want to design accelerator machines. So he had this great ability to apply the mathematics, the physics, in any field. And this process of his, it was active, in fact, from the motorcycle to the violation of parity in beta decay, following all his interests, that were very vast.

From the list of Frascati seminars in 1959 and Cabibbo's oral and written testimony, it can be clearly established that work on the electron-positron idea started in Rome in October 1959, if not also discussed between experimenters in Frascati.

---

[19] Raoul Gatto confirmed this episode in an e-mail exchange with Luisa Bonolis.

Fig. 10.9 Nicola Cabibbo in
Rome, early 1960s.
Photograph is courtesy of
Cabibbo's family and Luisa
Bonolis, who received it
directly from Nicola Cabibbo
on the occasion of the 2004
docu-film Bruno Touschek
and the art of physics

After Panofsky's seminar, discussions began in Rome among theorists, on the physical processes accessible through electron-positron collisions. Raoul Gatto, Touschek's colleague and friend, and Nicola Cabibbo, Fig. 10.9, Touschek's former student, began to study the vertex $\gamma \to 2\pi$ and $\gamma \to 3\pi$, which gave information on the pion form factor through electron-positron interactions, and they started calculations for a paper to submit for publication. The normal period for such a paper to be completed, namely for calculations to start, be finished and checked, and the text written and typed, would be two or three months. On February 17th, Cabibbo and Gatto's paper on "Pion Form Factors from possible high-energy experiments" reached the editorial office of *The Physical Review Letters* (PRL), in Brookhaven (Cabibbo and Gatto 1960). The timing of this submission implies that Cabibbo and Gatto must have started their calculations not long after Panofsky's seminar.

The revolutionary aspect of this article is in the study of the form factor in the so-called time-like region, which is only accessible through electron-positron annihilation. This was a totally new approach to the study of nuclear structure, not hitherto looked into. The creation of pions as products of electron-positron annihilation, could give a glimpse into the outstanding problem of particle physics, as pions were

Fig. 10.10 A young Francesco Calogero in 1958, photograph is personal communication to the Author, during the writing of this book, all rights reserved

considered to be carriers of the forces which keep nuclei together. The article was submitted and published about a month after its submission.

Cabibbo and Gatto's theoretical paper was not the only one submitted in February from the University of Rome to the *The Physical Review Letters*. In the same March 15th issue, another article was published mentioning the interest of electron-positron physics. Even more coincidentally, one of the two authors of this other paper, which arrived to the PRL Editorial Offices on February 5th, was Francesco Calogero, (Fig. 10.10) who had also been one of Touschek's students, having graduated together with Nicola Cabibbo in 1958.[20] The other author was the American theorist Laurie M. Brown, a Fulbright Research Scholar for 1959–1960, visiting University of Rome (Brown and Calogero 1960).[21] This article was entitled "The effect of pion-pion interaction in electromagnetic processes". While electron-positron collisions are not clearly discussed as such in the article, the interest in studying annihilation processes into pion pairs and access the pion form factor in the time-like regime is highlighted.

---

[20] Francesco Calogero (b.1935) is Emeritus Professor at Sapienza University of Rome. Recipient of the APS 2019 Dannie Heineman Prize for Mathematical Physics, Calogero served as Secretary-General of Pugwash Conferences on Science and World Affairs (1995 Nobel Peace Prize) from 1989 to 1997, and from 1997 to 2002 as chair of the Pugwash Council, of which he is still a member.

[21] L. M. Brown had received his PhD from Cornell University in 1951, with Richard Feynman as thesis advisor and was in Rome in 1958–1960, on leave of absence from Northwestern University, Evanston, Illinois.

It could not have been by pure coincidence that two of Bruno's former students started working on two separate papers dealing with electron-positron processes during the three months following Panofsky's seminar. From these two papers, both submitted in February 1960, and published in the same PRL issue, it appears that discussions and work had been taking place in Rome, during the winter 1959–1960, about what could be learnt from electron-positron collisions. Although Touschek is not mentioned in either of these two articles, it stands to reason that Bruno's question solicited the interest in electron-positron physics and he may have started discussions about such prospects, involving other Roman theorists as well, such as Marcello Cini, whose advice is acknowledged by Brown and Calogero.[22] Cini and Bruno were very close, having worked together (Cini and Touschek 1958), and Cini may have been involved in the exchange of ideas following Panofsky's seminar. On the other hand Bruno was very busy teaching, and may have had no time to start calculations. He had not yet given up his engagement in Pisa, and was now teaching two courses in Rome, one at the Scuola di Perfezionamento, one for the undergraduates. The latter was a completely new course on "Statistical Mechanics" for the fourth-year students, a very smart group of students, who kept him engaged with questions and required attentive preparation.[23] Among his students during that academic year there were Paolo Camiz, and Carlo Di Castro, who remembers Touschek's lectures in (Di Castro and Bonolis 2014).

> *Carlo Di Castro:* I became interested in statistical mechanics and in the theoretical physics of condensed matter, largely ignored as research topics in Rome, at a time end of the 1950s when everyone was engaged in the study of elementary particle physics. The course in statistical mechanics was given by Bruno Touschek, who was brilliant and thus certainly stimulating. Statistical mechanics was not his field, however. His course was based on the short book Statistical Thermodynamics by E. Schrödinger. He would also extemporise on specific topics, not teaching, strictly speaking, professional statistical mechanics, but rather how a theoretical physicist should approach problems with imagination, technique and enthusiasm.

Four years later, when Di Castro defended his thesis, Bruno Touschek was the second reader, the opponent, *contro-relatore* in Italian, who reviews the student's thesis. The thesis was excellent and Bruno returned his copy to Di Castro, 'With the compliments of the devil's advocate'.[24] In time, this course became famous for Bruno's perfect clarity and elegance in argumentation (Margaritondo 2021), inspiring the rise of a Roman school in theoretical physics.[25] Notes from this course would then take the

---

[22] Calogero and Cabibbo had graduated with Touschek in 1958, but had originally asked Cini for a thesis. According to Cabibbo, in the 2003 interview for the docu-film *Bruno Touschek and the art of physics*, Cini recommended them to do their thesis with Touschek, who was doing really interesting things.

[23] Letter to parents,12th December, 1959.

[24] December 2021 e-mail communication from Carlo Di Castro. Di Castro is Emeritus Professor at Sapienza University of Roma and a member of the Accademia dei Lincei. A summary of his scientific activity can be found in European Physical Journal History (2014), https://doi.org/10.1140/epjh/e2013-40043-5.

[25] Personal communication by G. Rossi. See also L. Maiani in the docu-film *Touschelk with AdA in Orsay*.

form of a book, written in 1970 with Giancarlo Rossi, who had graduated with him in 1966 (Touschek and Rossi 1970).

Apart from the two articles by the four theorists in Rome, whose parallel development indicates to have been started around the same time,[26] not much is known concerning what passed in Rome and Frascati between late October 1959, when electron–positron collisions had come up after Panofsky's seminar, and 16th January 1960, when, in a letter to his parents, Bruno mentions two proposals advanced to him by INFN. Were it not from the list of seminars in Frascati, it would probably have been difficult to even pinpoint the October 26th date, around which the beginning of electron–positron physics in the University of Rome can be placed.

As for the two proposals, mentioned by Bruno in January, the first one was for him to become Vice-President of INFN, which he calls an 'unbelievable' proposal. He mused to have perhaps become too stupid to do physics, and immediately dismissed it. It would have required regular visits to each one of the fifteen INFN Institutes spread throughout Italy, in addition to CERN, and he saw it would mean the end of him being active in research. The other proposal must have been in the air for quite some time, and he started thinking about it. It involved building a theoretical physics school in Frascati by the synchrotron. On personal grounds, it might lead him to move outside Rome, finding a house with a garden near the laboratory, which would be good for Francis, until he would be old enough to go to school. Elspeth would then have to learn to drive. There was however no immediate need to make a decision, which could be postponed to June, at the end of the academic year.[27]

Bruno had now reached a high standing as a theorist and by February 1960 the pressure grew on Bruno to take on some type of responsibility in the Frascati laboratories. They had been built under the helm of INFN, which provided scientific guidance and personnel. In 1959, INFN governing body, mostly Gilberto Bernardini and Edoardo Amaldi, together with Giorgio Salvini had been concerned not to let go waste the strong scientific position Italy had gained by the successful construction of the synchrotron. Now that the machine had ben built, time had come for physics to step-in, to have a view of what to look for. This meant not only to extract physics from the experiments the machine would allow to be performed, but to look forward to the future as well. Discussions along these lines find an echo in a conversation with Giorgio Salvini which took place on February 12th, 2007 in Frascati, during the preparation of a book about the Laboratories (Valente 2007).[28] Salvini said that, around 1957–1958, when it became clear that the synchrotron would soon be ready, as director of the Laboratories he urged to reserve office space for researchers from the University of Rome, believing it would increment their presence and involvement in the laboratory planning. He also understood that this meant not just engaging the

---

[26] As from e-mail correspondence with N. Cabibbo and R.Gatto.

[27] Letter to parents, January 16th, 1960.

[28] In 2007, Salvini, then 87 years old, promoted and organized the writing of a volume about the history of the Frascati National Laboratories, reaching up to modern days. He was very insistent that a full description of the birth and development of the Frascati theory group be included, as it was in fact done (Valente 2007, 165–177). Written notes were taken by this author, who participated to the volume preparation.

presence of experimenters, but that of the theorists as well. Raoul Gatto was the first to join, soon followed by Cabibbo. The obvious person to conquer to the nascent synchrotron efforts was of course Bruno Touschek, the most brilliant theorist in Rome at the time. Together with Amaldi, it was then decided to ask Bruno for a regular collaboration to the Laboratory programs.[29]

Such foresight is necessary when building modern accelerators which take years to build. In the case of the Frascati synchrotron, the entire enterprise had started in 1953 and was completed in six years, in 1959. Since then, the time between conception of an accelerator and running of particle beams has gone through a fivefold increase: the Large Hadron Collider, which had its first successful run in 2008, was envisioned at CERN around 1976, during early planning for LEP, the Large Electron Positron Collider. The design and length of the LEP tunnel were such so as to accommodate a proton-proton collider as well. Throughout the long periods of time between proposing and operating a particle accelerator, it is necessary to plan for yet unearthed physics phenomena which can be extracted when the machine will be ready. Thus, foresight and curiosity drive the planning of future accelerators, as research moves along during the years while an accelerator is built.

In 1959, the need to shape the future of Italian particle physics and reap the fruits from the synchrotron success was clear to Amaldi. Early in the 1950s, the inspiration for the construction of the synchrotron has come from Enrico Fermi, and, had he not passed away prematurely in 1954, he could have provided such guidance, but Fermi was not there anymore. Amaldi knew that one scientist in Rome had comparable theoretical knowledge and experimental insight: would Bruno accept the challenge to shape the future of Italian particle physics?

As Touschek wrote much later:[30]

The Birth of AdA A full regime of experimentation [at the electron synchrotron] was reached in the second half of 1959.

At the same time however, new preoccupations arose. All over the world newer and bigger machines were being built and planned and it was felt that if Frascati wanted to keep abreast something big and new had to be planned.

... A series of seminars was held in Frascati with idea of forming a feasible line of attack for entering phase II in the development of the laboratories. In one of them (Feb. 1960), I suggested that the synchrotron should be used for observing collisions between electrons and positrons.

While his former students worked on their articles, in addition to teaching Bruno was also occupied in settling aunt Ada's inheritance. This over, in January 1960 he

---

[29] Salvini, in his later reminiscences, places this decision around 1957–1958, but from Bruno's letters home, it appears that he started feeling a strong pressure to join Frascati only in 1959.

[30] From document entitled *A Brief Outline of the Story of AdA*, excerpts from a talk delivered by Touschek at Accademia dei Lincei - 24.5.74, reproduced courtesy of Rome Sapienza University, Physics Department Archives, https://archivisapienzasmfn.archiui.com, documents provided for purposes of study and research, all rights reserved.

started thinking seriously about the Frascati proposal. Later in his life, he would write:[31]

> I was ... attracted by the possibility of learning how a big
> enterprise like Frascati worked and particularly intrigued
> by the idea of having some contact with the technological
> side of the services, the techniques (vacuum, RF., magnets,
> liquid He(lium) and the various control mechanisms)
> necessary to the work of the big machine.

This later recollection is mirrored in a letter he wrote to his parents in March 1960, where he wrote: 'I have had enough of the theory for now',[32] indicating that he was looking for new research directions, Fig. 10.11.

Then, between the time Cabibbo and Gatto's article was received by PRL on February 17th and when it was published on March 15th, a revolution took place in Frascati. By that time, Touschek's suggestion to make an electron-positron experiment, and build an accelerator to do it, had been presented and accepted. A decision, which changed the future of particle physics, was taken.

10.4   The Making of AdA

In January 1960, Bruno was at a cross-point in his life. He had lost his theoretical mentor, Wolfgang Pauli, and the maternal aunt, who had been the closest link to his mother. Such personal losses are often behind the motivation to make important changes in the course of one's life. The time had arrived for Bruno to start a project he could call all of his own and where he could unify past and present knowledge. As he would write, this project put together 'all he had thought about it and much which others had suggested to him'. This sentence, belongs to the *incipit* of an undated text entitled "On the Storage Ring", likely to have been prepared soon after February 17th, the day he proposed an experiment in electron-positron collisions.[33] On that day, he plunged into the unthinkable (Rubbia 2004, 58), gambling to do something which was the only thing worth of his time and attention. This resolve is clear from the minutes of the meeting which took place in Frascati.

---

[31] Undated notes probably prepared for a book, after the talk Touschek delivered at the Accademia dei Lincei in 1974, reproduced courtesy of Rome Sapienza University, Physics Department Archives, https://archivisapienzasmfn.archiui.com at https://archivisapienzasmfn.archiui.com.

[32] *Ich hab' für den Moment von der Theorie genug*, letter dated as of 'last Monday' in March 1960.

[33] The undated typescript includes most material jotted down by hand in Touschek's notebook called S.R., *Storage Ring*, and dated 18.2.1960. It is likely that the typewritten version was properly prepared after the handwritten notes. An error in the formula he uses for the machine luminosity in S.R. is corrected in the typed text.

Fig. 10.11 Bruno Touschek at University of Rome, around 1960. On the blackboard behind him, one can see some formulae belonging to his theoretical work on the chiral transformations, and a symbolic waving good bye to this work. Family Documents, © Francis Touschek, all rights reserved

10.4.1   February 17th 1960: A Meeting for the Future of the Laboratory

On February 17th a meeting was called at the Frascati National Laboratories: on the agenda there was the plan to start a theoretical physics group in Frascati, (Amaldi 1981, 77), (Amman 1985).[34] Touschek was asked to give the introduction.

He began on a rather negative note about the proposed group. Did he already have in mind the direction where he was trying to lead his colleagues? The minutes show that he did. As this meeting traces the official birth of AdA's construction, it is

---

[34] (Amaldi 1981) in footnote 94) gives credit to Fernando Amman for telling him of the February 17th meeting, and to Icilio Agostini, INFN Chief Administrator, for preparing the Minutes, copy of which is available in Amaldi's papers in Sapienza University of Rome–Physics Department Archives at https://archivisapienzasmfn.archiui.com.

worth reproducing the minutes of the meeting. The following is a translation from the original Italian version:[35]

MEETING OF FEBRUARY 17, 1960

The meeting was dedicated to the program of setting up a dedicated theoretical physics group in Frascati. It started with an introduction by Touschek.

Touschek says that theorists generally care about the past and the future of experimental activity while the present is of little interest to them. The fact that experimental research is carried out in Frascati does not imply that there must necessarily be a theoretical group, since a theoretical group 'digests' the experimental results of many research centers and therefore the advantage of residing in an experimental research center is very small, for theorists.

On the other hand, if the purpose of the constitution of a theoretical group were to create a theoretical school, he would be against it; there is already the school in Rome: if the school in Rome is not adequate, it does not mean that one should have one in Frascati, while it is sure that creating one in Frascati would be at the expense of the school in Rome.

Theorists however are very interested in what Frascati intends to do in the near future. From reading the notes of the previous meetings, he had the impression of some uncertainty about the future activity of Frascati: he therefore tried to think of a 'future goal'. He reminds everyone that when it was decided to build an electron machine, it was because electrodynamic research is best done with electrons. But the study of electrodynamics implies the realization of very difficult experiments for the study of the 'form factor'.

Now, an experiment that is truly worth having, an experiment that would be truly first-order, and that would be capable of attracting theorists to Frascati (not only him, but also Gatto and certainly others) would be an experiment intended for the study of electron-positron collision. It is an experiment that Panofsky is thinking of doing:[36] it's all about getting there before him. The Frascati Synchrotron, with appropriate improvements or modifications, allows, in his opinion, to have this experiment realized. It is a question of making a program in this direction. First of all it would be necessary to work on improving the intensity, and probably it would be necessary to think about injecting with a linear accelerator instead of a Van de Graaff.

The points of a program of the type listed by him would therefore be:
1) Intensity
2) beam extraction
3) Positron acceleration.

This first part of Touschek's speech was followed by a full discussion during which:

Ghigo asserted that our synchrotron is not an easily convertible machine and that it is probably simpler to realize the experiment suggested by Touschek, by building an *ad hoc* 250 MeV.
Quercia said that the very kind of discussion that is taking place demonstrates the desirability of having theorists in Frascati. He then recalled that it is a question of establishing whether Frascati should be a 'service' [laboratory] and therefore centered all around the machine or rather something bigger.
Responding to Quercia and concluding his speech, Touschek specifies that in his opinion, in order to obtain what Quercia proposes about the 'stepping up' of Frascati, a theoretical group in Frascati is not necessary. It would suffice for a theorist to come 1 or 2 times a week, or maybe every afternoon and work on to get better exchanges with Rome. As for the

---

[35] Underlying of people's names follows the style of the original minutes.

[36] *Panofsky si propone di fare*, in the Italian version.

question of the 'Frascati service' he thinks that, even remaining in the theory of 'service', it is necessary that Frascati always remain at the forefront of [particle physics] research, and for this reason he has later proposed a program intended precisely to keep Frascati at the forefront of fundamental research.[37]

According to Bernardini there are two categories of people: those who know what to do and those who know what can be done. The first are the theorists, the second are the physicists and engineers of the type like the ones in Frascati. It is necessary to have a fruitful interaction between the two; but for this interaction a daily closeness is needed since it is a question of breaking a tradition of isolation between the two categories.

Salvini - Rome is not insufficient.

McDaniel - At Cornell there are 4 theorists; there is not much exchange between them and the experimenters, even though they are in the same building. The interests are different. It would be convenient to have the experimentalists here with a theoretical attitude and to leave the pure theorists in Rome.

Salvini asks McDaniel if Carlo Bernardini's theoretical program is good or not.

McDaniel believes it is difficult to form a theoretical physics group where there is no tradition for attraction of great theorists [38]

Ghigo asks if the theoretical group collaborates with the experimentalists on the program.

Amman says that the current theoretical-experimental situation must be changed, and it cannot be discussed in current terms.

Salvini confirms the above. He therefore still believes that the presence of a theoretical group in Frascati is still necessary, even if difficult [to realize]. A close theoretical-experimental collaboration is needed and this is proved by Touschek's own words.

Quercia more realistically, asks if, neglecting the metaphysical problems it is possible to have Touschek, Gatto or others in Frascati. Thus. it is not: program and then theorists, but theorists and then programs.

Touschek if one can plan for a large investment in Frascati (possibly to make other machines, etc.) the prospect may be interesting for him, otherwise he thinks that the Frascati-Rome interaction is sufficient.

Salvini the goal is what we are discussing.

Sacerdoti what was proposed by Touschek can be discussed in the budget as an alternative [to] new machines.

Salvini as regards Touschek's proposal, asks Touschek to discuss it with the machine group. He then asks whether or not one should try to arrive at a theoretical group.

Sacerdoti expresses his opinion that medium level theorists should be sought to be included in [the experimental] groups and then coordinated in a central organization.

McDaniel and Touschek observe that nothing is lost in trying to find theorists.

Touschek points out, however, that one should not expect too much from theorists in Frascati.

Querzoli Frascati is a type of laboratory unique in Italy and one of the few in Europe far from the University.

Thus, we experimenters can have a better view about the usefulness of theorists. The Frascati environment will probably be useful to the theorists because of the collaboration between theoreticians and experimentalists; Roma-Frascati probably doesn't work not for intrinsic defects but for the distance.

The theoretical group in Frascati is a question of life or death. He is against Sacerdoti's proposal.

---

[37] The minutes having clearly been prepared after the meeting, are confused in various points, i.e. what is meant by the twice mentioned Bernardini's program, or what does Salvini mean for the mentioned 'Touschek's program' proposed later?

[38] Boyce McDaniel from University of Cornell was visiting Frascati, at the time.

Salvini: based on what Touschek said, it will certainly be necessary to improve the collaboration with Rome, without prejudice to the theoretical group in Frascati.

Quercia asks what are the conditions that can push theorists to come to Frascati.

Salvini asks who agrees that there be a theoretical group in Frascati.

All agree.

Querzoli Touschek asked not so much that one program be made more than another, but that there be a development program. On the other hand, this is a condition for our stay in Frascati. What we ask of Touschek is to help us clarify the ideas on the development program.

Salvini will in any case pursue Touschek's proposal (O'Neill storage rings or other systems for e+ e- interaction)

He reads again Carlo Bernardini's program. Salvini will you accept the Bernardini's program or not?

Touschek: Okay, we can try.[39]

Salvini can we ask Touschek to increase his activity towards Frascati both for his proposal and for the creation of the theoretical group?

Touschek I have always declared that I am willing to come to Frascati even every day, if necessary. It is not an organizational problem.

Salvini the organizational problem is still very small compared to the total budget.

He then insists on the need for Touschek's collaboration in finding theorists.

All in favor of including the Carlo Bernardini's program in the four-year program.

The next meeting is scheduled for Wednesday 24/2/60 at 3 pm.

All the protagonists of the laboratory's future were present and approved the proposal.

The list of interventions includes those from Giorgio Ghigo and Giancarlo Sacerdoti, who designed AdA's magnet and followed its construction, Carlo Bernardini, the theorist to become the closest companion to Bruno in AdA's progress from Frascati to Orsay, Fernando Amman, who set his eyes on the future of AdA and built ADONE, Ruggero Querzoli, who built AdA's detectors with his student Giuseppe Di Giugno, who would be counting electrons in AdA in 1961 and later become a star in digital music,[40] Italo Federico Quercia (Spezia 2007) and Giorgio Salvini, succeeding each other as directors of the laboratory during those years. Others would soon join and contribute to AdA's construction, among them Gianfranco Corazza, in charge of the vacuum in AdA's doughnout, Mario Puglisi who had responsibility for the radio frequency set up. They could count on a team of highly skilled technicians, who had built the synchrotron and would be the backbone for AdA's success in Orsay.

Touschek's proposal for a totally new type of machine, so beautifully and easily accepted by the Frascati scientists, had come out of desperation.[41] He had felt the pressure for him to become what he called a *Haustheoretiker*, a house theorist, possibly losing independence and freedom.[42] A way out had to be sought. An

---

[39] From the minutes, it would appear that 'Bernardini's program' meant a long term presence of theorists, such as Touschek. In fact Salvini directly asks Touschek whether he would agree to 'Bernardini's program', basically asking for his commitment to be in Frascati on a regular basis. And to this, Touschek agrees.

[40] See Peppino Di Giugno, Genio Italiano. Macchine Digitali per la Musica in Tempo Reale, by A. Giordano, Esarmonia 2021.

[41] Letter to parents, March 28th 1960.

[42] The point is more explicitly made in the already mentioned notes prepared for a book about AdA and ADONE.

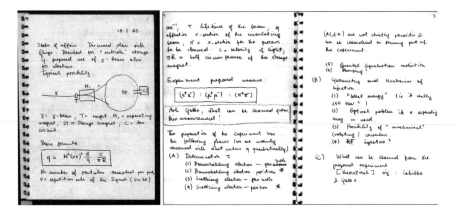

Fig. 10.12 First three pages of AdA's notebook, started on February 18th,1960, courtesy of Rome Sapienza University, Physics Department Archives, https://archivisapienzasmfn.archiui.com, documents provided for purposes of study and research, all rights reserved

involvement in the laboratory activities could entrap him in a situation where his intellectual growth and insight would be stymied. Like other times in his life, he jumped ahead and found a far out solution: had he not done it, in the middle of the war, when he had left Vienna for Germany, in order to be able to continue his studies in physics? Later, the move to Glasgow to obtain his doctorate may have been decided by others, the T-force in particular, but in 1952 the choice of leaving the UK for Italy had been his own, following a trail of survival in his desire to become a physicist in the fullest sense of the word. Was Bruno facing another desperate choice in his life? Through January and the first half of February, he had followed his friends' work. Then, on February 17th, he acted. He was not only moved by desperation, but also by the curiosity to see what an experiment with electron positron-collisions could reveal about the extraordinary world of physics. On that day, both Bruno and the Frascati laboratories stepped into the future, following a totally unknown path.

Things moved very rapidly after Touschek's proposal. Be it an O'Neill type storage rings or another system for electron-positron interactions, as Salvini has said, the idea had been accepted. Soon after the meeting, in the same afternoon, a working group was set up to prepare the actual proposal along the lines suggested by Giorgio Ghigo. Bruno's original idea of transforming the synchrotron into a positron-electron collider had of course been discarded. The synchrotron had just started functioning, and experimenters were eager for measurements to start with the detectors which they had built, collecting the data to be studied and reported in the articles they expected to write and present to their colleagues in Italy and abroad. There was no way these expectations could be postponed or dismissed for a hazardous adventure, never attempted before. Ghigo's proposal, supported by Sacerdoti, to build an *ad hoc* machine, which then became a proof-of-principle for electron-positron colliders, was the perfect way out. The dream of a bigger machine of course remained in Touschek's mind, but it could be taken up later, after AdA 's feasibility had been proved.

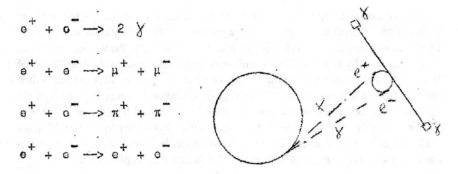

Fig. 10.13 A schematic drawing of the proposed experiment from minutes of March meetings, preserved by F. Amman, published in Internal note INFN-LNF-60-58, Discussioni Preliminari sull'A.d.A. prepared by G. Ghigo, © INFN – LNF

On the 18th, the day after the meeting, Bruno began what became known as AdA's note book, where he would jot down ideas, and the daily progress for the experiment. The first three pages of the note book are shown in Fig. 10.12. During the following two weeks, the team, consisting of Carlo Bernardini, Gianfranco Corazza, Giorgio Ghigo and Bruno Touschek, prepared a formal proposal, complete with processes to study, a time table and division of labour. An undated draft proposal in Bruno Touschek's handwriting, was prepared during these two weeks and presented on Monday, March 7th.[43] On that day, Bruno gave a seminar, where he proposed to build a storage ring for 250 MeV electrons produced by a γ-ray beam (from the synchrotron) impinging on a target, placed in the ring vacuum chamber. The layout and the processes to study are shown in Fig. 10.13. A similar set-up would be used for producing positrons. The proposal was considered sufficiently interesting to proceed with further studies about possible limitations on the beam lifetime. Thus the official date for AdA's birth is usually placed on March 7th, 1960.

Starting from March 7th, 1960, meeting followed meeting. As Touschek had said in February, the Americans were interested in building an electron-positron collider, but were not yet starting on this, being busy with constructing the electron-electron version. Thus, it could hopefully be a question of arriving to do the experiments before they did. No insurmountable problems having been identified, one week later, on March 14th, the proposal was formally approved by the laboratory scientists, with an initial expenditure of 8 Million Lire (Ghigo 1960). Touschek was given responsibility for the experiment, with Giorgio Ghigo in charge of technical and organizational questions, Carlo Bernardini coordinating theoretical work for the proposed accelerator. A preliminary realistic work plan was ready by March 16th.[44]

---

[43] Bruno Touschek's proposal from undated handwritten notes, likely to have been prepared for the meeting held on March 7th, 1960, is kept in Sapienza University of Rome–Physics Department Archives.

[44] As from minutes for March 7th, 14th and 16th discussions, dated March 22nd, 1960 by F. Amman, who gave it to C. Bernardini, personal communication by L. Bonolis. These notes were

The proposal, once approved by the laboratory scientific council, passed on to the INFN Board of Directors, where Touschek presented it for inclusion in the next Four Year Plan. Relating the events to his parents, Bruno noted that the session had began with an address by Gilberto Bernardini, outgoing INFN president (Amaldi would be the next one), who began asking Bruno the obviously rhetorical question: 'When will you stop proposing exclusively foolish experiments?'. Then, during all of the following day, Bruno's proposal had the floor. If Bruno had hoped that the project could be carried out by someone else, so that he could sit on the fence, watching the works and himself just having fun with the machine, it did not happen. The project was approved, and Bruno was in charge of funds and their deployment.[41]

Bruno Touschek. Excerpt from unpublished documents, around or after 1974. A storage ring is a system of magnets, which can keep charged particles in a closed orbit. That this can be done with magnets is due to the Lorentz force which acts perpendicularly to the direction of motion of the charged particle and also perpendicularly to the direction of the magnetic field. Disposed along the arc of a circle the Lorentz force can compensate the centrifugal force of the particle, which acts perpendicularly to the direction of motion and to the axis of rotation. In the arc of trajectory the centrifugal force points outward the Lorentz force inward and the two balance one another. The magnetic guiding field is not sufficient to keep particles (in particular electrons) in a stable orbit for any length of time. The reason is that the electrons undergo an accelerated motion in the bending magnets and accelerated particles emit radiation. If the energy loss by radiation would not be replaced the electron beam would rapidly spiral inwards and thus be lost. In order to keep the beam alive it is therefore necessary to replace its radiation losses. To do this one disposes a radiofrequency cavity (or more than one) in the electron orbit. The frequency of the RF-system has to be chosen as a multiple of the circulation frequency of the electrons (v/u ? c/u, where u is the length of the closed orbit). Electrons which pass the RF cavity at the right phase will then pick up an energy which just compensates their radiation losses. The replacement of these losses is stable. Small errors of phase are forgiven: they smooth themselves out. The principle of phase stability, which is at the bottom of this argument, was discovered by MacMillan in [1945]. It is the base on which synchrotrons are built.

---

then included in Ghigo's end of the year report of AdA's progress (Ghigo 1960), a copy of which was also sent to Amaldi by Ghigo's widow Mirella Ricci, *via* Giorgio Salvini, on November 23 1978, Edoardo Amaldi Papers in Rome Sapienza University, Physics Department Archives, https://archivisapienzasmfn.archiui.com.

The extraordinary thing in this story is the speed and interest which carried the new idea to be accepted and approved, as Salvini remembers in his 2013 interview:[45]

> *Salvini:* Touschek was keenest in building the machine. I asked an extra 20 million Lire for building AdA. It was immediately accepted. ...
> I have no direct merit in the AdA construction. AdA started with Touschek's interest, also the Russians, Budker, were very interested.

When Touschek had proposed an experiment to study electron positron collisions, he moved from a deep personal conviction that the plan was workable. This conviction was based on four reasons, which he outlined many years later, when preparing his talk at Accademia dei Lincei: the first reason was his profound knowledge of particle symmetries and his faith in the invariance under simultaneous Charge and Parity transformations, so called CP invariance. This property would make sure that if positrons moved in the same orbit as electrons they would meet, a necessary condition for them to clash. The argument is based on the particles having exactly the same mass, and circulating in the same ring. This vision shows how different Touschek's proposed experiment was with respect to Widerøe's ideas, expressed in the 1943 (1953) patent, Chap. 5. Touschek's was based on purely theoretical reasons and the existence of antimatter, i.e. of particles with same mass and opposite charge of those constituting ordinary matter. Another reason of interest for Touschek was based on the belief that true electrodynamics, which dictates interactions between electrons and the photons, 'cannot be indifferent to the existence of strong interactions', proof being the existence of a stable particle such as the proton which interacts with the electromagnetic field. One problem Touschek was aware of was of the challenge posed by having 'the first machine in which particles which do not naturally live in the world which surrounds us can be kept ad observed.' He was also aware of the costs of such machines, especially for Italy, still considered a poor country. To minimize costs, a single ring would be obviously cheaper that two rings, and using the already existing synchrotron would be even cheaper. As it turns out, the experimenters and machine scientists which had built the synchrotron, found the solution, a smaller machine, and the governing body, INFN, found the money.

10.4.2 The Construction

At this point, Bruno's regular life had to undergo the changes required by the urgency and cost of the new project. Panofsky's talks had let it be clear that the Americans would, at some time, engage in building electron-positron colliders, albeit not yet. Thus, to make the new effort worth of its cost in terms of both financial and human resources, there was no time to spare. The major component of the new machine was the magnet, which could not be built in Frascati, but needed to be made by a dedicated firm in Terni, in Umbria some 100 km drive from Rome. The firm promised

---

[45] See the docu-film Touschek with AdA in Orsay.

it could be ready by August, Giancarlo Sacerdoti and Touschek were to follow its construction according to the specific design prepared by Giorgio Ghigo. It is during those car trips to Terni that Bruno opened up with Giancarlo about his past before and during the war. In particular Bruno told him of things which had happened more than 20 years before and had remained impressed on his mind: how the atmosphere in Vienna had dramatically changed when Austria was annexed to Nazi Germany, and killings and violence were everywhere, (Sacerdoti 2004). Bruno also told him about moving to Germany, where Sommerfeld could count on his Prussian clan to protect him and allow him to continue to study.[46]

Touschek was fully involved in the making of AdA, following both the preparation of its parts, and worrying about the physics which could be extracted from it. Calculations were needed for the processes which one could expect to be observed with AdA, thus Cabibbo and Gatto went ahead to make more calculations beyond their relatively simple first paper.[47] This second work was going to become a landmark in electron positron physics, so much to be nicknamed *La Bibbia*, the Bible, by Frascati physicists. Long after, some forty years later, Cabibbo remembered the wonder of finding how many processes of interest had never been calculated before. Asked by L. Bonolis how he felt during that period, Cabibbo answered 'Calcolavamo e calcolavamo', namely calculations were following calculations (Cabibbo 2003). The completion and publication of the 'Bible' took a good part of the following year (Cabibbo and Gatto 1961). In the meantime, Touschek was eager to get a more precise estimate for the reactions originally proposed by Cabibbo and Gatto in their February article (Cabibbo and Gatto 1960), which required calculations beyond the first order in the QED coupling constant. Looking ahead to a problem that would fully absorb him a few years later, Bruno assigned the problem to one of his students, G. Putzolu, who graduated with him in July 1960 with a thesis on *Radiative corrections to Pion-Production in e^+e^- Collisions*, (Putzolu 1961).[48] The title of Putzolu's thesis points to Bruno's preoccupation to include higher order corrections in charged particle interactions and estimate their effects on the extraction of physical information from the collected data. This problem will become of great relevance with machines operating at higher energies and lead to Bruno's theoretical physics legacy through the creation of a theoretical physics school in Frascati in the late 1960's (Bonolis and Pancheri 2011, 36).

The first months of 1960 had been a period of absorbing and intense work for Bruno, and the mental and physical stress took over soon after the approval of the project. He fell sick with jaundice, and was forced to forgo visiting Frascati, under doctor's orders. But the pressure to move ahead with the construction of AdA, knowing that the Americans could be planning something similar, did not leave much

---

[46] Sacerdoti also writes that 'Among the new arms project there was one proposed by a stupid fool of a professor from Lipsia who was completely tied to the Nazi Party', referring to Schiebold's project mentioned in Chap. 3, and about Bruno being part of Widerøe's group.

[47] In a correspondance with L. Bonolis, Cabibbo called that first paper very simple, which could well have been true, but the idea wasn't.

[48] Soon after AdA's approval, this problem had been discussed by Cabibbo, Calogero and Putzolu (Ghigo 1960).

time. As soon as his doctor allowed, he was back in Frascati to follow the progress of the little machine. By that time a name had been given to be machine. In later years, Bruno wrote:[49]

```
... AdA and Adone. Both are mild puns. Originally AdA (born
in 1960) figured in my books as S.R. = storage ring. It was
later suggested - probably by Ghigo, I am not sure - to call
the thing AdA, which is short for 'Anello di Accumulazione',
which might be considered the Italian translation of storage
ring. My aunt Ada (which is short for Adele) had just died
and I was intrigued by the fact that I could now justly say:
'Ada is dead, long live AdA!'
```

The name of the collider has often been mentioned as one of Bruno's puns. In fact it had been proposed by others, as Bruno recounts, but the story has a resonance in that the machine sprang to life after Bruno's aunt had passed away. It also might suggest that Bruno's acceptance to spend part of his time in Frascati was an acknowledgment of how important Aunt Ada had been in his life.

The magnet was expected to be ready in August, and since Bruno had also to attend the usual Rochester conference, held in the US, to keep abreast with new developments, the usual summer vacations had to be shortened. In addition to follow AdA's progress, which was going along very well in Frascati, Bruno was continually invited to give talks, so that, unlike all the previous years, it was impossible to forecast what would become of the summer.[50]

In the spring, with the engagement in Frascati, Bruno's days took on a new weekly schedule. He had always lived in large metropolitan cities, Vienna, Berlin, Rome, and it is difficult to imagine that moving to a house with a garden in the countryside could appeal to him. On the other hand, Frascati's surroundings held many attractions given his love for swimming and the pleasant memories spent visiting Aunt Ada in her villa in the hills. Thus he developed a new schedule which he followed for many years to come. In the morning he would go to the university and meet his student Putzolu, whose thesis he was supervising. At 12 a drive out to Lake Albano was followed by swimming until 3 pm. After a late lunch in one of the small restaurants nearby, he would then go to the laboratory in Frascati, returning to Rome and be home around half past eight.

Work with AdA was progressing well. After the summer, and the arrival of the magnet from Terni, Bruno started drafting a paper on *The Frascati storage ring*, which was submitted to the *Nuovo Cimento* on November 7th and rapidly published (Bernardini et al. 1960b). It was the first of four history-making AdA papers, describing the project in detail.

Then came the time for the first measurements of the magnet, to start activating the current and test the beam. One of the crucial points about the scheme was its injection in the AdA ring and how to insert the positrons. The adopted solution was

---

[49] From Sapienza University of Rome–Physics Department Archives, file name 1-AT_11_92(8), Bruno Touschek Papers.

[50] Letters on March 28th, April 8th and an undated one, probably July, 1960.

Fig. 10.14 O'Neill's two rings proposed set-up, as drawn by Bruno Touschek in his 1974 Lincei talk, courtesy of Rome Sapienza University, Physics Department Archives, https://archivisapienzasmfn.archiui.com, documents provided for purposes of study and research, all rights reserved. At right, AdA's injection set-up, from LNF-60-046 (Bernardini et al. 1960a), © INFN–LNF, http://w3.lnf.infn.it/multimedia/

of course different from the tangent ring scheme proposed by O'Neill, schematically drawn by Bruno Touschek in Fig. 10.14, left panel. AdA's initial solution, shown in the right panel, called the *girarrosto*, the roasting spit, since AdA was simply turned over around a horizontal axis, was later abandoned (Bernardini 2004, 166).   The group was now confident that AdA could lead the way to higher energy physics, and Touschek started plans for a bigger and more powerful accelerator, one which would "probe the quantum vacuum", and discover new particles, through electron-positron annihilation. Following a memo prepared by Touschek in November 1960,[51] a document was prepared and submitted to INFN.[52]

10.4.3   Enthusiasm and Disappointments

Electrons and positrons circulating in AdA were produced from the materialization of γ rays, obtained directly from the synchrotron. For this reason AdA was placed initially next to the synchrotron, Fig. 10.15. In this photograph, taken in 1961, AdA appears to be almost like an intruder, or an afterthought, as indeed it was, parasitising next to the more impressive electron synchrotron, or almost like a phantom crawling in. Somehow, the little compact machine in the photograph projects a menacing feeling. Indeed AdA, as the model of future accelerators, would take over the old

---

[51] B. Touschek, *'ADONE' a Draft proposal for a colliding beam experiment* (typescript, Bruno Touschek Papers., Box 12, Folder 3.95.3, p. 3), Sapienza University of Rome–Physics Department Archives.

[52] On January 27, 1961, F. Amman, C. Bernardini, R. Gatto, G. Ghigo and B. Touschek presented the Internal Report "Storage ring for electrons and positrons 'ADONE'" (N. 68, Frascati National Laboratory). In February 1961 a study group was formally set up with the task of preparing a first estimate of the feasibility and costs of such a project.

Fig. 10.15 AdA, at left, in the Sycnchrotron Hall, and next to the Frascati electron synchrotron, ©LNF INFN, http://w3.lnf.infn.it/multimedia/

systems, and, a few years later, colliders built after AdA would make obsolete fixed target machines, such as the synchrotron.

The enthusiasm was high in Frascati, when, in February 1961, the magnet was turned on and electrons and positrons could be proven to circulate in the doughnut by observing the light signal they emitted, a phenomenon also known as synchrotron light radiation. The registration of the number of "electrons"[53] circulating in the machine and the length of time these electrons would "stay alive", were recorded, Fig. 10.16.

Spring 1961 saw more tests and measurements being done with AdA, while the theoretical work was completed by Raoul Gatto and Nicola Cabibbo (Cabibbo and Gatto 1961). The first AdA paper had arisen quite some interest and Touschek was invited to present the Frascati work on storage rings to the forthcoming International Conference, to be held in Geneva, June 5th to June 9th 1961.

But a problem arose. The number of electrons and positrons circulating in the ring was insufficient to prove that collisions, and hence annihilation, had taken place. An enormous step had been taken, but how to prove that it had been worth and continue in this direction? The Frascati team started preparing for the summer conferences, but there was a cloud over their enthusiasm.

The first act of the AdA story is now completed, (Amman 1989; Bernardini 2004). The next will be staged across the Alps.

---

[53] It was never clear whether the light observed was emitted by electrons or positrons.

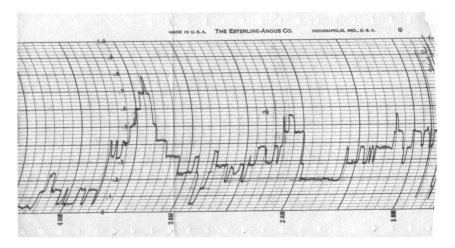

Fig. 10.16 The phototube record of February 27, 1961, showing steps that correspond to single electrons entering or leaving AdA, from L.Bonolis who photographed it, courtesy of Carlo Bernardini (personal collection)

10.5   Who Started First: Frascati or Novosibirsk?

The official date for AdA's conception can be placed around October 26th, the date of Panofsky's seminar in Frascati. From that time, theoretical work at University of Rome laid the grounds for Touschek's suggestion, on February 17th, 1960, to make an 'electron-positron experiment'. On that day, Frascati scientists, assembled to discuss the future of the laboratories, gave the go-ahead to the preparation of a proposal to build a small machine, a sort of prototype to study electron-positron collisions. A proposal was prepared and, on March 7th, 1960, the day Touschek presented a detailed plan to the synchrotron team of scientists, the construction of the first electron-positron storage ring was approved.

While there are plenty of records about AdA, not much was known about the activity beyond the Iron Curtain in those days. Giorgio Salvini, recalling AdA's beginnings, was always keen to give credit to the Russian group in Novosibirsk led by Gersh (Andrei Mikhailovich) I. Budker for their early interest in electron-positron collisions.[54] Indeed, in 1963 an electron-positron storage ring in advanced state of construction was shown to a selected group of Western visitors at the Novosibirsk laboratory (Budker and Naumov 1965; Marin 2009). Much later, an article by V. N. Baier, tells the story and gives the date of October 28th, 1959, for the day when Baier and Budker started thinking about electron-positron collisions (Baier 2006). While this is a very remarkable coincidence, since the idea would have arisen in the same

---

[54] A memorial of Gersh Budker's life (1918–1977) can be found at https://scfh.ru/en/papers/budker-in-four-perspectives/, SCIENCE First Hand: 29 Dec 2018, Volume 50, N3, based on Melik-Pashayeva's 1988 book *A.M. Budker in Four Perspectives*.

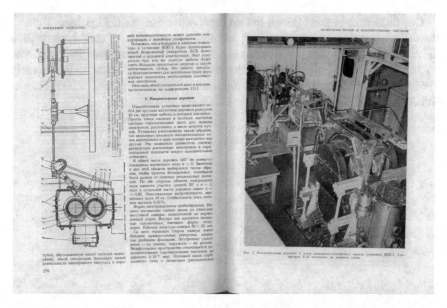

Fig. 10.17 The (vertical) Russian set up for the electron-electron collider VEP-1, from the Proceedings of the *4th International Conference on High-Energy Accelerators* held in Dubna in August 1963 (Kolomenskij et al. 1965)

days of Panofsky's seminars in Rome and Frascati, there is no reason to doubt that the Russian group started thinking about electron-positron collisions at around the same time than it happened in Italy. The Russian scientists had given and were giving prominent contributions to particle accelerator science and, by 1963, had an electron-electron tangent ring, VEP-1, in operation at the time of the 1963 Dubna conference (Kolomenskij et al. 1965). The Russian team was arranged initially as the Laboratory of New Acceleration Methods, of the Moscow Institute of Atomic Energy, under Budker's leaderships, and started to think about electron-electron collision, soon after the 1956 Geneva conference. Then, in 1958 the Laboratory was transformed into the Institute of Nuclear Physics and the team moved to Novosibirsk, in Siberia (Skrinsky 1995). The move delayed the construction of the ring, in addition, there was opposition among the Soviet Academicians concerning the electron-electron program, in particular by the great theoretician Isaak Pomeranchuk. After Pomeranchuk's visit to the new Institute, and his dismissal of possible interest in electron-electron collisions, Baier and Budker started thinking about electron-positron collisions, while the Novosibirsk team constructed VEP-1, Fig. 10.17, first officially presented at the 1963 Dubna Conference, about which more will be said in Chap. 12. The conceptual arrangement of two tangent rings for electron-electron collisions was similar to the on-going American project at Stanford, except that while the American rings were laid on a horizontal plane, the Russians put theirs on a vertical one! So, why didn't the Russian scientists arrive first to realize electron-positron collisions?

One difference is that Frascati moved rapidly to construct the machine. Bruno's theoretical physics insight and past accelerator experience combined with a formidable technical staff from the laboratory and unique theoretical work in Rome, so that it was in Italy that the first electron-positron storage came to life in early 1961, showing the way for other electron-positron colliders to follow, ACO in France, SPEAR in the USA, and VEPP-2 in the USSR (Baier 1963).

Chapter 11
Bruno Touschek and AdA: From Frascati to Orsay

...a visit to Frascati where intriguing things were happening.
...une visite à Frascati oú se passaient des choses qui intriguaient les esprits.
P. Marin, Éditions Dauphine, 2009

Abstract In June 1961, at a conference in Geneva, Bruno Touschek announced that Frascati had two electron-positron storage rings, AdA and ADONE. The announcement led two French physicists from the Orsay Laboratoire de l'Accélérateur Linéaire to come to Italy to see what was being constructed. A collaboration started, which brought AdA to France and, in two years time, led to the observation in Orsay of the first electron-positron collisions in a laboratory. The making of the collaboration between the two laboratories included visits between Orsay and Frascati, letters between Rome and Paris, and culminated with AdA leaving Frascati on July 4th, 1962 to cross the Alps on a truck, with the doughnut degassed to 10^{-9} mmHg through pumps powered by heavy batteries. This epoch-making trip and the exchanges which preceded it are recorded and presented through unpublished documents and interviews with some of its protagonists, Carlo Bernardini, François Lacoste, Jacques Haïssinski, Maurice Lévy.

When AdA had started functioning, in February 1961, there had been elation and great satisfaction, both in Rome and Frascati.[1] The little ring started functioning when the magnet had been turned on, and one could see electrons (or positrons?) to circulate in the doughnut by observing the light signal they emitted, as expected from the phenomenon of synchrotron light radiation. The number of "electrons" circulating in the machine and the length of time these electrons would "stay alive", was recorded. A recording similar to the one shown in Chap. 10 was sent by Touschek to Edoardo Amaldi on the morning following the second night of AdA's operation.

---

[1] See Bonolis and Pancheri (2018).

© The Author(s), under exclusive license to Springer Nature Switzerland AG 2022
G. Pancheri, *Bruno Touschek's Extraordinary Journey*, Springer Biographies,
https://doi.org/10.1007/978-3-031-03826-6_11

During the next night Amaldi went to see with his own eyes something that nobody had ever seen before: the synchrotron light emitted by a *single electron in orbit* which was visible to the naked eye through one of the portholes: "Bruno took an immense pleasure in showing this phenomenon which, to a certain extent, was commonplace, but at first sight appeared incredible. His enthusiasm was extreme [...]" (Amaldi 1981, 32). It also happened that Philip Dee, from Glasgow University, was visiting Rome with his wife in that particular time, and Bruno could show him AdA and its success.

But soon, the enthusiasm waned. Not enough electrons and positrons could be generated to sustain the probability of collisions. The main obstacle appeared to come from the slow injection mechanism from the electron synchrotron, whose bremsstrahlung photons, injected into the doughnut to hit a tantalum target, were too weak to produce a number of electrons and positrons sufficient to prove that collisions between them were taking place. Still the machine was operating, which by itself was a first success, a result never attained before, worth communicating to the particle physics community. In June, Touschek presented these results to the CERN *International Conference on Theoretical Aspects of Very High Energy Phenomena* (Bell et al. 1961).[2]

11.1 The 1961 Geneva Conference

Touschek's contribution appeared in the session on *Electromagnetic Interactions* which consisted of three talks, in the following order: one by Burton Richter, from Stanford University,[3] the second by Touschek, the third one by Raoul Gatto.[4] Gatto's talk was focused on the theoretical aspects of electron–positron physics, presenting the two papers authored with Nicola Cabibbo, the short one published in February 1960 (Cabibbo and Gatto 1960) and the lengthier one submitted to *The Physical Review* (recorded submission as of June 8th, 1961), just before leaving for the Conference (Cabibbo and Gatto 1961).

The contrast between the two first talks is striking. Richter seems to justify the choice NOT to start electron-positron experiments. In juxtaposition, Touschek is announcing they have already built a small prototype and have proposed to build a bigger ring. Let us see the two talks in their most important passages, at a distance of almost 60 years, with what, in Italian, is called *Il senno di poi*.[5]

---

[2] For these Proceedings, see also https://cds.cern.ch/record/280184.

[3] In 1976, the Nobel Prize in Physics 1976 was awarded jointly to Burton Richter and Samuel Chao Chung Ting "for their pioneering work in the discovery of a heavy elementary particle of a new kind", from https://www.nobelprize.org/prizes/physics/1976/summary/.s

[4] Raoul Gatto was at the time at University of Rome, and, two years later, would become Professor of Theoretical Physics at University of Florence.

[5] Wisdom, after the facts.

The session was started by Richter with a talk entitled *Colliding Beams Experiments* (Bell et al. 1961, 57)

Richter: Let me begin by saying that we all hope that this will be the last talk about what we are going to do when the experiment is ready. We hope that next year we can talk about what we have done.

He then continues by describing the two ring set-up for the proposed electron-electron collision experiment and the hoped for time schedule, and goes on discussing *Positron Experiments*, opening with a denial:

Richter: Whenever this subject has been brought up in the past, we have refused to commit ourselves about its prospects. I am not going to change this policy, but I would like to discuss the difficulties of the positron experiments a bit.

Having said this, he adds:

Richter: These problems are not the main reason for our long-standing silence on the experiment.

Basically, he insists that the electron experiments must first be shown to work well, before starting with positrons, namely:

Richter: Until we know what we can do in storing a beam, we cannot say anything about the positron experiment.

Then Touschek starts, with a simple concise sentence (Bell et al. 1961, 67)

Touschek: Frascati is developing two storage rings.

and then, after rapidly describing the ADONE project, which was still under design, he moves to describe the first project:

Touschek: The first project - AdA - was started in February 1960. It was clear from the beginning that this project would be a gamble, the calculated intensity of the machine being about a factor 500 less than what was needed for experimentation. It was nevertheless decided to go on with the project, mainly because it was hoped that experience in storage problems could be most rapidly gained in this fashion and that eventually ideas for increasing the intensity might be forthcoming.

Touschek's contribution included many technical drawings, as well as the phototube record of the number of electrons circulating in AdA, shown in the previous chapter. In Fig. 11.1 we show two photos extracted respectively from Richter's and Touschek's talk.In his talk, Touschek gives one of the clearest expositions of AdA's technical details, listing in his unique, and extraordinarily precise way, all the excellent reasons to start experimentation with electron-positron collisions. To complete this session, Gatto followed with a theoretical talk, where he makes a very prophetic statement (Bell et al. 1961, 76)

Gatto: High Energy electron-positron colliding beam experiments may become a field of spectacular development in high energy physics.

Fig. 4. Finished shells of Y-magnets at each end of interaction region.   Fig. 9. Tower with storage ring mounted for first test.

Fig. 11.1 Two figures shown at the 1961 Conference on Theoretical Aspects of Very High-Energy Phenomena. held 5—9 Jun 1961, CERN. At left the Y magnet at Stanford, with the region where the two electron orbits are tangent to each other to produce electron electron collisions from B. Richter's talk, at right AdA from Touschek's talk, (Bell et al. 1961)

11.2   And Then Came Pierre Marin and Georges Charpak:
Un vrai bijou

The conference at CERN and Touschek's talk represent a landmark in the development of electron-positron colliders.

Word spread that things were happening in the Laboratory on the Tusculum Hills. In the far away Russian city of Novosibirsk, beyond the Urals, at the Institute for Nuclear Physics, the scientists understood they were not alone in their work on electron-positron storage rings and increased their efforts (Baier 2006). Closer to Frascati, at CERN itself and in France, at the Laboratoire de l'Accélérateur Linéaire d'Orsay, *le LAL*, interest arose as to what the Italians were doing. France, Italy and Switzerland are close, travel between the Laboratories was frequent, and scientists could easily go and see by themselves what was new. And this is precisely what happened: two French scientists, one from Orsay, the other from CERN, went to Frascati to see with their own eyes what Touschek had announced.

At the time, in Orsay, a Linear Accelerator had been working since 1959, and the question of how to best exploit its discovery potential was object of debate. The team which had built the Linear Accelerator included George Bishop and Pierre Marin, who knew each other from Oxford and had come together to France in 1955, to work on the project. Marin (1927–2002) was then 34 years old and would become one of the main artifices of France's accelerator program. In 1961, with the linear accelerator by now successfully built and working, he was wondering which direction his research should take. When Marin asked around for interesting things to do in his research, he was told to go see what was happening in Frascati.

As Pierre Marin later wrote (Marin 2009, 46)[6]:

> *Marin:* Returning from a stay at CERN, after my thesis, I was searching for my own research directions, and G. R. Bishop suggested that I go to visit Frascati, where very intriguing things were happening. So, I went there, in the month of August, together with Georges Charpak, who was, at the time, collaborating with both CERN and American physicists on the measurement of the anomalous muon magnetic moment.[7]

The visit actually took place in July, as August is unbearably hot in and around the city of Rome.[8] Those who can, escape to the beaches, or go North, to the Dolomites, or, in those days, simply to the nearby hills, the so called Alban hills, from the name of the pre-Roman town of Alba Longa. The Alban Hills, *I colli Albani*, include a half-circle of volcanic hills, which, South East of Rome, limit the plains where the city lies. Among them, and closest to the city, is the Tusculum Hill, sloping down to the Frascati Laboratory mid-way to Rome, cooler and shady in the summer. In a hall next to the synchrotron Marin saw what Touschek had described in June, at the CERN conference (Marin 2009, 46).

> *Marin:* [...] a group of high caliber physicists [...] B. Touschek, C. Bernardini, G. Ghigo, F. Corazza, M. Puglisi, R. Querzoli and G. Di Giugno, [who] showed us with great pride a small machine, AdA, *un vrai bijou*.[9] [...]

The physicists, whom Marin and Charpak met in Frascati, had been part of the synchrotron team, with the exception of Touschek and Giuseppe Di Giugno, nicknamed Peppino, who had joined the AdA group, after graduating with Ruggero Querzoli in 1961 on a thesis about the synchrotron, Fig. 11.2. But, as Touschek and Bernardini knew, there were problems with their wonderful little machine. The group of Italian and French scientists started thinking about ways to make AdA produce some physics, beyond the great pride of having succeeded in accumulating electrons and positrons in some number. The estimates which Touschek had done in March of the previous year, and which were based on cross-section calculations by his colleagues Cabibbo and Gatto, showed that, even if some electrons and positrons were circulating in the ring, luminosity was too low and annihilation could not be proven. To go beyond and demonstrate the feasibility of this type of machines to do physics at high energy, the problem of injection had first to be solved. The Italian team had applied some very ingenious ideas but the rate remained orders of magnitudes smaller than

---

[6] Marin's book is presently out of print. A copy was given to the Author by Jacques Haïssinski, in 2009, during a visit to the Laboratoire de l'Accélérateur Linéaire. The book is in French, present translation is by A. Srivastava.

[7] Georges Charpak (1924–2010) was awarded the 1992 Nobel Prize in Physics 'for his invention and development of particle detectors, in particular the multi-wire proportional chamber', from https://www.nobelprize.org/prizes/physics/1992/summary/.

[8] Marin, in his telling almost forty years later, indicates the month of August for his visit. However from the letter sent by Laboratoire de l'Accélérateur Linéaire director, André Blanc-Lapierre, to Frascati director, Italo Federico Quercia, in December 1961, we learn that the visit had taken place in July. Given that Marin's memoirs were written some forty years after his visit, the indication in André Blanc-Lapierre's letter is more likely to be the correct one.

[9] A real gem.

Fig. 11.2 Ruggero Querzoli (at left) and his student, Giuseppe Di Giugno, in custom partty outfits, photograph courtesy of Giuseppe Di Giugno, all rights reserved

what a successful experiment would require. The problem was always the same: the beam of electrons in the synchrotron was too weak, it could not lead to a photon flux sufficient to create enough positron-electron pairs in AdA's inner target. Then, Pierre Marin observed that, in Orsay, the Linear Accelerator provided a well focussed 500 MeV electron beam, with an excellent intensity of 1 μA. Bernardini and Touschek looked at each other, moved slightly aside, talked for a few minutes, and then came back to Marin, Fig. 11.3. The question they posed was (Marin 2009, 47):

> *Touschek:* […] Do you think that LAL would agree to receive AdA?

And the answer was:

> *Marin:* A priori, the new director André Blanc-Lapierre would be quite open to welcome this sort of ideas.

Hopes were revived and unexpected perspectives opened. Marin and Charpak left. The seed of the future of AdA as a feasible way to high energy colliders had been planted. Marin carried back to Orsay a promise to follow up the idea of a collaboration and perhaps a transfer of people and machinery to France. In the months to follow, steps for a collaboration and the transfer of AdA from Frascati to Orsay were put in motion.

The operation was not trivial, as it involved moving a working accelerator, which was property of a Government institution, across two countries, a trip of about

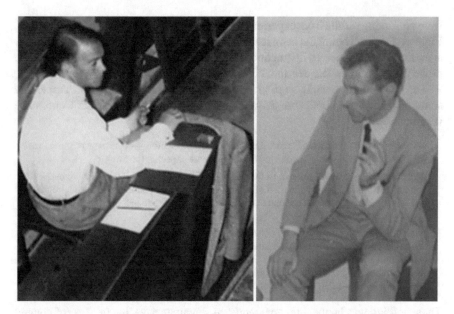

Fig. 11.3 Bruno Touschek, in early 1960s, courtesy of Giuseppe Di Giugno, and Pierre Marin, at right, courtesy of Jacques Haïssinski

1500 km, through custom and border controls. We should remember that in 1961, in Europe, there was no free circulation of goods and people and AdA was precious: it had costed some 20 million Lire, a large expenditure for pure research in Italy, at the time. The transfer therefore needed to be approved at the highest level. In addition, as the exchange of letters between the two Laboratories later showed, the collaboration involved a matter of scientific policy in Europe, something to which Edoardo Amaldi, one of CERN's founders, was particularly sensitive.

11.3 The 1961 Conference at Aix-en-Provence

As soon as the August vacations ended, both in Orsay and Frascati, activities started in earnest: the immediate scope was preparation of talks to present at the imminent conference in Aix-en-Provence, where the community involved in nuclear and particle physics, both experimentalists and theorists, was to gather in mid-September from 14th to 20th. The title of the conference, *The Aix-en-Provence International Conference of Elementary particles*, addressed directly, for the first time, the emergence of the field of elementary particle physics, singling it out from nuclear or accelerator or high energy physics, where it had been so far included. It was mostly a conference where up-to-date theoretical ideas would be debated in the plenary sessions, with experimental talks and other theory papers presented in parallel. The Italians were presenting results from experimentation with the Frascati electron synchrotron, Ruggero Querzoli being the team leader and giving the talk. From Orsay, Marin was attending, having received instructions from LAL director, André Blanc-Lapierre, to probe the Frascati colleagues about both AdA and ADONE.

The concluding remarks were given by Richard Feynman. It is worth repeating some of them here, applicable as well to what was happening at the Conference (Feynman 1962):

> *Feynman:* I want to ask what is most characteristic of the meeting — what new positions are we in at the present time — what kind of things do we expect in the future?

> At each meeting it always seems to me that very little progress is made. Nevertheless, if you look over any reasonable length of time, a few years say, you find a fantastic progress and it is hard to understand how that can happen at the same time that nothing is happening (Zeno's paradox).

> I think it is something like the way clouds change in the sky — They gradually fade out here and build up there and if you look later it is different. What happens in a meeting is that certain things which were brought up in the last meeting as suggestions come into focus as realities. They drag along with them other things about which a great deal is discussed and which will become realities in focus at the next meeting.

Thus, nothing was happening, apparently, as Feynman says, but he was (of course) right, as new realities would soon come into focus and "fantastic progress" would take place in due time.

Among the 90 talks, both plenary and parallel (Crémieu-Alcan et al. 1962), there was one by Raoul Gatto, who had co-signed ADONE's proposal in the previous month of February. His talk was in one of the parallel sessions and, as he had done in Geneva two months before, it addressed the discovery possibilities of electron-positron physics.

In addition to Gatto's talk, detailed discussions, about ADONE's prospects and AdA's results, took place between Pierre Marin from Laboratoire de l'Accélérateur Linéaire and Ruggero Querzoli from Frascati. Querzoli at the time was working on the synchrotron, but also on AdA, and had been one of the *physiciens du cru,*[10] whom Marin had met in July, when visiting AdA.

Feynman could not have known of such discussions, and no notice was given to Gatto's talk in his summary. This is partly understandable in light of the fact that the calculations by Cabibbo and Gatto were not using new techniques or envisioning new theoretical scenarios: they actually used the tools of QED and Feynman graphs to calculate the cross-sections of all known processes of interest in electron-positron physics, some of which were already present in the literature. The value of the paper was in the exhaustive study and its completeness. The great novelty of the paper, which had been submitted the previous July to *The Physical Review* (Cabibbo and Gatto 1961), was that it presented realistic calculations for processes that could now be observed and measured in presently foreseeable experimental set-ups, such as ADONE's.

---

[10] High caliber physicists, in English.

And so, while "nothing was happening", our heroes were building in the wings a totally new type of particle accelerator, and a revolution was being set in motion which would produce powerful tools of discovery.

11.4 Letters and Visits

On his return to Orsay, Marin prepared a detailed report to his director, accompanying it with a hand written note. The report, *Entrevue avec le Professor Querzoli de Frascati le 19-9-61* is divided into two parts, the first concerns AdA, the second is about ADONE.[11] About AdA, after describing the present status of the machine, successes, and limitations, such as only a moderate vacuum, a slow injection mechanisms etc., Marin listed the expectations of the Frascati group in case that a much better vacuum, i.e. 10^{-9} mmHg versus the actual 10^{-6}, could be reached. Among these, were the measurement of the annihilation process $e^+e^- \rightarrow 2\gamma$, studying the phenomenon of space charge and the effects of changing the beam energy. But then, at the end of this list of expectations, comes the crucial proposal. We transcribe it here in its original French version:

> *Marin:* Si les prévisions de calculs sont exactes, il semble possible de réaliser ce programme à Frascati […] S'il s'avérrait qu'il ne puisse être réalisé à Frascati, A.D.A. serait transporté à Orsay auprès de l'Accélérateur Linéaire.[12]

The report then addresses the new Frascati project, ADONE, an electron-positron storage ring, with a beam energy of 1.5 GeV. Marin informs his director that, although the project has not been officially approved, the Italians are rather confident it will be accepted, notwithstanding its much higher cost with respect to AdA. ADONE's budget was 2.5 billion lire, two orders of magnitude costlier than AdA's, with half of it to be spent on a powerful linear accelerator. This was a very ambitious project, which could be started after some further experimentation with AdA. Marin adds that Querzoli is very favourably inclined to have a French scientist visiting with the AdA team, in the coming months.

11.4.1 A Stormy Beginning

Three months passed however, and no further exchanges seem to have taken place in the immediate period following Marin's report. Then, at the end of December, a letter from André Blanc-Lapierre reached Italo Federico Quercia, then director of the

---

[11] Copies of the report and of the accompanying letter by Pierre Marin were obtained from LAL Archives, courtesy of Jacques Haïssinski.

[12] If the predictions of the calculation are exact, it seems possible to realise this program in Frascati …If it will happen that it cannot be realized in Frascati, A.D.A will be transported to Orsay, next to the Linear Accelerator.

Frascati Laboratories.[13] The letter, dated December 22nd, 1961, starts with a rather unexpected sentence: "Dear Professor Quercia, we are starting preliminary studies for a 1.3 GeV storage ring for positrons and electrons in Orsay", and then continues proposing that two or three people from Orsay come to Frascati to discuss some points about storage rings. André Blanc-Lapierre proposed that visits could start any time after the 23rd of January.[14]

This letter is very interesting, as it indicates that a French decision, to start planning for a storage ring at LAL, must have been taken between the time of Marin's report to the Orsay director in September (1962), and the December letter to Frascati, which was proposing a visit by some French scientists. The answer to this letter came from Edoardo Amaldi, then director of the Physics Institute of University of Rome, and INFN president. Amaldi's letter reached Orsay one month later, and was written in French, which, in those days, was the official language of diplomacy.[15] Indeed, André Blanc-Lapierre's letter had ruffled some feathers in Rome.

André Blanc-Lapierre's letter was probably received in Frascati during the end of the year vacation. As soon as the normal laboratory activities started again, Quercia called a meeting with Fernando Amman, Carlo Bernardini, Gianfranco Corazza and Bruno Touschek for Wednesday January 3rd. He showed them the letter from the French director and the decision was taken to answer positively to the French request, including agreeing to the proposed date of January 23rd for the visit to start.[16] Hopes for a collaboration which could solve AdA's injection problem and open the way to prove full feasibility of electron-positron storage rings, were renewed. Touschek, among the AdA team, was particularly anxious that a positive answer be sent immediately to the French colleagues. A letter by Quercia to Touschek, dated January 12th, less than 10 days after the meeting, implies some tension between Quercia and Touschek on this issue, since Quercia mentions a telephone call by Touschek "this morning", whose reason he "could not understand well".[16] We can only imagine that Touschek, when he saw that no letter to LAL had been sent after almost ten days from the meeting and more than two weeks from André Blanc-Lapierre's letter had passed, became upset, fearing that the French would be moving ahead with their proposed plan (for ACO, Anneau de Collisions d'Orsay), and that the visit would be delayed. He must have pressed the Laboratory director for an answer, in not too diplomatic terms. This is quite understandable, since he certainly knew that the future of AdA rested on the collaboration with LAL scientists and the use of their Linear Accelerator.

---

[13] See also Quercia's biography, *Italo Federico Quercia, note biografiche e documenti*, edited by Ugo Spezia, Collana di Storia della Scienza, Patrocini : SIF, ENEA, INFN, 2007.

[14] LAL Archives, courtesy of Jacques Haïssinski.

[15] LAL Archives, courtesy of Jacques Haïssinski.

[16] I. F. Quercia to B. Touschek, January 12th, 1962, Bruno Touschek Archive, Series 1, Folder. 4, Box 1, 1962–67 Correspondence, Sapienza University of Rome–Physics Department Archives in https://archivisapienzasmfn.archiui.com.

Touschek's phone call to Quercia did produce a reaction, and, on January 16th, Amaldi, answered to André Blanc-Lapierre.[17] The matter, indeed, did not simply involve a friendly agreement among scientists and two laboratories, as Touschek may have thought. Its complexity can be glimpsed from the fact that Quercia had sent a formal letter to answer Touschek's phone call. Since the University of Rome and the Frascati Laboratory were connected by a direct shuttle bus, which used to leave from the University of Rome every hour, it would have been simpler for Touschek and Quercia to just talk to each other: instead we have a formal letter, from the Laboratory director, in Frascati, to Touschek, in Rome. In fact, as one can see from the correspondence which followed, the matter needed to be thought over, with adequate consideration of scientific priority.

The reason for the delay in answering André Blanc-Lapierre's letter is apparent in the response which Edoardo Amaldi sent to André Blanc-Lapierre, four days later, on January 16. Amaldi starts his letter with the acknowledgment of the proposed collaboration, referring to Amman and Touschek for its implementation, but then he addresses the point of major interest for him, that of European accelerator strategy and Italian scientific priority. Unlike the previous one by André Blanc-Lapierre or anyone of those which followed, this is a formal letter, as proven by the fact that it was written in French, the language of diplomatic exchanges, not in English. We quote here from the letter:[14]

> *Amaldi:* [...] Pour ce qui concerne la construction d'un Accélérateur qui accumule dans le même anneau électrons et positrons, je désire vous faire savoir que le group dirigé par l'Ing. Amman est déjà arrivé à un degré plutôt avancé dans le projet d'une machine de ce genre pour 1,5 + 1,5 GeV [...] et que cet appareil constituera la partie essentielle de notre Second Plan Quinquennal [...].
>
> Nous croyons, pourtant, qu'il faudrait, en préparant des autres programmes, considérer tout cela, afin d'éviter des redoublements [...] Ceci est un cas particulier du problème plus général de la coordination [...] des plans de recherche de divers groupes nationaux surtout entre deux pays déjà si liés par des intérêt[s] communs, comme la France et l'Italie.[18]

He then continues with the authority given to him as one of the founders of CERN and protagonist of the European scientific reconstruction after the war:

> *Amaldi:* Un premier pas, dans ce sens, a été déjà fait avec la construction de l'European Accelerator Study Group, aux réunions duquel on a presenté le project italien dès Decembre 1960.[19]

---

[17] For a list of INFN Presidents, see https://www.presid.infn.it/index.php/it/10-articoli-del-sito/73-presidenti-dal-1954.

[18] Concerning the construction of an Accelerator that can store in the same ring electrons and positrons, I wish to let you know that the group directed by Eng. Amman has already reached an advanced stage in planning such machine with [energy] 1,5 + 1,5 GeV [...] and this apparatus will constitute the main part of our Second Five-Year Plan [...]. We believe, therefore, that, in starting other programs, one should consider all this, in order to avoid duplications [...] This is a particular case of the more general problem of coordination [...] of the research plans of different national groups, especially between two countries already so closely connected by common interests such as France and Italy.

[19] A first step in this direction has already been taken with the establishment of the European Accelerator Study Group, where the Italian project has been presented since December 1960.

This letter was received in Orsay, on January 24, and was followed by another letter by Amaldi to André Blanc-Lapierre, in which, having clearly stated Italy's priority, Amaldi welcomed the start of the collaboration. In this second letter, dated January 23rd, the exchanges between the two laboratories were blessed and specific plans for visits could start. Amaldi invited André Blanc-Lapierre to come to Frascati "accompanied, eventually, by another Member of your Group". The occasion was going to be an informal meeting held on the dates 7-8-9 of February, to discuss present results from the electron synchrotron, work at CERN by Italian groups, and, last but certainly not the least in Amaldi's priorities, reports on the ADONE project. This second letter reached Orsay on January 29th.[14] At his end, after his December letter, André Blanc-Lapierre had not remained idle: even before receiving the Italians' answer to his query, he had contacted the French Atomic Energy Commission, *Commissariat á l'Énergie Atomique (CEA)*, probably in the person of its President, Francis Perrin.[20] We learn that, on January 16th, André Blanc-Lapierre had proposed the new date of February 5th for a visit by three French scientists, F. Fer, Marin and Boris Milman, adding that one or two people from the French Atomic Energy Commission could be part of the expedition.[21] The inclusion of official visitors, not just scientists, indicates the importance the French scientific establishment attributed to the storage ring projects.

11.4.2   Visits and Meetings: Towards the Transfer of AdA

From February through June 1962, the collaboration was put in motion, resulting in reciprocal visits and (at least) one meeting in Geneva.

The first visit took place, as proposed by Amaldi, on the occasion of the Frascati *Congressino* held from 7th to 9th February 1962. Amaldi's introductory talk was dedicated for the major part to the ADONE project, which he clearly saw as crucial for the future of Italian and European high energy physics. André Blanc-Lapierre did not come, but four French scientists were welcomed by Amaldi in his opening address to the participants, H. Bruck, F. Fer, F. Guérin (M.le Guérin offers a welcome appearance of a woman in this story) and P. Marin.[22] In addition, the non Italian participants included F. Lefrançois, also from Orsay, Robert Wilson from Cornell, a UK scientist from Harwell, P.G. Murphy from Rutherford Laboratory, attending the meeting, Fig. 11.4. This may have created an opportunity for Touschek to visit

---

[20] Amaldi's letter reached Orsay only on January 24th, but the French Atomic Energy Commission, is mentioned in a 16th January letter written from Orsay to Quercia (on the same day as Amaldi's positive answer for a collaboration left Rome). We have not located this letter, but its existence is acknowledged, and its content described, in a January 23rd letter by Quercia to André Blanck-Lapierre—director to director.

[21] Marin and Milman had been part of the team which came from Oxford in 1955. F. Fer was among the participants in the Geneva 1961 Conference.

[22] The Report of the Meeting is available at http://www.lnf.infn.it/sis/preprint/detail-new.php?id=2980, as LNF-62 / 037.

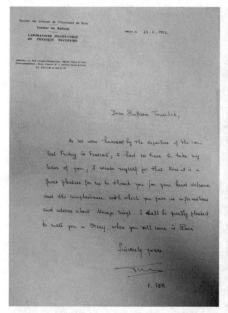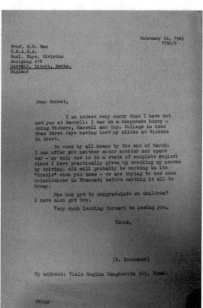

Fig. 11.4 February 16th, 1962 letter from F. Fer to Bruno Touschek thanking him after visiting Frascati at left, and February 24th, 1962 letter from Touschek to E. R. Rae, at the United Kingdom Atomic Energy Authority, indicating a visit to Harwell by Touschek, Sapienza University of Rome–Physics Department Archives, https://archivisapienzasmfn.archiui.com

the UK, later in February as mentioned in a letter from Touschek to Ernest Rae, at Harwell.[23]

By the end of February, Touschek was already confident that AdA would go to Orsay, at the same time he was also actively pursuing approval for the bigger ADONE project. Now that the "ice had been broken", as Touschek wrote in April to Blanc-Lapierre, Fig. 11.5, more visits and exchanges between Frascati and Orsay soon followed. The letters which passed between the two Laboratories during this period are very warm and friendly, as both sides were now eager to pursue the possible transfer of AdA and discuss the practical implications.

Interest in the Frascati projects was not limited to the French Laboratory. In February, Touschek was invited to attend a meeting at CERN organized by Kjell Johnsen, to examine large accelerator projects for the future of high energy physics.[24] Since 1961, a special Study Group on New Accelerators led by Johnsen had been established to work out the design study of the Intersecting Storage Rings, a

---

[23] For other contacts, such as a letter from O'Neill dated February 8th, 1962, see Sapienza University of Rome–Physics Department Archives, Bruno Touschek Papers Series 1, Folder. 4, Box 1, 1962–67 Correspondence.

[24] The subject of the meeting appears in the April 4th letter by André Blanc-Lapierre mentioned later in the text.

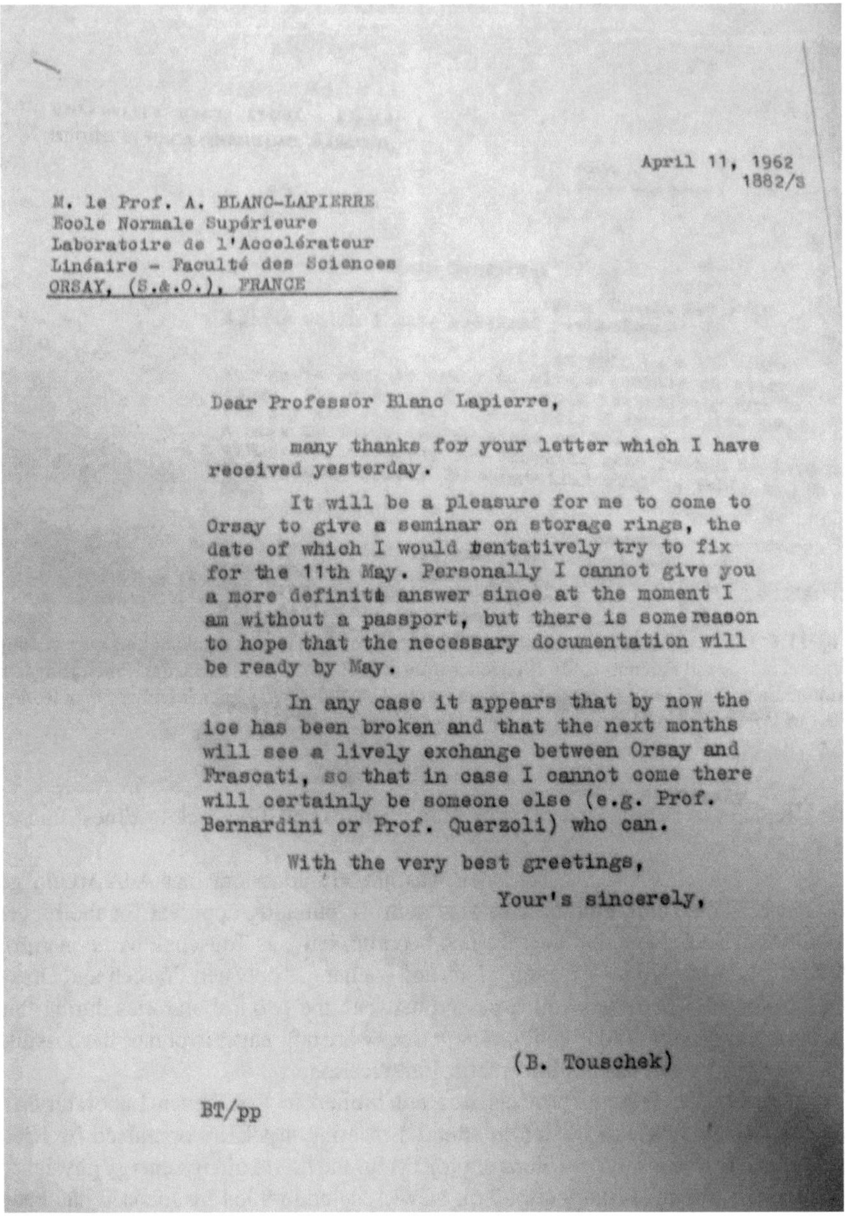

April 11, 1962
1882/S

M. le Prof. A. BLANC-LAPIERRE
Ecole Normale Supérieure
Laboratoire de l'Accélérateur
Linéaire - Faculté des Sciences
ORSAY, (S.&.O.), FRANCE

Dear Professor Blanc Lapierre,

many thanks for your letter which I have
received yesterday.

It will be a pleasure for me to come to
Orsay to give a seminar on storage rings, the
date of which I would tentatively try to fix
for the 11th May. Personally I cannot give you
a more definite answer since at the moment I
am without a passport, but there is some reason
to hope that the necessary documentation will
be ready by May.

In any case it appears that by now the
ice has been broken and that the next months
will see a lively exchange between Orsay and
Frascati, so that in case I cannot come there
will certainly be someone else (e.g. Prof.
Bernardini or Prof. Querzoli) who can.

With the very best greetings,

Your's sincerely,

(B. Touschek)

BT/pp

Fig. 11.5 Copy of the letter from Bruno Touschek to André Blanc-Lapierre, Bruno Touschek Papers in Sapienza University of Rome–Physics Department Archives, https://archivisapienzasmfn.archiui. com

proton-proton collider which would be the second large CERN machine (Russo 1996). Johnsen was clearly quite interested in Touschek's participation in the meeting. The invitation was very pressing, as we can see from Touschek's answer[25]:

Dear Dr. Johnsen,

your threat of ringing me has been transmitted by Amman and I am looking forward to its execution.

I shall certainly come for 3 or 4 days to participate at the enthusiast's meeting ...

With many greetings and looking forward to your call ...

André Blanc-Lapierre had been looking ahead to a possible encounter with Touschek in Geneva at the end of March, in the context of this meeting organized by Kjell Johnsen. However, notwithstanding his earlier positive response, for some reasons, Touschek did not attend the meeting. But Carlo Bernardini and Amman were there and the practical details of AdA's transfer to Orsay started to be discussed. André Blanc-Lapierre was sorry not to have met Touschek, and pressed him, on an April 4th letter, to come to Orsay and give a seminar.[23] Bruno Touschek welcomed the invitation, but had no passport at the time. Hoping to receive it in time, he chose the date of May 11th for his visit.[23]

It is uncertain whether Touschek was able to go to Paris and Orsay for the promised seminar. It would not have been his first visit, anyway. There exists a testimony of Touschek's visiting Paris and giving a seminar, from the theoretical physicist Maurice Lévy. In telling the episode, Lévy, professor at École Normale Supérieure who had been instrumental in calling André Blanc-Lapierre from Algiers to France in 1961, was unsure about the precise dating, tentatively placing it in the late 1950s.[26] Whatever the date of this episode, Lévy's anecdote brings Bruno's personality into a vivid light. This is what happened[27]:

Lévy: I knew Touschek by reputation and I had met him at several conferences. In Paris in the framework of our theory group, we had a weekly seminar where we invited people from all over and on several occasions we invited Touschek, who came and talked to us. Unfortunately I do not remember on which subject he talked. In fact he came at least two or three times. ...I have a small anecdote. In one of the visits, we had put him up in a small hotel on Boulevard Saint Michel and before retiring at night, he had put his shoes outside the door of his room, and when he opened the door on the next day, the shoes had disappeared. So, he had a problem, the owner of the hotel kept saying "this is a small hotel sir, we don't make shoes" and so on. Finally the proprietor of the hotel lent him a pair of shoes, which

---

[25] B.T. to K. Johnsen, 21 February 1962, as in footnote 23.

[26] This anecdote is reported in (Bernardini et al. 2015), and was originally told to the Author by Maurice Lévy during a interview in Paris, on May 24th, 2013. Lévy's memory of the episode as from the late 1950s could have partial confirmation in a January 16th, 1960 letter by Touschek to his family, where he mentions a trip to Paris and 'particularly beautiful shoes that they stole from me in Paris: from Funaro and very soft leather'. This trip to Paris appears to have been done for INFN official purposes, since he continues: 'When I reported the results of the Paris negotiations here at the 'Giunta' of the Research Council [INFN], a vote was taken so that I also get paid for my shoes and gloves.'

[27] From the interview with Maurice Lévy in May 2013, in Paris, for the docu-film Touschek with AdA in Orsay by E. Agapito, L. Bonolis and G. Pancheri.

were much too big for him, two or three sizes too big, and he went to a shop, at nine o'clock when the shops open, to get another pair of shoes for himself. He told us the story later on, [when he came to the Laboratory] with great sense of humour, Touschek was well known for his sense of humour.

The months to follow saw many lively exchanges, and detailed plans for AdA's transfer were made. One of the members of the team of the Linear Accelerator, François Lacoste, remembers visiting Frascati in the Spring of 1962[28]:

> *Lacoste:* I was witness to some initial contacts between Frascati and Orsay in '62 and I had occasion of one visit to Frascati with Boris Milman, where I met Bernardini and Touschek and we looked at some details about how it would be possible to bring AdA to Orsay.

11.5 How AdA Left Italy and Arrived in France

The transfer posed many technical problems, such as maintaining the vacuum in the doughnut while AdA traveled across the 1500 km and more between Frascati and Orsay. A major challenge in the storing of positrons was in fact the requirement of an extreme vacuum in the doughnut, to prevent scattering with the residual gas. To this aim, a legendary vacuum as low as 5×10^{-10} mmHg was reached in Frascati by Gianfranco Corazza and Angelo Vitale, nicknamed *Angelino*, a technician who specialized in outgassing in very low pressure vessels. Reaching such a low level vacuum required two or three months, and it was essential not to lose it during the transfer. As the vacuum was maintained by powerful devices constantly pumping residual gas from the doughnut, batteries would be needed to keep the pumps working and accompany AdA through Italy and the Alps into France and on to Orsay. Two trucks were hired for the transfer: a bigger one with AdA, sealed with its vacuum, with pumps and batteries, and a smaller one for other heavy equipment. Other lighter components could follow by plane.

There were also problems of a different nature: the head of the Italian National Committee for Nuclear Energy (CNEN), a government agency owning and overseeing the Frascati Laboratories, needed to be contacted and his agreement for AdA to leave Italy had to be obtained. Touschek, advised by Amaldi, wrote a letter and AdA was granted permission to go. Another possible problem involved the crossing of the border between Italy and France. The chosen route was across the Alps and the custom station was at Modane, near the Frejus.[29] Passing the French customs could be tricky, and had to be prepared in advance. Both Italian and French officers in the foreign ministries needed to be informed so as to act in case of difficulties.

When the time for the transfer came near, Touschek sent a letter to Francis Perrin, Haut Commissaire à l'Énergie Atomique au CEA, who had worked with Frédéric Joliot on nuclear chain reactions, and was very powerful and influential on all science

---

[28] From an interview with François Lacoste in May 2013, in Orsay, for the docu-film *Touschek with AdA in Orsay*.

[29] As the Frejus road tunnel did not yet exist, roads would go through the Moncenisio Pass.

matters in France. Perrin had been one of the founders of CERN, and knew Amaldi very well. Informal exchanges between them started as the date of transfer drew close. At the end of June, everything was finally ready.

Touschek's letter of June 28th, 1962, Fig. 11.6, reads:

Dear Professor Perrin,

I enclose a list of material for the second convoy Frascati-Orsay, which will presumably leave Rome on the 4th of July and should arrive in Paris on the 7th.

We very much hope that there will be no difficulties at the customs but, in case of emergency, we would much appreciate the help so kindly offered by you to Prof. E. Amaldi.

[...] it contains the vacuum chamber at 5×10^{-10} mm ...The ideal solution would be if some competent official at the Modane customs office could be informed before hand.

I will take the liberty of wiring you the exact (as near as possible) time at which the convoy can be expected to pass the frontier [...]

The story of AdA's transfer from Italy to France has been told on various occasions by its protagonists, Carlo Bernardini, who saw the convoy leave Frascati, and François Lacoste, who saw it arrive in Orsay.[30] When the time came for the transfer, the Frascati Laboratories called a trusted moving firm, in the person of 'legendary' Signor Grossi, *Il mitico signor Grossi*, as Bernardini calls him, often employed by the Laboratories for moving equipment to and from University of Rome, (Bernardini 2006). The 8 ton iron magnet inside which AdA's doughnut was placed, was put on the truck, with its set of batteries to keep the pumps working, and maintain the vacuum. Any mishap, namely any air that got into the doughnut would have meant months of extra work in Orsay to clean the chamber. The batteries had to be able to provide power for two, maximum three days, enough to cover the trip from Frascati to Orsay.

There was some worry that such weight could unbalance the truck and make it difficult to keep control. Signor Grossi did not think so, but Touschek was very concerned. So he jumped on the driver's seat and started driving the truck around the large square in front of the synchrotron building. Perhaps he just wished to try the driving, the result instead was to destroy a lamp post. Properly subdued, he let Signor Grossi take over and finally AdA left.

Travelling along the Italian Western sea-cost, the trucks then moved North towards the Alps, to reach the state border between Italy and France, in Modane, on the French side of the Alps. Angelo Vitale, the specialist of the vacuum, was travelling with the convoy. Customs checks were a very serious business at the time, so the customs officer asked: "What's inside?", meaning inside the bulky green round object, out of which protruded a short tube with a small round quartz window. This was the window through which scientists and their visitors in Frascati could see the synchrotron light emitted by the circulating electrons. What could Vitale say? The legend goes that he answered "There is nothing, just nothing", which was indeed the truth. But this was not sufficient to let AdA pass. So the help which Touschek had asked for, in his letter to Perrin, became necessary. Vitale called Corazza, who called Amaldi, who

---

[30] Carlo Bernardini, *Fisica vissuta*, 2006 Codice Edizioni, Torino, and François Lacoste interviewed by the Author, May 2013, Orsay.

June 28th, 1962
2015/S

Prof. F. PERRIN
Haute Commissaire à
l'Energie Atomique
69, rue de Varenne
P A R I S

Dear Professor Perrin,

I enclose a list of material for the second
convoy Frascati-Orsay, which will presumably leave
Rome on the 4th of July and should arrive in Paris
on the 7th.

We very much hope that there will no difficulties at the customs but, in case of emergency, we would
much appreciate the help so kindly offered by you
to Prof. E. Amaldi.

The present operation is most critical since
it contains the vacuum chamber at 5×10^{-10}mm. The
ideal solution would be if some competent official
at the Modane customs office could be informed before
hand.

I will take the liberty of wiring you the exact
(as near as possible) time at which the convoy can
be expected to pass the frontier.

With best thanks for your kindness and interest,

Yours sincerely,

(B. Touschek)

BT:pp

Fig. 11.6 Copy of the letter from Bruno Touschek to Francis Perrin about the transfer of AdA to Orsay, Bruno Touschek Papers, courtesy of Sapienza University of Rome–Physics Department Archives, https://archivisapienzasmfn.archiui.com, documents provided for purposes of study and research, all rights reserved

called Perrin, who called the French Minister of Interior, and from there, down the line of command to the French customs officer, until finally AdA was allowed to enter France (Bernardini 2006). The story as told by Lacoste is very similar:[31]

> *Lacoste:* The transfer of AdA to Orsay came during the summer of 1962 and I remember waiting for AdA and witnessing the arrival of AdA in Orsay.
>
> We had to wait a bit longer than we thought because AdA was stopped at the *frontière* by the French customs, who wanted to understand what they were bringing, what the team from Frascati was bringing and they were especially suspicious of what was inside AdA, and they wanted to look into, perhaps thinking that it could be drugs or whatever, into the vacuum ring, which was pumped during the trip because they had made the degassing, and so on, and it was vacuum of 10^{-8} and at that time good vacuum was very difficult to bring, so they had the pump working during the transit. So the customs wanted to open it and they only had a small window to ask [to be opened]. So they asked what the window was, and the Frascati people answered "It is in sapphire", which didn't improve the situation because sapphire is a jewel, for customs.[32] They had to wait and find a solution and, luckily, we had the intervention of Francis Perrin. ...

AdA arrived in Orsay, and was unloaded with its accompanying instrumentation. Then the empty trucks left, to go back to Italy. The image is still vividly recalled in Lacoste's words:

> *Lacoste:* I remember the trucks going back and, as the small truck was empty and the big truck also, they decided that it was easier for the driver of the truck to put the small truck on the big truck and I remember seeing them travel back that way.

AdA's "wedding trousseau", i.e. *il corredo*, as Carlo Bernardini used to call it in Italian, was completed a few days later with the remaining equipment being sent by plane. The list, detailed in a July 6th letter by Carlo Bernardini to François Lacoste, reads like a nursery rhyme[33]:

> Two Cherenkov glass counters ...
> one oscillograph ...
> one Movie camera ...
> 12 power supplies ...
> 30 modular units ...
> 6 scintillation counters
> ...

And then, after AdA and its outfits had arrived, the scientists and the technicians came. AdA was installed in Salle 500 next to the Linear Accelerator, and a new team, now composed of French and Italian scientists, moved on to the last leg of AdA's great adventure.[34] In Orsay, the Italian team joined forces with François Lacoste, Pierre

---

[31] François Lacoste comments here and in the next chapter are transcribed from May 2013 interviews in Orsay.

[32] According to Carlo Bernardini, one of the physicists who built AdA, the window was made of quartz, not sapphire (Bernardini 2003).

[33] Copy of this letter is courtesy of Mario Fascetti, the AdA technician who had built AdA's radiofrequency system, under Mario Puglisi's supervision.

[34] "Salle 500" means Salle 500 MeV, 500 Mev being the electron beam energy which was delivered by the Linac in this experimental hall.

Fig. 11.7 Pierre Marin (at center) from a still frame in a docu-film about Pierre Marin, produced by the Laboratoire de l'Accélérateur Linéaire, courtesy of Jacques Haïssinski. Appearing at right with white hair, one can glimpse Georges Charpak

Marin and a young graduate, Jacques Haïssinski, who would prepare his *Thèse d'État* on AdA, completed three years later, under the guidance of André Blanc-Lapierre.

The mission, which had been started by Pierre Marin and Georges Charpak in July 1961 to go and see what was happening in Frascati, had been accomplished: the *vrai bijou* was now in France, Fig. 11.7.

Chapter 12
Touschek with AdA in Orsay

> *The challenge of course consists in having the first machine in which particles which do not naturally live in the world which surrounds us can be kept and conserved.*
> Bruno Touschek, excerpts from a talk delivered at Accademia dei Lincei, 24-5-74, Sapienza University of Rome–Physics Department Archives.

Abstract After AdA's arrival in Orsay, the road to high energy particle physics with electron-positron colliders was made open by the better injection provided through the linear accelerator. The obstacles and successes of the two and a half years during which AdA was in the Laboratoire de l'Accélérateur Linéaire come alive from the transcripts of 2013 interviews with its protagonists, Carlo Bernardini, Peppino Di Giugno, Mario Fascetti, François Lacoste, and Jacques Haïssinski, whose doctoral thesis details all the experimental developments. Bernardini's published personal memories shine with the excitement of discovering the 'Touschek effect', which had a strong impact on similar projects in the US and USSR. In 1964, parallel work between Orsay, Frascati and Rome allowed to announce that collisions had taken place and that electron-positron colliders could become the tool for future discoveries in particle physics. Archival documents, and memories of that period highlight the theoretical physics work in Rome and the European dimension of AdA's final success.

In 1962, in Orsay, the parallel roads followed by France and Italy in the development of particle accelerators, met when AdA was downloaded from the truck and placed in Salle 500, next to the Linear accelerator.

The *Laboratoire de l'Accélérateur Linéaire d'Orsay*, where AdA arrived in July 1962, had been founded five years earlier, and was now ready to welcome the new arrivals from the Frascati Laboratories in Italy, its scientists and AdA, with *tout son attirail* (Marin 2009), namely its traveling outfit of working pumps, batteries, oscillograph, movie camera, and more. Both laboratories had come to life after the war, as the decision to found CERN was taken and, with it, the need arose to train new generations of European scientists in accelerator science. They also sprang from the dreams of three great pre-war scientists, Enrico Fermi on one side of the Alps, Frédéric Joliot, and Irène Joliot-Curie on the other (Pinault 2000; Bimbot 2007). They shared a vision, one of open spaces for students and researchers, to lodge new equipments, and overcome the strictures of University settings.

Thus, AdA's need for a stronger source of electrons and Laboratoire de l'Accélérateur Linéaire's search for new ideas and perspectives definitely met, when it was agreed to transport AdA to Orsay. The joining of minds and technologies between the Italian and the French laboratory transformed a partial success into a world record.

12.1   The Laboratoire de l'Accélérateur Linéaire

As it had happened in Frascati with the construction of the laboratories and the synchrotron, in Orsay a dream had also come true: after the war, driven by Frédéric Joliot and Irène Joliot-Curie's vision (Bimbot 2007), the Institut de Physique Nucléaire (IPN) had been created, in the old municipality of Orsay, in the Vallée de Chevreuse, which threads its way along the Yvette river, next to the Saclay plateau.

In 1942, Irène Joliot-Curie had started searching for an adequate location, where large installations could be available to students and researchers alike for experimentation in the new science of nuclear physics. She had toured the countryside, and, in 1942, Frédéric Joliot had written a letter to the Rector of the Académie de Paris suggesting to buy a large piece of land south of Paris, in view of a future extension of the University of Paris, not too far from the railway station of Gif-sur-Yvette. The piece of land that Joliot was thinking of was the one where the CEA is located today. After the 2nd World War, both Frédéric and Irène Joliot-Curie were eager to extend the Faculty of Sciences with Frédéric very much involved in the setting up of the CEA in the vicinity of Saclay, while Irène wanted to set up an "Institut de physique nucléaire et radioactivité", what later became the IPN.[1]

In 1950, a new generation of students and researchers was filling the existing urban laboratories in Paris, and the idea to create new spaces outside the city had gained momentum. A very convenient location for the construction of a laboratory

---

[1] The events leading to the choice of the Orsay site and the birth there of the IPN, were recollected by Frédéric Joliot in a conference at the 8th Lindau Nobel Laureate Meeting dedicated to Chemistry, on 1st July 1958, just two months before his death. The text of the Lindau conference *Le nouveau Centre de recherches fondamentales en physique nucléaire d'Orsay et la formation des chercheurs* was used in part by Joliot at the inauguration of the International Conference of Nuclear Physics held in July in Paris and later published in the journal *L'Âge Nucléaire*, 11, Juillet-Aôut 1958, p. 183 (Pinault 2000, 154). A version was also reproduced in Frédéric Joliot's collected works, a copy of which was kindly provided by Jacques Haïssinski. In these recollections, Frédéric Joliot recalls the role played by his wife, Irène Joliot-Curie, after the war, in establishing the development of a new institution in an adequate location in the surroundings south of Paris.

was found in Orsay, in the Vallée de Chevreuse, south of Paris. It had the advantage of being reached by the RER, the train from Paris, and to be close to existing institutions such as the Center for Nuclear Studies of the Atomic Energy Commission (C.E.A.) in Saclay, as well as to the Centre de Recherches of the C.N.R.S., and a piece of land of more than 100 ha was bought by the Faculty. To equip such new spaces for modern type nuclear physics research, Irène Joliot-Curie proposed the acquisition of a proton synchrocyclotron of 150 MeV, and building new office spaces. At the same time, the question to join CERN, posed to the French Parliament in 1954, highlighted the need for France to develop national resources in the new field of nuclear science and in 1954 the decision was taken to build both a proton synchrotron and a linear electron accelerator. The first machine was to be installed in Irène's Institute and its construction began first, while the linear accelerator, the LINAC, was to be installed in the future Laboratoire de l'Accélérateur Linéaire. When the decision to build a linear accelerator was taken by Yves Rocard, Directeur du Laboratoire de Physique de l'École Normale Supérieure (ENS) de la rue d'Ulm, in Paris, the question of an appropriate location was debated and the choice was to build it in Orsay, near the IPN. Unfortunately Irène Joliot-Curie did not have the joy to see her dream become reality. She passed away on 17 March 1956, soon to be followed by her husband, on 14 August 1958 (Bimbot 2007).

The Laboratoire de l'Accélérateur Linéaire had then been founded in 1957 and a linear accelerator became operational in 1959. The linear accelerator had been designed by Hubert Leboutet from the Thomson-CSF (Compagnie générale de télégraphie sans fil) and built by an equipe of scientists and technicians, directed by Jean-Loup Delcroix, (Blanc-Lapierre and Delcroix 1963), from the Laboratoire de Physique de l'ENS.[2] The first two directors of the Laboratory were Hans von Halban and André Blanc-Lapierre, who took over from Halban in 1962 and conducted the negotiations for the transfer of AdA.

Halban and André Blanc-Lapierre had both arrived at LAL from dramatic backgrounds.

When the decision was taken to build a linear accelerator and a laboratory around it, there came the decision to have a group of physicists to plan for possible experiments together with the necessary technical staff, and a director. The choice had fallen on Hans von Halban, the Austro-German collaborator of Frédéric Joliot, who, together with Lew Kowarski, had participated in 1939 to the experiments providing evidence that the number of free neutrons released in the fission of uranium is sufficient to induce fission in other nuclei, setting in motion a chain reaction. In 1941, as the German army was reaching Paris, Halban fled to the UK subtracting the precious barrels with heavy water, which could have otherwise helped the Germans in their atomic efforts.[3] In 1957, when LAL was being constructed, Halban was in Oxford.

---

[2] Hubert Leboutet collected his memoirs of the years 1950-60 in a 1994 publication of the group Histoire de Thales—AICPRAT, entitled *des ELECTRONS et des HOMMES—Les début des accélérateurs dans les années 50 et 60 à la CSF*, copy contributed by private communication from Jacques Haïssinski.

[3] This story is recalled in the 1948 movie Operation Swallow: The Battle for Heavy Water.

At that time, his collaboration with Joliot, and his work both in the UK and in Canada during the war, made him an excellent candidate for the direction of the new laboratory. He accepted to come to France and led the LAL experimental team which included Pierre Marin and François Lacoste. But the political climate was changing. He started having difficulties. He lacked the necessary support from the French scientific establishment, and could not obtain the financial and human resources needed for the future development of the laboratory. In 1961, he resigned from his position, for both health reasons and changed political atmosphere in France. His place was taken by André Blanc-Lapierre, who came from University of Algiers, where the war for independence was tragically disrupting the life of both France and Algeria. In a talk given at the 50th anniversary of the foundation of Laboratoire de l'Accélérateur Linéaire, the theoretician Maurice Lévy recalled his visit to Algiers in March 1961.[4] He himself having Algerian roots, Lévy had been happy to accept to give a series of lectures at the University of Algiers. He had been invited by André Blanc-Lapierre, professor at the University and director of the Institut de Physique Nucléaire, where a small proton cyclotron had been made available through the so called "Constantine Plan".[5]

When Lévy arrived in Algiers, he found the country he had known as a young man, engulfed in dramatic unrest and violence.[6] In many places one could see signs of the OAS, *Organisation Armée Secrète*, Secret Armed Organisation, a far-right French organization opposing Algeria's independence. A menacing feeling that something would soon happen and the situation turn for the worst was pervading life in the city. André Blanc-Lapierre was worried, not having clear prospects about alternative possibilities in France. When Lévy returned to Paris, the news of Halban's resignation reached him, and the idea of asking André Blanc-Lapierre to come to France and take over as LAL director arose. He talked this over with Yves Rocard. A director had to be found, and André Blanc-Lapierre had all the needed competences, such as being an expert in radio-frequencies, in addition to an excellent reputation as an applied mathematician. Worried about the future, when Maurice Lévy telephoned him in Algiers asking him to come and direct the Laboratoire de l'Accélérateur Linéaire, André Blanc-Lapierre was happy to accept and come to France, professor at the ENS and director of the laboratory in Orsay.

---

[4] See also 2013 interview with Maurice Lévy in Touschek with AdA in Orsay, and transcript of contribution to Symposium for 50 years of Laboratoire de l'Accélérateur Linéaire, document contributed to the Author, and *Le LAL a 50 ans*, Bulletin d'information interne du LAL—N. 43, 2006.

[5] The Constatine plan was a development plan for Algeria officially commenced on October 3, 1958, after the French President Charles de Gaulle announced it in the Algerian city of Constantine.

[6] See for instance the 1966 movie *The Battle of Algiers*, by Italian director Gillo Pontecorvo, younger brother of Bruno Pontecorvo.

12.2 AdA's Arrival and Installation in Orsay: Summer 1962

AdA was accompanied to Orsay by the two specialists of the vacuum, Gianfranco Corazza and Angelo Vitale, who had travelled with the two trucks and could be ready to take action if problems with the vacuum arose. When AdA arrived, it had been able to cover the 1500 km distance between Rome and Paris in two days and the vacuum in the doughnut had remained intact, to the great satisfaction of the Italian team who had so carefully prepared Ada for the trip. The scientists from both the Italian and the French side were in a great hurry to start the experimentation, as they knew the competition from the other side of the Atlantic would not be wasting time. Soon after the two trucks left Frascati, the remaining equipment needed for AdA's installation had been sent by plane, Fig. 12.1. Once the equipment reached Orsay, and the Italian technical team led by Giorgio Ghigo arrived by plane, the installation started in earnest, Fig. 12.2.

On the French side, the scientists who supervised the installation of AdA near the LINAC, through the month of July, were Pierre Marin and François Lacoste. The installation proved to be a non trivial affair, and accidents would occur as recalled in Haïssinski (1998, 17):

> AdA was greeted at Orsay by a small team composed of Pierre Marin and François
> Lacoste. Pierre Marin remembers quite vividly an incident which marked the installation of
> AdA and which could have turned out to be a dramatic one. The experimental hall where

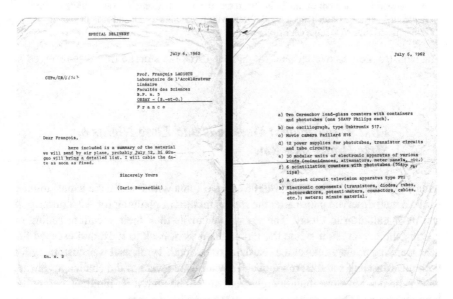

Fig. 12.1 Letter sent by Carlo Bernardini to François Lacoste to anticipate the arrival by plane of some equipment needed for AdA's installation and future experimentation, courtesy of Giuseppe Di Giugno

Fig. 12.2 At left, the LAL Main building, housing the Linear Accelerator, and, at right AdA installed in Salle 500 at LAL, both courtesy of Jacques Haïssinski

AdA was to operate was an intermediate energy hall, equipped with a special roof comprising a few water tanks having the shape of very large rectangular boxes which provided the proper radiation shielding. They could be moved horizontally in order to make room for a crane located above them. In the course of the AdA installation, it happened that, while the mechanical device used to support the ring was hanging on the crane hook, someone pressed a button which started the motion of the water tanks. These tanks were so heavy that when they reached the crane cable they just pulled the crane together with the AdA support. A member of the Italian team saw that the AdA support was heading towards a wall and screamed out. Thus alerted, Pierre Marin was able to run and stop the water tanks just in time to avoid a catastrophe.

The catastrophe was averted and by August 1962, the storage ring was installed at Orsay.

12.2.1   First Experiments: Weekends and Long Nights or Sixty Hours in a Row

After the July installation, August 1962 arrived, and with it came the usual summer break, a well deserved pause after the frantic months of planning for AdA's transport and its installation in Orsay. The French scientists took their vacation, sailing or visiting the countryside, while the Italian team went back to Italy, and escaped the heat spending a few weeks at the sea shore or in South Tyrol, as it was customary for both the Touschek's and Bernardini's. They all were aware of the challenges ahead, and of the need to store their energies, but also to be with their families, before the long absences to Orsay to be expected in the months to come.

   In Italy, the physics community was generally aware of the Frascati work but there often was a display of scepticism, which Touschek would counter by saying that "of

Fig. 12.3 Bologna 1962, September 9, 14th Meeting of the Italian Physical Society, Bruno Touschek seated in front, lower right, Carlo Bernardini, with light color jacket, upper left, Giuseppe Di Giugno, in light grey, at Carlo's right, in a photo courtesy of Giuseppe Di Giugno

course electrons and positrons have to meet, it is just a consequence of the CPT theorem, in fact", he would add "actually CP is enough!" (Bernardini 2004). Other objections, that interactions with the walls would separate the beams, were pushed aside, with a vigorous utterance, "Scheiße!". Such deep convictions came from his mind frame, as a theoretical physicist guided by faith in symmetry principles, also supported by his experience with electrons in Widerøe's betatron. The Frascati and the Orsay team shared his enthusiasm.

In early September, before joining the AdA team in Orsay, where AdA was now ready for the new set-up, Bruno Touschek, Carlo Bernardini and the young Giuseppe Di Giugno attended the yearly Italian Physical Society meeting in Bologna, where Touschek presented an update of AdA's progress. We show them in Fig. 12.3, in a photograph taken during the Conference: Touschek, in front seat, is seen at lower right, tanned from his vacation, intense and focused. He knows he will soon begin the final leg of a journey into the unknown.

By mid-September the team had reassembled in Orsay, ready to begin the new line of experiments. Pierre Marin remembers the AdA period, which stretched from July 1962 until the final runs in '63 and '64, as *fievreuse*, feverish. Both the French and the Italians knew they were on the verge of a potentially epochal breakthrough in accelerator physics. As Touschek has said in February 1960, in Frascati, when he proposed the construction of AdA, they were not alone in the world to think of electron-positron collisions.[7] They knew they could not waste any time on the face of the competition from the other side of the Atlantic, by the American teams of which they were all aware and probably frightened. The outcome was not assured either, namely to demonstrate that collisions had taken place was not a given, as the discovery of the *Touschek effect* in winter of 1963 would very soon show.[8]

Once in Orsay, as soon as the experimentation started, Touschek and Bernardini understood that a unique opportunity had been dealt to them. The enthusiasm rose, while the hopes, which every run entailed, brought a further step in understanding the working of the accelerator. The Italian team, scientists and technicians, would come for a series of runs, which took place during week-ends, when the other French users of the LINAC would let them the use of the beam and they could experiment injecting the electrons in AdA. No personal testimony by Touschek for such period is available so far, except for a well known drawing, the *Magnetic Discussion*, Fig. 12.4, which may be attributed to this period, according to Carlo Bernardini (Bernardini 2006).

A detailed record of the results was kept by Touschek, in two large notebooks, where he reported all the calculations he was doing to understand the behaviour of the particles in the beam, and the many hypotheses and checks following the measures. A copy of the first AdA logbook exists, in Bruno Touschek's papers, together with the original of the second logbook, whereas a further notebook was unfortunately lost.[9] To these notebooks, we shall often refer to pinpoint some dates through this chapter.

The original team welcoming AdA in Orsay was constituted by Pierre Marin and François Lacoste, who has vivid memories of that period.[10]

François Lacoste: I participated in the start of AdA in Orsay, we were in a small room which doesn't exist anymore. We were using the electron beam of the 500 MeV station to produce gamma rays which were targeted inside AdA to produce electrons and positrons. The storage rate was quite small so we needed about 48 hours to have sufficient number of electrons. You could count electrons one by one by the differences in the light [emitted by the accelerated electrons].

In November, the French team was joined by Jacques Haïssinski, a young doctoral student, whose father, Moïse Haïssinski, had worked in the Radium Institute, in Paris,

---

[7] Minutes of the Frascati meeting, February 17th, 1960.

[8] When exactly the team discovered the effect, which will be discussed later, is not well established. Carlo Bernardini mentions "a night in 1963" in Bernardini (2004) and confirms it in Bernardini (2006), 70, where he mentions it as *una famosa notte del 63*, a famous night in 1963. Since the paper was certainly written by mid-March, it can be expected that the "night" was some time in January or early February 1963.

[9] The original of the first AdA logbook, with starting date 4 April 1960, is in possession of Touschek's son Francis Touschek, with the full copy to be consulted in Bruno Touschek Papers(Box 11, Folder 89), together with the Notebook "Ada II" (Box 11, Folder 91) and a smaller Notebook ("Quaderno di AdA", Box 11, Folder 90), in Sapienza University of Rome–Physics Department Archives.

[10] Extracted from recorded interviews for the movie Touschek with AdA in Orsay by E. Agapito, L. Bonolis, and G. Pancheri. Interviews were recorded at Laboratoire de l'Accélérateur Linéaire, in Orsay, May 2013 .

MAGNETIC DISCUSSION

directed by Marie Curie. Jacques Haïssinski had worked with François Lacoste on experimentation at the Laboratoire de Physique de l' École Normale Supérieure, equipped with two accelerators, and, after a fellowship period in Stanford, at HEPL (WW Hansen Experimental Physics Laboratory), had come back to France for his military service, due to terminate in fall 1962. Thus he was not present when AdA arrived, but joined the team after the summer, being offered a position as a doctoral student[11]:

Jacques Haïssinski: I have to say that while I was at Stanford, a few month before leaving, I had an opportunity to see the beginning of the construction of an e^-e^- ring, and therefore I knew a little bit about the possibilities of such machines, but it took many years in fact for the Stanford machines to take some data. Nevertheless I had some idea what colliding beams were.

At the end [of this period] I was offered a position by André Blanc-Lapierre, to come as expected and work here. He asked me if I was willing to join the team which had just

---

[11] From Jacques Haïssinski's interview for the movie Touschek with AdA in Orsay by E. Agapito, L. Bonolis, and G. Pancheri. Interviews were held at Laboratoire de l'Accélérateur Linéaire, in Orsay, May 2013.

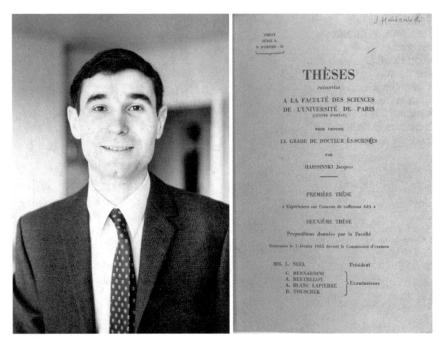

Fig. 12.5 Jacques Haïssinski in 1968 at left, and the frontispiece of his 1965 *Thèse d'État* at right, courtesy of Jacques Haïssinski

started to work on AdA, and of course I was very much attracted by this program, again it was frontier physics …

I came here, beginning of November of '62, and AdA was already there, of course.

Jacques Haïssinski worked on AdA's measurements until 1964 and his *Thése d'État*, which leads to the highest education level in France's system, constitutes the best document of AdA's achievements, from both the scientific and the historical point of view, Fig. 12.5.

Jacques Haïssinski's testimonies were given during interviews in May 2013, which took place in the so-called *Igloo*, a round building where the original ring ACO is maintained as part of the museum Sciences ACO. ACO stands for Anneau des Collisions d'Orsay. It was a circular electron positron storage accelerator, which was designed and built in the years '63–'65 under the supervision of Pierre Marin. Its design benefited particularly from the experience that the French team got with AdA.

Jacques Haïssinski: When I joined the AdA collaboration, AdA was already installed in the experimental hall, a few tests had been made. First thing I was in charge was to try and upgrade the line which transfers the electrons between the linear accelerator and the ring in order to increase the injection rate. So that was my first work. The injection rate was much higher than what it was in Italy. But it would still take many hours, between ten [to] fifteen hours to fill the ring with electron or positrons. One has to remember that the goal of bringing AdA to Orsay was to try and to study collective effects, effects that take place when there are many particles which are in the beam. In particular the main goal was to

Fig. 12.6 Jacques Haïssinski working at AdA's control station in Orsay, at left (photo courtesy Jacques Haïssinski), and, at right, Gianfranco Corazza, taking a temporary rest, sleeping in the SIMCA car the Italian team used to rent, in Orsay, courtesy of G. Corazza

check that the two beams, the electron particles and the positron bunch, were crossing and overlapping precisely in order to get the maximum rate of collisions. And that took two years, two and a half of experimentation here at Orsay. The so called data taking runs used to take place during the week-ends, they would start on Friday evening and end up next Monday morning, which means that we were working sixty hours in a row, non stop. And of course this was a bit stressing and tiring, but I think we were all very enthusiastic and most of the members of the team would stay in the control room most of the time. Sometimes of course there was some possibility of sleep far away, but most people were still there most of the time.

Two photographs from those times are shown in Fig. 12.6. As soon as experimentation started, the injections of electrons and positrons in the AdA ring with the LINAC, which was a much stronger source than the Frascati synchrotron, had showed a marked improvement, and raised high hopes for the next step in experimentation. The technique used was the same as in Frascati, namely, electrons and positrons were first generated inside the doughnut, by a photon hitting an internal target. Then one type of charges, say the electrons, would be bent by AdA's magnetic field into the expected circular orbit, while the other, oppositely charged, particle would just be wasted away. Once a sufficient number of, say, electrons had accumulated, AdA would be flipped over like a pancake (Haïssinski 1998, 21), namely rotated around a horizontal axis and the magnetic field would change direction. It would then be the turn of the electrons to waste away, while the positrons would now be directed to move into the same circular orbit as the electrons, but circulating in the opposite direction.

This effect is the one represented graphically by Touschek's drawing of Fig. 12.4, which represents the classical law of "the left hand", which established how a magnetic field acts on a charged particle moving perpendicular to it. The mechanism flipping AdA, turning it upside down, was called *il girarrosto*, in Italian, the roasting spit in English.

Jacques Haïssinski[12]: As I said, it used to take a long time to store the particles, and of course during that time Bruno Touschek was monitoring very precisely the rate at which

---

[12] May 2013, interview with Jacques Haïssinski, in Orsay.

particles were stored and when he felt that the beam was not up to the best possibility, I remember that he would use a watch and put numbers, checking the rate.

Was Bruno repeating a routine followed twenty years before, in the Müller factory, in Germany, during the dark years of the war, when trying to get Widerøe's betatron to work? Jacques Haïssinski continues:

> *Jacques Haïssinski:* After a while we had to understand that something had to be done because he [Touschek] was getting more and more unhappy because probably the people operating the accelerator were falling asleep or at least dozing and then I would run to the other control room to the linear accelerator, which was 100 meters away, and spend, sometimes, hours trying to get the engineers or the technicians motivated to get the best possible beam that AdA could benefit from.

> During the year and a half in which the data was taking, it was not always easy, there has been a number of surprises, and in physics surprises are difficulties to overcome …When the ring was flipped over to change the injection from electrons to positrons, sometimes, the beam which was stored was lost, it took us a while to understand what was going on. By sheer luck Bruno Touschek and also Pierre Marin, I remember, both checked what was going on and they noticed that when they were flipping the ring, there were some magnetic dust particles within the doughnut which were falling across the beam and which would eliminate all the particles.

This difficulty with losing the beam when changing electrons with positrons during the injection, is echoed in François Lacoste's words:[10]

> *François Lacoste:* The long storage time was a problem because we also would lose electrons now and then, either through short stops of the electrical supply by the EDF [Elecricité de France] which was not stable or also when we flipped from the electron to the positron side and the small powders could fall and would destroy the beam. You had to start all over again and I remember a technician of the Frascati [team] who would [say] …every time we lost the beam: "elettroni tutti morti". And this I still remember because it was very frustrating.

Then as Bernardini (2004) also mentions, the puzzle could be solved when

> *Carlo Bernardini:* …by sheer chance, Bruno and Pierre Marin were looking through the porthole in the donut searching for the malefic elfs who were destroying the beam and found them in the form of fluttering diamagnetic dust left over from the welding of the donut and moving under gravity along the magnetic field lines and passing through the beam.

Giorgio Ghigo is quoted by Pierre Marin for having suggested the solution which was finally adopted: Marin calls it *la parade*, the perry, in a sport-like term which clearly reflects the spirit of the team who was fighting a battle against time, to be the first to see the particles collide.

Giorgio Ghigo had the brilliant intuition that the same effect of inverting the direction of magnetic field (by flipping AdA upside down) could be obtained by changing the direction by which the positrons would enter it! To this effect, after the electrons had already been stored in the ring, AdA was first translated away, then rotated by 180 degrees around a vertical axis, and finally translated back to the point of arrival of the LINAC beam, so that the relative orientation of the positron velocity and the magnetic field would now send the positrons to be bent so as to

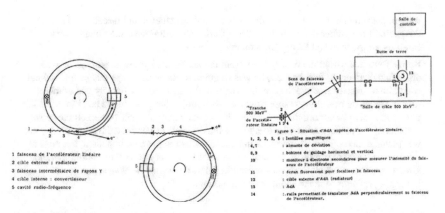

Fig. 12.7 Figures 4 and 5 from Jacques Haïssinski's thesis show the positioning of AdA with respect to the linear accelerator and the translating platform at right, with mechanism of injection of electrons and positrons, at left. Courtesy of Jacques Haïssinski

enter the same orbit as the electrons, but in opposite direction. "For obvious group-theoretical reasons…", commented Touschek (Bernardini 2003, 5).[13] Gianfranco Corazza found a clever solution to realize the new scheme proposed by Ghigo. With the help of Antonio Marra he had a translating platform built allowing the two faces of the target to be exposed to the LINAC beam, so that the magnetic dust, the "small powder" remained stuck to the bottom of the donut. It was, as Pierre Marin says, a very elegant solution, Fig. 12.7 from Jacques Haïssinski's thesis. Later on, when discussing storage rings, and AdA's injection mechanism in particular, in his presentation to the 1963 Spring Meeting of the American Physical Society, Gerard O'Neill would admiringly call it simple, but very efficient.

> *Jacques Haïssinski:* Another difficulty that we ran across, not a major one, nevertheless it took some time to repair it, was that the detectors which were used to look at the annihilation or the collision at least between electrons and positrons, they were big blocks of lead glass which used to detect the Čerenkov light produced by the photons coming from the beam collisions. Because these detectors were kept in the experimental hall where the primary beam from the LINAC was coming in, there was very high radioactivity and this radioactivity progressively blackened totally the two Čerenkov, so we had to restore the transparency of these blocks and this took some time. That was not a major problem, but nevertheless it was one of the hurdles we had to overcome.[14]
>
> The data taking runs, all started on Friday evening. The very first measurement which had to be done was to calibrate the instruments, that is to measure carefully the intensity of the particles which were circulating in the beams. This was essential, since what we were looking for were collective effects, which depend on the 'number' of particles stored; there

---

[13] As Bruno wrote to his parents in June 1943, he had studied group theory because he would have liked to definitely dominate this area as completely as possible after the war.

[14] The solution was to heat up the Čerenkov detector, a heavy glass cylinder, just below the softening temperature, and then gradually cool it to avoid internal tensions that might create fractures. The Čerenkov was brought to Frascati and Corazza took care of this delicate procedure which took about 48 h (Corazza 2008).

were perhaps hundred times higher than what could be achieved in Frascati. And during this period I remember Bruno Touschek and Carlo Bernardini were watching us very carefully all the time and taking notes on the log book.

Bruno Touschek would rather stay away from the knobs, the hardware, perhaps he was dissuaded from doing that, because he was so quick in changing things that people did not want him to interfere too much with the program. But he never missed any one of these runs in Orsay, he was always there from the beginning to the end, always had input making plots and …always putting things forward. It was essentially Ruggero Querzoli the senior member of the collaboration who was telling him to leave us continue the program, which was planned. Sometimes of course Bruno Touschek would get a bit impatient, but most of the time he had some smile on his face, because he had a very good sense of humour, he always saw things which were very special in his own way and he always had some interesting remarks to make about what was going on.

12.2.2 Italians in Paris

The arrival by plane of the technical core of the AdA team is humorously recalled by one of the Frascati technicians, Mario Fascetti, in the docu-film Touschek with AdA in Orsay.[15] Fascetti recalls their arrival at Orly, with many baggages, and reminisced that Giorgio Ghigo, upon arrival, rented a Renault for the drive to Orsay. But, as Ghigo started the engine of the Renault, emotion and inexperience gave way: he engaged the car in the third gear, the car started and immediately jumped to a violent stop. The luggage, not so well secured on the top of the car, fell off rolling on the highway. Luckily, as Fascetti says, nothing happened.

As for the other members of the team, Carlo Bernardini remembers[16]:

> *Carlo Bernardini:* We used to fly to Orsay from Rome all the week-ends. The colleagues, our friends in Orsay, dedicated all their week-ends to us, the families were not so happy about these absences during the week-ends.

They would leave Rome on Friday evening and return either on Sunday or Monday night. Bernardini, who had teaching obligations, would take the evening flight back to Rome, a Viscount aircraft, which would then continue towards Bangui and Brazzaville. Touschek preferred the train, taking the overnight trip between Rome Termini Station and the Gare de Lyon, in Paris.

In those days, the cultural differences between the two teams were much stronger than those we can see now.

> *François Lacoste:* The organisation of the Frascati team was much more military than for us: the technicians would call Touschek and Bernardini "Dottore", which was not at all our custom, as we would - all the physicists between them, and the technicians- call us by our last names.

The Italian team in Orsay included senior scientists—Bruno Touschek, Carlo Bernardini, Giorgio Ghigo, Ruggero Querzoli and Gianfranco Corazza—as well as a young

---

[15] For the Italian version see Touschek con AdA a Orsay.

[16] June 2013 interviews in Frascati for the docu-film Touschek with adA in Orsay.

post-doc, Giuseppe (Peppino) Di Giugno, and four technicians: Bruno Ilio, Angelo Vitale, Mario Fascetti and Giorgio Cocco. While most of the scientists had travelled abroad to conferences and the like, the Paris transfer was a wholly new adventure for the young Peppino Di Giugno and the technicians.[17] Fascetti, remembers that, before leaving Frascati, the Laboratory director, Italo Federico Quercia, bought for them tickets for a show in one of the famous Parisian strip-tease theatres, the *Café Mayol*: not quite the Moulin Rouge, but still enough for the young technicians to enjoy something not existing in Italy, at the time. Fascetti kept the (cancelled) ticket to this day, as a memento of the greatest adventure of his life, together with photos taken in Orsay and Paris, two of which, dated as of July 1962, are shown in Fig. 12.8. Among Fascetti's recollections, there is one example of the enthusiasm and drive of those days well worth remembering. In an interview recorded in June 2013, Fascetti recalls a week-end when the whole experiment almost failed[18]:

Mario Fascetti: I was always available to make sure that the high frequency apparatus would be working in all its parts. But, once, an accident happened. It was Friday. We were allowed to use the linear accelerator only on Saturdays and Sundays. Next to the radio frequency station, there was a tall stack of heavy co-axial cables, two or three meter high. I was in the laboratory, but not there, and someone passed by, brushed against the stack, which broke, and all the cables, kilos and kilos of it, fell from high on top of the radio frequency apparatus, damaging it heavily. It was a panicky situation. That meant to stop experimentation with Ada and fix it. But how? We could only work on Saturdays and Sundays [with the beam], and I was the only one who could fix it, as I had worked to build it [in Frascati]. I tried a call to Rome, and speak with my boss, ingegner Puglisi, but could not find him. Then, it came to my mind that I should be able to do it, by myself: 'I have built it after all', I told myself. So I asked one of the French scientists, either Marin or perhaps Lacoste, and asked for two of their technicians to help me and repair the damage. This was a very difficult thing to do, since, as I remember well, the French personnel used to disappear as as soon as their work hours were off, at 4 or 5 in the afternoon. But somehow they were able to convince two people ...That night we worked and the morning after the accelerator was working and available to all the experimenters.

The beauty of the *Ville Lumière* did not escape the attention of the young technicians of the team, but they saw its marvels with a special twist. When asked by Giorgio what they thought of the Eiffel tower they had just visited, they rapsodized about the iron bolts which kept it together.[19] A snapshot taken during a lunch break, has Giorgio Ghigo and Gianfranco Corazza at lunch with Mario Fascetti and other technicians of the AdA team, Fig. 12.9. For the youngest of the team, Peppino Di Giugno, it was a life forming experience. Reflecting on it in 2013,[16] he compared working on AdA with present day high energy physics research, striking a note which had already resonated in Frédéric Joliot's last thoughts, and would be present in Touschek's, as well, when, at the end of his life, he was at CERN in 1978.

---

[17] Di Giugno would later leave physics for electroacoustics and digital sound, and work with Pierre Boulez, Luciano Berio, Robert Moog. See a recent book about Di Giugno's work in digital music, https://www.esarmonia.it/negozio/libri/didattica/peppino-di-giugno-genio-italiano-macchine-digitali-per-la-musica-in-tempo-reale/.

[18] The original interview in Italian was translated in English by the Author.

[19] As related by Giorgio Ghigo's son Andrea in *Touschek with AdA in Orsay*.

Fig. 12.8 Members of the Italian team in Orsay in 1962, with the left panel showing Giorgio Cocco, Bruno Ilio, Gianfranco Corazza and Angelino Vitale with AdA in the background, while the right panel shows Angelino Vitale, Peppino Di Giugno (at center) and Mario Fascetti (at right), photoes courtesy of Mario Fascetti

Fig. 12.9 At lunch, in Paris, one distinguishes Mario Fascetti, third from left, Gianfranco Corazza, head of table, Giorgio Ghigo, first from right. Photo courtesy of Mario Fascetti, all rights reserved

Peppino Di Giugno:[18] I always say that I consider myself to have been extremely lucky to have been part of the AdA group. To be part of such group is one thing which can happen only once in life …Around 1975 I left high energy physics, mostly because it had become like a factory, and today I could not work at CERN with two or three thousand people.

Fig. 12.10 At left, Giuseppe Di Giugno with AdA in Ferrara in 1991, on the occasion of an INFN exhibition about particle physics, courtesy of Di Giugno. At right, a recent photo of François Lacoste from https://www.deasyl.com/english/, and Pancheri and Bonolis (2018), 31, reproduced with permission

Paris reserved many surprises to the young Italian, as when he recollects the time he visited his friend François Lacoste at home:

> *Peppino Di Giugno:* One evening, one of the French collaborators, François, invited me for dinner at his home. He arrived with an enormous car, I forgot what it was, a Cadillac or a Bentley, perhaps a Rolls Royce. He lived in a palace on the Champs Élisées. I was stupefied: more than a palace, it was a museum, with columns, waiters, paintings right and left. While we were eating, I was looking around, and saw an embalmed crocodile on the floor, and all around us paintings of crocodiles, and also a marble crocodile. I asked him "how come you have so many crocodiles in this house?" and he answered that this was the house symbol.
>
> I had clearly not known who was our friend Lacoste!

In Fig. 12.10 we show photos of the two friends in later years.[20]

12.3 The Winter of 1962–63

François Lacoste helped in installing AdA and collaborated during the first runs of the machine, but soon he left particle physics to work in aerospace industries. The team missed him, but in the meantime Jacques Haïssinski had joined the AdA group. As Bernardini says, "luckily we could count on Jacques", who took over with great dedication and enthusiasm. His *Thèse d'État* at the end of three years

---

[20] Di Giugno's photograph was taken during the INFN public exhibition *Dai quark alle Galassie*, held in Ferrara, April 13–May 4, 1991.

of experimentation with the only existing electron-positron ring in the world at the time, represents the best document of the great AdA adventure.

By December 1962, the new arrangement for injection was ready for operation (Touschek 1963, 117). Bernardini was very proud of having invented a novel injection procedure including an amplitude modulation of the radio frequency during the now very short LINAC pulse, which increased the accumulation rate by no less than two orders of magnitude (Bernardini 2006, 70). At this point, the puzzle of the magnetic dust, which killed the beam—*elettroni tutti morti*—when AdA was flipped over, had been understood and solved, and the injection rate should have been good enough to allow the team to address the measurement of the interactions between the beams. Good enough, but not sufficient to see production of new particles, and Touschek proposed to look for a process which had a higher probability to occur, and hence needed a lesser number of stored particles. It was an old friend of Touschek, namely annihilation into two γ rays, which he had considered from the very beginning, on the day he had first sketched his ideas about AdA.[21]

12.3.1 Looking for Two γ Rays

On December 21st, 1962, the first run to look for $e^+e^- \rightarrow 2\gamma$ started, as recorded in Fig. 12.11, from Touschek's log book.

Experiments in particle physics are a complex ensemble of hardware and software: there is a logical construct to identify the events one looks for, then the actual building of the electronic instruments which will collect the particle's signals, and the digital realization of the idea, fed into a computer program. It was 1962, nobody had yet built an apparatus for a colliding beam experiment, which could detect two γ rays, exiting the accelerator in opposite directions in coincidence with each other and with the beams circulating in the storage ring. The AdA team did it. They could count on the exceptional digital vision of Giuseppe Di Giugno, and on Ruggero Querzoli's capacities as an experimentalist who had built one of the particle detectors for the Frascati synchrotron. The 2–γ experimental apparatus for AdA was built in Frascati and assembled in Orsay, and was ready by December, with the logical circuit devised by Peppino Di Giugno and realized by Querzoli, Fig. 12.12.[22]

No firm conclusions could be drawn from the measurements, as there were inconsistencies and contradictions between expectations and observations, although the

---

[21] Touschek had proposed that the process of annihilation into two γ rays be used as the standard process against which to measure all the others, and it is conceivable that term luminosity—because the two gammas would play the role of a unit such as the *lumen*—was invented by him. The concept used to calculate the rate of interaction in colliding beam experiments was introduced by G. O'Neill in his first works on storage rings (O'Neill 1956).

[22] Di Giugno personal communication to the Author, July 2018.

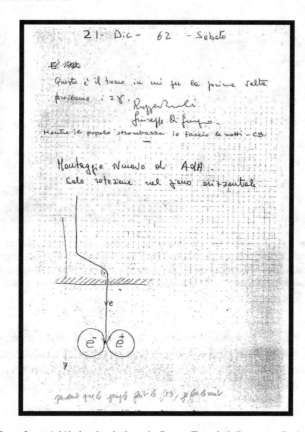

Fig. 12.11 Copy from AdA's log book, kept in Bruno Touschek Papers at Sapienza University of Rome–Physics Department Archives, https://archivisapienzasmfn.archiui.com, stating that on December 21st, 1961 "This is the run in which for the first time we shall try [to look] for 2γ", signed by Ruggero Querzoli and Giuseppe Di Giugno. This is followed by a few lines by Carlo Bernardini (CB), "While the people are trumpeting, I work through the nights", and a sketch of the new AdA set up with rotation only on the horizontal plane. At the very bottom, one can see the French translation of Carlo's lines, i.e. "*pendant que le peuple fait la fête, je fais la nuit*", most likely by Jacques Haïssinski

number of stored particles appeared now to be sufficient to observe annihilation into two γ rays. Touschek was concerned that this could be an effect of *radiation damping*, a phenomenon he was very familiar with since his betatron times.[23]

---

[23] A trace of Touschek's work on radiation damping in a betatron done during the war, can be found in a letter to Arnold Sommerfeld from Kellinghusen, dated September 28, 1945 (Deutsches Museum Archive, Munich, NL080,013). A German version, "Zur Frage der Strahlungsdämpfung im Betatron," is among Rolf Widerøe's papers at the Library of the Wissenschaftshistorische Sammlungen of Eidgenössische Technische Hochschule. An undated English version, written in Göttingen (The Effect of Radiation-Damping in the Betatron), apparently prepared for submission to the *Physical Review*, is in BTA, Box 4, Folder 15. Interestingly, in the last two lines Touschek added: "Thanks are due to Dr. Widerøe for discussions on this problem, which has been treated by the author during March, April 1945", *that is when he was Gestapo's prisoner*. A further, but slightly different version

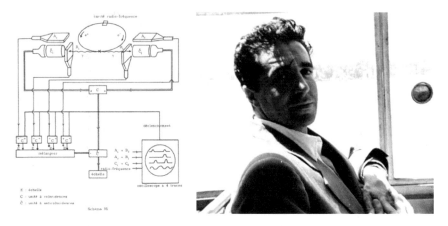

Fig. 12.12 The logical scheme for measuring the process *electron* + *positron* → 2γ, from Jacques Haïssinski's thesis. The γ rays exiting AdA (center top) are depicted as two wavy lines labelled γ, to be detected by an assembly of electronics and detectors. The apparatus, was constructed by Giuseppe Di Giugno and Ruggero Querzoli, seen in the right panel, photograph courtesy of Giuseppe Di Giugno, all rights reserved

However, as it often happens, it was not the spectre of well known phenomena which bothered AdA. Instead, it was a combination of totally new effects, which in a few months would lead to Touschek's greatest insight and to the main legacy of the AdA group, the understanding of how to plan for, and build, higher energies particle colliders.

12.3.2 The AdA Effect

The turning point of AdA's path towards becoming a milestone in particle physics, took place in early 1963. The process of electron-positron annihilation into two gammas had been searched for, but the results of the experiment did not correspond to the expectations. There arose the suspicion that the volume of interaction was not correctly estimated and that the volume occupied by electrons and positrons in their respective bunches could be larger than originally thought.[24]

As the year 1963 came by and rolled into its winter months, the team started focusing on the beam life time and its size, and the correlation between the number of injected particles and the beam life time were measured and studied. This is when the surprises arrived.

(Radiation Damping and the Betatron) can be retrieved at the Archive of the Defense Technical Information Center, https://apps.dtic.mil/docs/citations/ADA801166. This one is dated Göttingen, September 28, 1945, the same date of the letter written to Sommerfeld.

[24] For the same number of particles in each beam, the number of collisions which can be experimentally observed decreases if the volume occupied by the particles increases.

Fig. 12.13 The Café de la Gare, where Touschek often went to to reflect alone and look for solutions to AdA's problems, photographed by L. Bonolis and the Author in 2013

They were in Orsay, like every week-end, ready for an exceptional charge in AdA, a luminosity test in the best possible conditions. The charge of the first beam began at dinner time. Everything was working. The injection speed kept constant during three-four hours. They had three monitors: a monitor showing the injection speed from the synchrotron light detected by the phototube, a monitor showing the LINAC intensity from the control room, a monitor for the gas pressure in the vacuum chamber measured by the Alpert vacuometer. Shortly after midnight they had 10^7 electrons circulating in the vacuum chamber, a number compatible with the life time at the measured pressure, which was slightly over 10^{-9} Torr. They were already dreaming to reach the goal of 10^9 circulating electrons which would allow them to observe some meaningful process. But as soon as they reached 10^7 stored electrons, the injection speed appeared to be conspicuously slowing down, as recalled by Carlo Bernardini[25]:

Bernardini: After four or five hours of steady charging, we saw, plotting the data, that some kind of saturation was beginning, "as if the beam lifetime depended on the stored current". This was the immediate perception, confirmed by a simple plot. Bruno went crazy. Corazza and myself had immediately offered what appeared to be the most trivial explanation: such an intense synchrotron radiation is extracting residual gas from the walls of the donut. The intensity of radiation depends on particle's number, therefore the scattering lifetime depends too. It looked plausible; but no change of pressure was registered from the Alpert vacuometer, even if the sensitivity was adequate.
Bruno became very pensive, but then we saw he had a flash of thoughts and said: "I will think about it; try to make some measurements at lower energies of AdA." We were working at 220 MeV. He left and went to the Café de la Gare, as usual when he was preoccupied.

The Café de la Gare, where Touschek often went to reflect alone and look for solutions to AdA's problems, is still there, in Orsay, just in front of the RER station of Gif-sur-Yvette, Fig. 12.13. Two drawings by Touschek, perhaps reminiscent of that night and the intense period of attempting to understand AdA's data, can be seen in Fig. 12.14.

---

[25] Carlo Bernardini, personal communication to L. Bonolis, December 9, 2009, also Bernardini (2006).

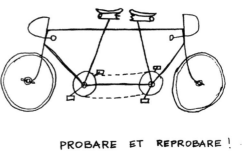

PROBARE ET REPROBARE !

Fig. 12.14 Two drawings by Bruno Touschek, likely to have been done during the intense 1962–63 winter. Family Documents, © Francis Touschek, all rights reserved

Bernardini: (Bernardini 2003, 6)[26] One night, in '63, we understood that the ring held no more secrets for us. We had noticed a decrease in the beam life time with increasing number of stored particles. It was a dramatic observation, a death sentence for storage rings. Bruno left at 5:30 in the morning and went to the Café de la Gare for a drink of his favourite Rosé Sec with the workers arriving from Paris [with the first morning train].[27] He was back at 6:30, extremely excited: he had understood that transverse momentum had been transferred into longitudinal motion through the [so-called] Møller scattering between electrons in the same bunch. [dots] Bruno was both worried and satisfied: the problem was a serious one for AdA, but he had calculated the energy dependence of what would later be called the *Touschek effect* and seen that it would not prevent higher energy machines to work. His capacity to calculate the effect in one hour's time left us admiring and astonished.

The data were already there; we got more in the next few days. There was an astonishing agreement between measurements and Touschek's plot on the tablecloth of the Café de la Gare. Bruno had computed by hand all details but asked me to check the scale factor, which I did immediately in the late morning. Scale, exponents etc. were perfectly working. The reason why Bruno was calmer was that the energy dependence did not anticipate a disaster for higher energy rings like ADONE: this was a low energy effect. But we had the opportunity to see it with AdA just because the vertical size of the beam was much smaller than predicted with multiple scattering.

Jacques Haïssinski remembers well what happened that night[28]:

Haïssinski: And then of course one surprise was the fact that the more particles were stored in the AdA ring, and the shorter was the lifetime. That is to say, we were losing the particles at a faster rate when …and this was totally unexpected and it's Bruno Touschek

---

[26] The original text was translated from Italian by the Author.

[27] In other written versions or interviews, Bernardini gives an earlier hour for Touschek's leaving the laboratory. We thank L. Oliver from Orsay for pointing out that RER train service would not run before 6 a.m.

[28] May 2013 interviews in Orsay.

who understood what was going on. It was a typical collective process which was taking place within the bunches: particles belonging to the same bunches were making oscillations, they are moving one with respect to the other, and some of these collisions are accompanied by a transfer of energy, there is some relativistic effect which plays into this process and, at the end, the two particles, which have collided, are just lost.

And so, first, what Touschek understood is that it was an internal process like that, and then, quite rapidly during the night, after having understood that this was the basis of the process, he made the calculations and showed that this was exactly what was observed. So, later on, this took the name of the *Touschek effect* and it was one of the major features which had to be taken into account in the design of the storage rings which were built after AdA.

Di Giugno also has a clear memory of those fateful hours[29]:

Di Giugno: When Touschek then explained how it worked, it was amazing how simple the explanation was. With pen and pencil, he showed how this fact depended on energy, and there was one factor here and another there, and it was amazingly simple…

This was an example of what they used to call Bruno's serendipity:

Bernardini: He had such a precise idea of what was happening in the beam, as if he could see the electrons with his eyes.

The *AdA effect*, as it was initially called, could have been devastating for the operation of the larger machine, ADONE. Luckily, what Touschek's calculations showed—and measurements immediately confirmed—the effect diminishes rapidly with energy: both ADONE and ACO appeared to be on the safe side! "Bruno was walking frantically from a wall to another in the Lab, like a billiard ball. He was terribly excited: it doesn't happen very often to be happy because one has understood the why of a misfortune…" (Bernardini 2006, 71).

The results of the measurements and observations by the AdA team during the months up to February 1963, were summarized by Pierre Marin in (2009), 49, where he also acknowledged that "AdA's contribution was of capital importance for Orsay, which did not possess any culture of circular accelerators whatsoever, but it was also fundamental to the whole line of lepton storage rings which followed."

12.3.3 After the Discovery of the Touschek Effect

Soon after the discovery and the understanding of the effect, the team prepared the third article on AdA's experimentation which was submitted to *The Physical Review Letters*, on April 1st, 1963, where it was published just a month later: "Eventually, we believe that the consistency of the data we dispose of at present is fairly good and encourages in proceeding further with storage ring machines since these data do not show any failure 'in principle' of the main idea" (Bernardini et al. 1963, 13).

At around the same time, Touschek wrote to Burton Richter in Stanford to inform the Princeton-Stanford group of what they had found.[30]

---

[29] May 2013 interview, translated from Italian by the Author.

[30] Bruno Touschek Papers, Box 1, Sapienza University of Rome–Physics Department Archives.

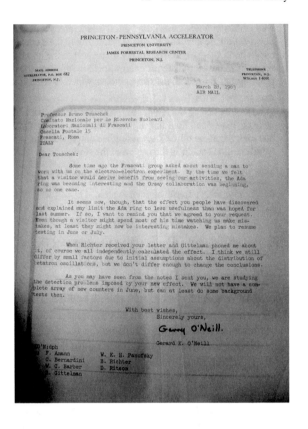

The Frascati and the American group had been in very friendly and collaborative terms at least since January 1962, when O'Neill had congratulated Ghigo and Bernardini on their initial successes with AdA. At the time, O'Neill had written: "…many congratulations on your success in storing a beam with so long a life time. I look forward with much interest in hearing the details", concluding with "Very best wishes to all of you and we hope to hear more good news".[31] The tone of this letter is very different from the one received by Touschek after he informed Richter in March 1963 of the discovery of the *AdA effect*. By that time, the AdA team was submitting the article, which would be rapidly accepted and published in just a month.

The answer Touschek received from Gerard O'Neill, on March 28th 1963, was not encouraging. It was rather cold, even dismissive both of what the Franco-Italian team had discovered, namely the Touschek effect, and pessimistic of the value of AdA's future prospects, given the limits imposed by the new effect on the luminosity at AdA's energy, Fig. 12.15.

About the Touschek effect, O'Neill writes that "It seems now that the effect you people have discovered and explained may limit the AdA ring to less usefulness than

---

[31] In 1962 O'Neill came to Europe, visiting both Frascati and Orsay, as from the unpublished report entitled *Summary of visits to Frascati and Orsay* (O'Neill 1962).

was hoped for last summer." And then he adds that they all calculated the effect, after receiving Touschek's letter, and basically agree on the conclusions.

While it is true that at this point AdA could not aspire to observe electron-positron annihilation into new particles, the importance of AdA's studies on beam dynamics and its impact on planning future machines did not escape the American team. A reflection on both the above points can be found in the talk which O'Neill gave at the end of April, in Washington D.C., at the Spring Meeting of the American Physical Society, 22–25 of April (O'Neill 1963d).[32] The importance of what the team had discovered is highlighted by the fact that, in this talk, O'Neill already calls it the *Touschek effect*, thus extrapolating it from the limitations of a small, low energy collider such as AdA. After describing how the Princeton-Stanford project was born and its present stage of operation, O'Neill opens a new paragraph by saying that "In 1960, a group at Frascati, the Italian National Laboratory, became interested in colliding beams. They built an electron-positron simple storage ring called AdA [...] This group [...] has made rapid progress and has passed some of the important milestones in colliding beam technique sooner than we have. The differences are mainly in the areas of vacuum technique and the method of injection." About the method of injection, O'Neill is clearly impressed by the way injection is accomplished, he calls it an "inefficient, but simple method" and admits that the sacrifice in efficiency has the great advantage that nothing can go wrong and that its simplicity makes it possible for the vacuum in AdA to be good. He acknowledges the importance of the effect discovered by the Orsay-Frascati team, calling it *Touschek effect*, the name used today, although in an article published on *Science* in summer 1963 O'Neill called it also the "Frascati effect". In this article, O'Neill mentioned it as the third of "three problems, not foreseen earlier, which will limit but (we now think) not prevent the carrying out of electron-electron colliding-beam experiments" (O'Neill 1963a, 683–684).

12.3.4 Single Bremsstrahlung: The Calculation That Helped to Confirm That Collisions Had Taken Place

O'Neill's March letter, with its put-down of AdA's prospects, had been sent in copy to the Stanford scientists, Wolfgang (Pief) Panofsky, Burt Richter and David Ritson, as well as to Bernard Gittelman, O'Neill's Princeton colleague. It had also been sent to Fernando Amman and Carlo Bernardini, the Italian members of the ADONE and AdA team. The large number of recipients, who would be reading the letter in copy, suggests the attempt, by the US colleagues, to establish some official status on the experimentation with the date of March 1963, which could eventually give credibility to Richter's later claim that AdA had been no more than a "scientific curiosity" (Richter 1997, 169).

---

[32] See https://www.osti.gov/biblio/4875549-storage-rings.

Anyone would have been discouraged in reading such letter, which reflected the opinion of so powerful competitors: not so Bruno Touschek and the intrepid Franco-Italian team. If AdA's luminosity was still too low to observe annihilation into new particles and even annihilation into two γ rays, there was another process which had a higher cross-section and which could demonstrate that the beams would collide, namely elastic electron-positron scattering.[33] But electron-positron scattering, namely $e^+e^- \to e^+e^-$ would be very difficult to distinguish from various background processes. On the other hand, according to Quantum Electrodynamics, such a reaction is always accompanied by photon emission, the so called *bremsstrahlung* process which corresponds to the reaction $e^+e^- \to e^+e^- + \gamma$.

It became clear that one could record the correlations between the photons emitted by electrons or positrons and the intensity of the particle beams, and the team prepared the first of the four runs which would finally establish the first observation of electron-positron collisions. These measurements would allow calculation of how many particles were produced when the photon was recorded. But was this number in agreement with the expectations from theory? The calculation of a similar process—bremsstrahlung from electron-electron scattering, namely for $e^-e^- \to e^-e^-\gamma$—was present in the literature, but no calculation of the cross-section for the case of electron-positron scattering was known to have been done as yet: it had not been included in the *Physical Review* paper by Cabibbo and Gatto about electron-positron processes (Cabibbo and Gatto 1961), nor in the more recent Baier's article, March-June 1963 (Baier 1963), which had followed it.[34]

Thus, to prove that electron-positron scattering had taken place, a calculation needed to be done, fast. And indeed it was done, between Rome and Frascati, as we shall see next.

It had in fact happened that, around that time, early 1963, two top physics students at University of Rome, Guido Altarelli and Franco Buccella, were close to finishing their course of studies and had been looking for a thesis in theoretical physics.[35] Early in 1963, Franco Buccella approached Raoul Gatto, already highly regarded and soon to become Professor at University of Florence, while Altarelli went to see Bruno Touschek.[36]

Gatto suggested Buccella start looking, rather generally, at the calculation of "radiative corrections" to electron-positron annihilation, and directed him to study the first four chapters of the book by Jauch and Rohrlich, on Quantum Electrodynamics (Jauch and Rohrlich 1955b). Gatto and Touschek were very close, and Touschek had

---

[33] First attempts to observe the process $e^+e^- \to 2\gamma$ had started on Saturday, December 21st, 1962, but no conclusive evidence had been obtained (Haïssinski 1998, 17).

[34] https://doi.org/10.1070/PU1963v005n06ABEH003471.

[35] The course of physics studies in Italian universities lasted four years and led to a *Laurea in Fisica*, with the title of Doctor in Physics, upon completion of an original research article or of a review paper, such that it could be published in peer reviewed international journals.

[36] Guido Altareli (1941–2015), Professor of Theoretical Physics at University of Rome, was also head of CERN Theory group in the years 2000–2004. One of his most influential works is the article entitled *Asymptotic freedom in parton language* (Altarelli and Parisi 1977), co-authored with 2021 Physics Nobel Prize winner Giorgio Parisi.

certainly been sharing with Gatto his concern about photon emission and extraction of physics from electron machine performances. He was also thinking of ADONE and collective radiative effects, such as those he had studied in Glasgow with Walther Thirring (Thirring and Touschek 1951), a problem, generally referred to as 'radiative corrections'.

Buccella remembers[37]:

> *Buccella:* I started reading the book, but, after going through the first four chapters, it was for me totally incomprehensible. I was in my fourth year of physics studies, still had to finish a number of exams, and I did not have any clue as to what I was supposed to calculate nor how to start doing the calculation, so I postponed the thesis to concentrate on the exams.

Around the same time, the other student, Guido Altarelli, had gone to talk to Bruno Touschek and asked for a thesis, as his friend Buccella had done with Gatto. Altarelli either did not understand or was not pleased with what Touschek proposed, and began arguing with him. During the discussion, Touschek suddenly jumped up and rushed to a chair, where his jacket, which had been hanging next to a small heater, had started going on fire. Altarelli then decided this professor was not his favourite and turned to Gatto for a less exciting thesis adventure.[38]

The presence of a heater in Touschek's office, places the episode in the winter. It may thus have been February or early March 1963 when Altarelli went to see Gatto.

By this time, Touschek had made an order of magnitude estimate to prove that the cross-section for single bremsstrahlung was consistent with AdA's observation, but he knew a more precise proof had to be done.[39] In early Spring, it was clear what was needed for AdA and when Altarelli asked Gatto for a thesis, the problem of calculating single bremsstrahlung emission in elastic electron-positron scattering was assigned, with Buccella also joining in, given the expected complexity of the calculation. Buccella and Altarelli, Fig. 12.16, were friends since childhood, Altarelli very tall and lean, with a sarcastic vein, Buccella shorter and handsome, more naive, and gifted with a great sense of humour.

---

[37] Private communication by Buccella to the Author, May 9th, 2018. Franco Buccella (b. 1941) is a former Professor of Theoretical Physics at University of Naples.

[38] This episode, by private communication from Altarelli to the Author in 2008, was published in the Italian journal *Analysis*, issue 2/3—June/September 2008, in an article about Frascati e la fisica teorica: da AdA a DAΦNE. Guido Altarelli later became Professor of Theoretical Physics at University of Rome, and Head of CERN Theory group. He passed away on 30 September 2015.

[39] This was discussed by Touschek in his presentation at The Brookhaven Summer Study Meeting on accelerators, which took place in June 1963 (Touschek 1963).

Fig. 12.16 Left panel, Franco Buccella at the time of his thesis, 1963, personal contribution to the Author, and Guido Altarelli, photo courtesy of Monica Pepe-Altarelli

12.4   The Summer of 1963

Summer arrived, and with it the summer meetings, following the well established tradition to present the year's work in various worldwide venues. In just two years' time, storage rings had now become an important topic in accelerator physics, and the AdA team was ready to present their work. Two were the main events ahead of the community of accelerator physicists: the first was a 5 weeks long Summer Study in Brookhaven, in the USA from 10th June to 19th July, and the second the traditional accelerator conference to be held in Dubna, in the USSR, in August. The first near New York, the second not far from Moscow.

12.4.1   Summer Studies in Brookhaven

Touschek was invited to lecture at Brookhaven National Laboratory, in Upton, near New York, where a Summer Study on Storage Rings, Accelerators and Experimentation at Super High-Energies had been organized. In his talk, entitled *The Italian Storage Rings* he described the status of experimentation with AdA and then the situation of ADONE, the bigger project (Touschek 1963). Touschek relates that although they had observed about 10 genuine beam-beam events, they did not observe annihilation into two gammas, as the two Čerenkov counters did not signal the expected coincidences. Given the expectations, both for the theoretical cross-section and the registered number of particles in the beam, they had to conclude that the cross-sectional area of the beams, the other factor from which depends the luminosity, was larger than expected, being at least 20 times higher. He then put forward the hypothesis that the few observed events, the genuine 10 beam-beam events, were probably due to radiative electron-positron scattering. To substantiate his hypothesis, he gave an order of magnitude estimate of both single and double bremsstrahlung processes. He then stressed the fact that "Life time measurements are of paramount importance when considering the feasibility of coincidence experiments."

The question of the volume of the beam is amply discussed in Touschek's contribution, but it would seem only from a theoretical point of view, namely that if the dimensions had been of the *natural* type, one should have seen annihilation into two gammas. But since they did not see that, the observed gammas had to be radiative events, which had a bigger cross-section, and the culprit were the dimensions of the beam. He then proceeds to relate the life time measurements which had taken place that year. He says that "the first accurate measurements were made in February 1963". Various discrepancies about the life time are discussed and finally he says that such discrepancies can be eliminated if one assumes the beam is considerably higher than "natural" and that this increase in height is due to some unwanted coupling between radial and vertical betatron oscillations … "Accurate photometric measurements carried out by Marin and Haïssinski show a beam height of less than $150\,\mu$. Measurements of beam height now in preparation involve the observation of forward-scattered electrons instead of the annihilation reaction." (Touschek 1963, 197)

Gerald O'Neill was also present and made two different presentations (O'Neill 1963c, b), where he recalled that the Princeton-Stanford storage rings of electrons against electrons "was designed to provide a sensitive quantum-electrodynamic test" for which the energy was chosen to be 500 MeV. He also remarked that a comparison with the AdA storage ring showed "a rather fortunate difference in parameters which has made these rings complementary." Surprisingly, Lawrence W. Jones, who gave the summary talk, did not mention AdA in the twenty-pages overview on the experimental utilization of colliding beams, not even in the brief historical introduction outlining US proposals and attempts before 1957 and "main advances" since then Jones (1963).

12.4.2 Dubna and the New Institute in Novosibirsk: *Une grande première*

The summer of 1963 is also when the AdA team learnt that beyond the Iron Curtain, and further away, beyond the Urals, in the Novosibirsk Laboratory of the Siberian branch of the USSR Academy of Sciences, a team of Russian scientists, led by Gersh Itskovich Budker, unbeknownst to everyone in the West, had started the construction of an electron-positron colliding beam accelerator.

A posteriori, the Russian work did not directly challenge AdA's primacy in proving that collisions had taken place, but later on it was the Russian electron-positron storage rings, VEPP-2, and the French ACO which produced the first important particle physics results, when they were first to study the properties of the so-called vector mesons, ρ, ω and ϕ, new particles which added important mile-stones to the understanding of the building blocks of matters, the quarks.

The announcement was made at the Dubna conference, which ran from 21st to the 27th August, (Kolomenskij et al. 1965). and went initially unnoticed.

In 1963, this Conference represented the highlight of the conference season for accelerator physicists. Its proceedings can now be found at the site http://inspirehep. net/record/19356, but for quite a long time, they were practically unavailable to the Western world, having been written in Russian. Bernardini presented the team's work on the beam life-time in storage rings (Bernardini et al. 1965), Henri Bruck, from Orsay, presented the ACO project. Pierre Marin was also attending the conference, but their memories are rather different. While Bernardini hardly mentions the Russian progress in storage rings, (Bernardini 2006), Marin's impressions are quite detailed, (Marin 2009, 52). In particular, he remembers the difficulty for the Western participants to understand any Russian. As a result, the only guide to the talks was the list of abstracts, and not much was understood of Budker's talk on the Russian progress in building colliders in his Institute in Novosibirsk, where he had moved. And then, at the end of the conference Budker invited a selected group among the foreign participants to visit INP, the Institute for Nuclear Physics in Siberia. Fernando Amman and Carlo Bernardini were among the 25 scientists invited to travel to the new Laboratory, but Bernardini's only mention of this trip is about a forced stop-over of the plane in Tomsk and he says nothing about what they saw in Novosibirsk (Bernardini 2006, 72). And still, as Pierre Marin puts it in his memoirs (Marin 2009, 52),

C'était une grande première.

This was indeed a grandiose debut, a first night to which they had been invited from all over the world. It was an amazing visit to a totally new laboratory, which had been built in just a few years, moving people and equipment from Moscow to the new location of Novosibirsk, beyond the Urals, where the Siberian Branch of the Soviet Academy had established a great laboratory under the direction of Budker. The astonished visitors, who had not really understood what Budker had said in Russian during his presentation in Dubna, saw two colliders in advanced stage of construction, VEPP-1, electrons against electrons, and VEPP-2, electrons against positrons, an eye-opener of what the Russian scientists had been doing, without the West even knowing anything of it.[40] For the French scientists, who had just began the construction of ACO, whose first elements were still being finished in the machine shop, it was a shock to be facing the electron-positron collider VEPP-2, with beam energy of 700 MeV (higher than ACO's), in an advanced stage of construction, Fig. 12.17.

---

[40] VEP is the Russian name for colliding electron beams and VEPP for colliding electron-positron beams.

Fig. 12.17 VEPP-2, from the 1963 Dubna Conference Proceedings, (Kolomenskij et al. 1965)

The Conference Proceedings were in Russian, and were sent abroad, as it was customary, to the participants libraries. The Americans translated it in English a couple of years later, but the volume in English was not widely circulated and the individual contributions were posted on inspirehep.net only recently. The whereabouts of the Proceedings in France are related by Pierre Marin in his book, but it will be worth recalling them, later in this chapter, from a letter Marin wrote to Bernardini on December 26th, in the year 2000. Enclosed with the letter was a copy of the translation in French of Budker's talk of Dubna '63, whose first page we show in Fig. 12.18.

12.5   Observing Collisions

The summer conferences being over, back to Orsay, the work with AdA resumed in earnest. The report on the beam life time had been published and presented both at Brookhaven and in the USSR. It was now time to proceed to implement its consequences, and attack the next obstacle, the demonstration that although annihilation was out of reach, still AdA could prove that collisions were taking place.

Having grasped the geometry and dynamics of the colliding beams of electrons and positrons, the team was now on the finishing line, but in order to get the final

Fig. 12.18 At left, the first page of Budker's talk at Dubna '63, in the French translation, courtesy of Jacques Haïssinski, and at right Gersh Budker's photo from http://sesc.nsu.ru/famed/index.php? view=detail&id=83&option=com_joomgallery&Itemid=145

prize, namely to be accepted in the history of science as the first to witness, track and record electron-positron collisions in a laboratory, the team had to convince the world: only then, they could move ahead to continue their new dreams, ACO for the French and ADONE for the Italians.

To complete the work, the two teams had to start precise measurements of the process which could be observed with AdA, and confront them with theoretical expectations. Touschek had already a fair idea of the order of magnitude of the result, but precise formulas had to be given and numbers checked. Indeed, as the fall of 1963 started, the needed calculation for the process they could claim to have observed, $e^+e^- \rightarrow e^+e^- + \gamma$, became available. The two students in Rome had spent the months of July and August, working on their thesis, skipping almost entirely the usually sacred summer vacations, as Franco Buccella remembers. There were many obstacles, theoretical and computational, some of them almost intractable, but they persevered, and indeed the article they would later write (Altarelli and Buccella 1964) is still a classic text, amply used and quoted. They worked through the summer of 1963, taking only one week vacations. Touschek helped them not to lose heart, giving them now and then some insight on why the calculation appeared so difficult, and indeed it was, and, at times, hinting at the possible solution. Finally, they found their way to the result through an approximation which is still used and they could then start writing the thesis. The thesis was defended, with full honours, at the University of Rome, in November 1963. Thus, the theoretical results were ready for the time the AdA team in Orsay would start the final experimentation in December 1963, when the AdA team positioned themselves in earnest, building statistics.

Touschek was also looking ahead to ADONE. Now that the discovery of the *Touschek effect* had cleared the way for higher energy colliders, the problem of radiative corrections to future ADONE experiments began worrying him.

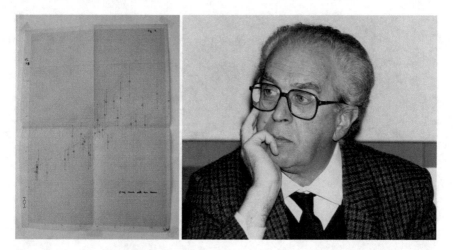

Fig. 12.19 The synthesis of all the data for gamma rays count collected when two beams were circulating in AdA, from Touschek's recorded plot, in the original plot photographed by L. Bonolis from Carlo Bernardini's personal papers, data published as (Fig. 9) in Bernardini et al. (1964). At right, Carlo Bernardini in later years, courtesy of L. Bonolis, personal copy from C. Bernardini, all rights reserved

12.5.1 December 1963: Starting the Final Runs

The first of four final runs took place in December 1963. They had to observe the correlation of gamma rays with the counting of positrons and electrons, and check that the number of such events was in accordance with their acquired knowledge about the volume of the bunches and the calculation of the elementary cross-section being done by Altarelli and Buccella. The second run took place in January 1964, a third one in February and the fourth one in April 1964. The results collected during the four data runs, Fig. 12.19, clearly indicated correlation between the counts from both sides. In May, enough evidence had been accumulated, figures started to be drawn, the article was written. In Frascati and Rome, the calculation of single hard bremsstrahlung in electron positron scattering was finished, and the comparison between the theoretical calculation and AdA's data showed that AdA's observations were consistent with expectations that electrons and positrons had indeed made collisions. This result was of course also based on the knowledge the AdA group had acquired of the beam dimensions and their internal dynamics, controlled by the Touschek effect.

Thus the team did not lose heart and, in the end, managed to demonstrate that the collisions were taking place and that the overlap between the electron and the positron bunches was complete. In the first half of 1964, an article with the results from the four runs was prepared, to be circulated by ordinary mail, in *preprint* form among the interested community of accelerator physicists, and submitted, in July

Fig. 12.20 At left, Jacques Haïssinski is shown in the ACO control room, kept as it was then, with the 1:4 copy of AdA, on the occasion of the Bruno Touschek Memorial Lectures, September 9–13, 2013, celebrating fifty years since the 1963 observation of electron-positron collisions, courtesy of Jacques Haïssinski. At right one can see AdA on the Frascati Laboratories grounds, 2013 ©INFN, © INFN–LNF, http://w3.lnf.infn.it/multimedia/

1964, to the *Nuovo Cimento* (Bernardini et al. 1964) with the title *Measurement of the rate of interactions between stored electrons and positrons.*

In 2013 Jacques Haïssinski, Fig. 12.20 summarised the results of AdA's experimentation[41]:

> *Haïssinski:* Les mesures finales qui ont été faites ont été des mesures qui ont porté sur le nombre de collisions par seconde entre les électrons et les positrons. C'était la première fois au monde que l'on montrait que les particules, effectivement, intéragissaient et entraient en collision les unes avec les autres et donc, ça montrait que, on peut dire, l'essentiel, pas tout, bien sûr, on a fait d'autres découvertes par la suite, mais l'essentiel, quand même, des caractéristiques de ce type de machines était validé et permettait de penser que les générations ultérieures [de ce type de machine] séraient utilisables pour faire de la physique des particules à très haute énergie.[42]

---

[41] May 2013 interview with Jacques Haïssinski in Orsay.

[42] The final measurements which were done dealt on the rate of collisions among electrons and positrons. It was the first time in the world that one could show that the particles effectively interacted and collided against each other, and this allowed to think that these machines could be used to do [experiments] in high energy physics. There were other discoveries, of course, but the essential point was that the characteristics of this type of machines were validated and that successive generations [of these machines] could be used for very high energy physics.

12.6   What Was Happening Beyond the Iron Curtain

With the publication of the results of the four runs, during 1963 and 1964, AdA's adventure was completed. It had proved that this type of machine could work, and the AdA team had been the first to show this. But had AdA been the first ever particle collider to function? And had been Touschek the first to propose storing electrons and positrons in a single ring in order to make them collide and probe the QED vacuum? Had he been the first to realize the idea, together with a group of exceptional young scientists and technicians, from Orsay and Frascati? It seems to us, at the end of this chapter, that we can answer in the affirmative. This was an incredible feat, a victory of David against Goliath, which sprang from the hard times of the war through which Touschek, and the rest of Europe, had survived.

The story from the point of view of the scientists beyond the Iron Curtain shows how close the race had been. Budker's team at the Laboratory of New Acceleration Methods of the Institute of Atomic Energy of the USSR Academy of Sciences in Moscow (in 1960 named *Kurchatov Institute*, after its founder Igor V. Kurchatov) had focused since 1957 on colliding beams, in particular on physics development and design of the electron-electron collider VEP-1. VEP-1 construction started in 1958–1959, and was the first accelerator implementing colliding beams, after AdA. As recalled by Alexander N. Skrinsky, this activity had started "just upon D. Kerst's proton collider suggestion and G. O'Neill's proposal to use radiation damping for electron beams storing and compression." (Skrinsky 1995, 14). Then, in 1958, Budker's Laboratory was transformed into the Institute of Nuclear Physics (INP) of the Siberian Branch of Academy of Sciences. Between 1961–1962 the whole INP moved to Novosibirsk, where the first circulating beam was obtained with VEP-1 in August 1963 and the first luminosity was detected in May 1964 (Levichev et al. 2018, 407). In the meantime, they had begun to work on the idea of electron-positron collisions, which would later materialize in the VEPP-2 collider. As seen before, the collider activity of INP was presented for the first time at the International Conference on High-Energy Accelerators held in Dubna in August 1963 (Budker and Naumov 1965).[43]

We recall here some comments and impressions by Pierre Marin about the Dubna Conference from a letter sent to Carlo Bernardini in the year 2000, reproduced in Pancheri and Bonolis (2018), Fig. 28:

Dear Carlo,

following our recent telephone conversation, please find below some comments on the 1963 Dubna Conference on large accelerators in relation to the origin of e^+e^- collider ideas in the Soviet Union.

First a few facts:

---

[43] For an outline of accelerator developments in Novosibirsk and a presentation of research work with VEP-1 and VEPP-2 see Budker (1966), Budker et al. (1966), Auslender et al. (1966). For an early account and presentation of the Russian projects made by Budker himself, see Budker (1967).

– as you know the conference took place in August. The participants were delivered a Bulletin with the abstracts in Russian, accompanied by a short English translation. Nobody among the French delegates understood Russian and we were only able to grasp a bare minimum of Budker's talk

– the proceedings were available only in 1964 and Henri Bruck had a translation made by CEA in France. This probably arrived at Orsay towards the end of 1964

– as you remember, Budker organized a visit of IPN [Institute for Nuclear Physics] at Novosibirsk just after the Conference and I have a clear recollection of seeing VEPP-2, as described in Budker's talk, that is, well advanced but not yet assembled

– in 1964 at Orsay, we were busy with the construction of ACO and the preparation of the first experimental set-up, together with our collaboration in AdA.[44] There was little time to pay attention to the origin of e^+e^- in the USSR and the secrecy about the Kurchatov Institute at Novosibirsk was not a favourable situation to ask for questions

– at this point I should mention that a group of 25 French nuclear physicists, I was one of them, was invited in 1957 by the Soviet Academy of Science to visit installations in Moscow, Leningrad and Kiev. We paid an extremely short visit to the Kurchatov Institute, but nothing came out due to the extreme secrecy existing at the time (Dubna and Leningrad were not open)

– two years ago, during one of usual stays of W. Baier at LAL, I had asked about the e^+e^- origins in the USSR. He said that this was his ideas and as soon as started talking on the subject at the Kurchatov, all the work on e^-e^- stopped immediately and they concentrated on e^+e^- collisions. At that occasion, he mentioned that Budker had gone, one day, to the Academy [of Sciences] to propose the idea. He was opposed an argument he could not immediately answer whereas Baier was able to. The day after, Budker went back to the Academy and got the authorization to start on the new field. When questioned about the existence of written documents, notes, minutes …of the Academy, he said there might be something but difficult to reach now. A year after, Baier suggested that people involved at that time in electron storage rings and still active should meet and have a thorough discussion of the subject

– I should mention that the translation of Budker's talk was kept in the archives of the Orsay Storage Rings Group at Orsay until now, but when questioned neither H. Zyngier, M. Sommer nor Jacques Haïssinski had any recollection of it. It is only at the celebration of the Bruno Touschek Symposium in 1998, at Frascati (Haïssinski 1998), that our interest (I am talking of J.H. and myself) was strongly revived on the subject. To my great confusion, I must say that it is at the occasion of a wild sorting out of old archives at Orsay, that a few days ago I came across a copy of the translation of Budker's talk. It was in the verge of being thrown away by the new generation of people …!

Here above, are the only facts I was able to recollect since the 1998 Symposium.

A few other comments:

---

[44] Efforts were made in view of preparing experiences and detectors specifically dedicated to the new research topics to be explored with colliding beams (Augustin et al. 1965). The group directed by A. Blanc-Lapierre included: (1) physicists and engineers of the Laboratoire de l'Accélérateur Linéaire d'Orsay and of the Centre d'Études Nucléaires de Saclay who participated in the project and construction of ACO as well as in tests (M.M.G. Arzelier, J.E. Augustin, R.A. Beck, R. Belbéoch, M. Bergher, H. Bruck, G. Gendreau, P. Gratreau, J. Haïssinski, R. Jolivot, G. Leleux, P. Marin, B. Milman, F. Rumpf, E. Sommer, H. Zyngier); (2) Physicists participating to the realization of the experimental device at ACO (MM. J.E. Augustin, J. Buon, J. Haïssinski, F. Laplanche, P. Marin, F. Rumpf, E. Silva) (Marin 1966).

- VEPPII was probably not assembled and not injected at the Kurchatov. Touschek life time would have been a striking evidence with positrons at such a machine at low energy, even with low currents. We also know that the AdA findings on this were new to the Siberian group in 1963.
- the move from Kurchatov to Novosibirsk resulted probably to a large interruption of their activities. In 1963 the institute at Novosibirsk was already huge. Despite the large available manpower it may have taken several years to achieve it. Remember also that electron-positron collisions were only one of their concerns.
- Baier's sayings reported above are in agreement with Budker's talk. However, looking at the references concerning the reported activities of Baier, some dated back to 1959, they only deal with electron electron collisions. This adds to the mystery of the period.
- finally one can ask what are the original contributions due respectively to Budker and to Baier, accelerator ideas on the one hand and theoretical vision on the other hand? For the time being it is only conjectural

Sincerely yours
Pierre

In the year 2006, the well known theoretical physicist Vladimir N. Baier published a memoir which tells his side of the electron-positron collider adventure. In his memoir, Vladimir Baier tells of the difficulties and delays encountered by the Russian group led by Budker, delays due also to the moving to Novosibirsk, including people and instrumentation. He also recalled that "Because of pathological secrecy adopted at that time in USSR, all activity in Kurchatov Institute was considered as 'for restricted use only' and the special permission for publication in open journals or proceedings was necessary for each article" (Baier 2006, 5). Baier included an interesting date for the day in which he started working on electron-positron colliders: 28 October 1959, around the same time we can ascribe to Panofsky's seminar in Frascati on 26th of October. Although the beginning of electron-positron activity in the USSR was likely to have started after Panofksy's presentation in Kiev in July, reinforced by the one ay CERN in September, Baier's date does not appear anywhere else. On the other hand, the starting point for the events which led to Touschek's proposal in February 1960, is clearly defined by Panofsky's visit to Rome in late October 1959, and seminar in Frascati, as seen in Chap. 10.

12.7   After AdA

When the AdA team published their fourth and last work on AdA's experimentation (Bernardini et al. 1964), four and a half years had passed since the day when Touschek had proposed to build AdA. While AdA's last article was being published in December 1964, everywhere in the world new projects for colliding beam facilities were already in place, in Italy with ADONE, in the USA with SPEAR, in the USSR with VEPP-2, in France with ACO, at CERN with the Intersecting Storage Ring for protons (Bryant 1983).[45]

Both ADONE and ACO had been envisioned and planned before AdA moved to Orsay. In December 1960, as AdA had been constructed and was near to enter in operation in February 1961, Touschek had proposed to build an electron-positron collider which could reach such energy to explore annihilation into pairs of particles and antiparticles of all the kinds known at the time: $\mu's$ and $\pi's$, kaons and protons and neutrons. ADONE would need to be much bigger, and costlier, but it was rapidly approved and, in 1963, construction of the ADONE building had started, in an empty lot across the street from the land where the electron synchrotron and AdA had been built (Amman et al. 1965).

In France, the construction of ACO was approved in 1963 (Bruck and Group 1965) and the collider became operational in 1967. As presented in Dubna, ACO's original value of $E_{max}/beam$ was 450 MeV, but later on this energy was pushed up to 540 MeV, probably to ensure comfortable conditions to produce the ϕ meson and study its properties (Augustin et al. 1973).

After 1964, the roads followed by the protagonists of AdA's story moved apart. Giorgio Ghigo passed away not long after the end of AdA's adventure. Bruno Touschek and Carlo Bernardini returned to Italy, fully immersed in planning for ADONE and on how to extract physics results from this machine. Giuseppe Di Giugno had to leave for his military service, and, during his absence, the second AdA log book, kept in his office in University of Naples, was thrashed. Di Giugno is the only one of the AdA team who left the academic life of particle physics: he profused his genius for electronics into music and was one of the early founders of today's digital world in music.[46]

As for the French team, Jacques Haïssinski finished his thesis and started working on ACO, joining Pierre Marin, already head of the project. Thus, in Orsay, through experimentation with AdA, the world of colliders had been opened and with it the road to the larger French accelerators, and their future exploration with synchrotron light, through SUPERACO, and, in modern days, the ESFR in Grenoble and SOLEIL on the Saclay plateau, south of Paris, above the Vallée de Chevreuse.[47]

---

[45] CERN's interest in storage rings for protons had an early start. Proposals for tangential rings for protons were presented at the end of 1960, and it is in 1962 that the actual proposal for the ISR colliding protons against protons was presented. The first proton-proton collisions were recorded on 27 January 1971. CERN's interest in storing antiprotons appeared also very early. In 1983, Peter Bryant wrote a brief account of the events leading to antiprotons in the ISR, on behalf of then dissolved CERN ISR division. Bryant recounts that the storage of anti-protons in the ISR was discussed as early as 1962, even before the ISR were constructed. Physics runs with proton on antiprotons in the ISR took place starting in April 1981 until December 1982, at maximum 62.7 GeV c.m. energy.

[46] See https://medias.ircam.fr/x7b9990, http://www.lucianoberio.org/node/36288 for some memories from Di Giugno's first approach to electronics in modern musics. See also http://120years. net/sogitec-4x-synthesiser-giuseppe-di-giugno-france-1981/.

[47] A highlight of Orsay's road to sinchrotron light radiation studies and applications can be found in the docu-film Soixante années d'exploration de la matière avec des accélérateurs de particules, by E. Agapito and G. Pancheri, with the collaboration of Jacques Haïssinski.

12.7.1 How Touschek Found the Way to Prove That Electron-Positron Colliders Would Work

What was Bruno's role in AdA's final success in Orsay? Not just in driving people's work or sitting at Café de la Gare, when preoccupied. The whole sequence of events, from the February 17th, 1960 proposal to the publication of AdA's last paper in 1964, is clearly driven by Bruno's insight and determination. When he discovered and calculated the *Touschek* effect, he understood that AdA would not be able to observe any of the processes he has listed in his original proposal. Having seen that the effect would not be as important for higher energy machines, he could have just given up, and waited for ADONE to work. But this was not his way. He had started with AdA, and had to prove the feasibility of the *vrai bijou*. His knowledge and understanding of electrodynamics suggested the solution. A proof that electrons and positrons annihilate, is given by photon bremsstrahlung, radiation emission, before or after the annihilation.[48] The process could be compared with the observed rate of electron-positron elastic scattering accompanied by an observable photon. This is when he started wondering how to calculate the process. He spoke with Gatto about it. And Gatto assigned the problem as a graduation thesis to two of the brightest students of an already exceptional year, Guido Altarelli and Franco Buccella. By November 1963, when they graduated with full honours, the feat was done. Early in 1964 data were collected in Orsay, wisdom about the size of the beams was put together with data and with the cross-section calculation, and AdA's proof-of-concept was done.

Once more, the gamble had paid off. Now, it was time to focus on ADONE.

---

[48] A semi-classical description of what happens is given in a later paper (Etim et al. 1967a). In semi-classical terms, if annihilation takes place, it means that the initial state present at time $t = -\infty$ disappears at time $t = 0$, to reappear at $t = +\infty$. During the transition, the initial state charged particles are slowed down to zero and then reappear to be accelerated again to final velocities. This makes the transition to be equivalent to a deceleration and acceleration process during which radiation is emitted.

Part III
Maturity

Chapter 13
ADONE: A Legacy of Wins and Losses

All those, among us, who had the privilege of knowing Bruno,
will never forget him, and we have an immense debt of gratitude
to his intelligence, generosity and friendliness.
Raoul Gatto Bruno Touschek Memorial Lectures, 1987 Frascati

Abstract ADONE's construction, from its approval in 1963 to the time which led to the first collisions in 1969, includes some of the best years in Touschek's life. During these years Touschek saw some of his dreams come true, as electron-positron collisions were recognized everywhere as an important tool to probe the world of particle physics, and saw ADONE becoming reality. While following its construction, he built in Frascati a theoretical physics group, which developed his ideas about quantum resummation and, later, carried them on to Quantum ChromoDynamics. ADONE started working in spring 1969 and its early successes in discovering multiple hadron production confirmed Bruno's faith that new and unexpected final states would be produced in the annihilation of matter against antimatter. A few years later, in 1974, the discovery of the particle J/Ψ, with a conventional type accelerator and a collider built after AdA's success, was simultaneously announced by two US teams and confirmed by Frascati a few days later, after a dramatic search. The road to the experimental establishment of the Standard Model of Elementary Particles was open.

In the last two chapters, Touschek's familial voice was silent. The letters he sent to his father through most of his life, and on which much of his story has been narrated here, are not available for the period of AdA and ADONE's construction. After November 6th, 1960, the letters speak again only in 1969, between May 20th 1969 and February 18th, 1971. After Bruno's father's passed away in 1971, the whole set of letters was then sent to Bruno by his step-mother. The missing letters were probably kept in one of the four boxes, which Elspeth showed to Luisa Bonolis and the Author in 2009, but were not photographed back then, and have been lost after she passed away. At the same time, large archival and bibliographic material, of both historical and personal value, has been published about these years by a number of its protagonists. From these voices, the story of Bruno's last most productive years can be told. Touschek's own view of the world around him is, however, far from being absent. This view is present in a large number of sketches and drawings, which give Bruno's satirical or sarcastic description of an important period at the University of Rome, from mid 1960's until he went to CERN in 1977 (Amaldi 1981, 58–69). From there, on March 8th 1978, a CERN car accompanied him to Austria, where he passed away on May 25th, 1978.

13.1 ADONE and Its Golden Years

In February 1960 Touschek had advanced the proposal to construct an electron-positron collider and see what interesting physics could be produced in the annihilation. The collider, whose construction had been approved by the Frascati management in March, was smaller and less powerful than what he had originally envisioned, but AdA, a scaled down version to become a proof-of-principle, was going along well and promised to be finished in a few months. So much so, that, only eight months after AdA's construction had been approved, he felt confident that the machine would work and ready to propose what he had wanted to do from the very beginning: a collider capable of doing physics. Thus, even before the first electron beam had circulated in AdA, and much before AdA proved the feasibility of storing electrons and positrons in a single ring and allow detection of matter-antimatter collisions in the laboratory, Touschek started plans for a bigger accelerator, with higher particle density and higher energy per beam.

In early November 1960, he was aware of the interest AdA was raising, and of the possible competition from across the Atlantic Ocean. He had been invited to give some seminars in Berlin and Hamburg, and, before taking the trip, Bruno thought ahead. AdA would be waiting for him in Frascati, and intense weeks of work would now arrive, with AdA's magnet to be measured in November and December. Things could be expected to get really serious in January, when the first attempts for electrons to circulate in the ring would start. Through his friends, he had learnt that the Frascati scientists were really first in this field and his judgement was that the American competition would not bother them for a year.[1] But the final objective for him was to do physics, not just to build a prototype, much as this would be ground-breaking. No time could be wasted, plans for the next move had to be set in motion.

To start with, on November 9th, 1960 he put down his ideas for an accelerator to be called ADONE, meaning a bigger AdA, but also something as beautiful as an Adonis, the Greek god paradigm of beauty. As shown in Fig. 13.1, his mind turned to the physics which could be extracted from collisions at a substantially higher energy that the one in AdA. Thus he thought of proposing a beam energy such that all known particles could be created in pairs through the initial electron-positron

---

[1] Letter to parents on November 6th, 1960.

113

A D O N E - a Draft Proposal for a
Colliding Beam Experiment.

B.Touschek,
Rome, 9.Nov.60.

It is proposed to construct a synchrotron
like machine capable of accelerating simultaneously
electrons and positrons in identical orbits. The suggested maximum energy is 1.5 Gev for the electrons as well
as the positrons. This energy allows one to produce pairs
of all the so called 'elementary particles' so far known,
with the exception of the neutrino, which only becomes
accessible via a weak interaction channel.

It is assumed that experiments in which there
are only two particles in the final state are most easy
to interpret. There are 16 such reactions, namely:

(1) 2γ . This is the only reaction in which
the real intermediate state is 'quasi real' and in which
therefore there should be no 'radiative corrections'.
This reaction should serve as a 'monitor'. The cross-section is $2.6 \cdot 10^{-31}$ cm².

(2) e^+,e^- . This reaction will show strong
angular variations and may require 'good geometry'. It
would give information on the brkdown of electrodynamics
at distances corresponding to about 1/3 the Comptonwavelength of the proton.

(3) μ^+,μ^- . Test of electrodynamics in 'bad
geometry'. May also serve as an indication of the fundamental difference between electrons and muons.

(4) $\pi^+\pi^-$ reveals the interaction between
pions in odd parity states.

(5) $2\pi^0$: charge exchange interaction for pion-pion scattering.

(6) K^+K^- : interaction of K-mesons in odd
parity states.

(7) $\overline{K}^0,K^0$: Charge exchange interaction between
K-mesons.

(8) $p,\bar{p}$: interaction of proton and antiproton
in even parity odd charge parity states.

(9) $n,\bar{n}$: same as (8) but for the charge
exchange reaction.

(10) through (15). Interactions simple or
with charge exchange of hyperons.

-3-

A very preliminary design study of the
magnet leads to the following requirements:
A major problem in accelerating electrons is
to compensate their radiation losses. In order to safeguard a small radiative decay of the beam it is necessary
to use strong focussing. Also, it does not seem to be
convenient to make the magnet very small. We are now
studying the design of a strong focussing DO magnet with
a momentum compaction factor of 0.04 (the same as in
Livingston's machine) consisting of 8 straight sections
of 2m length and 8 curved sections with a radius of
curvature of 10m. In such a design the radiation losses
are 45 kvolts per turn and a radiofrequency working in
the fourth harmonic (i.e. at a wave length of about 20m)
and with a peak voltage of 200 kvolt should be quite
sufficient to make the radiative lifetime of the beam
practically infinite.

The cross section of the circulating beam
in an ideal machine (complete vacuum, no alignment errors)
would then be 2.10^{-7} cm².

If the radiofrequency is placed in one of
the straight sections most of the beam-beam reactions
should take place in the straight sections (owing to
the bunching of the charges). To have a luminosity of
1/sec and straight section of the monitoring reaction
one would have to have

$$N_+N_- = 10^{15}$$

(N_+,N_- = number of circulating positrons,electrons).
It is quite possible to inject about 10^{11} electrons
so that only 10^7 positrons would be needed. It does
seem possible to reach this figure if positrons can be
injected accumulatively in about 10 linac pulses.

The present proposal has been discussed in
Frascati. It appears that it would involve an expense
of between 3 and $5 \cdot 10^5$ lit. It was the agreed opinion
of these meetings that if the work now in progress on
ADA shows real promise (we will know more about this
in Frebruary 1961) the development of ADONE should be
considered as compulsory, particularly in consideration
of the fact, that the work at the moment going on in
Frascati puts us into a position of great advantage.

Fig. 13.1 First and third page of memo prepared by Touschek on November 9th, 1960, from Touschek's papers, reproduced courtesy of Sapienza University of Rome–Physics Department Archives, https://archivisapienzasmfn.archiui.com, documents provided for purposes of study and research, all rights reserved

annihilation, and with sufficient leeway to study in detail their production process. It was November 1960: AdA, whose energy could be high enough to produce pions and muons, was still in the construction stage. The new machine should go beyond. Bruno's mind went through all the known particles, and tried to see what could be found beyond, and which interesting physics processes could be investigated with the future machine. He thought of the heaviest of the elementary particles, neutrons (n) and protons (p), whose mass is close in value, $\approx 1\,\mathrm{GeV}$, and saw that the new machine should have at least a center-of-mass energy of $2\,\mathrm{GeV}$. Arguing that it could be interesting to study the threshold behaviour in the creation of nucleon-antinucleon pairs, as well that of pairs of some unstable known particles, such as the Hyperons with mass around $1.2\,\mathrm{GeV}$, Bruno proposed a center-of-mass energy of $3\,\mathrm{GeV}$. Such energy appeared, and was at the time, a far away dream.

As we shall see, this choice for the maximum energy of the new machine he was proposing to be built, was to be dramatically regretted on November 11th 1974. However, at the time, aiming for an energy of $3\,\mathrm{GeV}$ in the center of mass was as good as aiming at $4\,\mathrm{GeV}$, certainly cheaper anyway. This was an important consideration,

Fig. 13.2 Bruno Touschek and Edoardo Amaldi, date unknown, reproduced courtesy of Sapienza University of Rome–Physics Department Archives, https://archivisapienzasmfn.archiui.com, documents provided for purposes of study and research, all rights reserved

since the cost of such machine appeared be much higher than anything Italy had started so far, close to $3 - 5$ Billion Lire. As no one had yet thought of building an electron collider of such high energy, namely there were no competitors yet in sight, the ADONE project would still put Frascati in a position of great advantage, as Bruno wrote in his memo.

Bruno had been discussing this plan in Frascati, and the memo may have been prompted by Amaldi, who wanted to present the ADONE project at the December meeting of the European Committee for Future Accelerators.[2] With an eye toward possible European competitors, Amaldi was laying down Italy's priority claims for electron-positron colliders, and preparing the grounds for the national approval of the project costs. Amaldi had shaped post-war European and Italian physics reconstruction, leading the European particle adventure, both as protagonist and *deus ex-machina*, the unseen stage director, Fig. 13.2. Together with Bruno, younger by 13 years, he created the conditions for the new physics to come with particle colliders.

By January 1961, the general lines of the project had been prepared. Touschek presented them in an internal LNF note, together with Fernando Amman, Carlo

---

[2] He did so and mentioned it in his January 16th, 1961 letter to André Blanc-Lapierre, Laboratoire de l'Accélérateur Linéaire Archives, courtesy of Jacques Haïssinski.

Bernardini, Raoul Gatto and Giorgio Ghigo.[3] Soon after, a Study Group was officially formed and, by the end of the year, the note had become an official proposal for the realization of an electron-positron storage ring of 1.5 GeV beam energy.[4] Three of the original proposers are absent from the authors' list, Touschek and Carlo Bernardini being at that stage fully involved in AdA's progress, while Raoul Gatto had gone back to his mission as a theoretical physicist. He was preparing, with Nicola Cabibbo, the extensive article on e^+e^- physics, which became known as *la Bibbia* among Frascati physicists, (Cabibbo and Gatto 1961). By this time, AdA had been presented at CERN, arising the interest of Pierre Marin and Georges Charpak from Orsay, and encouraging Budker and his collaborators in Novosibirsk to continue the construction of VEPP-2 (Baier 1963, 2006). David Ritson, from MIT, was in Frascati as well, and was one of the proposal's authors, as were Claudio Pellegrini and Mario Bassetti, who would be instrumental to ADONE's successful operation.[5]

While AdA was in Orsay, parallel preparatory work for ADONE progressed. Its future was in everybody's mind when the luminosity in AdA had suddenly dropped on the fateful 1963 February night: could ADONE suffer the same fate? Was AdA's problem signaling an insuperable limit to how many positrons and electrons could be stored in the same ring? That would have made it impossible to observe collisions with ADONE, and the road to higher energies through electron-positron collisions would be closed. It was not an improbable occurrence. Failure is often met in science, and routes have to be changed. Indeed, for a brief time, the physics hoped to be seen emerging from particles annihilation, had seemed beyond reach. On that night, for a few hours, hopes had been temporarily crashed. As we know, this story had a swift happy ending. By early morning, Touschek had understood the origin of the effect, and could explain to the anxious Franco-Italian team that the problem would be lessened at higher energies, namely ADONE was safe, though AdA was not.

13.1.1 1963: A Storm Temporarily Avoided

At the end of 1963, while AdA was still in Orsay, the ADONE group was ready for a status report (Amman 1985). One of the major decisions, which had been taken, concerned the acquisition of a LINAC, the linear accelerator needed for positron injection with sufficient intensity and energy. Offers from industry were solicited, comparison of costs and performance indicated that the American company Varian Associates could provide the best solution. But, while all this was taking place, a storm was gathering over the laboratories and the institution which had been

---

[3] F. Amman et al., "ANELLO DI ACCUMULAZIONE PER ELETTRONI. ADONE", LNF-61/005 (27.1.61).

[4] Amman et al., LNF-61-065.

[5] Claudio Pellegrini (b.1935), Emeritus Professor at University of California at Los Angeles, was awarded the 2008 US President Fermi Medal for pioneering work on X-ray free electron lasers and collective effects in relativistic particle beams.

financing constructions and owned the land, the National Committee for Nuclear Energy, *Comitato Nazionale per l'Energia Nucleare* in Italian, CNEN for short.[6] Together with the National Research Council (CNR), CNEN had funded the costs for construction and development of the Frascati National Laboratories, as a pure research adventure, in the context of its statutory mission for the development and pacific exploitation of nuclear energy. Felice Ippolito, CNEN Secretary General, was a strong supporter of the nationalization of the Italian electricity system and believed nuclear energy could be the future for Italian electricity production. This was a dangerous path to follow, and Ippolito soon came under fire. To quote what Fernando Amman, ADONE's project director, wrote in 1985 (Amman 1985):

> *Fernando Amman:* In the summer of 1963, the atmosphere at CNEN was becoming more and more choking: it was necessary to close the LINAC [LINear ACCelerator] contract before the storm that was in the air broke out on CNEN, in order to avoid the probable consequences of the retreat, at leat temporarily, of the National Research Council, and very long delay for ADONE. That July 29th [the day when the contract with Varian, the firm providing the LINAC, was signed] came as a great relief for all of us.
>
> Less than a week passed and an article on a weekly magazine by [Giuseppe] Saragat (few years later he became President of the Italian Republic) marked the launching of an absurd attack to Ippolito that brought him to jail.

It was a big national scandal, which signalled the first of future attacks on the Italian nuclear energy program (Baracca et al. 2017). Amaldi and the Rome physics department were first in line to defend Ippolito, with frequent discussions and strategy meetings taking place mostly in Bruno's home. Emilio Colombo, the Minister of Industry, which controlled CNEN, was also often present (Bernardini 2006, 73). Bruno, as always the sarcastic witness, left many drawings about this period, Fig. 13.3.

During one of these evenings at Bruno's home, Amaldi, Salvini and Carlo Bernardini were animatedly discussing how to help Ippolito. At one point, the conversation turned to lighter subjects. Bruno, wanting to stress that the issue was urgent, in fact even tragic, burst out in Italian : "A parte cio', signora Lincoln, com'era lo spettacolo?"[7] Touschek's question, addressed in particular to Salvini, underlined the absurdity of making futile comments when the tragedy was unfolding nearby (Bernardini et al. 2015, 287). This is why, in other drawings, or comments with friends, Touschek referred to Salvini as "la vedova allegra", *the merry widow* in English.

It was in fact a tragedy. Ippolito (1915–1997), professor of Geology at University of Naples at the time of the newspaper campaign again him, was arrested on March 3rd, 1964. Sentenced to 11 years of prison and to pay a heavy monetary fine, he was also stripped of his civil rights. After two years of prison, he was pardoned by the President of the Italian Republic, the same Giuseppe Saragat who had launched the journalistic attach against him. Ippolito was reinstated as Professor in University of Naples, later authoring many books of political and economic studies content. He is

---

[6] When originally established in 1952, CNEN had been called CNRN, *Comitato Nazionale per le Ricerche Nucleari.*

[7] In English, "By the way, Mrs. Lincoln, how was the show?"

Fig. 13.3 A drawings by Bruno Touschek, with a caricature of Edoardo Amaldi under bars, echoes Amaldi's public defense of Ippolito, and fears of Amaldi being under attack as well. The lower right dedication reads "With promise to bring you there the spaghetti". Family Documents, © Francis Touschek, all rights reserved

seen in Fig. 13.4, together with other protagonists of Touschek's story, assembled in Frascati for Salvini's 70th birthday. Among them, one sees Nicola Cabibbo, who had been one of Touschek's first students in 1957 and was President of INFN, at the time of the photograph. Next to him, there is Enzo Iarocci, director of the Frascati Laboratories, and himself future INFN president. Together they developed, and carried to conclusion, the project for a new generation electron-positron collider, called DAFNE, Double Accelerator for Nice Experiments (Bonolis et al. 2021).

Luckily for ADONE, the order for the LINAC was signed in July 1963. The cost for other main orders were placed between 1963 and 1964, while construction started on the large unoccupied site across Strada del Sincrotrone. Bruno returned full time to Rome, to teach, do research in theoretical physics and follow ADONE's progress day by day. However ADONE was an extremely complex machine, of a scale completely different from AdA, and Bruno could let Amman, Claudio Pellegrini, Mario Bassetti and others solve the rising difficulties, which in fact they did. By that time the laboratory on the hill had grown and matured. Building the synchrotron, envisaging and then constructing AdA, had groomed a generation of accelerator physicists and engineers among the first in the world and, in December 1967, the first beam was injected in ADONE (Amman 1985).

Fig. 13.4 Seated in first row, from the right: Nicola Cabibbo, Enzo Iarocci, Gilberto Bernardini, Felice Ippolito, Matthew Sands, Fernando Amman, Giorgio Salvini, Costanza Salvini Catenacci and the Author, 1990, © INFN–LNF, http://w3.lnf.infn.it/multimedia/

While the machine was under construction, Bruno and his friend Carlo Bernardini had started worrying about how to extract meaningful physics from it.[8]

---

[8] Bruno's vision in proposing first AdA and then ADONE can be expressed through two of Einstein's formulae: $E = Mc^2$, the relativistic connection between a particle mass M and its rest energy, and the photoelectric effect relation $E = h\nu$, between the energy needed to create a particle and its characteristic frequency $\nu = \omega/(2\pi)$. Combining the two formulae, in particle-antiparticle annihilation, a total energy $E = 2E_{beam}$ is made available to create a state resonating with angular frequency $\omega = Mc^2/\hbar$. This is how Bernardini referred to Bruno's description of electron-positron machines capable of probing the frequencies on which the quantum vacuum oscillates (Bernardini 2006, 62). Bruno's own view appears as a remark appended to a talk presented at a conference in Pisa, in 1976 (Conversi 1976b, 89): *"Touschek: ...the Fourier component of the dielectric constant $\epsilon(\omega)$ is directly proportional to the total cross section $\sigma_{tot}(E)$, in which $E = \hbar\omega$ is the total c.m. energy of the annihilation process between electrons and positrons. Colliding beam experiments therefore explore the [polarization of the] vacuum all along the real axis."* Touschek's spirit comes immediately alive by the footnote to this remark, which, in Bruno's characteristic style, is indicated as "This is what B.T. wrote after the conference: it is what he wished he had said". Bruno put down many other interesting observations in his post-conference remark, one of them that the future of high energy physics would see the unification of the three interactions between elementary particles, as was in fact observed a few years later (Amaldi et al. 1991).

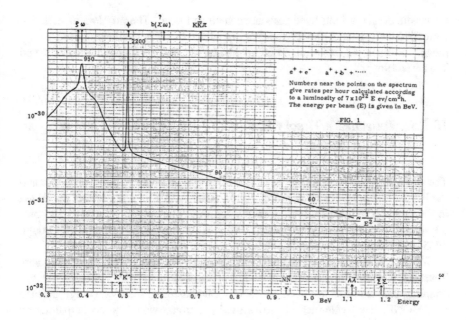

Fig. 13.5 Luminosity and energy plot of the energy scanning proposed by Carlo Bernardini for ADONE, from (Bernardini 1965), © INFN–LNF, http://w3.lnf.infn.it/multimedia/. The peaks correspond to the signals from the resonances which are created in the annihilation at the corresponding c.m. energy, $E_{c.m.} = 2E_{beam}$

13.1.2 Bernardini's Idea to Look for Vector Mesons

In 1965, Carlo Bernardini was as keen as Bruno to devise interesting experiments to perform at ADONE. In an unpublished note (Bernardini 1965), he proposed to use the wide range of center of mass energies explorable at ADONE to look for a special kind of particles, the so-called "Vector bosons", which could be produced in electron-positron annihilation. Three such particles had been discovered at the time, all unstable. Two of them, called ρ and ω, were short lived, and decayed in π-mesons, the other, called ϕ, had a longer life time and was mostly decaying into a K-meson pair. If the beam energy is increased by small steps at the time, a procedure which became later known as "energy scanning", more and more energy would become available in the center of mass and one could see the formation of the resonances and their decay, as shown in the plot of intensity of the signal *vs.* beam energy in Fig. 13.5. To observe such curves would have been possible only with electron-positron machines and this was, *a priori*, a very interesting experiment to perform. One glitch about such experiment was that the observation of a curve such as shown for the ϕ-meson in Fig. 13.5, required a high resolution of the machine energy and the calculation of QED infrared radiative corrections of a type not yet developed for production of a long lived resonance such as the ϕ.[9] The problem of radiative correction had arisen. Although the experiment was never run in the proposed energy range, Touschek put this problem on his mental list, and approached it in (Etim and Touschek 1966).[10]

13.1.3   Touschek's Theory Group and His Legacy of Photon Resummation

The early '60s had been a time of exhilaration and adrenaline, both for Bruno and the Frascati laboratories (Bernardini 2006). After AdA's final paper was published in December 1964, a new period started with the preparation of ADONE's commissioning. For Bruno it was a happy and creative time which would last through 1965–1966 and 1967, until mid 1968.

While AdA was reaching its final goal to be proof-of-concept for future colliders, Touschek worried that processes of higher order in $\alpha_{e.m.}$, the electromagnetic (e.m.) coupling constant, had not been adequately calculated. At the high energy reached by ADONE, the accelerated electrons (and positrons) would be accompanied by concomitant radiation effects, whose calculation appeared to be a necessity and required a special effort, when planning for new discoveries. Remembering how AdA had been born through the need to plan ahead for the laboratory, Touschek finally accepted the task which Salvini had tried invain for him to take on in 1960: the creation of a theory group in the Frascati laboratories. He succeeded in doing so, perhaps beyond his own expectations.

Thus, even before his return from Orsay, he turned his attention to ADONE, along two directions: devising meaningful experiments and preparing to correct experimental results for concomitant radiative effects. A crop of extremely brilliant students was graduating in physics during those years, among them Giovanni Gallavotti.[11] Bruno assigned him a thesis on ADONE's radiative corrections, and Gallavotti graduated on the same November 1963 session as Altarelli and Buccella, who graduated with Gatto with the calculation of single photon emission, instrumental in proving that collisions had taken place in AdA.

After Altarelli and Buccella's work, another step would be to calculate double photon emission. It was the typical subject thesis for excellent students, and Bruno

---

[9] According to Heisenberg's uncertainty principle, the very narrow shape of the curve, as a function of the energy available for the particle production, corresponds to a long decay-time.

[10] Once the main formalism had been developed, he suggested a through calculation to the Author, who completed it in 1969, after she had left and moved to Boston (Pancheri 1969). I did suggest to have him and Etim as co-authors, but Touschek did not accept. The formalism developed in this paper was later extended to extract the life-time of the J/Ψ resonance (Greco et al. 1975), through the coherent state approach developed by Mario Greco and Giancarlo Rossi (Greco and Rossi 1967), and inspired by Touschek.

[11] Giovanni Gallavotti (b.1941) Emeritus Professor of Theoretical Physics at Rome Sapienza University, was awarded the 2007 Boltzmann Medal for his contributions to statistical mechanics.

assigned this calculation to Paolo Di Vecchia, then 22 years old.[12] Bruno however did not have the time, nor, perhaps, the inclination, to follow the increasingly difficult calculation details and the student 's progress was understandably slow.

At the same time, the ADONE machine group was in need of calculating the same higher order electromagnetic effects, in order to increase precision for monitoring the machine performance. The calculation of double photon emission was assigned to Mario Greco, who had just graduated from the University of Rome and had been awarded a fellowship from CNEN to study particle physics processes at Frascati. Mario had graduated with Benedetto De Tollis on the effect of the vector mesons ρ and ω in pion photoproduction, useful for the ADONE project.[13] He arrived in Frascati in early 1965 and was well advanced in his work when Touschek came back from Orsay. Touschek was very pleased when he was told of Mario's work, so much so that he startled the young researcher by going to his office to congratulate him on how his calculations (on the double bremsstrahlung process) were progressing (Greco 2018). Mario was then delegated to follow Di Vecchia's thesis work to its completion (Greco and Di Vecchia 1967).

Mario became one of Touschek's long time friends.They shared the passion for theoretical physics, but they also spent time together playing tennis in the Laboratories court, swimming and fishing in Positano or in the murky waters of the Albano lake. Bruno had always been a keen swimmer, in Hamburg during the early times of his work with Widerøe, later in Scotland in the icy waters of Loch Lomond, in the lovely Tyrolean lakes around Flecken. In Italy, he would swim at Monte Circeo and Sperlonga, just South of Rome, or in Livorno, on his way back to Rome from Milan or Turin. At a certain point, he even thought of having a swimming pool built in the Laboratories, on the roof the office building next to the synchrotron. He was rapidly discouraged by a simple calculation of the weight which the roof could not support. There was almost no sporting activity he did not feel obliged to try. He skied on the Alps at 3500 m altitude, he had practiced fencing in Rome and Vienna as a boy, and then again in Hamburg, during the war, when the heavy bombing of the city had not yet started, and he still was relatively carefree.

Towards the end of 1965, the Frascati theory group started taking shape. In the wake of ADONE's growing importance, a number of fellowships were made available for new graduates to work in Frascati with Bruno. When I joined the laboratories in May 1966, in addition to Mario Greco the group included Paolo Di Vecchia, Giancarlo Rossi and Etim Gabriel Etim, a Nigerian student doing his thesis with Bruno, holder of a fellowship from ENI, *Ente Nazionale Idrocarburi*, the Italian Oil Institute. The extreme complexity of the double bremsstrahlung calculation had convinced Bruno that perturbation theory could not be a viable tool to pursue further precision, in addition to being plagued by order-by-order infinities, which required going always one step higher in the order of the electromagnetic coupling. The work done with his friend Walter Thirring about the infrared catastrophe came back to

---

[12] Paolo Di Vecchia (b. 1942) is a string theorist, Nordita professor emeritus.

[13] Mario Greco (b. 1941) was a Frascati researcher before becoming Professor of Theoretical Physics at University of Pavia, later moving to Roma3 University, presently INFN Eminent Scientist.

his mind, and when Etim had approached him asking for a thesis, he was asked to research the current literature about infrared radiative corrections.

The subject was not new. The main theorem about the infinities plaguing the effect of radiation had been proved before the war (Bloch and Nordsieck 1937). With the post-war advent of QED, the theoretical difficulties rising from the infrared divergence had been solved by Schwinger (Schwinger 1949), examined by Brown and Feynman (Brown and Feynman 1952), and a full calculation order by order in perturbation theory had been obtained (Jauch and Rohrlich 1955a; Yennie et al. 1961), but the application of these calculations to electron-positron experiments at energies as high as ACO, ADONE or VEPP-2, was not obvious. Other existing more pragmatic approaches (Tsai 1965) were unsatisfactory in Bruno's view, and did not appeal to Bruno's elegant mind, well exemplified by his invincible faith that "electrons and positrons MUST meet because of the CPT theorem" (Rubbia 2004; Amaldi 2017). After the calculation of single bremsstrahlung, and double bremsstrahlung, further precision could only come from a different approach. Bruno was aware that his best work could come from first principle considerations. But he also had interest and faith in the power of experimentation and observation. In this sense he was a true follower of Heisenberg's approach to physics (Heisenberg 1944). He turned his attention to soft photons, those which escape detection because their energy is smaller than the experimental resolution, and which flood the detectors with near zero energy in infinite number.

A closed form expression for the summation of infrared photons found in the scientific literature, puzzled Bruno (Lomon 1956, 1959; Eriksson 1961): the corrections were given through a compact expression which appealed to his strong sense of beauty, and he wondered whether it should be possible to derive it from analyticity principles. There were other approaches to the problem, but he did not favour the perturbative treatments, and was suspicious of the way the infrared catastrophe was circumvented. He decided something had to be done, after a conversation with Ugo Amaldi, Edoardo's son and a brilliant young researcher at Istituto Superiore di Sanita'. One day, Bruno had called him on the phone, worried that the preparation of ADONE's experiments were not proceeding with a speed corresponding to the machine progress. Ugo had agreed to start some planning for an experiment measuring in detail the properties of the ϕ−resonance, whose observation had also been the object of Carlo Bernardini's interest. The question of radiative corrections came up and, when they met again, Bruno inquired about Ugo's progress. "What method did you use to calculate them?" asked Bruno. "The Kessler method", answered Ugo (Amaldi 2017, 154). This treatment, well known at the time (Kessler 1962), was however unknown to Bruno or, in his vision, insufficient to deal with the complexity of the problem. Dismissing it, with one of his lapidary comments, i.e. "I only know the Kessler sisters" (beautiful twin dancers in a very popular TV show), Bruno went on to formulate his own approach.

The Kessler anecdote is typical of Touschek's approach to a new problem. He had developed his own way during the war years, after being expelled from University of Vienna for racial reasons, and had to continue to study at Paul Urban's home. In Hamburg he had taught himself to deal with electronic equipments, building klystrons and circuit breakers, in Berlin he had to learn how to apply his theoretical background to the workings of a betatron. All through his life, Bruno had been his own teacher, and learnt to be self-reliant. In February 1960, this capacity had further developed in the mental attitude which led him write in his first notes about AdA:[14]

```
... The following is a very sketchy proposal for the
construction of a storage ring in Frascati. No literature
has been consulted in its preparation, since this invariably
slows down progress in the first stage, necessary though it
may be in the consecutive stages of the development. I shall
present here all I have thought about it and much, which
others have suggested to me and to anticipate the question:
No, I have not properly read O'Neil [sic], but I hope that
someone will.
```

In 1960 Bruno had not properly read O'Neill's work (O'Neill 1956), neither had he read Paul Kessler's when Ugo mentioned it to him, but he gathered together what he had been thinking for a long time about radiation effects and the infrared problem, including the Bloch and Nordsieck theorem, and Heisenberg's view of physical phenomena and the observer's role. Bruno's focus was now on soft photons, radiation emitted in large amounts of individual photons of such low energy that they cannot be properly counted nor distinguished one from the other. Hidden by the experimental resolution, individual soft photons can escape detection, but still obfuscate extraction of data. The first approach to the problem was laid out in the unpublished note with Etim (Etim and Touschek 1966). This first work has an interesting title, which shows Bruno's sense of humor: "A proposal for the administration of radiative corrections", a pun on the word 'administration', which could point to administering a medicine to cure the measurements of the disease of emitting soft photons, or to prepare the rules needed to extract significant physics.

When the author of this book joined the laboratories in May 1966, in Touschek's mind the problem was fully developed and a plan had emerged. In his words, "The picture of an experimentalist as one counting soft photons is not entirely realistic: existing perturbation theory works in a representation in which the number of photons is diagonal and the emission of additional photons requires a further step in the perturbation procedure. The experimenter on the other hand does not see single photons but rather an unbalance of energy and momentum between the incident and emerging particles." His vision included approaching the analyticity properties of the unobservable soft photon emission, the re-defintion of particle states as a coherent superpositions of soft photon fields, and the application of radiative corrections to electron-positron experiments, to test both the validity of QED and the study of resonant states production, one of the long sought objectives for ADONE's construction (Bernardini 1965).

Three seminal papers stemming from Bruno's vision rapidly followed (Etim et al. 1967b), (Greco and Rossi 1967), (Pancheri 1969). Through these papers, Bruno left

---

[14] Undated typed text, presumably prepared after the 17th February, 1960 meeting, Bruno Touschek Papers papers in Sapienza University of Rome–Physics Department Archives.

Fig. 13.6 From left, Etim G. Etim in Frascati during the 1987 Bruno Touschek Memorial Lectures (BTML), photographed by the Author, and Bruno Touschek with Italo Federico Quercia and Ruggero Querzoli (at center), in 1967, during ADONE's costruction, © INFN – LNF, http://w3.lnf.infn. it/multimedia/. At right, the Author in 1966, all rights reserved

a major legacy to the young researchers of his group. In later years, they expanded Touschek's vision beyond the initial formulation, successfully applying it to narrow resonance production, and carrying the idea of soft quantum resummation to studies in Quantum Cromodynamics (QCD). Although he suggested the work which could be done through the 1967 and 1969 papers, Touschek co-authored only one of them, in the unlikely authors' trio shown in Fig. 13.6. As for the other two papers, he suggested the problem but did not sign the work. The absence of his name shows his characteristic generosity, how he would open a field and let others develop it, without expecting or pretending to be an author as well. Conversely, he would include his collaborators as fully fledged authors, even when the vast majority of the work was his own. This is certainly true for the radiative correction paper (Etim et al. 1967a).

In those years, other Rome graduates joined the Frascati theory group, among them Pucci Di Stefano, who had done her thesis with Francesco Calogero's supervision, and later joined Gatto's group in Florence.[15] A long time member of the group was Aurelio Grillo, who graduated with Touschek in 1967.[16] Touschek's group attracted foreign visitors, such as Yogendra Srivastava, Fig. 13.7, with whom long lasting collaborations were established, while other brilliant graduates from Rome joined the group in the 1970's, among them Fabrizio Palumbo and Calogero Natoli, enlarging the group's research interests to nuclear physics and condensed matter studies.

After the discovery of QCD, it was natural for Touschek's group to extend to strong interactions his ideas about resummation (Pancheri-Srivastava and Srivastava 1977; Greco et al. 1978; Greco 2019). An extension of Touschek's work also appears in

---

[15] Maria Grazia, *Pucci*, Di Stefano (1941–2004) was one of the most brilliant physics students at University of Rome, and is warmly remembered in (Maiani and Bonolis 2017b).

[16] Aurelio Grillo (1945–2017), after many years as a researcher at Frascati, joined the INFN Gran Sasso Laboratory, where he was head of the theoretical physics group until his sudden death after suffering an asthma attack on the laboratory grounds.

Fig. 13.7 Yogendra Srivastava (at left), visiting from Northeastern University, Boston, and Mario Greco in the Frascati Laboratories, around 1970, photographed by the Author, all rights reserved, and a present day view of the Frascati National Laboratories with the ADONE cupola at the center and the Tusculum hills in the background, © INFN-LNF, http://w3.lnf.infn.it/multimedia/

an influential paper by Giorgio Parisi, who had graduated with Nicola Cabibbo, and was done when Parisi was a postdoctoral researcher in Frascati (Parisi and Petronzio 1979).[17]

13.1.4 Touschek's Contribution to ADONE

Fernando Amman, who participated to the construction of the synchrotron and AdA's initial works in Frascati, co-signed the 1960 official proposal for ADONE and became its machine director, gave explicit testimonials of Bruno's contributions. On 25 May 1980, he wrote to Amaldi:[18]

Amman:

[…] I would like to share with you some thoughts about the "roaring years" of the storage rings, from 1960 to 1970.

It may appear strange that Bruno, who had so profoundly and directly engaged in AdA's realization, is seen as having had an only marginal role in ADONE's: this may be so if

---

[17] Giorgio Parisi (b.1948) was awarded the 2021 Nobel Prize in Physics 'for the discovery of the interplay of disorder and fluctuations in physical systems from atomic to planetary scales', sharing one half of it with Syukuro Manabe and Klaus Hasselmann. President of the Accademia dei Lincei from 2018 to 2021, Parisi had previously received many awards for his work in quantum field theory and statistical mechanics.

[18] Letter from F. Amman to E. Amaldi, 25 May 1980, original document in Italian, in Sapienza University of Rome–Physics Department Archives, Archivio Amaldi, Fasc. (folder) 4, S.F. (subfolder) 5 - Scatola (box) 524. Translated from Italian by the Author, reproduced courtesy of Sapienza University of Rome–Physics Department Archives, https://archivisapienzasmfn.archiui.com, documents provided for purposes of study and research, all rights reserved.

looking at the official documents and publications, but it is not how things went, at least in my opinion.

From the very beginning, since 1960, the scientists working on AdA and on the early ADONE project were closely connected, through Bruno, Carlo Bernardini and Gianfranco Corazza: the first experimentations on the rings, mostly of technological nature, developed through a collaboration with hardly distinguishable roles and very close contacts through the early period of ADONE's planning (1963-64), which ran in parallel with AdA's measurements.

[…]

Bruno lived intensively all of ADONE's phases, from the project to the construction (1965–1967), to the difficult year when instabilities developed in ADONE (1968). We did not often get together in those days, as he feared to waste my time, as Carlo Bernardini told me. Bruno was mostly following the technical work through Gianfranco Corazza, [but] he was also in awe of ADONE's dimensions, which led him to see it as an "industrial" type enterprise. At the same time, he knew ADONE was his own creation, born from his original idea [carried on through his determination and vision, Author's Note]. Every time a problem arose, and he felt to be able to help, he was there and gave his contribution. Such was the case of the transverse instability, which had been dealt with in (Laslett et al. 1965) for the continuous beam case, and which he solved for the bunched beam case of storage rings with Ferlenghi and Pellegrini (Ferlenghi et al. 1966).

He gave much time and effort to the Committee for Experiments at ADONE, *Commissione Esperienze con ADONE*, asserting his own ideas and stimulating discussions, although he was left with a negative feeling about it, a precognition of the problems later seen to arise when ADONE's experimentation started.

What needs to be underlined is the unity of the AdA-ADONE enterprise, beyond the official documents: a unity which came about through […] Bruno, Carlo Bernardini, Gianfranco Corazza, Ruggero Querzoli (and myself).

[…]

Bruno was the initiator, the continuity link between AdA and ADONE during the ten golden years of the laboratory, the person who had a great idea and let others turn it to reality; I believe that his scientific and personal qualities have been essential [dots] to attain the success, if success it has been.

Amman's words, "that Bruno [dots] is seen as having had an only marginal role in ADONE's", reflect—tragedy or merit?—Touschek's often undeclared contribution to the scientific work he inspired and directed in his collaborators. Indeed, the crucial contribution of Bruno's backstage role is not always acknowledged. He knew this early in his life, when he wrote to his father from Glasgow about lack of recognition of his own contribution to his students' work, but this occasional regret was never made public, and it only appears in the family letters. His participation is likely to have been at the heart of Widerøe's betatron success, but he never claimed it as his own, although trace of pride in his contribution can be found in the 1943 and 1944 letters. In AdA's and ADONE's case, the situation is different. For AdA, he is universally acknowledged as the main protagonist, both in Frascati and through the Orsay adventure's high points, as it would later be for his contribution to the formation of students and collaborators. Less so in ADONE's case, where his contribution during its construction is partially shadowed by the growing complexity of the machine. However, all through the 1960s, this capacity of Bruno's to inspire and move towards ground breaking physics goals is at the heart of ADONE's successful construction

Fig. 13.8 At left, Piergiorgio Picozza, Bruno Touschek and Italo Federico Quercia, LNF director 1961–1963 and 1970–1973, with an unknown visitor on the ADONE floor, at right Amman (left) in discussions with Touschek (out of camera), Italo Federico Quercia and Ruggero Querzoli, LNF director 1967–1970. © INFN–LNF, http://w3.lnf.infn.it/multimedia/

and planning, which ensued from the close and warm collaboration between Amman, Touschek, Quercia and Querzoli. Some of Touschek's best photographs come from the central period of ADONE's construction, the years 1966–1967 the 'golden years of the laboratories', as in Fig. 13.8, where Touschek is seen with Piergiorgio Picozza, who graduated with Italo Federico Quercia.[19]

13.2 ADONE: Problems and First Results

Clouds had been gathering, in those years, over the laboratories and Italy.

In 1966, the University of Rome had been occupied by the students, in the wake of what was happening elsewhere, notably at Berkeley. But it was still a friendly occupation: directed against the Rector and other authorities, accused to be tolerant of right wing violence within the campus, the 1966 occupation of the physics department did not divide faculty and students, and life returned to normal after a few months. It was just a prelude of much to come. Spring 1968 saw students rebel with violent clashes with the police and political unrest breaking out everywhere. It took many years for Italy to overcome the disruptions and polarization which marked the 1970's. Those years became known as *anni di piombo* in Italian, literally meaning years made of lead or as heavy as lead, of which bullets are made, and the killings which took place in those years. Teachers, magistrates, students, sometimes on both sides of the political spectrum, ordinary people, were killed. In 1978, Aldo Moro, Italy's Prime Minister, was assassinated before peace could come back to the country.

University life also paid a price. For a number of years, fear of political disruptions made it difficult to hold seminars or scientific meetings in the Rome Physics

---

[19] Piergiorgio Picozza (b. 1941) is Professor Emeritus of the University of Rome Tor Vergata.

Institute.[20] Bruno's bitterness towards the students during those years is well remembered by many of his colleagues. When it became routine that a foreign scientist, most often an American, would not be allowed by the self-installed left-wing student committee to give his scientific presentation in the physics institute, Bruno likened such prohibition to the discrimination applied before World War II in Austria and Germany towards Jewish or socialist leaning professors. The leftist movement was particularly active in the humanities department, where it promoted by decree the existence of the so-called *trenta politico*, or *trenta di gruppo*, which meant that an A mark (corresponding to 30/30) could be assigned to a group of students, even though perhaps only one of them deserved it. Such practice was less frequent in the physics institute, but disruptions during lectures or examinations frequently occurred there as well. When a student called Bruno a "nazi baron" for his uncompromising behaviour, he felt profoundly offended. He had suffered discrimination and abuse of power during his youth and could not avoid feeling shock at what was happening in the universities. His drawings during those years were copious and vividly reproduce the atmosphere in the institute, Fig. 13.9. They were often drawn on scrap pieces of paper, abandoned on his desk and collected by friends, or crushed as useless in his pockets, but retrieved and ironed out by his wife Elspeth. Tempers were high, professors would not be sacred any more, nor were the students.

Ugo Amaldi, then a young researcher at the Istituto Superiore di Sanità, remembers the shock when he heard that his father Edoardo, exasperated by being disturbed while giving exams, had physically attacked Franco Buccella, the same brilliant student whose thesis work, together with Guido Altarelli, had allowed proving that annihilation in AdA had taken place. Indeed, the accident was a shock both for Franco and for the professor (Amaldi 2017, 158). Franco had been very active in taking the students side against the academic establishment. Two years earlier, during the first occupation of the University, he had led the group of students presenting their demands to Giorgio Salvini, in charge of negotiations. Giorgio Salvini's first reply had been: "Show me your student card [showing his grades]". When Franco, head of the delegation, could point to a perfect A+ score, Salvini could not refuse to listen to him.[21]

In 1968, Edoardo Amaldi's reaction took place during an examination session, in the physics institute main lecture hall.[22] Amaldi was opposed to give the infamous 'political score', 30/30, which the student assembly had voted should be

---

[20] A well known incident took place in 1972, on the occasion of a talk to be given by Sidney Drell, head of the theoretical physics division at Stanford, and author of a famous book on particle physics (Bjorken and Drell 1965), a mandatory reading for particle physics students and researchers. As Drell started to speak, Franco Buccella invited the audience to leave the seminar room in protest, because Drell was a member of the so called *Jason Committee* advising President Nixon on the Vietnam war. I was present, as were some foreign visitors, among them Yogendra Srivastava and Robert Jaffe, and did not leave, and others also stayed and listened to the talk, but many others followed Buccella's solicitation.

[21] Personal recollection.

[22] The large cavernous hall, previously named 'Experimental Physics Hall', is now named after him as 'Edoardo Amaldi Hall'.

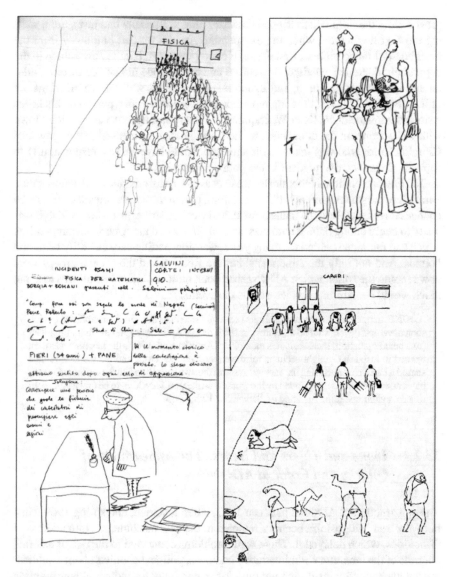

Fig. 13.9 A few of Bruno Touschek's drawings about university student unrest in late '60s and early '70s. Family Documents, © Francis Touschek, all rights reserved

the norm. Franco was not a student anymore, and had by now become assistant professor to Amaldi's chair, hence a member of the examination board. But his political views were with the student movement and their view of examinations as a tool of oppression. He walked over to the blackboard and started writing "Down with exams, selection tool of the dominant classes". This is by now a rather innocuous remark, and, even in 1968, it was not a particularly offensive comment, but it was a provocation, the drop which tripped the vase. Franco Buccella had just started writing the first few words, when the exasperated Amaldi pushed him away from the blackboard. Franco hit back in defense. Nobody was hurt, but the atmosphere in the department darkened. Mediation occurred and the episode did not go further. It did remain in everybody's mind, and Bruno made one of his well known drawings out of it. In Fig. 13.9 the last drawing remembers the episode with a person on his knees writing the upside down letter W, graphically corresponding to the expression 'down with'. The name on the door refers to the distinguished condensed matter physicist Giorgio Careri, who was Institute director and would have had the responsibility to formally denounce the incident to the police authorities.

The Frascati National laboratories was not avoid of disturbances and delays either. Paralleling the student unrest in the universities, part of the laboratories staff requested changes unacceptable to the management, and went on strike, just when ADONE was ready to begin taking data. Bruno was devastated. Indeed the American competition, which had remained behind for many years, was now getting very close, by leaps and bounds. And not only the Americans, but the Russian and the French groups were now reaping the fruits, which ADONE had been built to obtain. Not all was lost, but much would be. Amman's voice details the events: [23]

> ADONE started functioning with a single beam on December 8th, 1967. Instabilities [which soon developed] were studied and cured in 1968, and in April-May 1969 the first luminosity [two beam operation] measurements took place (Amman et al. 1969); the machine was then opened to instal the straight sections for the experiment detectors, and the workers protest started when we were closing the vacuum system (this work was finished during the following days, activities already suspended by the workers strike), on May 30th, to be precise. A more or less regular operation restarted on September 19th, 1969.

13.2.1 Lions and Tigers and Bears: The Appearance of Quarks and Color at ADONE

The construction of ADONE between 1963, when it started, and spring 1969 when both electron and positron beams circulated in the new machine, ran into all kind of difficulties, which delayed it. There were problems connected to the rise of external controls of the laboratory administrative practices, political conflicts about establishing a nuclear energy institution not only for research but for industrial consumption as well, a development opposed by both electrical and oil companies, from within and from outside Italy, and then financial difficulties, rising with the increasing size of the ADONE project. To top it all, while the electron beam was injected in December 1967 (Amman 1985), instabilities developed with the positron beam, and then a labor strike stopped everything.

---

[23] Letter from F. Amman to E. Amaldi, 25th 1980, translated by the Author, as in [18], reproduced courtesy of Sapienza University of Rome–Physics Department Archives, https://archivisapienzasmfn.archiui.com, documents provided for purposes of study and research.

At the same time, Frascati was no more the only laboratory in the world building electron-positron colliders. In addition to a number of particle colliders under constructions and close to completion, in 1968 two electron-positron colliders had started operating and had produced physics results. Had it not been for the strike which paralized Frascati, ADONE would have been with them, or perhaps even preceding them. ACO, whose construction had been approved in 1963, saw its first beams circulating in 1967, with physics results published in 1968 (Augustin et al. 1968). ACO's initial energy reach was only 1.1 GeV in the center of mass, but sufficient to study the properties of $\rho-$ and $\omega-$mesons. In Novosibirsk, VEPP-2 had also started working by this time, publishing physics results already in 1967 (Auslander et al. 1967).

Both the French and the USSR rings operated at quite a lower energy than ADONE and their discovery potential was correspondingly limited, but the really dangerous competition would come from Stanford, where projects for the construction of an electron-positron accelerator with comparable energy to ADONE's had been under way since 1965 (Rees 1986). So that, while the Americans were adjusting their project, with an eye to ADONE's progress, in Frascati Touschek, Amman and Pellegrini were 'hunting for instabilities and relative cures' (Amman 1985).

In Spring 1969, when electrons and positrons were both circulating in ADONE, Bruno had become very famous. Construction of electron-positron rings was under way in all major world laboratories, while at CERN the Intersecting Storage Rings for protons had started operations (Hübner 2012; Bryant 1992). Bruno was invited to give seminars in Europe and elsewhere, and organized the Varenna Summer School on "Physics with intersecting storage rings", to be held in Villa Monastero from 16th to 26th June 1969. All the major accelerator physicists were invited to give lectures, Fig. 13.10.

In Frascati, the laboratory was ready to start measurements. Bruno had hoped that ADONE would open a new world for particle physics, and it did. There were four detectors placed around the ring, searching for the different type of particles expected to be produced in the annihilation. Two of them had been designed and built to observe final states with particles such as photons, or pairs of electrically charged particles, muons, electron and positrons, pions and kaons, all specified as interesting in Touschek's original memo. Pions and kaons are generically labelled as 'hadrons', namely particles heavier than electrons, labelled *leptons*, from the Greek adjective for light. Hadrons have both weak, electromagnetic and strong interactions, whereas leptons do not interact through the strong force, and their class includes μ's, τ's and neutrinos. In electron-positron annihilation, a virtual photon is created, materializing with the same strength in either a pair of hadrons or leptons. Apart from the different mass values of these particles, the production cross-section for hadrons was expected to be the same as the one for leptons. But, there was a surprise, appearing as soon as ADONE was turned on: hadrons were produced more frequently than expected. The comparison between results from these two different types of experiments, Fig. 13.11, shows how, soon after its start, ADONE signaled an anomalous production of hadronic particles, registered by the detectors surrounding the ring.

Fig. 13.10 Course group-shot from the Proceedings of the International School of Physics "Enrico Fermi", Course XLVI, "Physics with intersecting storage rings", edited by B. Touschek (1969). Bruno Touschek is sixth from left in first row between Matthew Sands (at his right) and Gerard O'Neill. Jacques Haïssinski and Vladimir Baier next to O'Neill. Claudio Pellegrini is third from right in second row, Pedro Waloschek appears at the upper left. © SIF, reproduced with permission

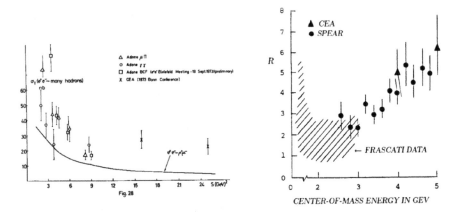

Fig. 13.11 Data in the left plot indicate the intensity of hadronic type particle production (Mencuccini 1974; Conversi 1976b), compared with theoretical QED expectations (full line) (Zichichi 1976). At right a compilation of multiple hadron production data from (Richter 1984)

The Frascati data were collected between 1969 and 1973, and showed that hadron production through the same annihilation process was different from the known expectation from QED. Something new was clearly happening at ADONE's energies. When these measurements first appeared, in 1969, electron-positron collisions were exploring the vacuum at energies no other collider had yet reached, and the results were a surprise which took time to understand. Touschek was elated.[24] Something

---

[24] Luciano Maiani's comment during the 2020 Cabibbo Memorial event in Frascati laboratories.

new was appearing as ADONE's energy was moving away from the known region below 2 GeV towards higher center of mass energies.

The observed phenomenon of multiple hadron production was totally unexpected. It was later confirmed by observations at the Cambridge Electron Accelerator and SLAC. The full significance of this discovery was not immediately obvious, but Touschek's intuition was right: something new was making its appearance. This is what had been driving his push for matter and antimatter to smash through accelerators, which could extract the unknown and discover new physics. ADONE was showing that electron-positron colliders were the tool to go further in probing the structure of the infinitesimal world. As it happened, the world of quarks, unobservable particles, fitted with a new quantum number called *color*, which manifest themselves only in hadronic form, was making an official debut. Theorists had suggested their existence since 1964 and there had been strong hints of new phenomena, signaled from SLAC experiments in the so-called *Deep Inelastic Scattering* region, but ADONE shed on them a strong light, demonstrating the capacity to disentangle unknown processes from the expected background. That ADONE could be a very clean tool was clear from these early results, which anticipated that particle colliders would ultimately take over conventional fixed target machines, as a better high energy precision tool.

13.2.2 Intermezzo: Bruno During the Years 1969–1973

While ADONE was showing its worth, Touschek's fame grew, as did, of course, the competition to ADONE from other colliders. In 1969 Bruno was invited by Harvard University to give the 1969–1970 Loeb Lectures, four lectures about electron-positron colliding beams. These lectures were a remarkable recognition of his contribution to the field. In the same academic year 1969–1970, the five Loeb Lecturers included Robert Hofstadter and Peter Kapitza, both Nobel prize winners.

The Harvard invitation was an acceptance of Bruno's ground-breaking role in the rising field of colliding beam machines. His role was well known in Cambridge, where, under the joint sponsorship of Harvard University and the Massachusetts Institute of Technology (MIT), the construction of an electron-electron collider had morphed into that of an electron-positron accelerator (CEA), which would start operating in 1972. As for Bruno, after directing the 1969 Varenna School on electron-positron collider, he felt that the subject was not interesting him anymore. This is something which had happened to him before: losing interest in a subject once he had mastered it, was not new to him. It had happened after the betatron had started working, and, in part, after his theoretical work through 1957 and 1959. He was often ahead of his time, and in need of new challenges. When the invitation from Harvard came, he was obviously aware of Harvard's fame as a major American university, but was not awed. In fact, to bring it down to a normal level, he drily noted to his father that Harvard had gone through its student riots, just like every other university.

This could be relevant to this trip, as he sarcastically wrote to his parents, 'in case I long for it [the student unrest] in the fortnight'.

He had thought to use the Christmas holidays to prepare the four or five lectures to deliver in Cambridge. But the holidays went by with family life and a new dog, a cocker spaniel, which reminded him of the one he had in Glasgow many years before, Fig 13.12. The real point, that the topic no longer interested him, was mixed to his not being sufficiently impressed by the prestige of this visit: he had met and known personally some of the great physicists of his time, he himself was one. But one does not refuse an invitation by Harvard University, especially when it comes with a lofty *honorarium*.[25] It cannot be excluded that this invitation was the recognition that he was a front-runner for bigger honors, even for the Nobel Prize. But the lectures were not a success. As contemporary work at Harvard shows, theoretical interest at the famous university was focused on topics which Bruno had remained away from, and viceversa.[26] By this time, electron-positron colliding beams were mainstream from the point of view of machine innovation, and their greatest discovery yet to come. When this arrived, in 1974, the major credit went to others.

Bruno arrived in Boston, where he remained two weeks, on February 21st, 1970. I was in Boston at the time, as was another of the young people from Bruno's old Frascati group, Paolo Di Vecchia, spending one year at MIT, the Massachusetts Institute of Technology. We went to Bruno's lectures and afterwards went to meet him at a pub in Harvard Square, but the three of us had moved away from the Frascati days, when ADONE was still in the making. The sparkle of a vacation in Positano, which we had spent together with Bruno and his family in September 1966, dimmed in the icy days of the Cambridge winter, and the encounter was a disappointment.

Returning to Rome, Bruno was busy with his teaching duties and finishing the book on statistical mechanics, with Giancarlo Rossi.[27] Since 1969, and through 1973, Bruno was appointed Aggregated Professor, a position not exactly like that of full professor, which was overdue, but at least such that he did not have to give up his Austrian citizenship. He was appointed in November 1969 and started teaching the course of "Mathematical methods in Physics" (Amaldi 1981, 23). In 1972 he was nominated a foreign member of the National Academy, the *Accademia dei Lincei*, in recognition of his contributions to science and teaching.

Finally in 1973, the Italian law changed. Bruno could transfer from his *Professore Aggregato* position to *Extraordinary Professor*. Because of his Austrian citizenship, he had been excluded from this position, which should naturally have come to him many years earlier. But he could yet not become a full-fledged professor. According to still existing provisions, an Extraordinary Professor only becomes Ordinary Professor after three years of service, thus enjoying higher salary and other benefits. There were

---

[25] Letter to parents on February 19th, 1970.

[26] The article where a new quark, the *charm*, was proposed, had its birth in early 1970 at Harvard, through Sheldon Glashow, John Iliopoulos from École Normale Supérieure, and Luciano Maiani from University of Rome (Glashow et al. 1970).

[27] Giancarlo Rossi (b.1943) is Honorary Professor of Physics at Rome University of Tor Vergata, and has given contributions to theoretical Field Theory with focus on Lattice QCD simulations, on Statistical Mechanics and Biophysics.

Fig. 13.12 Bruno Touschek with his cocker spaniel, named Lola, around 1970, Family Documents, © Francis Touschek, all rights reserved

forms to be filled, innumerable copies of his many papers to be submitted, as many copies of his CurriculumVitae, and, after all this had been done, an examination in front of a committee of other professors, nominated by the Ministry. Bruno could not subject himself to all the burocratic work, his strength was affected by some health problems, his time was taken by physics, and family obligations. It was left to his friends and colleagues to collect and collate all the papers. The Ministry then appointed the Commission to approve the nomination. One of the members of the Commission was Nicola Cabibbo, who had done his thesis with Bruno in 1957. In later years, in an interview for the docu-film *Bruno Touschek and the Art of Physics*, Cabibbo remembered his embarrassment in having to examine his teacher.

13.2.3 The November Revolution: Winners and Losers

The full power of electron-positron collider showed itself in 1974 with the discovery of a particle made of new quarks. What became one of the most notable discoveries in particle physics had its roots in conventional projectile on fixed-target accelerators, but the combination of energy reach and precision attainable by the colliders changed the world of particle physics.

13.2.3.1   A Discovery From the Other Side of the Atlantic

In 1968, as ADONE was suffering interruptions and strikes in the laboratory, an experiment from Brookhaven showed a curious data distribution. The experiment, which was directed by Leon Lederman, future director of Fermi Fermi National Accelerator Laboratory, aimed to measure the scattering of a beam of protons from a hydrogen target and the observation of the energy dependence of the probability to produce a $\mu^+\mu^-$ pair.[28] When the data were analyzed, an anomalous increase in this probability was observed in an energy region between 2 and 4 GeV (Christenson et al. 1970). This was a region which had been anticipated by some theorists to be void of new particle production, and was called 'the desert'. At that time, the particle 'zoo' of the mid 1950's had been enriched by many newcomers, all of which could be classified as built from quarks, invisible entities coming in three types: *up, down* and *strange*. By studying the possible quantum number combinations, one could explain some features of the resonant states which had been observed, but particles with masses above 2 GeV had not been found. This was the reason of believing in the 'desert'. Lederman's experiment indicated otherwise. The key which later brought the solution, came from purely theoretical calculations. First suggested in 1964 (Bjorken and Glashow 1964), the idea that a fourth quark may be needed took some time to be fully accepted. Full theoretical consistency was proved in 1970 (Glashow et al. 1970) when Sheldon Glashow and two young visiting physicists, Luciano Maiani from University of Rome and John Iliopoulos from École Normale Supérieure in Paris, showed that a fourth quark was needed to make the quark model theoretically safe from infinities.[29] How to obtain experimental proof of their idea? They spoke with colleagues at Harvard and with both experimentalists and theorists at MIT, a few blocks from Harvard Square, down towards the Charles River (Maiani and Bonolis 2017a).

During this time, Samuel Ting, from MIT, had come back from Europe, and was thinking about possible experiments to search for new vector mesons. In his original planning Ting and his collaborators had focused on performing the experiment at different accelerators, such as the Alternating Gradient Synchrotron (AGS) in Brookhaven National Laboratory (BNL) or the Intersecting Storage Rings at CERN (with protons). As for ADONE, while already functioning since 1969, it did not offer a sufficiently high energy reach for the exploration (only 3 GeV), while SPEAR, the Stanford Positron Electron Asymmetric Ring, in 1971 had not yet started operation. After looking at the various possibilities, the decision fell on Brookhaven, in Long island. Between the initial idea for the experiment, in 1971, and the data taking in 1974, three years went by, through approval for funding, construction and various accidents which delayed the experiment.

---

[28] Leon Lederman (1922–2018) was awarded the 1988 Nobel Prize in Physics 'for the neutrino beam method and the demonstration of the doublet structure of the leptons through the discovery of the muon neutrino', jointly with Melvin Schwartz and Jack Steinberger.

[29] Sheldon Glashow shared with Abdus Salam and Steven Weinberg the 1979 Nobel Prize in Physics for 'their contribution to the theory of the unified weak and electromagnetic interaction between elementary particles, including, *inter alia*, the prediction of the weak neutral current.'

Fig. 13.13 The original data plot of the October 1974 excitation curve of the J/Ψ resonance, a photograph taken by the Author in June 2018, in Ulrich Becker's office in the now dismantled Cyclotron Building on Vassar Street, in Cambridge, Massachusetts, courtesy of Ulrich Becker, all rights reserved

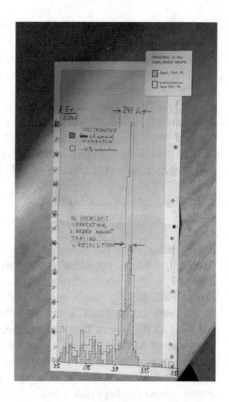

By the end of August 1974, most of the initial teething problems of the experiment had been conquered and data analysis could start. Ulrich Becker, one of Ting's close collaborators, in a conversation with the Author, remembered his excitement in October 1974, when he saw an exceptionally large amount of electron-positron pairs being recorded as a peak in the invariant mass distribution around 3 GeV, Fig. 13.13.[30] It was the signal that a new particle decaying into an electron and a positron had been created (Ting 1976a).

Through late October and early November, the excitement at BNL and MIT grew, and Ting started writing a paper to report their findings (Ting 1976b). While Ting was preparing the article, on the other side of the United States, in Stanford, a similar search had been going on at SPEAR. Since 1972, experimentalists at SPEAR had searched for confirmation of Frascati data of multiple hadron production, which still constituted a puzzle for many physicists. In summer 1974 a step-by-step search, changing the machine energy by as little as 1 MeV at the time, like a real precision tool, had started, under Burton Richter's lead (Richter 1984). When the energy went

---

[30] Ulrich Becker (1938–2020) was Emeritus Professor in the MIT Physics Department. He gave important contributions to particle physics, including to the discovery of the J/Ψ particle, for which Burton Richter and Samuel Ting were awarded the 1976 Nobel Prize in Physics. He was also a major contributor to the Alpha Magnetic Spectrometer (AMS) on the International Space Station, as from https://news.mit.edu/2020/ulrich-becker-mit-professor-emeritus-physics-dies-0409.

near 3.1 GeV, hadron production jumped by a factor 100, muon pairs by 20, electron pairs by a factor 2.

As word reached the East Coast of what was happening across the country in Stanford, Ting decided to finish the article he had been preparing. On November 10th, he left the article to be submitted to the *Physical Review Letters* editorial office in Upton, in Brookhaven, and flew to Stanford, where he had been scheduled to give a seminar. He did another thing before leaving: he asked one of his collaborators, Sau Lan Wu, to call their Italian colleagues and inform them of the discovery.[31] The story of the joint press conference announcing the discovery of the new particle is well known. Less so what happened in Frascati after Sau Lan Wu's call.[32]

13.2.3.2   Phone Calls from America

In November 1974, the director of the Frascati National Laboratory was Giorgio Bellettini, a young, but already well known experimental particle physicist, from University of Pisa. He had distinguished himself in a series of important experiments at the CERN ISR.[33] Ting's interest in the possibility to perform his experiment at the ISR and join Bellettini's group had led to frequent contacts between Bellettini and Sau Lan Wu, one of Ting's closest collaborators. In fact, in 1974, the BNL group was aware of ADONE's searches and Bellettini knew of the ongoing BNL work, but was unaware of the latest developments. [34]

The news of the extraordinary data plot produced by the BNL experiment after the summer, had been kept rather secret, at least with respect to European laboratories. Thus, Frascati was caught unaware, when a phone call from New York reached Bellettini on the night of November 11th.[35] In Bellettini's memory the conversation went something like this:

> *Sau Lan Wu:* "We have found a new particle!"
> *Bellettini:* "What's so special about it?"
> *Sau Lan Wu:* "It is very heavy [massive] and very narrow [long lived] and decays in an electron-positron pair."
> *Bellettini:* "How heavy?"
> *Sau Lan Wu:* "About 3.1 GeV."

She also said they should start searching for it at ADONE. It was an absolute surprise, and the search in Frascati started immediately. The catch was that, in order to

---

[31] In (Amaldi 2017, 158), Ugo Amaldi remembers to have received Wu's call on November 10, as had Bellettini. At the time Ugo was head of INFN National Committee for experimental high energy physics.

[32] In his Nobel Lecture, Ting also mentions that Frascati was informed on November 11.

[33] Bellettini would later lead one of two experiments at the $p\bar{p}$ collider at FermiLab in Chicago, which in 1995 found the *top* quark, the last building block of the particle world in the Standard Model, (Abe et al. 1995; Abachi et al. 1995).

[34] Bellettini's e-mail communication to the Author on February 11th, 2021.

[35] G. Bellettini's talk at the 2013 Bruno Touschek Memorial lectures.

find this particle, ADONE's energy had to be raised, beyond the nominal machine possibilities, which had been fixed since 1960 to be 3 GeV. Until then ADONE had been operating around a c.m. energy of 2.8 GeV, and to increase beyond 3 GeV was a risk which needed the authorization from Fernando Amman, the director of the ADONE program, since pushing the machine beyond the energy design could bring irreparable damage. Such accidents were known to happen, sometimes resulting in fires because of high electrical tensions, or explosions, because of dangerous chemicals in the machine hall. *A posteriori* regrets arose not to have asked the accelerator group to increase the energy right away. The decision was taken by Giorgio Salvini, who was at the time a member the $\gamma\gamma 2$ detector group, and had both the experience and the authority to do it. He had created the laboratories and directed the construction of the synchrotron, had made the crucial decision to approve the AdA proposal, and had been INFN president during the building of ADONE (1966–1969):

> *Salvini*: I was informed by Bellettini and Wu that a resonance had been found beyond 3 GeV. And, therefore, there was no hope for ADONE. Because ADONE reached a limit of 3 GeV. When I got more information, I learned that the resonance was just over 3 Gev; on the other hand, they were very hectic days, because there was this news of the resonance; and we could not observe it except by pushing the machine [beyond the design limits]. Amman wasn't there so I found myself deciding about our machine, because instead of Amman there could only be me. [36]

To force the machine to go beyond its design limits, one needed to know how much the resonance could be found away from 3.0 GeV: 3.2 or 3.5 or beyond? The lee-way for forcing the machine and go higher in energy was no more than 3 to 5% of the design energy, and to go higher would have been too risky. It is at this point that another phone call from the US broke the impasse and made Salvini decide.

This time, the phone call came from the West Coast, and it was Mario Greco who made it. Greco, on his way to visit one of his collaborators in Mexico city, had been invited to Stanford by Sidney Drell, head of the theoretical physics group. Greco and Drell had become good friends during Drell's sabbatical year in Italy, spent between Frascati and Rome. When Greco arrived at SLAC, directly from the hotel, and still in the lift taking him up to the Theory Division, he heard the news of the J/Ψ discovery. Immediately grasping that Frascati could also see the resonance, if ADONE could just be pushed a bit higher in energy, Mario asked to be allowed to make a phone call to Bellettini in Frascati. He did so from Drell's office and gave the value of the resonance peak to be around 3.1 GeV, i.e. around 3100 MeV (Greco 2018).[37]

Beam energy in ADONE was then slowly increased one MeV at the time, looking for hadronic events, normally one in long intervals of time. But when the energy of 3098 MeV was reached, the frequency, at which the counters gave signal, exploded. Rushing to the machine control room, the astonished researchers saw the electronic counters blinking like like a Christmas tree.[38]

---

[36] June 2013 interview with G. Salvini, recorded and translated by the Author from the original Italian transcription.

[37] See also M. Greco and G. Pancheri in *Analysis-online*, Vol.2/3 (2008).

[38] February 13th, 2021, e-mail communication by Rinaldo Baldini, one of ADONE's experimenters in the $\gamma\gamma 2$ group led by Corrado Mencuccini.

Salvini: When we risked forcing the machine, we found it the next day, I remember we were in the laboratories, when that gamma gamma detector, that gave us, generously, a tap every three minutes to say he could make it, at some point [went like] Pim ! Pim ! Pim ! At first, I thought we were drunk, then I said: 'No, that was the resonance, of course'.

But this is a dramatic story, because if ...the resonance, instead of the J/Ψ at 3100 [MeV], had been at 2900, 2800, we would have seen it [right away]. We had already explored up to 3000.

...in fact, we found the resonance at 3.1 Gev. It was a curse for ADONE: ADONE would certainly have found it if it had been [designed to go] beyond 3 Gev. Instead, ...up to 3 GeV there was nothing. And, at 3.1 GeV, there it was. When I pushed the machine, we did find it. And, in fact, we published the results together [with the Americans].[39]

The new particle was called J/Ψ, both to continue in the tradition of giving Greek names to resonant particles, as well as to go along with the Latin alphabet, as for then undiscovered W and Z bosons, the carrier of the weak force. While the choice of the name as Ψ is credited to Richter, Ting assigned the name J (Ting 1976b, [334]).[40]

It was clear that this was a discovery which would shake the world of particle physics. The dramatic jumping of the signals from the electronic counters such as it had never been seen before, had to come from a new world, as it was indeed: a bound state made of a new quark, the *charm* and its antiparticle, had been discovered.

At ADONE, the resonance was found on November 13th. Data were collected and analyzed in no-time, errors estimated and plots drawn. On the 18th, the article was dictated by Salvini on the phone to the office of *The Physical Review Letters* in Upton, where the two articles by the BNL and SLAC groups had already been submitted, respectively on the 12th and 13th.[41] The three articles were published in the same issue of the journal, with an exceptional editorial note (Aubert et al. 1974; Augustin et al. 1974; Bacci et al. 1974).

And where was Bruno during those days? No one seems to remember his presence in Frascati. But his interest, in what ADONE was doing, had not waved. Not long after the discovery, he urged the young members of his group to apply radiative corrections to the J/Ψ excitation curve, which they did, once more inspired by his vision, (Greco et al. 1975).

In those years, Touschek's interest had shifted to teaching and bringing science to high-schools. He was still musing about particle physics, and In 1974, his last physics published paper appeared, (Touschek 1974), with the affiliation *Garvens S.p.A*, the firm he had inherited after aunt Ada's death in 1959, Sect. 9.1. At Accademia dei Lincei, he started organizing a series of lectures by prominent physicists, such as P.A.M Dirac, or Rolf Widerøe, Fig. 13.14. By 1976, he started having health problems. He also saw that the scale of experimental efforts and the concomitant rivalries

---

[39] June 2013 interview with G. Salvini, recorded and translated in Italian by the Author.

[40] Different explanations exist, such that the letter J just follows K, the name of strange mesons, as from https://en.wikipedia.org/wiki/J/psi_meson#The_name.

[41] The list of authors of the ADONE paper is full of typing errors, later corrected in an *erratum*. It is a reflection of Salvini's personality that his name appears as G.S., as if it were obvious to the American secretary, typing the article during the phone call, that G. S. stood for Salvini's name.

Fig. 13.14 Bruno Touschek at left, Paul Dirac in the center, Beniamino Segre (President of the Accademia) and Marcello Conversi last at right, at the Accademia dei Lincei on the occasion of Dirac's lecture, on April 15th, 1975. Courtesy of Sapienza University of Rome–Physics Department Archives, https://archivisapienzasmfn.archiui.com, documents provided for purposes of study and research

had changed the world of particle physics in a way he disliked.[42] His interest in up-to-date theoretical developments had also waned.

13.2.3.3 Why Bruno Missed the Prize

The importance of the discovery of the J/Ψ was recognized by the scientific community with the award of the 1976 Nobel Prize to Samuel Ting and Burton Richter, the leaders of the experiments at BNL and SLAC, for 'their pioneering work in the discovery of a heavy elementary particle of a new kind'. Had the Italian physics community been differently organized, they could have shared the prize: unlike at SPEAR or Brookhaven, four distinct groups were operating at ADONE, each one with a different spokesman. There was no scientifically dominant personality recognized as such by all the four groups, and old, if not ancient, rivalries between

---

[42] See his note, 'This is what Touschek wrote after the conference: this is what he wished he had said', following Conversi's contribution in (Conversi 1976a, [89]).

regions and universities were still keeping their scientists apart. On the other hand, physics was moving into a different *modus operandi*, the one which became known as 'big science'. A central organization, with a clear hierarchy structure, was needed to obtain the funds, avoid repetitions and plan the different parts of a detector. While Italy had been ahead in constructing ADONE, its past still prevented the unification of resources and efforts. Thus it would have been impossible to decide if one single individual could share of the Nobel prize for the J/Ψ discovery. The only one could have been Bruno Touschek. So, why didn't he get it?

The prize had acknowledged "pioneering work", and it is astonishing that Touschek's name does not appear in any of the official Nobel Prize documents. He is mentioned neither in Ting's lecture (Ting 1976b), nor in Richter's (Richter 1976). Nor does the official press release say anything of the scientist who had proposed and built the first electron-positron ring, carried through the proof-of-concept, discovered a crucial effect affecting all storage rings, and, as early as 1960, had the courage and insight to envision ADONE. If one reads the official Nobel Prize documents, Frascati is mentioned, but Touschek's contribution to the field does not exist. This book has been started to remind the physics community of his genius.

But Bruno's genius had no affiliations. No nation could fully claim him as his own, no ethnic group either. He shunned political correctness. His caustic remarks often hit the mark. Thus the dice were against him, and when the recognition of storage ring successes came, he was left out.

13.3   Bruno's Last Year at CERN

In 1977, Bruno's health was declining (Amaldi 1981, 1). Worried about his family's future, and also interested in CERN's new programs, he welcomed an offer as 'Senior Visiting Scientist' at CERN, where the design of a powerful electron–positron collider, the LEP, was under way, and, in October, he moved to Geneva. Bruno did not like the direction high energy physics was following, with large collaborations organized like an industrial enterprise, but other plans, closer to his way of looking at physics, were also discussed. One in particular appealed to him, the making of a proton-antiproton collider, the $S\bar{p}pS$, which, five years later, would lead Carlo Rubbia and the UA1 team to the experimental discovery of the carriers of the electroweak force, the so called W and Z bosons (Arnison et al. 1983b).[43] Rubbia remembers long discussions with Bruno about the project, first at CERN and later at the Hôpital de La Tour (Rubbia 2004). Bruno, 'sharp and lucid as ever', although already quite ill, started writing a paper about stochastic cooling, which became his last incomplete scientific work, published posthumously (Touschek 1979).

As spring 1978 approached, conscious of the coming end, Bruno arranged his return to Austria, Fig. 13.15. A CERN car brought him from Hôpital de La Tour, in

---

[43] The Nobel Prize in Physics 1984 was awarded jointly to Carlo Rubbia and Simon van der Meer 'for their decisive contributions to the large project, which led to the discovery of the field particles W and Z, communicators of weak interaction.' from https://www.nobelprize.org/prizes/physics/1984/summary/.

Fig. 13.15 Bruno Touschek
in 1978, at Hôpital de La
Tour, in Geneva, Sapienza
University of Rome–Physics
Department Archives,
https://archivisapienzasmfn.
archiui.com, documents
provided for purposes of
study and research

Geneva, to Igls, near Innsbruck. The extraordinary journey which had taken him from
his first encounter with racial discrimination to his realization as a great scientist,
was over, and when his life prematurely ended, he was back to his country of birth,
in Austria, where he passed away on May 25th, 1978 (Amaldi 1981, 1–2).

Chapter 14
Epilogue: A Personal Memory

I first entered the grounds of the Frascati National Laboratories in May 1966. I had graduated in physics from University of Rome in February, with a thesis on the "Coalescence and decay of photons on nuclei", an unlikely process to be measured at the time, but a good training in Quantum Electrodynamics (QED) calculations, of special interests in Rome and Frascati. In April, back from a post graduation ski vacation in the Dolomites, I heard from two colleagues, Giancarlo Rossi and Paolo Di Vecchia, that there was a possible opening as a research fellow with a new theoretical physics group led by Bruno Touschek in the Frascati National Laboratories. Very hesitatingly, I went to knock on Touschek's door on the second floor of the Physics Institute in Rome. At the time, first and second year students were allowed to access only the first floor of the Institute, where the main Lecture halls were located. Professor offices were on the second floor and enormous by later standards. To access them, one had to be at least senior students. As I entered, Touschek was discussing some calculation with Giancarlo Rossi. I asked about the possibility of joining the new group, he was very kind and promised an answer. A few weeks later I was told I had been accepted: my thesis advisor, Benedetto De Tollis, had vouched for my capacity to do good and independent work. This is how my life-long attachment to Touschek began, 55 years ago.

My first encounter with him had been as a teacher, when I was attending his lectures on Statistical Mechanics, a famous course which trained and inspired many different student generations (Margaritondo 2021). I met him for the last time in summer 1976, on the *Alpen Express*, a train which crossed the Alps from Innsbruck to Rome, through the Brenner pass.

I was returning to Rome from a vacation in one of the Trentino valleys, Val di Non, where my father's family came. Nine years before, I had left Italy and moved to Boston, in November 1967, leaving Frascati and Rome. We used to return to Italy very often, for extended periods of time, for summer vacations or sabbaticals, which we spent mostly in the Frascati Laboratories, on leave from Northeastern University

G. Pancheri, *Bruno Touschek's Extraordinary Journey*, Springer Biographies,
https://doi.org/10.1007/978-3-031-03826-6_14

where I used to teach part-time and my husband, Yogendra (Yogi) Srivastava, a theoretical physicist, was Professor of Physics.

On that day on the train, I was in the corridor watching the mountain range disappear as we headed South, when I heard a loud voice so uniquely accented that it could only be Bruno's. I remember feeling a great emotion. Since I had left Italy, I had seen him in very few occasions. The most recent encounter had been in Frascati soon after the discovery of the J/Ψ, in December 1974 or early January 1975. In a meeting, where Frascati J/Ψ data where presented and discussed, he had urged Mario Greco, Yogi and myself to apply infrared radiative corrections to the experimental results and extract the natural life-time of the new particle. Once more, his physics vision was right on the issue, we did an extensive work, obtaining a very precise estimate, which remained for a while the most accurate. Bruno was very pleased to see his ideas become alive, the curve following so closely the data that he doubted the estimated experimental errors could be as small as reported.[1]

In summer 1975, I had returned to Boston and had not seen him again until the day we met on the train. However, in the years which followed my leaving Bruno's group and move to the US, I had never forgotten the ideas about radiation emission and the resummation method he had formulated. Since the time I had left Frascati on November 30th 1967, I had been thinking about the way to continue on the resumma-tion problem and done some further work which had made me think that Touschek's (resummation) method echoed a strong interaction behaviour, a power law result known as *reggeization* (Pancheri-Srivastava 1973). In 1975, after the parenthesis on the J/Ψ, I went back to one problem left unsolved in the old work with Bruno, a problem on which we had spent days and nights, in 1966. It concerned applying soft photon resummation in momentum space, but our quest for a compact formula akin to the one for energy radiation, had led nowhere. In the paper we discussed at length the analyticity properties of the momentum distribution, but obtaining a closed form turned out to be impossible. Bruno however was seldom wrong in his physics intu-ition, and his interest on tackling the problem in momentum space had pinpointed a feature which would later dominate in Quantum Chromodynamics (Greco et al. 1978; Parisi and Petronzio 1979; Greco 2019).

One of the conclusions of the work with Bruno had been that soft photon resum-mation in momentum space was not required in QED, given the smallness of the coupling constant. After finishing the work on the J/Ψ, which was based on QED, I turned again my attention to apply Bruno's resummation in momentum space for strong interaction particle phenomenology. This problem occupied my research time thought 1975 and early in '76, until some interesting results had appeared. All along I had been driven by the desire to show Bruno how far his method could go, and that I had not forgotten his teaching. As often happens to those who leave a place, there is a lingering feeling of having betrayed the trust of the ones we left behind, as we moved away. Applying Touschek's resummation ideas to strong interactions was the way to keep alive Bruno's teaching. By summer 1976, a work along those lines,

---

[1] He thought that, statistically, some experimental points should be above and others below the theoretical curve.

with my husband Yogi, had been completed and we had started writing it (Pancheri-Srivastava and Srivastava 1977). My main drive had been to tell it to Bruno, and this is what I began doing, on that day in the train. But he was not interested, so we spoke of many other things, his family, and political unrest in Italy, until the train reached Rome, and, then, I did not see him again.

Some time later, in 1977, I learnt through Mario Greco that he had been very ill, in Rome Polyclinic Hospital, and, in March 1978, that he was in a hospital in Geneva (Greco 2018). I immediately sent him a short note, wishing him to recover and communicating the birth of my son a few months before. The note I sent was among the papers Elspeth showed to me and Luisa Bonolis, when we went to visit her in 2009. Once more, I felt a great emotion at seeing this note among his papers. And then Elspeth said something I had never expected: "If you had not left...". I did not ask why she said so, and soon regretted it, as she unfortunately passed away a few years later.

I am now finishing this book and I think again to what Elspeth told me. It was not my leaving which inspired Elspeth's remarks, but her memory of a time when Bruno was at a high point of his intellectual strength. Reflecting now, on Bruno's life, from the perspective of the present book, I can see that the two years 1966 and 1967 may have been the best in his life. He had been the initiator of a field of research, electron-positron physics, which was now developing everywhere, and was close to see it become reality in ADONE, the beautiful machine. He was famous in the world of particle physics, and was surrounded by a group of bright young researchers working on his ideas. The machine he had invented and the physics he hoped would come out, were soon to appear. His health was still good, family life serene.

Today, as I look back, I also see that I had left before everything changed, just in time to have been part of Bruno's group and his last few golden years.

Chapter 15
Postface

This book is clearly incomplete. Much is still to be told about the life of the extraordinary human being who was Bruno Touschek, the circumstances which forged his life and the impact of his work. Since 1953, when he arrived in Rome, his teaching and vision in Quantum Field Theory contributed to the formation of a generation of great theoretical physicists, themselves the initiators of novel venues of knowledge. In Frascati, the main Auditorium of the National Laboratories is named after him, and AdA, the first electron-positron collider, is still there, under a canopy, seen by over 5000 yearly visitors. A few hundred meter downhill, the ADONE building houses DAFNE, the electron-positron collider in operation since the year 2000 to study violations of CP invariance, Touschek's guiding principle in his belief that electrons and positrons would meet when circulating in the same ring, under the same magnetic field. Across the Alps, upgrading of the Large Hadron Collider is under way, while the world of particle physics is looking further ahead to build 100 km long electron-positron colliders, at CERN itself and in China.

Recently, on December 2–4, 2021, a Symposium was held in Rome, at Sapienza University and the Lincei National Academy, and in Frascati, at the INFN National Laboratories. The Symposium celebrated hundred years after Touschek's birth, and his influence on both accelerator and theoretical physics came alive through the testimonies of many scientists who had known him as a teacher or a colleague. The Proceedings of the Symposium are under preparation and will be published with the title *Bruno Touschek 100 years - Memorial Symposium 2021*.[1]

Much more is still to be done to bring Touschek's legacy to younger generations and to the public at large. Most of his correspondence, both private and public, is to be studied and published. In this book, the over 200 letters Touschek sent to his father since his teen years when he was away from Vienna, have been used as a guide through his life, but a full transcription and translation is waiting to

---

[1] Edited by L. Bonolis, L. Maiani and G. Pancheri and published by Springer Nature, the Symposium Proceedings are expected to appear in 2022.

433
G. Pancheri, *Bruno Touschek's Extraordinary Journey*, Springer Biographies,
https://doi.org/10.1007/978-3-031-03826-6_15

Fig. 15.1 Some incomplete drawings, among those Elspeth Touschek used to recover from Bruno's pockets, after his coming home from the university. Touschek's captions read as '3rd Meeting' (top left), and 'he is not from our faculty' (top right). Family Documents, © Francis Touschek, all rights reserved

be published. These letters give a day-by-day account of Bruno's odyssey during the war, from Vienna to Munich, Hamburg, Berlin, and his escape from the final deportation to the Kiel concentration camp. They hold great human and historical interest, as one more testimonial to Europe's dramatic past. Other aspects of Bruno's life, related to his maternal and paternal family, are yet to be explored, beyond this book's content. Not all of his drawings has been published. A collection organized around their contemporary context will illustrate life at the University of Rome during the transition from the 1950's reconstruction to the turbulent times of the 1968 student movement, and beyond. Finally, questions still remains, which may find an answer only in five or ten years from now: why did Touschek miss the Nobel Prize and was he ever nominated for it? This could have happened, either soon after ADONE started operating, discovering multiple hadronic production, or later, on the occasion of the discovery of the J/Ψ. In time, the interested researcher will find the answers.

In the meanwhile, in the tradition of Amaldi's biography, this book closes with a choice of Touschek's unpublished drawings, Fig. 15.1.

References

G. Aad and ATLAS Collaboration, Observation of a new particle in the search for the Standard Model Higgs boson with the ATLAS detector at the LHC. Phys. Lett. B, **716**(1), (2012). https://doi.org/10.1016/j.physletb.2012.08.020

S. Abachi et al., Observation of the top quark. Phys. Rev. Lett. **74**, 2632–2637 (1995). https://doi.org/10.1103/PhysRevLett.74.2632

F. Abe et al., Observation of top quark production in $\bar{p}p$ collisions. Phys. Rev. Lett. **74**, 2626–2631 (1995). https://doi.org/10.1103/PhysRevLett.74.2626

Academy of Science USSR and IUPAP, (ed.), *Proceedings, 9th International Conference on High Energy Physics, v. 1–2 (ICHEP59), Kiev, USSR, July 15–25, 1959*, (Moscow, 1960). Academy of Science USSR/International Union of Pure and Applied Physics. http://inspirehep.net/record/1280969

A. Alberigi, F. Amman, C. Bernardini, U. Bizzarri, G. Bologna, G. Corazza, G. Diambrini, G. Ghigo, A. Massarotti, G. P. Murtas, M. Puglisi, I. F. Quercia, R. Querzoli, G. Sacerdoti, G. Salvini, G. Sanna, P. G. Sona, R. Toschi, A. Turrin, M. Ageno, and E. Persico, Operation at 1000 MeV of the Frascati Electronsynchrotron. Il Nuovo Cimento, **11**(2), 311–312, (1959). ISSN 1827–6121. https://doi.org/10.1007/BF02859729

G. Altarelli, F. Buccella, Single photon emission in high-energy $e^+ - e^-$ collisions. Il Nuovo Cimento **34**(5), 1337–1346 (1964). https://doi.org/10.1007/BF02748859

G. Altarelli, G. Parisi, Asymptotic freedom in parton Language. Nucl. Phys. B **126**, 298–318 (1977). https://doi.org/10.1016/0550-3213(77)90384-4

E. Amaldi, The Years of Reconstruction. Part I. Scientia **114**, 51–68 (1979)

E. Amaldi, The Years of Reconstruction. Part II. Scientia **114**, 439–451 (1979)

E. Amaldi, *The Bruno Touschek legacy (Vienna 1921 - Innsbruck 1978)*. Number 81–19 in CERN Yellow Reports: Monographs. (CERN, Geneva, 1981). 10.5170/CERN-1981-019. https://cds.cern.ch/record/135949/files/CERN-81-19.pdf

E. Amaldi, *L'Eredita' di Bruno Touschek*. (Societa' Italiana di FIsica, 1982)

E. Amaldi, *Da via Panisperna all'America*. (Editori Riuniti, 1997). URL ISBN:8835943272

E. Amaldi, *The Adventurous Life of Friedrich Georg Houtermans, Physicist (1903–1966)*. (Springer-Verlag, Berlin Heidelberg, 2012). URL https://doi.org/10.1007/978-3-642-32855-8

E. Amaldi, E. Fabri, T.F. Hoang, W.O. Lock, L. Scarsi, B. Touschek, B. Vitale, Report of the committee on τ-mesons. Il Nuovo Cimento **12**(S2), 419–432 (1954). https://doi.org/10.1007/BF02781542

U. Amaldi, Remembering Bruno Touschek. In *Bruno Touschek Memorial Lectures*, vol. 33 of *Frascati Physics Series*. ed. by M. Greco and G. Pancheri. (INFN-Laboratori Nazionali di Frascati, 2004), pp. 89–92. URL http://www.lnf.infn.it/sis/frascatiseries/Volume33/volume33.pdf

U. Amaldi, *Ricordi e nostalgie del Laboratorio di Fisica dell'ISS*, vol. 12. (Istituto Superiore di Sanita', 2017)

U. Amaldi, W. de Boer, H. Furstenau, Comparison of grand unified theories with electroweak and strong coupling constants measured at LEP. Phys. Lett. B **260**, 447–455 (1991). https://doi.org/10.1016/0370-2693(91)91641-8

F. Amman, The Early Times of Electron Colliders. Rivista di Storia della Scienza **2**, 130–151 (1985)

F. Amman, The early times of electron colliders, in *The Restructuring of Physical Sciences in Europe and the United States*. ed. by M. De Maria, M. Grilli, F. Sebastiani. (1945–1960). (World Scientific, Singapore, 1989), pp. 449–476

F. Amman, R. Andreani, M. Bassetti, C. Bernardini, A. Cattoni, R. Cerchia, V. Chimenti, G. Corazza, E. Ferlenghi, and L. Mango, Status report on the 1.5 Gev electron positron storage ring—adone. In *Proceedings, 4th International Conference on High-Energy Accelerators, HEACC 1963, v.1–3: Dubna, USSR, August 21 - August 27 1963*. ed. by A. A. Kolomenskij and A. B. Kuznetsov. (Oak Ridge, TN, 1965), pp. 309–327. NTIS. URL http://inspirehep.net/record/918674/files/HEACC63_I_314-338.pdf

F. Amman et al., Two-beam operation of the 1.5 GeV electron-positron storage ring ADONE. Lettere al Nuovo Cimento, **1**, 729–737 (1969). https://doi.org/10.15161/oar.it/1448378666.75

Anonymous, Edinburgh conference on elementary particles. Nature, **165**(4185), 54–56 (1950). https://doi.org/10.1038/165054a0

G. Arnison et al., Charged particle multiplicity distributions in proton anti-proton collisions at 540-GeV center-of-mass energy. Phys. Lett. B **123**, 108–114 (1983). https://doi.org/10.1016/0370-2693(83)90969-3

G. Arnison et al., Experimental observation of isolated large transverse energy electrons with associated missing energy at $\sqrt{s} = 540$ GeV. Phys. Lett. B **122**, 103–116 (1983). https://doi.org/10.1016/0370-2693(83)91177-2

A. Aruta, Shocking waves at the museum: The bini-cerletti electro-shock apparatus. Med. Hist. **55**, 407–412 (2011). https://doi.org/10.1017/s0025727300005482

M.G. Ash, Denazifying scientists and science, in *Technology Transfer out of Germany after 1945*, ed. by M. Judt, B. Ciesla. Studies in the History of Science, Technology and Medicine, vol. 2 (Harwood Academic Publishers, Amsterdam, 1996), pp. 61–80

J.J. Aubert et al., Experimental observation of a heavy particle. J. Phys. Rev. Lett. **33**, 1404–1406 (1974). https://doi.org/10.1103/PhysRevLett.33.1404

J.E. Augustin, P. Marin, and F. Rumpf, Traces et parcours d'electrons, muons et pions dans une chambre a etincelles a plaques epaisses. Nucl. Instrum. Methods **36**, 213–225 (1965). URL https://doi.org/10.1016/0029-554X(65)90427-1

J.E. Augustin, J.C. Bizot, J. Buon, J. Haissinski, D. Lalanne, P.C. Marin, J. Perez-y Jorba, F. Rumpf, E. Silva, and S. Tavernier, Study of electron-positron annihilation into $pi^+ pi^-$ at 775 MeV with the orsay storage ring. Phys. Rev. Lett. **20**, 126–129 (1968). https://doi.org/10.1103/PhysRevLett.20.126

J.E. Augustin, A. Courau, B. Dudelzak, F. Fulda, G. Grosdidier, Haissinski, J.L. Masnou, R. Riskalla, F. Rumpf, and E. Silva, Evidence for the phi-meson contribution to vacuum polarization obtained with the orsay e^+e^- & colliding-beam ring. Phys. Rev. Lett. **30**, 462–464 (1973). URL https://doi.org/10.1103/Phys.Rev.Lett.30.462

J.E. Augustin et al., Discovery of a Narrow Resonance in e^+e^- Annihilation. Phys. Rev. Lett. **33**, 1406–1408 (1974). https://doi.org/10.1103/PhysRevLett.33.1406. [Adv. Exp. Phys. 5, 141 (1976)]

V.L. Auslander, G.I. Budker, J.N. Pestov, V.A. Sidorov, A.N. Skrinsky, A.G. Khabakhpashev, Investigation of the ρ-meson resonance with electron-positron colliding beams. Phys. Lett. **25B**(6), 433–435 (1967). https://doi.org/10.1016/0370-2693(67)90169-4

V.L. Auslender, G.A. Blinov, G.I. Budker, M.M. Karliner, A.V. Kiselev, A.A. Livshits, S.I. Mishnev, A.A. Naumov, V.S. Panasyuk, Y.N. Pestov, V.A. Sidorov, G.I. Sil'vestrov, A.N. Skrinskii, A.G. Khabakhpashev, and I.A. Shekhtman, Accelerator development in novosibirsk: Iii. status of the positron-electron vepp-2 storage ring. J. Nucl. Energy. Part C, Plasma Phys. Accel. Thermonuclear Res. **8**(6), 683–688 (1966). https://doi.org/10.1088/0368-3281/8/6/307.http://stacks.iop.org/0368-3281/8/i=6/a=307

C. Bacci et al., Preliminary result of frascati (ADONE) on the nature of a new 3.1-GeV particle produced in e^+e^- annihilation. Phys. Rev. Lett. 33, 1408 (1974). https://doi.org/10.1103/PhysRevLett.33.1408, https://doi.org/10.1103/PhysRevLett.33.1649. [Erratum: Phys. Rev. Lett. 33, 1649 (1974)]

V.N. Baier, High-energy interactions betweens electrons and positrons. Soviet Phys. Uspekhi **5**(6), 976–997 (1963). URL http://stacks.iop.org/0038-5670/5/i=6/a=R07

V.N. Baier, Forty years of acting electron-positron colliders. arXiv:hep-ph/0611201, Nov. 2006. URL https://arxiv.org/abs/hep-ph/0611201

A. Baldin and I. Semenyushkin, Twenty years of the synchrophasotron of the JINR High-Energy Physics Laboratory. Sov. At. Energy **43**, 1146–1147 (1977). URL https://doi.org/10.1007/BF01117960

A. Baracca, G. Ferrari, and R. Renzetti, The "go-stop-go" of Italian civil nuclear programs, beset by lack of strategic planning, exploitation for personal gain and unscrupulous political conspiracies: 1946–1987, 2017. URL https://arxiv.org/abs/1709.05195

G. Battimelli, I. Gambaro, Da via Panisperna a Frascati: gli acceleratori mai realizzati. Quaderni di Storia della Fisica **1**, 319–333 (1997)

G. Battimelli, M. De Maria, G. Paoloni, *Le carte di Bruno Touschek* (Università La Sapienza, Roma, 1989)

G. Battimelli, M. De Maria, and G. Paoloni, *L'Istituto Nazionale di Fisica Nucleare Storia di una Comunità di Ricerca.* (Laterza, 2001)

W. Beindorf, Einphasig gespeister meßphasenschieber aus drehtransformator und drehspannungsscheider. Archiv für Elektrotechnik, **37**(11), 542–554 (1943)

J.S. Bell, F. Cerulus, T. Ericson, J. Nilsson, and H. Rollnik, (ed.), *International Conference on Theoretical Aspects of Very High-energy Phenomena. 5–9 Jun 1961, CERN, Geneva, Switzerland.* (Geneva, 1961). CERN, CERN. URL https://cds.cern.ch/record/280184

S. Beller, *Vienna and the Jews, 1867–1938: A Cultural History.* (Cambridge University Press, 1989). ISBN 978-0521407274

V.R. Berghahn, Technology, reparations, and the export of industrial culture. Problems of the German-American relationship, 1900–1950, in *Technology Transfer out of Germany after 1945*, ed. by M. Judt, B. Ciesla. Studies in the History of Science, Technology and Medicine, vol. 2 (Harwood Academic Publishers, Amsterdam, 1996), pp. 1–10

C. Bernardini, Vector Boson Hunting with ADONE. Technical Report 145 (LNF-65/47), Frascati National Laboratory, December 27 1965. URL http://www.lnf.infn.it/sis/preprint/detail-new.php?id=2752

C. Bernardini, AdA: the smallest e^+e^- ring. In *The Restructuring of Physical Sciences in Europe and the United States, 1945–1960*, ed. by M. De Maria, M. Grilli, and F. Sebastiani. (World Scientific, Singapore, 1989), p. 444

C. Bernardini, Bruno Touschek and AdA. In *Adone, a milestone on the particle way*, ed. by V. Valente. (INFN, Frascati, 1997), p. 1

C. Bernardini, Remembering Bruno Touschek, his work and personality. In *Bruno Touschek and the birth of e^+e^- physics*, vol. XIII, ed. by G. Isidori. (Frascati National Laboratories, 1998), pp. 9–16

C. Bernardini, La nascita degli anelli di accumulazione per elettroni e positroni. In *Atti del XXII Congresso Nazionale di Storia della Fisica e dell'Astronomia, Università degli Studi di Genova, Genova-Chiavari, 6–8 giugno 2002*, ed. by M. Leone, A. Paoletti, and N. Robotti. (Napoli, 2003). IISF. http://www.sisfa.org/wp-content/uploads/2013/03/002-BERNARDINI-DEFINITIVO.pdf

C. Bernardini, AdA: the first electron-positron collider. Phys. Perspect. **6**(2), 156–183 (2004). URL https://link.springer.com/article/10.1007/s00016-003-0202-y

C. Bernardini. *Fisica vissuta*. Codice, 2006. ISBN 9788875780517. URL https://books.google.com/books?id=hYhpAAAACAAJ

C. Bernardini and L. Bonolis, ed. by *Enrico Fermi. His Work and Legacy*. (Springer-Verlag, Berlin Heidelberg, 2004). URL https://www.springer.com/de/book/9783540221418

C. Bernardini, G. Corazza, G. Ghigo, and B. Touschek, The Frascati Storage Ring, 1960a. URL http://www.lnf.infn.it/sis/preprint/detail-new.php?id=3180

C. Bernardini, G.F. Corazza, G. Ghigo, B. Touschek, The Frascati storage ring. Il Nuovo Cimento **18**(6), 1293–1295 (1960). https://doi.org/10.1007/BF02733192

C. Bernardini, G. F. Corazza, G. Di Giugno, G. Ghigo, R. Querzoli, J. Haissinski, P. Marin, and B. Touschek, Lifetime and beam size in a storage ring. Phys. Rev. Lett. **10**(9), 407–409, (1963). URL https://doi.org/10.1103/PhysRevLett.10.407

C. Bernardini, G. Corazza, G. Di Giugno, J. Haissinski, P. Marin, R. Querzoli, B. Touschek, Measurements of the rate of interaction between stored electrons and positrons. Il Nuovo Cimento **34**(6), 1473–1493 (1964). https://doi.org/10.1007/BF02750550.https://link.springer.com/article/10.1007

C. Bernardini, G. F. Corazza, G. Di Giugno, G. Ghigo, R. Querzoli, J. Haissinski, P. Marin, and B. Touschek, Lifetime and beam size in electron storage rings. In *Proceedings, 4th International Conference on High-Energy Accelerators, HEACC 1963, v.1–3: Dubna, USSR, August 21 - August 27 1963*, ed. by A.A. Kolomenskij and A.B. Kuznetsov. (Oak Ridge, TN, 1965), pp. 411–415, NTIS. URL http://inspirehep.net/record/918681/files/HEACC63_I_416-421.pdf

C. Bernardini, G. Pancheri, C. Pellegrini, Bruno Touschek: From betatrons to electron-positron colliders. Rev. Accel. Sci. Technol. **08**, 269–290 (2015). https://doi.org/10.1142/S1793626815300133

J. Bernstein, *Hitler's Uranium Club: The Secret recordings at Farm Hall*. (Springer Verlag, New York, 2001). URL https://doi.org/10.1007/978-1-4757-5412-4

J. Bernstein, Heisenberg in Poland. Am. J. Phys. **72**, (2004). https://doi.org/10.1119/1.1630333

H. Bethe, The German Uranium project. Phys. Today **53**(7), 34 (2000). https://doi.org/10.1063/1.1292473

H.A. Bethe, M.E. Rose, The maximum energy obtainable from the cyclotron. Phys. Rev. **52**, 1254–1255 (1937). https://doi.org/10.1103/PhysRev.52.1254.2. URL https://link.aps.org/doi/10.1103/PhysRev.52.1254.2

H.J. Bhabha, R.H. Fowler, The scattering of positrons by electrons with exchange on Dirac's theory of the positron. Proc. R. Soc. Lond. **154**, 195–206 (1936). https://doi.org/10.1098/rspa.1936.0046

S.K. Bhattacharjee, 1959 VARENNA summer school on "weak interactions". Curr. Sci. **29**(2), 48–50 (1960). ISSN 00113891. https://doi.org/10.2307/24212700. URL http://www.jstor.org/stable/24212700

R. Bimbot, Les années Joliot. La naissance du Laboratoire de physique nucléaire d'Orsay (1956–1958). La revue pour l'histoire du CNRS, **16**, 41 (2007). URL http://journals.openedition.org/histoire-cnrs/1597

J.D. Bjorken and S. D. Drell, Relativistic quantum fields. 1965

J.D. Bjorken, S.L. Glashow, Elementary particles and SU(4). Phys. Lett. **11**, 255–257 (1964). https://doi.org/10.1016/0031-9163(64)90433-0

A. Blanc-Lapierre, J.-L. Delcroix, The Orsay linear accelerator. Onde Électrique **43**, 597–598 (1963)

F. Bloch, A. Nordsieck, Note on the radiation field of the electron. Phys. Rev. **52**(2), 54–59 (1937). https://doi.org/10.1103/PhysRev.52.54

L. Bonolis, Bruno Touschek vs. Machine Builders: AdA, the first matter-antimatter collider. La Rivista del Nuovo Cimento, **28**(11), 1–60 (2005). https://doi.org/10.1393/ncr/i2005-10006-x

L. Bonolis, *Maestri e allievi nella fisica italiana del Novecento* (Compositori, Pavia, 2008)

L. Bonolis, Bruno Rossi and the racial laws of fascist Italy. Phys. Perspect. **13**(1), 58–90 (2011). https://doi.org/10.1007/s00016-010-0035-4

L. Bonolis, International scientific cooperation during the 1930s. bruno rossi and the development of the status of cosmic rays into a branch of physics. Ann. Sci. **71**(3), 355–409 (2014). https://doi.org/10.1080/00033790.2013.827074. PMID: 24908796

L. Bonolis, Bruno Touschek Remembered. 1921–2021. Bibliog. Sources **10** (2021)

L. Bonolis, M.G. Melchionni, *Fisici Italiani del tempo presente* (Storie di vita e di pensiero. Marsilio, Venezia, 2003)

L. Bonolis, G. Pancheri, Bruno Touschek: particle physicist and father of the e^+e^- collider. Europ. Phys. J. H **36**(1), 1–61 (2011) https://doi.org/10.1140/epjh/e2011-10044-1.https://link.springer.com/article/10.1140/epjh/e2011-10044-1

L. Bonolis and G. Pancheri, Bruno Touschek and AdA: from Frascati to Orsay. *arXiv e-prints*, arXiv: 1805.09434, May 2018. https://arxiv.org/abs/1805.09434

L. Bonolis and G. Pancheri, Bruno Touschek in Germany after the War: 1945–1946. *arXiv e-prints*, October 2019. URL https://arxiv.org/abs/1910.09075

L. Bonolis, F. Bossi, and G. Pancheri, The frascati national laboratories. Il Nuovo Saggiatore, **37**, 47–60 (2021). URL https://www.ilnuovosaggiatore.sif.it/issue/65

M. Born, *Atomic Physics*, 5th edn. (Blackie & Sons, Glasgow, UK, 1951)

M. Born, L. Infeld, Foundations of the new field theory. Proc. R. Soc. Lond. **A144**(852), 425–451 (1934). https://doi.org/10.1098/rspa.1934.0059

M. Born and L. Infeld, On the quantization of the new field equations. II. Proc. R. Soc. Lond. **150**(869), 141–166 (1935). https://doi.org/10.1098/rspa.1935.0093

W. Bosley, J.D. Craggs, W.F. Nash, R.M. Payne, X-Radiation from a 20-Mev. Betatron. Nature **161**(4104), 1022–1023 (1948). https://doi.org/10.1038/1611022a0

L. Brown, M. Dresden, L. Hoddeson, (ed.) *Pions to Quarks: Particle Physics in the 1950s* (Cambridge University Press, United States, 1989). 9780521309844. https://doi.org/10.1017/CBO9780511563942

L.M. Brown, F. Calogero, The effect of pion-pion interaction in electromagnetic processes. Phys. Rev. Lett. **4**, 315–317 (1960). https://doi.org/10.1103/PhysRevLett.4.315

L.M. Brown, R.P. Feynman, Radiative corrections to compton scattering. Phys. Rev. **85** 231–244 (1952). https://doi.org/10.1103/PhysRev.85.231. URL https://link.aps.org/doi/10.1103/PhysRev.85.231

H. Bruck and S.-R. Group, The Orsay project of a storage ring for electrons and positrons of 450 Mev maximum energy. In *Proceedings, 4th International Conference on High-Energy Accelerators, HEACC 1963, v.1–3: Dubna, USSR, August 21 - August 27 1963*, ed. by A.A. Kolomenskij and A.B. Kuznetsov. (Oak Ridge, TN, 1965). pp. 365–371. NTIS, NTIS. URL http://inspirehep.net/record/918676

S. Brush, History of the Lenz-Ising Model, Rev. Mod. Phys. **39**, 883–893 (1967). https://doi.org/10.1103/RevModPhys.39.883

T. Brustad, Rolf Widerøe: Why is the originator of the science of particle accelerators so neglected, particularly in his home country? Acta Oncol. **37**(6), 603–614 (1998). https://doi.org/10.1080/028418698430313

P.J. Bryant, Antiprotons in the ISR. IEEE Trans. Nucl. Sci. **30**(4), 2047–2049 (1983). https://doi.org/10.1109/TNS.1983.4332713

P.J. Bryant, *A brief history and review of accelerators* (Course on General Accelerator Physics, In CERN Accelerator School, 1992)

G. I. Budker, Accelerator development in novosibirsk: I. general review. J. Nucl. Ener. Plasma Phys. Accel. Thermonuclear Res. **8**(6), 675 (1966). URL http://stacks.iop.org/0368-3281/8/i=6/a=305

G.I. Budker, Accelerators with colliding particle beams. Sov. Phys. Uspekhi, **9**(4), 534–542 (1967). URL http://iopscience.iop.org/article/10.1070/PU1967v009n04ABEH003010

G.I. Budker and A.A. Naumov, Studies of colliding electron-electron, positron-electron and proton-proton beams. In *Proceedings, 4th International Conference on High-Energy Accelerators, HEACC 1963, v. 1–3: Dubna, USSR, August 21 - August 27 1963*, ed. by A.A. Kolomenskij and A.B. Kuznetsov, pp. 334–363, (Oak Ridge, TN, 1965. NTIS). URL http://inspirehep.net/record/918675

G.I. Budker, N.A. Kushnirenko, A.A. Naumov, A.P. Onuchin, S.G. Popov, V.A. Sidorov, A.N. Skrinskii, and G.M. Tumaikin, Accelerator development in novosibirsk: Ii. status of work on the vep -1 electron storage machine. J. Nucl. Energy. Plasma Phys. Accel. Thermonuclear Res. **8**(6), 676–683 (1966). URL http://stacks.iop.org/0368-3281/8/i=6/a=306

U. Busch, W.W. Bautz, The development of betatrons in the siemens centers of berlin and erlangen. Z. Med. Phys. **15**, 87–100 (2005)

C.C. Butler, Recollections of patrick blackett 1945–1970. Notes Rec. R. Soc. Lond. **53**(1), 143–156 (1999, 2020). URL www.jstor.org/stable/531934

N. Cabibbo, e^+e^- Physics—a View from Frascati in 1960's. In *Adone a Milestone on the Particle Way*, ed. by V. Valente, Frascati Physics Series. (INFN Frascati National Laboratories, 1997), p. 219

N. Cabibbo, *Fisici Italiani del tempo presente. Storie di vita e di pensiero. Interview with Nicola Cabibbo*. (Marsilio, 2003), PP. 45–66

N. Cabibbo, R. Gatto, Pion form factors from possible high-energy electron-positron experiments. Phys. Rev. Lett. **4**, 313–314 (1960). https://doi.org/10.1103/PhysRevLett.4.313

N. Cabibbo, R. Gatto, Electron positron colliding beam experiments. Phys. Rev. **124**, 1577–1595 (1961). https://doi.org/10.1103/PhysRev.124.1577

P. Camiz, *Un anno a Rovere (1943–1944)* (Ginevra Bentivoglio Editori A, Editore, 2018)

R. Casalbuoni and D. Dominici, The teacher of the gattini (kittens), 10, 2018. URL https://arxiv.org/abs/1810.06413

D. Cassidy, *Uncertainty* (Freeman, The Life and Science of Werner Heisenberg. W. H, 1993)

D. Cassidy, Controlling German science, I.U.S. and allied forces in Germany, 1945–1947. Hist. Stud. Phys. Biolog. Sci. **24**(2), 197–235 (1994). https://doi.org/10.2307/27757723. URL http://hsns.ucpress.edu/content/24/2/197

D. Cassidy, Controlling German science, II. Bizonal occupation and the struggle over west German science policy, 1946–1949. Hist. Stud. Phys. Biol. Sci. **26**(2), 197–239 (1996). https://doi.org/10.2307/27757762. URL http://www.jstor.org/stable/27757762

D. Cassidy, Farm Hall and the German Atomic Project of World War II. (Springer International Publishing, 2017). https://doi.org/10.1007/978-3-319-59578-8

R. Cavendish, Hitler and Mussolini meet in Rome. Hist. Today, **58** (2008). URL https://www.historytoday.com/archive/hitler-and-mussolini-meet-rome

M.B. Ceolin, The discreet charm of the nuclear emulsion era. Ann. Rev. Nucl. Part. Sci. **52**(1), 1–21 (2002). https://doi.org/10.1146/annurev.nucl.52.050102.090730

J. Chadwick, The existence of a neutron. Proc. R. Soc. Lond. Ser. **136**(830), 692–708 (1932). https://doi.org/10.1098/rspa.1932.0112

P. Chadwick. I.N. Sneddon, O.B.E. 8 December 1919 – 4 November 2000. Biogr. Mem. Fellows R. Soc. **48**, 417–437 (2002). URL www.jstor.org/stable/3650270

S. Chatrchyan et al., Observation of a New Boson at a Mass of 125 GeV with the CMS Experiment at the LHC. Phys. Lett. B **716**, 30–61 (2012). https://doi.org/10.1016/j.physletb.2012.08.021

J.H. Christenson, G.S. Hicks, L.M. Lederman, P.J. Limon, B.G. Pope, E. Zavattini, Observation of massive muon pairs in hadron collisions. Phys. Rev. Lett. **25**, 1523–1526 (1970). https://doi.org/10.1103/PhysRevLett.25.1523

J. Christopher, *The Race for Hitler's X-Planes: Britain's 1945 Mission to Capture Secret Luftwaffe Technology. Britain's 1945 Mission to Capture Secret Luftwaffe Technology*. (Spellmount, 2013)

M. Cini and B. Touschek, The relativistic limit of the theory of spin 1/2 particles. Il Nuovo Cimento, **7**(3), 422–423 (1958). URL https://doi.org/10.1007/BF02747708

M. Cini, G. Morpurgo, B. Touschek, A non-perturbation treatment of scattering and the "Wentzel-Example". Il Nuovo Cimento **11**(3), 316–317 (1954). https://doi.org/10.1007/BF02781399

F. Close, *Half-Life: The Divided Life of Bruno Pontecorvo* (Physicist or Spy. Basic Books, New York, 2015)

F. Close, *Antimatter*. (Oxford Landmark Science. Oxford University Press, 2018). ISBN 978-0-19-883191-4

F. Close, *Trinity* (The Treachery and Pursuit of the Most Dangerous Spy in History, Penguin Random House UK, 2019)

J.D. Cockcroft, E.T.S. Walton, Disintegration of Lithium by Swift Protons. Nature **129**(3261), 649 (1932). https://doi.org/10.1038/129649a0

G. Consolmagno, *The Heavens Proclaim. Astronomy and the Vatican*. LEV Libreria Editrice Vaticana (Vatican Publishing House, 2009). URL https://www.vaticanobservatory.org/education/the-heavens-proclaim-astronomy-and-the-vatican/

M. Conversi, e^+e^- Physics. In *In *Pisa 1976, Proceedings, Frontier Problems In High Energy Physics*, 65–90 and Preprint - CONVERSI, M. (76,REC.OCT.) 28p*, 1976a

M. Conversi, e^+e^- Physics, **9** 1976b

M. Conversi, From the Discovery of the Mesotron to that of its Leptonic Nature, in *40 Years of Particle Physics*. ed. by B. Foster, P.H. Fowler. *Proceedings of the International Conference to Celebrate the 40th Anniversary of the Discoveries of the pi- and V- particles, held at the University of Bristol, 22–24 July 1987*. (Adam Hilger, Bristol and Piladelphia, 1988), pp. 1–20

M. Conversi and O. Piccioni, Sulla disintegrazione dei mesoni lenti. Il Nuovo Cimento (1943–1954), **2**(1), 71 (1944a). https://doi.org/10.1007/BF02903046

M. Conversi and O. Piccioni, Misura diretta della vita media dei mesoni frenati. Nuovo Cimento **2**, 40–70 (1944b). URL https://doi.org/10.1007/BF02903045

M. Conversi and O. Piccioni, On the mean life of slow mesons. Phys. Rev. **70**(11–12), 859–873 (1946). URL https://doi.org/10.1103/PhysRev.70.859

M. Conversi, E. Pancini, O. Piccioni, On the decay process of positive and negative mesons. Phys. Rev. **68**(9–10), 232 (1945) 10.1103/PhysRev.68.232. URL https://link.aps.org/doi/10.1103/PhysRev.68.232

M. Conversi, E. Pancini, and O. Piccioni, On the disintegration of negative mesons. Phys. Rev. **71**(3), 209–210 (1947). https://doi.org/10.1103/PhysRev.71.209. URL https://link.aps.org/doi/10.1103/PhysRev.71.209

G. Corazza, *Maestri e allievi nella fisica italiana del Novecento*, Chap. 8, pp. 259–286. (La Goliardica Pavese, 2008)

B. Cork, G.R. Lambertson, O. Piccioni, W.A. Wenzel, Antineutrons produced from antiprotons in charge-exchange collisions. Phys. Rev. **104**, 1193–1197 (1956)

E. Crémieu-Alcan, P. Falk-Vairant, and O. Lebey, (ed.), *Proceedings, Conférence Internationale d'Aix-en-Provence sur les Particules Elémentaires*, vol. 2, (Gif-sur-Yvette, France, 1962). CEN, CEN. URL http://www-spires.fnal.gov/spires/find/books/www?cl=QC721.IN83::1961

I. Curie, F. Joliot, Un nouveau type de radioactivité. Comptes rendus hebdomadaries des sances de lAcadmie des Sciences **198**, 254–256 (1934)

S.C. Curran, Philip Ivor Dee. 8 April 1904 – 17 April 1983. Biogr. Mem. Fellows R. Soc. **30**, 140–166 (1984). URL https://www.jstor.org/stable/769823

S.C. Curran, J. Angus, A.L. Cockcroft, Beta spectrum of tritium. Nature **162**(4112), 302–303 (1948). https://doi.org/10.1038/162302a0

P.F. Dahl, Rolf Wideröe: Progenitor of Particle Accelerators. Technical Report SSCL-SR-1186, Superconducting Super Collider Laboratory, Dallas, Texas, March 1992. URL http://lss.fnal.gov/archive/other/ssc/sscl-sr-1186.pdf

E. De Waal, *The Hare with Amber Eyes: a Hidden Inheritance* (Farrar, Straus and Giroux, New York, 2012)

C. Di Castro, L. Bonolis, The beginnings of theoretical condensed matter physics in rome: a personal remembrance. EPJH **39**, 3–36 (2014). https://doi.org/10.1140/epjh/e2013-40043-5

D. Dickson, Germany's 75 years of free enterprise science. Science, **234**(4778), 811–812 (1986). ISSN 0036-8075, 1095–9203. https://doi.org/10.1126/science.234.4778.811. URL http://science.sciencemag.org/content/234/4778/811

C. Eberle, A.C. Grayling, among the dead cities: The history and moral legacy of the WWII Bombing of civilians in Germany and Japan. Ethics, **117**(2), 356–363 (2007). https://doi.org/10.1086/510700

M. Eckert, *Arnold Sommerfeld: Science, Life and Turbulent Times (1868–1951* (Springer, New York, 2013). ISBN 978-1-4614-7461-6

M. Eckert, Sommerfeld's *Atombau und Spektrallinien*. In *A History of Quantum Physics through its Textbooks*, ed. by M. Badino and J. Navarro, (2013b). URL https://www.mprl-series.mpg.de/studies/2/7/index.html

R.J. Eden, The analytic behaviour of Heisenberg's S matrix. Nature **165**(4185), 54–56 (1949). https://doi.org/10.1038/165054a0

F.R. Elder, A.M. Gurewitsch, R.V. Langmuir, and H.C. Pollock, A 70 Mev synchrotron. J. Appl. Phys. **18**(9), 810–818 (1947). 2020/03/05. https://doi.org/10.1063/1.1697845

R. Elliott and J.H. Sanders, Maurice henry lecorney pryce. 24 January 1913 - 24 July 2003. Biogr. Mem. Fellows R. Soc. **51**, 355–366 (2005). https://doi.org/10.1098/rsbm.2005.0023

K.E. Eriksson, On radiative corrections due to soft photons. Il Nuovo Cimento (1955–1965), **19**(5), 1010–1028 (1961). URL https://doi.org/10.1007/BF02731243

S. Esposito, Ettore Majorana, Unveiled Genius and Endless Mysteries. (Springer International Publishing AG, 2017). https://doi.org/10.1007/978-3-319-54319-2

G. Etim, G. Pancheri, and B. Touschek, The infra-red radiative corrections for colliding beam (electrons and positrons) experiments. Il Nuovo Cimento B, **51**(2), 276–302 (1967a). URL http://inspirehep.net/record/1940376

G.E. Etim and B. Touschek, A proposal for the administration of radiative corrections, 1966. URL http://www.lnf.infn.it/sis/preprint/detail-new.php?id=2655

G.E. Etim, G. Pancheri, B. Touschek, The infra-red radiative corrections to colliding beam (electrons and positrons) experiments. Nuovo Cim. B **51**, 276–302 (1967)

E. Fabri, B.F. Touschek, La vita media del mesone τ. Il Nuovo Cimento **11**(1), 96–97 (1954). https://doi.org/10.1007/BF02780875

U. Fentsahm, Der "Evakuierungsmarsch" von Hamburg-Fuhlsbüttel nach Kiel-Hassee (12.–15. April 1945). Informationen zur Schleswig-Holsteinischen Zeitgeschichte, **44**, 66–105 (2004)

E. Ferlenghi, C. f Pellegrini, and B. Touschek, The transverse resistive wall instability of extremely relativistic beams of electrons and positrons. Il Nuovo Cimento B, **44B** (1966)

E. Fermi, E. Amaldi, B. Pontecorvo, F. Rasetti, and E. Segre', Azione di Sostanze Idrogenate sulla Radioattivita' Provocata da Neutroni. La Ricerca Scientifica, **5** (1934)

E. Fermi, E. Teller, V. Weisskopf, The decay of negative mesotrons in matter. Phys. Rev. **71**(5), 314–315 (1947). https://doi.org/10.1103/PhysRev.71.314

B. Ferretti, The absorption of slow mesons by an atomic nucleus. In *Report of an International Conference on Fundamental Particles and Low Temperatures, held at the Cavendish Laboratory, Cambridge, on 22 /27 July 1946*, vol. 1. (Fundamental Particles, London, 1947), pp. 75–77

B. Ferretti, Sulla cattura atomica dei mesoni lenti. Il Nuovo Cimento (1943–1954), **5**(4), 325–365 (1948). https://doi.org/10.1007/BF02784466

B. Ferretti, Sulla diagonalizzazione della hamiltoniana nella teoria dei campi d'onda e sulla teoria dei sistemi chiusi. Il Nuovo Cimento (1943–1954), **8**(2), 108–131 (1951). https://doi.org/10.1007/BF02773048

B. Ferretti, R. Peierls, Radiation damping theory and the propagation of light. Nature **160**(4068), 531–532 (1947). https://doi.org/10.1038/160531a0

R.P. Feynman, R.R., Leighton, and M. Sands, *The Feynman Lectures on Physics*. (California Institute of Technology, 1964–1966). URL https://www.feynmanlectures.caltech.edu

R.P. Feynman, Conclusions of the conference. In *Proceedings, Conférence Internationale d'Aix-en-Provence sur les Particules Elémentaires: Aix-en-Provence, France, Sept 14–20, 1961*, vol. 2, ed. by E. Crémeiu-Alcan, P. Falk-Vairant, and O. Lebey, (Gif-sur-Yvette, France, 1962), pp. 205–210. CEN. URL http://inspirehep.net/record/1377883/files/Pages_from_C61-09-14_205.pdf

R.P. Feynman and R. Leighton, *Surely You are Joking, Mr. Feynman! (Adventures of a Curious Character*. (Norton & Co., New York, 1985)

W. Fletcher, Sir Samuel Crowe Curran. 23 May 1912–25 February 1998. Biogr. Mem. Fellows R. Soc. **45**, 95–109 (1999). https://doi.org/10.1098/rsbm.1999.0008. URL https://royalsocietypublishing.org/doi/abs/10.1098/rsbm.1999.0008

B. Ford, *Secret Weapons: Death Rays* (Bloomsbury Publishing, Doodlebugs and Churchills Golden Goose, 2013)

C. Frank (ed.), *Operazione Epsilon* (Le trascrizioni di Farm Hall. Selene Edizioni, Milano, 1994)

G. Fraser, *The Quark Machines: How Europe Fought the Particle Physics War*. Institute of Physics Pub., 1997. ISBN 9780367806552. URL https://books.google.it/books?id=BAZCzQEACAAJ

M. Frayn, R. Butler, *Copenhagen* (Bloomsbury Publishing, Student Editions, 2017)

V.J. Frenkel, Professor Friedrich Houtermans—Arbeit, Leben, Schicksal. Biographie eines Physikers des zwanzigsten Jahrhunderts. Preprint 414, Max Planck Institute for the History of Science, Berlin, 2011. URL https://www.mpiwg-berlin.mpg.de/sites/default/files/Preprints/P414.pdf

D.W. Fry, J.W. Gallop, F.K. Goward, J. Dain, 30-Mev. Electron synchrotron. Nature **161**(4092), 504–506 (1948). https://doi.org/10.1038/161504a0

L. Gariboldi, Constance Charlotte Dilworth. Il Nuovo Saggiatore, **20**, 16–21 (2004). URL https://www.ilnuovosaggiatore.sif.it/issue/40

R. Gatto, Memories of Bruno Touschek. In *Bruno Touschek Memorial Lectures*, vol. 33 of *Frascati Physics Series*, ed. by M. Greco and G. Pancheri, (INFN-Laboratori Nazionali di Frascati, 2004), pp. 69–75. URL http://www.lnf.infn.it/sis/frascatiseries/Volume33/volume33.pdf

G. Ghigo, Discussioni Preliminari sull'A.d.A., 1960. URL http://www.lnf.infn.it/sis/preprint/detail-new.php?id=3189

J. Gimbel, Deutsche Wissenschaftler in Britischem Gewahrsam. Ein Erfahrungsbericht aus dem Jahre 1946 über das Lager Wimbledon. In *Vierteljahrshefte für Zeitgeschichte*, vol. 3, ed. by K. Bracher and H.-P. Schwarz, (R. Oldenbourg Verlag, München, 1990a), pp. 459–483. URL https://www.ifz-muenchen.de/heftarchiv/1990_3.pdf

J. Gimbel, *Science, Technology and Reparations. Exploitation and Plunder in Postwar Germany*. (Stanford University Press, Stanford, CA, 1990b). ISBN 978-0-8047-1761-8

G. Giovanni Battimelli, F. Buccella, and P. Napolitano, Raoul Gatto, a great Italian scientist and teacher in theoretical elementary particle physics. pp. 145–169, December 2019. https://doi.org/10.1393/qsf/i2019-10065-7

S.L. Glashow, J. Iliopoulos, L. Maiani, Weak interactions with lepton-hadron symmetry. Phys. Rev. D **2**, 1285–1292 (1970). https://doi.org/10.1103/PhysRevD.2.1285

M. Goeppert-Mayer, Double Beta-disintegration. Phys. Rev. **48**, 512–516 (1935). https://doi.org/10.1103/PhysRev.48.512.URL https://link.aps.org/doi/10.1103/PhysRev.48.512

M. Goeppert Mayer, On closed shells in Nuclei. Phys. Rev. **74**, 235–239 (1948). https://doi.org/10.1103/PhysRev.74.235. URL https://link.aps.org/doi/10.1103/PhysRev.74.235

M. Goeppert Mayer, On closed shells in Nuclei. II. Phys. Rev. **75**(12): 1969–1970 (1949). https://doi.org/10.1103/PhysRev.75.1969. URL https://link.aps.org/doi/10.1103/PhysRev.75.1969

K. Gottstein and A. Chao, Heisenberg and Bohr—another view. Science, **295**, 2211 (2002). https://doi.org/10.1126/science.295.5563.2211b

S.A. Goudsmit, *Alsos* (History of modern physics and astronomy. AIP Press, Woodbury, NY, 1996). 978-1-56396-415-2

F. Goward and D. Barnes, Experimental 8 Mev. synchrotron for electron acceleration. Nature, **158** (1946). URL https://doi.org/10.1038/158413a0

M. Grandolfo, F. Napolitani, S. Risica, and E. Tabet, vol. 12. (Istituto Superiore di Sanita', 2017)

A. Grayling, *Among the Dead Cities*. (Bloomsbury, London, 2006). URL https://doi.org/10.1086/510700

M. Greco, *Fisica e Avventure*. (Exhorma, 2018). URL http://www.exhormaedizioni.com

M. Greco, *Coherent States in Gauge Theories and Applications in Collider Physics*. (World Scientific, Singapore, 2019). https://doi.org/10.1142/9789811213908_fmatter

M. Greco, P. Di Vecchia, Double photon emission in e^+e^- collisions. Nuovo Cim. **50**, 319 (1967). https://doi.org/10.1007/BF02827740

M. Greco and G. Pancheri, *1987 Bruno Touschek Memorial Lectures*, vol. XXXIII of *Frascati Physics Series*, (Frascati, 2004), ed. by INFN Frascati National Laboratories. URL http://www.lnf.infn.it/sis/frascatiseries/Volume33/volume33.pdf

M. Greco, G. Rossi, A note on the infrared divergence. Nuovo Cim. **50**, 168 (1967). https://doi.org/10.1007/BF02820731

M. Greco, G. Pancheri-Srivastava, Y. Srivastava, Radiative corrections for colliding beam resonances. Nucl. Phys. B **101**, 234–262 (1975). https://doi.org/10.1016/0550-3213(75)90304-1

M. Greco, F. Palumbo, G. Pancheri-Srivastava, Y. Srivastava, Coherent state approach to the infrared behavior of nonabelian Gauge theories. Phys. Lett. **77B**, 282–286 (1978). https://doi.org/10.1016/0370-2693(78)90707-4

O. Greenberg, Discovery of the color degree of freedom in particle physics: A personal perspective. Particle Physics on the Eve of LHC (2009). https://doi.org/10.1142/9789812837592_0002

F. Guerra and N. Robotti, The beginning of a great adventure: Bruno Pontecorvo in Rome and Paris. Nuovo Cimento C, **37**(5), 39–53 (2014). URL https://en.sif.it/journals/sif/ncc/econtents/2014/037/05/article/8

J.C. Gunn and H.S.W. Massey, Interaction of mesons with a potential field. Proc. R. Soc. Lond. Ser. A. Math. Phys. Sci. **193**(1035), 559–579 (1948). https://doi.org/10.1098/rspa.1948.0062. URL https://royalsocietypublishing.org/doi/abs/10.1098/rspa.1948.0062

K. Hahn, W. Kleen, and A. Woeldike, HEINRICH HÖRLEIN/GÜNTHER JOBST 60 Jahre/HERMANN SCHÜLER 60 Jahre. Physikalische Blätter, **10**, 320–322 (1954). URL https://onlinelibrary.wiley.com/doi/10.1002/phbl.19540100705

J. Haïssinski, From AdA to ACO. Reminiscences of Bruno Touschek. In *Bruno Touschek and the birth of e^+e^- physics*, volume XIII of *Frascati Physics Series*, ed. by G. Isidori, (INFN Laboratori Nazionali di Frascati, 1998), pp. 17–31

C. Hall, *British Exploitation of German Science and Technology, 1943–1949* (Routledge, New York, 2019)

O. Haxel, J.H.D. Jensen, and H.E. Suess, On the "magic numbers" in nuclear structure. Phys. Rev. **75**(11), 1766 (1949). https://doi.org/10.1103/PhysRev.75.1766.2. URL https://link.aps.org/doi/10.1103/PhysRev.75.1766.2

J. Hecht, *Lasers, Death Rays, and the Long, Strange Quest for the Ultimate Weapon.* (Prometheus Books, 2019). 1633884600. URL https://cds.cern.ch/record/2316355

J.L. Heilbron and R.W. Seidel, *Lawrence and His Laboratory Lawrence and His Laboratory: A History of the Lawrence Berkeley Laboratory*, vol. I. (University of California Press, 1990)

T. Heinze, O. Hallonsten, and S. Heinecke, From periphery to center: Synchrotron radiation at desy, part i: 1962–1977. Hist. Stud. Nat. Sci. **45**(3), 447–492 (2015). ISSN 19391811, 1939182X. URL https://www.jstor.org/stable/10.1525/hsns.2015.45.3.447

W. Heisenberg, Die "beobachtbaren Größen" in der Theorie der Elementarteilchen. Zeitschrift fü Physik **120**(7), 513–538 (1943). https://doi.org/10.1007/BF01329800

W. Heisenberg, Die beobachtbaren Größen in der Theorie der Elementarteilchen. II. Zeitschrift für Physik **120**, 673–702 (1943). https://doi.org/10.1007/BF01336936

W. Heisenberg, Die beobachtbaren Größen in der Theorie der Elementarteilchen. III. Zeitschrift für Physik **123**(1), 93–112 (1944). https://doi.org/10.1007/BF01375146

W. Heisenberg, Mesonenerzeugung als Stosswellenproblem. Z. Phys. **133**, 65 (1952). https://doi.org/10.1007/BF01948683

W. Heisenberg, Professor Max Born. Nature **225**(5233), 669–671 (1970). https://doi.org/10.1038/225669a0

W. Heisenberg, *Physics and Beyond: Encounters and Conversations* (Harper and Row, New York, 1971)

W. Heisenberg, *Gesammelte Werke. Collected Works.* (Piper & Co. Verlag and Springer-Verlag, 1984–1989)

W. Heisenberg and E. Heisenberg, *My Dear Li: Correspondence, 1937–1946.* (Yale University Press, 2016)

W. Heitler, The quantum theory of damping as a proposal for Heisenberg's S-matrix. In *Report on an International Conference on Fundamental Particles and Low Temperatures, held at the Cavendish Laboratory, Cambridge, on 22–27 July 1946*, vol. 1. ed. by T.P. Society, (Fundamental Particles, Taylor & Francis, 1947), pp. 189–194

W. Heitler, *The Quantum Theory of Radiation* (Dover, New York, Third edition, 1984)

K. Hentschel, G. Rammer, Physicists at the University of Göttingen, 1945–1955. Phys. Perspect. **3**(2), 189–209 (2001). https://doi.org/10.1007/PL00000529

A. Hermann, L. Belloni, J. Krige, U. Mersits, and D. Pestre, *History of CERN. Launching the European organization for nuclear research*, vol. 1. (North-Holland, 1987)

P. Higgs, My Life as a Boson. 2010

R. Hofstadter, On nucleon structure. In *9th International Annual Conference on High Energy Physics*, 1960

R.P. Hopkins, The historiography of the allied bombing campaign of Germany. In *Electronic Theses and Dissertations. Paper 2003*, 2008. URL https://dc.etsu.edu/etd/2003

N. Hu, On the application of Heisenberg's theory of S-Matrix to the problems of resonance scattering and reactions in nuclear physics. Phys. Rev. **74**(2),131–140 (1948). https://doi.org/10.1103/PhysRev.74.131. URL https://link.aps.org/doi/10.1103/PhysRev.74.131

K. Hübner, The CERN intersecting storage rings (ISR). Europ. Phys. J. H **36**(4), 509–522 (2012). https://doi.org/10.1140/epjh/e2011-20058-8

G. Ising, Prinzip einer Methode zur Herstellung von Kanalstrahlen hoher Voltzahl. Arkiv fr Matematik, Astronomi och Fysik, **18** (1924)

G. Ising, Högspänningsmetoder för atomsprängning. *Kosmos, Annuary of the Swedish Phys. Ass.*, 1933

D. Iwanenko, I. Pomeranchuk, On the maximal energy attainable in a Betatron. Phys. Rev. **65**, 343 (1944). https://doi.org/10.1103/PhysRev.65.343

A. Jacobsen, *Operation Paperclip* (The Secret Intelligence Program to Bring Nazi Scientists to America. Little, Brown and Company, New York, NY, 2014). 978-0-316-22104-7 978-0-316-23982-0

J.M. Jauch and F. Rohrlich, *The Theory of Photons and Electrons*. (Addison-Wesley Educational Publishers Inc, 1955a). ISBN ISBN-13: 978-0201033007

J.M. Jauch and F. Rohrlich, *Theory of Photons and Electrons*. (Addison-Wesley Educational Publishers Inc, 1955b)

F. Joliot, I. Curie, Artificial production of a new kind of radio-element. Nature **133**, 201–202 (1934)

L.W. Jones, Physics to be done with colliding beams. In *Proceedings, 1963 Summer Study on Storage Rings, Accelerators and Experimentation at Super-High Energies*, vol. C63-06-10, ed. by J. Bittner, (1963), pp. 253–276. URL http://inspirehep.net/record/48297/files/C630610-p253.PDF

D. Kaiser, *Drawing Theories Apart: The Dispersion of Feynman Diagrams in Postwar Physics* (The University of Chicago Press, Chicago and London, 2005)

H.F. Kaiser, European electron induction accelerators. J. Appl. Phys. **18**(1), 1–18 (1947). https://doi.org/10.1063/1.1697549

N. Kemmer and R. Schlapp, Max Born, 1882–1970. In *Biographical Memoirs of Fellows of the Royal Society*, vol. 17, (Royal Society, 1971), pp. 17–52. URL http://doi.org/10.1098/rsbm.1971.0002

D.W. Kerst, Acceleration of electrons by magnetic induction. Phys. Rev. **58**, 841 (1940). https://doi.org/10.1103/PhysRev.58.841

D.W. Kerst, The acceleration of electrons by magnetic induction. Phys. Rev. **60**, 47–53 (1941). https://doi.org/10.1103/PhysRev.60.47

D.W. Kerst, R. Serber, Electronic orbits in the induction accelerator. Phys. Rev. **60**, 53–58 (1941) https://doi.org/10.1103/PhysRev.60.53. URL https://link.aps.org/doi/10.1103/PhysRev.60.53

D.W. Kerst, F.T. Cole, H.R. Crane, L.W. Jones, L.J. Laslett, T. Ohkawa, A.M. Sessler, K.R. Symon, K.M. Terwilliger, N.V. Nilsen, Attainment of very high-energy by means of intersecting beams of particles. Phys. Rev. **102**, 590–591 (1956). https://doi.org/10.1103/PhysRev.102.590. [115(1997)]

P. Kessler, La méthode des processus quasi réels en physique des hautes énérgies. In *1st Aix en Provence International Conference on Elementary Particles*, vol. 1, (1962) pp. 191–196

P. Kienle, In memory of Bjørn Wiik. Nucl. Phys. News **9**(2), 31–33 (1999). https://doi.org/10.1080/10506899909411122

O. Klein, Mesons and Nucleons. Nature **161**(4101), 897–899 (1948). https://doi.org/10.1038/161897a0

R. Kollath, G. Schumann, Untersuchungen an einem 15-MV-Betatron. Zeitschrift für Naturforschung A **2**(11–12), 634–642 (1947). https://doi.org/10.1515/zna-1947-11-1205

A.A. Kolomenskij, K.A.B., and A.N. Lebedev, (ed.), *Proceedings, 4th International Conference on High-Energy Accelerators, HEACC 1963, v.1–3 : Dubna, USSR, August 21 - August 27 1963*, (Oak Ridge, TN, 1965). NTIS, NTIS. URL http://inspirehep.net/record/19356

J. Krige, The installation of high-energy accelerators in Britain after the War. Big Equipment but not "Big Science," in *The Restructuring of Physical Sciences in Europe and the United States*. ed. by M. De Maria, M. Grilli, F. Sebastiani, 1945–1960. (World Scientific, Singapore, 1989), pp. 488–497

J. Krige, *American Hegemony and the Postwar Reconstruction of Science in Europe* (MIT Press, Cambridge, MA, 2006)

L.J. Laslett, V.K. Neil, and A.M. Sessler, Transverse resistive instabilities of intense coasting beams in particle accelerators. Rev. Sci. Instr. **36** (1965). https://doi.org/10.1063/1.1719595

C.M.G. Lattes, H. Muirhead, G.P.S. Occhialini, and C.F. Powell, Processes involving charged mesons. Nature, **159**(4047), 694–697, (1947). URL https://doi.org/10.1038/159694a0

E.O. Lawrence and M. Livingston, The production of high speed protons without the use of high voltages. Phys. Rev. **38**, 834 (1931). URL http://prola.aps.org/abstract/PR/v40/i1/p19_1

E.O. Lawrence and M. Livingston, The production of high speed light ions without the use of high voltages. Phys. Rev. **40**, 19–35 (1932). URL http://prola.aps.org/abstract/PR/v40/i1/p19_1

J. D. Lawson, Early Synchrotrons in Britain, and Early Work for CERN. Technical Report, CERN, 1997. URL http://cds.cern.ch/record/340513

L. Ledermann, Life in physics and the crucial sense of wonder. CERN Courier, **49**(8), 22–24 (2009) URL https://cds.cern.ch/record/1734431

S. Lee, *Sir Rudolf Peierls. Selected Private and Scientific Correspondence. Vol. 1–2.* (World Scientific, Singapore, 2007). https://doi.org/10.1142/5941

S. Lee, (ed.), *Sir Rudolf Peierls. Selected Private and Scientific Correspondence*, Vol. 2. (World Scientific, Singapore, 2009)

T.D. Lee and C. Yang, Question of parity conservation in weak interactions. Phys. Rev. **104**(1), 254–258 (1956). URL https://link.aps.org/doi/10.1103/PhysRev.104.254

E. Levichev, A.N. Skrinsky, G.M. Tumaikin, and Y.M. Shatunov, Electron–Positron beam collision studies at the Budker Institute of Nuclear Physics. Sov. Phys.-Uspekhi, **61**(5), 405–423 (2018). URL http://stacks.iop.org/1063-7869/61/i=5/a=405

C. Llewellyn Smith, Genesis of the Large Hadron Collider. Philos. Trans. R. Soc. Lond.: Math. Phys. Eng. Sci. **373** (2032). 20140037 (2015). URL https://doi.org/10.1098/rsta.2014.0037

E.L. Lomon, The joining of infra-red and ultra-violet calculations. Nucl. Phys. **1**(2), 101 – 111 (1956). ISSN 0029-5582. https://doi.org/10.1016/0029-5582(56)90061-X. URL http://www.sciencedirect.com/science/article/pii/002955825690061X

E.L. Lomon, Radiative corrections for nearly elastic scattering. Phys. Rev. **113**, 726–727 (1959) https://doi.org/10.1103/PhysRev.113.726.URL https://link.aps.org/doi/10.1103/PhysRev.113.726

S. Longden, *T-Force. The forgotten Heroes of 1945*. (Constable and Robinson, London, 2009)

B.A.C. Lovell, Patrick Maynard Stuart Blackett, Baron Blackett, of Chelsea, 18 November 1897-13 July 1974. Biogr. Mem. Fellows Roy. Soc. **21**, 1–115 (1975). https://doi.org/10.1098/rsbm.1975.0001

G. Luders, On the equivalence of invariance under time reversal and under particle-antiparticle conjugation for relativistic field theories. Kong. Dan. Vid. Sel. Mat. Fys. Med. **28N5**(5), 1–17 (1954)

M. Mafai, Il lungo freddo: storia di Bruno Pontecorvo, lo scienziato che scelse l'URSS. (Arnoldo Mondadori Editori, Milan, 1992)

L. Maiani, L. Bonolis, The charm of theoretical physics (1958–1993). Eur. Phys. J. **H42**(4–5), 611–661 (2017). https://doi.org/10.1140/epjh/e2017-80040-9

L. Maiani and L. Bonolis, The LHC timeline: a personal recollection (1980–2012). Europ. Phys. J. H **42**(4–5), 475–505 (2017b). URL https://doi.org/10.1140/epjh/e2017-80052-8

E. Maiorana, Teoria simmetrica dell'elettrone e del positrone. Il Nuovo Cimento, **14** (1937). URL https://doi.org/10.1007/BF02961314

G. Margaritondo, A lezione da bruno touschek. Quaderni di Storia della Fisica, **29**, 2021

P. Marin, L'Anneau de collisions a électrons et positrons d'Orsay. Le Journal de Physique. Colloques, **27**C4(11–12), C4–54 (1966). URL https://doi.org/10.1051/jphyscol:1966405

P. Marin, *Un demi-siècle d'accélérateurs de particules* (Éditions du Dauphin, Paris, 2009)

E.M. McMillan, The synchrotron—a proposed high energy particle accelerator. Phys. Rev. **68**(5–6), 143–144 (1945). https://doi.org/10.1103/PhysRev.68.143. URL https://link.aps.org/doi/10.1103/PhysRev.68.143

M.A. McPartland, *The Farm Hall Scientists. The United States, Britain, and Germany in the New Atomic Age, 1945–1946*. Dissertation, The George Washington University, Washington, DC, 2013. URL https://search.proquest.com/pqdtglobal/docview/1436978271/abstract/DDC811FAFC464047PQ/2

C. Mencuccini, $e^+e^-e^+e^-$ Colliding Beam Experiment, 1974. URL http://www.lnf.infn.it/sis/preprint/detail-new.php?id=2023

U. Mersits, From cosmic-ray and nuclear physics to high-energy physics, in *History of CERN*. ed. by A. Hermann, J. Krige, U. Mersits, D. Pestre. Launching the European Organization for Nuclear Research, vol. 1. (North-Holland, Amsterdam, 1987), pp. 3–52

W. Moore, *Schrödinger: life and thought*. (Cambridge University Press, 1989)

G. Morpurgo, B.F. Touschek, Remarks on the validity of the tamm-dancoff method. Il Nuovo Cimento **10**(12), 1681–1694 (1953). https://doi.org/10.1007/BF02781663

G. Morpurgo, B. Touschek, L.A. Radicati, On time reversal. Il Nuovo Cimento **12**(5), 677–698 (1954). https://doi.org/10.1007/BF02781835

G. Morpurgo, L.A. Radicati, and B. Touschek, Time reversal in quantized theories. In *Proceedings of the 1954 Glasgow Conference on Nuclear and Meson Physics. International Union of Pure and Applied Physics. 13–17 July, 1954*, ed. by E. Bellamy and R. Moorhouse (Pergamon Press, London, New York, 1955)

I.F. Offenberger, *The Jews of Nazi Vienna, 1938–1945*. (Palgrave Macmillan, Cham, 2017). URL https://link.springer.com/book/10.10072F978-3-319-49358-9

G.K. O'Neill, Storage ring synchrotron: device for high energy physics research. Phys. Rev. **102**(5), 1418–1419 (1956). URL https://doi.org/10.1103/PhysRev.102.1418

G.K. O'Neill, Summary Of Visits To Frascati And Orsay. Technical Report PRINT-64-258, SLAC, 1962. URL http://inspirehep.net/record/1097443/export/hx

G.K. O'Neill, Storage rings. Science, **141**(3582), 679–686 (1963a). URL http://science.sciencemag.org/content/141/3582/679

G.K. O'Neill, Vertical instabilities in electron storage rings. Part II. Some experimental results from the stanford electron storage ring. In *Proceedings, 1963 Summer Study on Storage Rings, Accelerators and Experimentation at Super-High Energies*, vol. C63-06-10, ed. by J. Bittner, (1963b), pp. 375–378. URL http://inspirehep.net/record/48304/files/C630610-p375.PDF

G.K. O'Neill, Storage-ring work at stanford. In *Proceedings, 1963 Summer Study on Storage Rings, Accelerators and Experimentation at Super-High Energies*, vol. C63-06-10, ed. by J. Bittner, (1963c), pp. 209–227. http://inspirehep.net/record/48292/files/C630610-p209.PDF

G.K. O'Neill, Storage Rings. Conference CONF-46-33; AED-Conf-63-048-59, Princeton University, NJ (United States), 15 April 1963d. https://www.osti.gov/biblio/4875549-storage-rings

J. Oppenheimer, C. Yang, C. Wentzel, R. Marshak, R. Dalitz, M. Gell-Mann, M. Markov, and B. D'Espagnat, Theoretical Interpretation of New Particles. In *6th Annual Rochester Conference on High Energy Nuclear Physics*, (1956), pp. VIII.1–36

M. Osietzki, Kernphysikalische Grossgeräte wzischen naturwissenschaftlicher Forschung, Industrie und Politik: Zur Entwicklung der ersten deutschen Teilchenbeschleuniger bei Siemens 1935–1945. Technikgeschichte **55**, 25–46 (1988)

G. Pancheri, Infra-red radiative corrections for resonant processes. Nuovo Cim. A **60**, 321–329 (1969). https://doi.org/10.1007/BF02757350

G. Pancheri, Bruno Touschek and the Frascati theory group. In *Bruno Touschek Memorial Lectures*, vol. 33 of *Frascati Physics Series*, ed. by M. Greco and G. Pancheri, pp. 101–103. (INFN-Laboratori Nazionali di Frascati, 2004). http://www.lnf.infn.it/sis/frascatiseries/Volume33/volume33.pdf

G. Pancheri and L. Bonolis, Touschek with AdA in Orsay and the first direct observation of electron-positron collisions. *arXiv e-prints*, arXiv:1910.09075, 2018. https://arxiv.org/abs/1812.11847

G. Pancheri and L. Bonolis, Bruno Touschek in Glasgow. The making of a theoretical physicist. *arXiv e-prints*, 2020. https://arxiv.org/abs/2005.04942

G. Pancheri-Srivastava, Reggeization of the photon in quantum electrodynamics. Phys. Lett. **44B**, 109–111 (1973). https://doi.org/10.1016/0370-2693(73)90314-6

G. Pancheri-Srivastava, Y. Srivastava, Transverse-momentum distribution from Bloch-Nordsieck method. Phys. Rev. D **15**, 2915 (1977). https://doi.org/10.1103/PhysRevD.15.2915

W.K.H. Panofsky, Electromagnetic interaction and nucleon structure. In *Proceedings, 9th International Conference on High Energy Physics, v.1–2 (ICHEP59): Kiev, USSR, Jul 15–25, 1959*, vol. 1, ed. by A. of Science USSR, I. U. of Pure, and A. Physics, (Moscow, 1960), pp. 378–409, URL http://inspirehep.net/record/44195/files/c59-07-15-p378.pdf

G. Parisi, R. Petronzio, Small transverse momentum distributions in hard processes. Nucl. Phys. B **154**, 427–440 (1979). https://doi.org/10.1016/0550-3213(79)90040-3

B.T. Pash, *The Alsos Mission* (Award House, New York, NY, 1969)

W. Pauli, *Exclusion Principle, Lorentz Group and Reflection of Space-Time and Charge*, (McGraw-Hill, New York, 1955), pp. 30–51

R. Peierls, E. Segre, B. Rossi, E. Amaldi, R. Marshak, and E. Teller, Anti-nucleons. In *6th Annual Rochester Conference on High Energy Nuclear Physics*, (1956), pp. VII.–26

R.E. Peierls, *Birds of Passage* (Princeton University Press, Recollections of a Physicist, 1985)

H. Pender and S. Warren, *Electric Circuits and Fields*. (McGraw-Hill book Company, Incorporated, 1943). URL https://books.google.it/books?id=d2shAAAAMAAJ

O. Piccioni, The discovery of the leptonic property, in *Present Trends, Concepts and Instruments of Particle Physics*. ed. by G. Baroni, L. Maiani, G. Salvini. Symposium in Honour of Marcello Conversi's 70th Birthday (Roma, 3–4 November 1987), vol. 15. (Italian Physical Society, Bologna, 1988), pp. 171–193

M. Pinault, *Frédéric Joliot-Curie* (Odile Jacob, Paris, 2000)

W. Plessas, Paul Urban: promoter of modern theoretical physics. In *Few-Body Problems in Physics '95*, vol. 8. ed. by R. Guardiola, (Springer, Vienna, 1995). https://doi.org/10.1007/978-3-7091-9427-0_1

J.A. Poirier, D.M. Bernstein, J. Pine, Scattering of 200-mev positrons by electrons. Phys. Rev. **117**(2), 557–565 (1960) https://doi.org/10.1103/PhysRev.117.557.URL https://link.aps.org/doi/10.1103/PhysRev.117.557

B. Pontecorvo, Nuclear Capture of Mesons and the Meson Decay. Phys. Rev. **72**, 246–247 (1947) https://doi.org/10.1103/PhysRev.72.246. URL https://link.aps.org/doi/10.1103/PhysRev.72.246

B. Pontecorvo, *Enrico Fermi : ricordi di allievi ed amici*. (Edizioni Studio Tesi, 1993)

G. Puppi, Sui mesoni dei raggi cosmici. Il Nuovo Cimento **5**(6), 587–588 (1948). https://doi.org/10.1007/BF02780913

G. Putzolu, Radiative Corrections to Pion-Production in e^+e^- Collisions. Nuovo Cimento, **20** (1961)

L.A. Radicati, B. Touschek, On the equivalence theorem for the massless neutrino. Il Nuovo Cimento **5**(6), 1693–1699 (1957). https://doi.org/10.1007/BF02856061

M. Raggam-Blesch, Privileged under Nazi-Rule: The Fate of Three Intermarried Families in Vienna. J. Genocide Res. **21**, 378–397 (2019). URL https://doi.org/10.1080/14623528.2019.1634908

H. Rechenberg, The early S-matrix theory and its propagation (1942–1952). In *Pions to Quarks. Particle Physics in the 1950s*, ed. by L. Brown, M. Dresden, and L. Hoddeson, (Cambridge University Press, Cambridge, 1989), pp. 551–570

J. Rees, Colliding beam storage rings—a brief history. SLAC Beam Line S **9**, 1–8 (1986)

E. Regenstreif, (ed.), *CERN Symposium on High-Energy Accelerators and Pion Physics. Proceedings, 1st International Conference on High-Energy Accelerators, HEACC 1956, v.1–2*, (Geneva, 1956). CERN, CERN. URL http://inspirehep.net/record/19326

R.A. Ricci, Remembering Gilberto Bernardini. Europhys. News **26**, 117 (1995). URL https://www.europhysicsnews.org/articles/epn/pdf/1995/05/epn19952605p117.pdf

B. Richter, From the PSI to charm—the experiments of 1975 and 1976 (1976). URL https://www.nobelprize.org/prizes/physics/1976/richter/lecture/

B. Richter, Colliding beams at stanford. Machines and experiments. SLAC Beam Line, **1984N7**, 1–16 (1984)

B. Richter, The rise of colliding beams. In *3rd International Symposium on the History of Particle Physics: The Rise of the Standard Model*, (1992), pp. 261–284

B. Richter, The rise of colliding beams. In *The Rise of the Standard Model. Particle Physics in the 1960s and 1970s*, ed. by L. Hoddeson, L. Brown, M. Riordan, and M. Dresden, (Cambridge University Press, 1997), pp. 261–284

G.D. Rochester, C.C. Butler, Evidence for the existence of new unstable elementary particles. Nature **160**(4077), 855–857 (1947). https://doi.org/10.1038/160855a0

W. Rößler, *Eine kleine Nachtphysik. Geschichten aus der Physik.* (Birkäuser, Basel, 2007). URL https://doi.org/10.1007/978-3-7643-7744-1

L. Roth, Francesco Severi. J. Lond. Math. Soc. **s1–38**, 282–307 (1963). https://doi.org/10.1112/jlms/s1-38.1.282

C. Rubbia, Edoardo Amaldi, 5 September 1908 – 5 December 1989. Biogr. Mem. Fellows R. Soc. **37**, 1–31 (1991). https://doi.org/10.1098/rsbm.1991.0001. URL https://royalsocietypublishing.org/doi/abs/10.1098/rsbm.1991.0001

C. Rubbia, The role of Bruno Touschek in the realization of the proton antiproton collider. In *Bruno Touschek Memorial Lectures*, vol. 33 of *Frascati Physics Series*, ed. by M. Greco and G. Pancheri, (INFN-Laboratori Nazionali di Frascati, 2004), pp. 57–60. URL http://www.lnf.infn.it/sis/frascatiseries/Volume33/volume33.pdf

A. Russo, Science and industry in Italy between the two world wars. Hist. Stud. Phys. Biol. Sci. 281–320 (1986). https://doi.org/10.2307/27757567

A. Russo, The intersecting storage rings. In *The construction and operation of CERN's second large machine and a survey of its experimental programme*, ed. by J. Krige, *History of CERN*, vol. III, (Elsevier, 1996), pp. 97–170

G. Sacerdoti, Remembering Bruno Touschek. In *Bruno Touschek Memorial Lectures*, vol. 33 of *Frascati Physics Series*, ed. by M. Greco and G. Pancheri, (INFN-Laboratori Nazionali di Frascati, 2004), pp. 93–95. URL http://www.lnf.infn.it/sis/frascatiseries/Volume33/volume33.pdf

G. Salvini, *L'elettrosincrotrone e i Laboratori di Frascati.* (Nicola Zanichelli, 1962)

G. Salvini, From AdA to Tristan and LEP. In *Bruno Touschek Memorial Lectures*, vol. 33 of *Frascati Physics Series*, ed. by M. Greco and G. Pancheri, (INFN-Laboratori Nazionali di Frascati, 2004), pp. 61–68. URL http://www.lnf.infn.it/sis/frascatiseries/Volume33/volume33.pdf

G. Salvini, *L'UOMO UN INSIEME APERTO* (Mondadori Universitá, LA MIA VITA DA FISICO. Mondadori Education, 2010). ISBN:978-8861840317

M. Sands, *The Making of An Accelerator Physicist.* (Cambridge University Press, 1989)

M. Sands and B. Touschek, Alignment errors in the strong-focusing synchrotron. Il Nuovo Cimento, **10**(5), 604–613 (1953). URL https://doi.org/10.1007/BF02815285

G. Sardanashvily, In memoriam: Dmitri Ivanenko. (2016). URL https://arXiv.org/abs/1607.03828

S.S. Schweber, *QED and the Men Who Made It: Dyson* (Schwinger, and Tomonaga. Princeton University Press, Feynman, 1994)

J. Schwinger, On radiative corrections to electron scattering. Phys. Rev. **75**, 898–899 (1949) https://doi.org/10.1103/PhysRev.75.898. URL https://link.aps.org/doi/10.1103/PhysRev.75.898

J. Schwinger, The theory of quantized fields. i. Phys. Rev. **82**, 914–927 (1951). https://doi.org/10.1103/PhysRev.82.914. URL https://link.aps.org/doi/10.1103/PhysRev.82.914

E. Segrè, *A Mind Always in Motion. The Autobiography of Emilio Segrè*. University of California Press, Berkeley, 1993. URL https://publishing.cdlib.org/ucpressebooks/view?docId=ft700007rb;query=;brand=ucpress

W. Shirer, *Berlin Diary* (Knopf, Alfred A, 1941)

W. Shirer, *The Rise and Fall of the Third Reich. A history of Nazi Germany*. Simon & Schuster, 1950

A. Skrinsky, Accelerator field development at Novosibirsk (History, Status, Prospects). In *Proceedings of the 1995 Particle Accelerator Conference, 1–5 May 1995, Dallas, Texas, USA.*, vol. 1, (IEEE, 1995), pp. 14–16. https://doi.org/10.1109/PAC.1995.504558. URL http://ieeexplore.ieee.org/document/504558/

I.N. Sneddon and B.F. Touschek, The excitation of Nuclei by electrons. Proc. R. Soc. Lond. Math. Phys. Sci. **193**(1034), 344–356 (1948a). 2019/04/04. https://doi.org/10.1098/rspa.1948.0050

I.N. Sneddon, B.F. Touschek, A note on the calculation of the spacing of energy levels in a Heavy Nucleus. Math. Proc. Cambridge Philos. Soc. **44**(3), 391–403 (1948). https://doi.org/10.1017/S0305004100024397

I.N. Sneddon, B.F. Touschek, Nuclear models. Nature **161**(4080), 61 (1948). https://doi.org/10.1038/161061a0

I.N. Sneddon, B.F. Touschek, Production of mesons by electrons. Nature **163**(4144), 524 (1949). https://doi.org/10.1038/163524a0

A. Sørheim, *Besatt av en drøm* (Historien om Rolf Widerøe. Forlaget Historie & Kultur AS, Oslo, 2015)

A. Sørheim, *Obsessed by a dream. The Physicist Rolf Widerøe—a Giant in the History of Accelerators.* (Springer, Cham, 2020). https://doi.org/10.1007/978-3-030-26338-6

A. Speer, *Inside the Third Reich: memoirs* (The MacMillan Company, New York, 1970)

U. Spezia, *Italo Federico Quercia: note biografiche e documenti.* 21mo Secolo S.r.l., 2007

M. Steenbeck, Beschleunigung von Elektronen durch elektrische Wirbelfelder. Naturwiss. **31**, 234–235 (1943)

D. Stratigakos, *Hitler's Northern Utopia: Building the new order in occupied Norway.* (Princeton University Press, 2020). ISBN ISBN:9780691198217

The Physical Society and Cavendish Laboratory, (ed.), *Report of an International Conference on Fundamental Particles and Low Temperatures, held at the Cavendish Laboratory, Cambridge, UK, on 22–27 Jul 1946*, (London, 1947). Physical Society and Cavendish Laboratory, Cambridge, England, The Physical Society

W. Thirring, *Lust am Forschen* (Seifert Verlag GmbH, Wien, 2008)

W.E. Thirring and B. Touschek, A covariant formulation of the Bloch-Nordsieck method. Philos. Mag. **42**(326), 244–249 (1951). URL https://doi.org/10.1080/14786445108561260

S. Ting, Discovery of massive neutral vector mesons: one researcher's personal account—Discovery story. Adv. Exp. Phys. **5**, 115 (1976)

S. Ting, The Discovery of the J Particle: a Personal Recollection. Nobel Lecture, (1976b). URL https://www.nobelprize.org/prizes/physics/1976/ting/lecture/

J. Tiomno, J.A. Wheeler, Energy spectrum of electrons from meson decay. Rev. Mod. Phys. **21**, 144–152 (1949) https://doi.org/10.1103/RevModPhys.21.144.URL https://link.aps.org/doi/10.1103/RevModPhys.21.144

B. Touschek, Excitation of Nuclei by electrons. Nature **160**(4067), 500 (1947). https://doi.org/10.1038/160500a0

B. Touschek, Zum analytischen Verhalten Schrödinger'scher Wellenfunktionen. Zeitschrift für Physik, **125**(4–6), 293–297 (1948a). URL https://doi.org/10.1007/BF01454900

B. Touschek, Zur Theorie des doppelten β-Zerfalls. Zeitschrift für Physikr Physik, **125**(1–3), 108–132 (1948b). URL https://doi.org/10.1007/BF01337622

B. Touschek, Note on Peng's treatment of the divergency difficulties in quantized field theories. Math. Proc. Cambridge Philos. Soc. **44**(2), 301–303 (1948). https://doi.org/10.1017/S0305004100024294

B. Touschek, Das synchrotron. Acta Physica Austriaca **3**, 146–155 (1949)

B. Touschek, CXVIII. A Perturbation Treatment of Closed States in Quantized Field Theories. Lond. Edinb. Dublin Philos. Mag. J. Sci. **42**(333), 1178–1184 (1951). https://doi.org/10.1080/14786445108561363

B. Touschek, Parity conservation and the Mass of the Neutrino. Il Nuovo Cimento, **5**(3), 754–755 (1957a). URL https://doi.org/10.1007/BF02835605

B. Touschek, The mass of the Neutrino and the non-conservation of parity. Il Nuovo Cimento, **5**(5), 1281–1291 (1957b). URL https://doi.org/10.1007/BF02731633

B. Touschek, A note on the Pauli transformation. Il Nuovo Cimento **13**(2), 394–404 (1959). https://doi.org/10.1007/BF02732949

B. Touschek, Remarks on the neutrino gauge group. In *Proceedings, 9th International Conference on High Energy Physics, v.1–2 (ICHEP59): Kiev, USSR, Jul 15–25, 1959*, vol. 2, ed. by Academy of Science USSR and IUPAP, (Moscow, 1960), pp. 117–125. Academy of Science USSR/International Union of Pure and Applied Physics. URL http://inspirehep.net/record/1280995/files/c59-07-15-p117.pdf

B. Touschek, The Italian storage rings. In *Proceedings, 1963 Summer Study on Storage Rings, Accelerators and Experimentation at Super-High Energies. 10 June - 19 July, 1963, Brookhaven National Laboratory, Upton, NY*, vol. C63-06-10, ed. by J. Bittner, (1963), pp. 171–208. URL http://lss.fnal.gov/conf/C630610/p171.pdf

B. Touschek, What is high energy? Phys. Lett. B **51**, 184–186 (1974). https://doi.org/10.1016/0370-2693(74)90211-1

B. Touschek, An Analysis of Stochastic Cooling, January 1979. URL http://www.lnf.infn.it/sis/preprint/detail-new.php?id=1631

B. Touschek, G. Rossi, *Meccanica statistica* (Boringhieri, Torino, 1970)

B.F. Touschek, An estimate for the position of the lowest Dipole-Level of a Nucleus. Lond. Edinb. Dublin Philos. Mag. J. Sci. **41**(319), 849–851 (1950). https://doi.org/10.1080/14786445008561016

Y.-S. Tsai, Radiative corrections to colliding beam experiments, (1965)

S. Turchetti, *Il caso Pontecorvo*. (Sironi Editore, 2007)

S. Turchetti, *The Pontecorvo Affair: A Cold War Defection and Nuclear Physics* (University of Chicago Press, Chicago, 2012)

V. Valente, (ed.), *Adone, a milestone on the particle way*, vol. VIII of *Frascati Physics Series*. (INFN, 1997)

V. Valente, *Strada del Sincrotrone Km 12*. (Istituto Nazionale di Fisica Nucleare, 2007). ISBN 978-88-88610-15-3

W. Vansant, *Bombing Nazi Germany*. (Zephir Press, 2013)

V. Veksler, Concerning some new methods of acceleration of relativistic particles. Phys. Rev. **69**(5–6),244 (1946). https://doi.org/10.1103/PhysRev.69.244. URL https://link.aps.org/doi/10.1103/PhysRev.69.244

R. Vergara Caffarelli, Ricordando Pontecorvo. Il Nuovo Saggiatore **20**, 8–15 (2004). URL https://www.ilnuovosaggiatore.sif.it/issue/40

F. Vissani, What is matter according to particle physics and why try to observe its creation in lab, 2021. URL https://arxiv.org/pdf/2103.02642.pdf

H. von Halban, F. Joliot, L. Kowarski, Liberation of Neutrons in the nuclear explosion of Uranium. Nature **143**(3620), 470–471 (1939). https://doi.org/10.1038/143470a0

H. von Halban, F. Joliot, L. Kowarski, Number of Neutrons liberated in the nuclear fission of Uranium. Nature **143**(3625), 680 (1939). https://doi.org/10.1038/143680a0

G.-A. Voss, Rolf Widerøe. Phys. Today, **50**(8), 79–80 (1997). URL https://doi.org/10.1063/1.881866

M.C. von Reichenbach, Richard Gans: The First Quantum Physicist in Latin America. Phys. Perspect. **11**, 302–317 (2009). https://doi.org/10.1007/s00016-008-0416-0

H. Waenke, Hans E. Suess. Biogr. Mem. Natl. Acad. Sci. **87**, 354–372 (2005). URL https://www.nap.edu/read/11522/chapter/20371

P. Waloschek, *The Infancy of Particle Accelerators. Life and work of Rolf Widerøe (edited by P. Waloschek).* (Friedr. Vieweg & Sons Verlagsgesellschaft (Braunschweig and Wiesbaden), Braunschweig, Germany, 1994). URL https://doi.org/10.1007/978-3-663-05244-9_7

P. Waloschek, *Todesstrahlen als Lebensretter:Tatsachenberichte aus dem Dritten Reich.* (Atelier OpaL Productions, 2004)

P. Waloschek, *Death-Rays as Life-Savers in the Third Reich.* (DESY, 2012). URL http://www-library.desy.de/preparch/books/death-rays.pdf

J. Walsh, Max planck society. Filling a Gap in German research. Science, **160**(3833), 1209–1210 (1968). ISSN 0036-8075, 1095-9203. https://doi.org/10.1126/science.160.3833.1209. URL http://science.sciencemag.org/content/160/3833/1209

E.T.S. Walton, The production of high speed electrons by indirect means. Proc. Camb. Phil. Soc. **25**(Pt IV), 469–481 (1929)

V. Weisskopf, Hans Kopfermann (1895–1963). Nucl. Phys. **52**, IN1–IN2 (1964). https://doi.org/10.1016/0029-5582(64)90684-4

W. Westendorp, E. Charlton, A 100 Million Volt induction electron accelerator. J. Appl. Phys. **16**, 581 (1945). https://doi.org/10.1063/1.1707508

G. Wick, Rend. Accad. dei Licei, **21**, 1935

R. Widerøe, Ueber ein neues Prinzip zur Herstellung hoher Spannungen. Archiv für Elektrotechnik **21**(4), 387–406 (1928) https://doi.org/10.1007/BF01656341. URL https://doi.org/10.1007/978-3-662-40440-9_14

R. Widerøe, European induction accelerators. J. Appl. Phys. **18** (1947). https://doi.org/10.1063/1.1697841

R. Widerøe, Some memories and dreams from the childhood of particle accelerators. Europhys. News, **15**(2), 9–11 (1984). URL https://doi.org/10.1051/epn/19841502009

R. Widerøe, Über ein neues prinzip zur herstellung hoher spannungen. Archiv für Elektrotechnik **21**(4), 387–406 (1928). https://doi.org/10.1007/BF01656341

E. Wigner, *Gruppentheorie und ihre Anwendungen auf die Quantenmechanik der Atomspektren* (Vieweg Verlag, Braunschweig, 1931)

E. Wigner, *Group Theory and its Application to the Quantum Mechanics of Atomic Spectra.* (New York: Academic Press, Translation by J.J. Griffin, 1959)

B.H. Wiik, Rolf Widerøe and the development of particle accelerators. Acta Oncol. **37**(6), 615–625 (1998). https://doi.org/10.1080/028418698430322

R. Wilson, *Physics is Fun. Memoirs of a Life in Physics.* (Richard Wilson, Department of Physics, Harvard University, 2011)

E. Wolf, Recollections of Max Born. Astrophys. Space Sci. **227** (1995). https://doi.org/10.1007/BF00678085

C. Wu, E. Ambler, R. Hayward, D. Hoppes, R. Hudson, Experimental test of parity conservation in β decay. Phys. Rev. **105**, 1413–1414 (1957). https://doi.org/10.1103/PhysRev.105.1413

D.R. Yennie, S.C. Frautschi, H. Suura, The infrared divergence phenomena and high-energy processes. Ann. Phys. **13**(3), 379–452 (1961). https://doi.org/10.1016/0003-4916(61)90151-8

A. Zichichi, Selected topics in strong interactions, QED and unsolved problems . In *In *Pisa 1976, Proceedings, Frontier Problems In High Energy Physics*, 109–164* (1976)

D. Zimmerman, The society for the protection of science and learning and the politicization of British science in the 1930s. Minerva, **44**(1), 25–45 (2006). URL https://doi.org/10.1007/s11024-005-5405-8

T. Ziolkowski, Hauptmann Iphigenie in Delphi: A travesty? Ger. Rev. Lit. Cult. Theor. **34**(2), 105–123 (1959). https://doi.org/10.1080/19306962.1959.11786964

S. Zweig, *The World of Yesterday.* (The Viking Press, 1943)

Index

© The Editor(s) (if applicable) and The Author(s), under exclusive license
to Springer Nature Switzerland AG 2022
G. Pancheri, *Bruno Touschek's Extraordinary Journey*, Springer Biographies,
https://doi.org/10.1007/978-3-031-03826-6

Printed in the United States
by Baker & Taylor Publisher Services